Nuclear Energy Conversion

Nuclear Energy Conversion

M. M. EL-WAKIL

Professor of Mechanical and Nuclear Engineering

University of Wisconsin

The American Nuclear Society
La Grange Park, Illinois

Library of Congress Catalog Card Number 78-6169

International Standard Book Number 0-89448-015-4

Copyright © 1978 by The American Nuclear Society, 555 North Kensington Avenue, La Grange Park, Illinois 60525, USA

First Printing 1971 by International Textbook Company

Preface

It is a source of pride and satisfaction for me that the American Nuclear Society (ANS) has chosen this volume, *Nuclear Energy Conversion*, and its companion volume, *Nuclear Heat Transport*, as their first venture in textbook publishing. The two volumes were originally published by the International Textbook Company, which is no longer in existence. The two volumes have found acceptance in their role of filling needed classroom texts in the areas of heat generation and thermo-fluid design of nuclear reactor cores and in nuclear power systems.

The ANS is an organization that has as its main objectives the encouragement of research and scholarship, the dissemination of information, and the cooperation with government and educational institutions in scientific and engineering matters relating and allied to the atomic nucleus. It is therefore actively engaged in nuclear education via the publication of scientific and technical journals and monographs. On the university and other educational levels, it has been active in organizing student chapters, meetings, and publications.

It is therefore a natural extension of ANS activities that they get into textbook publishing. I am therefore gratified that ANS has seen fit to assume publication of these two volumes and look forward to a fruitful association in the service of university and other students and practicing engineers who may stand to benefit from these two books.

M. M. El-Wakil

May 1978
Madison, Wisconsin

Publisher's Foreword

The American Nuclear Society (ANS) has long pursued the general publishing objective of accelerating the peaceful uses of nuclear energy by publishing books directed to and advancing the professional interests of engineers and scientists in the nuclear field. Volumes in the continuing ANS monograph series, although not prepared primarily as textbooks, are being used by educators as source books for teaching in areas of current technology.

With publication of *Nuclear Energy Conversion* and its companion volume, *Nuclear Heat Transport*, ANS is making nuclear engineering textbooks—written for this purpose alone—available for the first time to the nuclear community.

American Nuclear Society-published books are directed to the interests of scientists and engineers concerned with the application of nuclear energy. With the addition of Dr. El-Wakil's texts, ANS broadens its commitment to provide detailed information in specialized technical areas to educators and their students.

The American Nuclear Society regards these publishing activities and their contributions to the growing achievements in nuclear energy application as an essential obligation to ANS membership and to the nuclear field.

Norman H. Jacobson
Manager, ANS Publications

Contents

CHAPTER 4.   BOILING-REACTOR POWER PLANTS . . . 94

CHAPTER 5.   PRESSURIZED-WATER REACTORS . . . 145

CHAPTER 6.   PRESSURIZED-WATER REACTOR
POWER PLANTS . . . 167

Contents

Contents

chapter 1

Survey of
Nuclear Power Systems

1-1. INTRODUCTION

Man's primary sources of energy originate in reactions involving the atom or its nucleus. This applies to such varied sources as the energy obtained by combining atoms of carbon or hydrogen with atoms of oxygen (combustion), and the energy received from the sun due to the combination or fusion of four nuclei of the hydrogen atom at high temperatures (thermonuclear). The latter gives rise to a multitude of secondary energy sources on earth such as the plants and organisms that eventually turn into fossil fuels.

On earth, abundant energy due to reactions involving the nucleus, *nuclear energy,* has been obtained by fissioning heavy nuclei into two or more lighter nuclei each, and, to a lesser extent, by making use of the decay energy of some radioactive nuclei. Also, research is in progress to find practical means for the generation of large quantities of energy by fusing two light nuclei into a heavier one.

This book deals mainly with the conversion of nuclear energy to electrical energy. The purpose of this chapter is to survey the origin and types of nuclear power systems, and therefore to help place in their proper perspective the various topics that will be discussed in the book.

1-2. NUCLEAR ENERGY CONVERSION

The reactions involving the atom are the *chemical* reactions in which two or more atoms combine (or separate) to form new substances. An example of the chemical reaction is the ordinary process of combustion in which one atom of carbon combines with two atoms of oxygen to form carbon dioxide. Chemical reactions involve the binding (or separation) of entire atoms by sharing or exchanging their orbital electrons. The nuclei of the participating atoms are unaffected. The total mass of the materials entering the reaction undergoes negligible change (decrease in exothermic or increase in endothermic reactions) after the reaction is completed. The energy liberated (or absorbed) is due to the binding (or separation) energy of the participating atoms.

The reactions involving the nucleus are the *nuclear* reactions which involve changes in the nucleus of the atoms as well as in the number of their orbital electrons. Nuclear reactions are of many types. Those that produce energy on a large scale, however, are of the fusion or fission type. In *fusion,* two or more light nuclei are fused into a heavier and more stable nucleus. The new nucleus has a mass slightly less than the sum of the masses of the original nuclei. In *fission,* a relatively heavy nucleus is fissioned, or split, into two or more lighter nuclei whose combined mass is less than that of the original nucleus. Other less productive, but nonetheless useful, forms of nuclear energy are due to the *radioactive decay* of radioisotopes. Radioisotopes may be found in nature or artificially produced as fission products or by neutron irradiation.

Table 1-1 gives orders of magnitude of the energies produced per reaction from each of the above-mentioned reaction types. By reaction it is meant the combustion of one molecule of hydrogen or one atom of carbon, the fusion of four nuclei of hydrogen or two nuclei of deuterium, etc., the fission of one uranium nucleus, and the single radioactive decay of one radioisotope.

TABLE 1-1

Type of Reaction	Approximate Energy per Reaction
Chemical	3-4 ev
Fusion	3-18 Mev
Fission	200 Mev
Radioactivity	1-5 Mev

There are many forms of energy that can be converted to one another. Energy conversion can occur in single or multiple steps. Table 1-2, the *energy conversion matrix,* shows in block form the types of energy that can, with present technology, be converted to others, and the processes or devices by which conversion is accomplished. Figure 1-1 shows the types of power systems required for given outputs and duration times with limited fuel supply, as in outer or remote space.

The conversion of nuclear energy to electrical energy is the theme of this book. This conversion can be done in the following ways:

(a) In a single step, by the use of direct collection devices, also called *nuclear batteries* (horizontal row, Table 1-2).

(b) In two steps, from nuclear to thermal by fusion or fission, then

Energy Conversion Matrix

To

Convert	Gravitational	Kinetic	Thermal	Chemical	Nuclear	Electrical	Electromagnetic
Gravitational		Falling objects					
Kinetic	Rising objects (rockets)		Friction (brakes)	Dissociation by radiolysis		Electrical generator. MHD	Accelerating charge (cyclotron). Phosphor
Thermal		Thermal expansion (turbines). Internal combustion engines		Phase change (boiling). Dissociation		Thermoelectricity. Thermionics. Thermomagnetism. Ferro-electricity	Thermal radiation
Chemical		Muscle	Combustion			Battery. Fuel cell	Chemiluminescence (fireflies)
Nuclear		Radioactivity (α particles, A-bomb)	Fission. Fusion	Radiation catalysis (hydrazine plant). Ionization (cloud chamber)		Direct collection devices. Nuclear battery	Gamma reactions (Co^{60} source, A-bomb)
Electrical		Electric motors. Electrostriction (sonar transmitter)	Resistance heating	Electrolysis (aluminum production). Battery charging			Electromagnetic radiation. Electroluminescense
Electromagnetic		Solar cell. Radiometer	Absorption of thermal radiation	Photosynthesis (plants). Photochemistry (photographic film)	γ-neutron reactions ($Be^9 + \gamma \rightarrow Be^8 + n$)	Photoelectricity (light meter). Radio antenna. Solar cell	

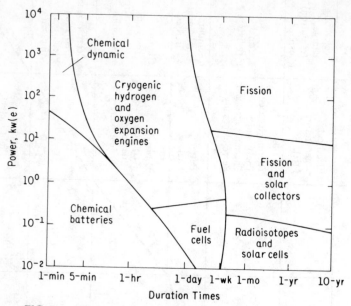

FIG. 1-1. Power systems required for given power outputs and
duration times with limited fuel supply.

(go vertically to diagonal then horizontally) from thermal to
electrical by thermoelectricity or thermionics.

(c) In three steps, from nuclear to thermal by fussion or fission,
from thermal to kinetic by thermal expansion of heated fluid in a
turbine, and finally from kinetic to electrical by an electrical
generator.

This book will deal with the conversion of nuclear energy to electrical
energy by the above three routes.   The primary emphasis will be on (c),
the fission-turbine-generator route, Chapters 3-12, since it is the one that
is capable of producing the much-needed world power requirements on a
large scale.   The other two routes, (a) and (b), produce much smaller
amounts of energy for specialized uses, such as for remote earth, outer
space, biomedical applications and others.   They will be discussed in
Chapters 13-15.   The fusion process is in the very early stages of development and is covered in Chapter 16.   The economics of fission power
will be discussed in Chapter 17.

The remainder of this chapter is an introduction to and survey of the
field of nuclear power systems.

1-3.   ENERGY FROM NUCLEAR FUSION

In the sun and stars the reactions are essentially those in which four nuclei of hydrogen fuse and form one helium nucleus according to

$$4_1H^1 \longrightarrow {}_2He^4 + 2\beta^+ \qquad (1-1)$$

The reactions are called *thermonuclear* because extremely high temperatures are required to trigger them.   On earth, although the utilization of fission preceded that of fusion in both weapons and power generation, the basic fusion reaction was discovered first.   In the 1920's research on particle accelerators produced the first man-made fusion reaction.

Man-made fusion is accomplished by fusing two, instead of four, nuclei.   Table 1-3 lists some of the possible reactions and the energies

TABLE 1-3

No	Fusion Reaction		Energy per Reaction, Mev
	Reactants	Products	
1	D + D	T + p	4
2	T + D	He^4 + n	17.6
3	D + D	He^3 + n	3.2
4	He^3 + D	He^4 + p	18.3

produced by them.   Such reactions are possible because: (a) there is a much greater probability of two particles colliding than of four, and (b) the 4-hydrogen reaction that occurs in the sun and stars requires, on the average, billions of years of completion, whereas the D-D reaction, for example, requires only a minute fraction of second.

All natural waters on earth contain about 1 part in 6,666 of *heavy water,* D_2O, which contains deuterium.   Deuterium is therefore plentiful and could supply man's fuel needs indefinitely.   Heavy water is concentrated by electrolytic, distillation or chemical exchange methods. Deuterium can be distilled from liquid hydrogen.

To accomplish the feat of producing useful energy by fusion, basically three performance criteria must be met:

(a) The *temperature* of the fuel must be heated to between 10^8 and 5×10^9 °K, corresponding to between 10^4 and 5×10^5 ev respectively.   This is due to the fact that, to cause fusion, it is necessary to accelerate positively charged nuclei to high speeds so that collision between them takes place despite the electrical repulsive forces between them.   The necessary kinetic energies are produced by raising their temperatures.   These temperatures transform the fuel gas into a plasma (a mixture of electrons, positive ions, and neutral particles).

(b) The *density* of the resulting plasma must be of the order of 10^{15} ions/cm³ (still corresponding to a high vacuum at room temperature).

(c) The *confinement time* of the ions at these temperatures and density must be of the order of tenths of a second.

There is a trade-off between density and confinement time. If the densities are lower than 10^{15}, for example, the confinement time must be longer, the product of the two should be about 10^{14} The above criteria pose severe physical problems. No container material can stand the high temperatures. The devices that have been constructed so far have heated and confined the plasma, away from the container walls, by strong magnetic fields.

Advances in attaining the above goals have been made in recent years, particularly in the United States, the United Kingdom, and the Soviet Union. The advances have been in attaining the individual goals, but not in meeting the combination of all criteria in the same device. The Soviet Union's Tokamak T-3 device has shown the best performance to date (1970). There is now unprecedented international cooperation in the field of fusion power, and after years of pessimism, there is much optimism that a practicable fusion power device will be a reality in about 25 years.

1-4. THE FUSION REACTOR

A fusion reactor using the deuterium-tritium fuel cycle (No. 2, Table 1-3) may look like that shown in Fig. 1-2. The plasma is confined in the center of the reactor by the magnetic coils on the outside. The neutrons

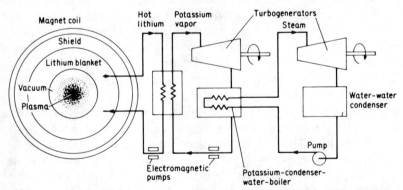

FIG. 1-2. Elements of a fusion reactor power plant with deuterium-tritium fuel cycle and potassium-water dual fluid working cycle.

emanating from the D-T reaction are absorbed in a liquid lithium blanket surrounding the plasma according to the reactions:

$$_3Li^6 + _0n^1 \longrightarrow _1T^3 + _2He^4 + 4.8 \text{ Mev} \tag{1-2}$$

$$_3Li^7 + _0n^1 \longrightarrow _1T^3 + _2He^4 + _0n^1 - 2.5 \text{ Mev} \tag{1-3}$$

Lithium contains about 7.5 percent Li^6 and 92.5 percent Li^7, but has a much higher n,α cross section for Li^6. The above reactions breed tritium, some of which is reused in the fuel cycle. The net energy generated in the lithium blanket is transported to a potassium Rankine cycle. Liquid potassium is vaporized in a heat exchanger and used to drive a turbogenerator. The exhaust potassium is condensed in a potassium-condenser-water-boiler which produces steam for use in a conventional steam Rankine cycle.

Part of the electrical power produced in the two turbogenerators is used to provide the needed intense magnetic field as well as supply station auxiliaries such as vacuum and liquid-metal pumps, etc.

Chapter 16 covers the current fusion technology in more detail including other fuel cycles and concepts such as the *fussion torch* or those employing direct conversion methods.

1-5. ENERGY FROM NUCLEAR FISSION*

Unlike fusion, which involves particles of similar electrical charge and therefore requires high kinetic energies to initiate reaction, fission can be caused by a neutral particle, the neutron. The neutron can strike and fission a heavy nucleus at high, moderate, or low speeds, without being repulsed. Fission can also be caused by other particles. Neutron bombardment, however, is the only practical way of obtaining a sustained reaction, since two or three neutrons are usually released for each one engaging in fission. These keep the reaction going.

The neutron was identified by Chadwick in 1932. Much work and speculation about fission took place between 1932 and 1939, including the work of Meitner and Hahn in Germany. The latter found, in one experiment in which neutrons were captured by uranium nuclei, that barium was produced. This was the first identification of a fission fragment and the first proof of fission.

It was not until December 2, 1942, however, that Enrico Fermi and his associates succeeded in producing continuous or sustained fission in an atomic *pile*. The pile was erected under the stands of the sports stadium of the University of Chicago. It was named *pile* because it consisted

* Sections 1-5 through 1-8 may be bypassed by those with a background in reactor physics.

of a large number of uranium and graphite blocks piled on top of each other.   The Fermi *chain reaction* was the event that signaled the dawn of the nuclear age.

The two isotopes U^{238} and Th^{232}, found in nature, are called *fertile* materials because fissionable materials are manufactured from them. There are three fissionable isotopes (by neutrons of all energies).   These are U^{235}, Pu^{239}, and U^{233}.   The first is found in natural uranium.   It can be used in this state or in an *enriched* form; i.e., its percentage is increased beyond that in natural uranium (0.711 percent).   It can also be used in chemical combination or alloyed with other materials.   Plutonium 239 is manufactured by nuclear reactions from U^{238}.   Uranium 233 is similarly manufactured from Th^{232}.

When a neutron hits a fissionable nucleus in the right way, it causes it to split into two (rarely three or four) lighter nuclei, usually not of equal mass, called *fission fragments*.   Two or three neutrons are ejected, mostly at the time of fission.   Many isotopes belonging to some 35 elements have been identified as fission fragments.   Fission fragments are transformed by *radioactive decay* into other isotopes.   These and the fission fragments are called *fission products*.

An example of fission is one in which barium 137 and krypton 97 are fission fragments, and two neutrons are ejected.

$$_0n^1 + {}_{92}U^{235} \longrightarrow {}_{56}Ba^{137} + {}_{36}Kr^{97} + 2{}_0n^1 \tag{1-4}$$

The sum of the reactant masses is greater than the sum of the product masses by 0.2080 amu.   This is equivalent to about 194 Mev, 8.6×10^{-18} kwhr or 2.9×10^{-14} Btu of energy.   On a mass basis, the *complete* fissioning of 1 g_m of U^{235} produces about 5×10^{23} Mev or 2.2×10^4 kwhr. The complete fissioning of 1 lb_m of U^{235} produces about 3.4×10^{10} Btu, or the same amount of energy produced by about 2.5 million lb_m of coal.

The complete fissioning of all U^{235} nuclei in a fuel mass is, however, impossible because many of the fission products capture neutrons in nonfission reactions.   In time, the number of neutrons so captured becomes great enough, because of the accumulation of the products, that the fission chain can no longer be sustained.   Depending upon fuel enrichment, this happens when only a small percentage (often less than 1 percent) of the fissionable nuclei in the fuel has been consumed. Further use of this *poisoned* fuel can be made only by removing the fission products and reprocessing.

Thus a fuel mass is capable of producing much smaller quantities of energy than indicated above.   This is referred to by the term *burnup*. A common unit for burnup is the Mw-day/ton.   This is the amount of energy in megawatt-days produced from each metric ton (1,000 kg_m) of *fuel*.   Fuel is defined as all uranium, plutonium, and thorium isotopes, not including alloying or other chemical compounds or mixtures, in the fuel charge. The term *fuel material* is used to refer to the entire fuel charge including chemical, alloy, or mixture used, but not cladding or other structural materials.

1-6. THE CHAIN REACTION

An important parameter in fission is the number of newly born neutrons, also called *fission neutrons,* in a single fission reaction per neutron engaging and thus lost in such a reaction. For U^{235} this number is, on an average, 2.47. In a reactor where controlled and sustained energy production is desired, conserving neutrons is a vital matter.

There are mainly two reasons why not all the fission neutrons cause further fission. The first is the nonfission capture or absorption of some neutrons by the fission products; by nonfissionable nuclei in the fuel, structural material, coolant, and moderator; by the fissionable fuel itself; and, in the case of research and test reactors, by materials deliberately inserted in the reactor *core* (the volume within the reactor occupied by the active fuel) for test purposes.

The second reason is that a certain percentage of neutrons escape, or *leak* out, from the core. The smaller the surface-volume ratio of the core, i.e., the larger its size, the lower this percentage. Other things being equal, the core size must be increased to the point where a chain reaction is possible. This size is called the *critical size,* since a smaller core would be incapable of sustaining a chain reaction. The mass of the fuel in such a core is called the *critical mass.*

In a reactor using uranium, 100/2.47 or about 40.5 of each 100 fission neutrons must ultimately engage in fission to keep the reactor critical. However, only about 84 percent of the neutrons that get absorbed in U^{235} cause fission, the remainder reacting with it to produce U^{236}, an isotope of no particular importance. Consequently, a total of about 40.5/0.84 or 48 neutrons must be absorbed in U^{235}. This leaves a maximum of about 52 that may be allowed to leak out of the core and to become absorbed in other core materials.

1-7. NEUTRON ENERGIES AND MODERATION

Fission neutrons possess a wide range of speeds, averaging about one-tenth the speed of light, and correspondingly high kinetic energies. They collide with various core-material nuclei and slow down. Thus all neutrons flying about in a core possess kinetic energies that vary between several Mev down to minute fractions of an electron volt. They are classified as *fast, intermediate*, and *slow*. The lowest kinetic energies that the neutrons may reach are equivalent to those of the adjoining molecules and atoms. Since these energies are a function of the temperature of the medium, neutrons in this state are called *thermal,* a special category of slow neutrons.

The fertile nuclei U^{238} and Th^{232} can be fissioned but only with fast neutrons. The fissionable nuclei U^{233}, U^{235}, and Pu^{239} fission with neutrons of all energies. In the thermal range, however, the *probability* of fission is higher, the slower the neutrons (the $1/V$ law). A physical explanation of this is that a neutron has a better chance of reacting with the nucleus if it is slow and consequently spends more time in the vicinity of the nucleus.

If a mass of natural uranium (0.7 percent U^{235}, 99.3 percent U^{238}) is used in a reactor core, the probability of fissioning the abundant U^{238} nuclei diminishes rapidly as the newly born neutrons slow down. Many of these neutrons are then captured by the same U^{238} nuclei in nonfission reactions, the probability of which increases sharply as the neutrons reach intermediate energies. This is called *resonance absorption.* The few neutrons that escape this capture and slow down further acquire a high probability of fission with U^{235}. Some fission, but the number of the new neutrons plus those produced by the fast fission of U^{238} is much less than those starting the cycle. Thus a critical mass cannot be made from natural (or low-enriched) uranium alone.

In order to overcome the above difficulty, the neutrons must be slowed down past the intermediate or resonance energy range in a material that does not excessively absorb them and that has good slowing-down properties. To do this, the fuel is divided into small elements, such as plates, rods, hollow cylinders, pins, etc. The space between the elements is filled with a material with the required properties. Such a material is called a *moderator.* Thus a large proportion of the fission neutrons escape from the fuel and into the moderator before resonance energies are attained. In the moderator they are slowed down past these energies, and upon reentry into the fuel they have largely been *thermalized* (slowed down to thermal energies) and are ready to fission U^{235}. This is the process of *moderation.* It is obvious that the moderator-fuel ratio in a core is an important design parameter.

Figure 4-11 shows a typical fuel subassembly containing a 7×7 array of rod-type fuel elements. The interspace is filled with a moderator such as water. The elements are covered by a material such as zirconium which protects the fuel against chemical reaction with the water and prevents gaseous fission products from escaping. Such a cover is called *cladding,* jacket or can. For some fuels it also acts as a structural support.

A good moderator slows down a neutron after a small number of *collisions* with its nuclei. The size of these nuclei should therefore be about the same as that of the neutron. As with billiard balls, one is slowed down more effectively if it hits another billiard ball than if it hits a much heavier object, such as a bowling ball, or a much lighter one, such as a marble.

Hydrogen is an excellent moderator because its nucleus is a proton, approximately the same size as a neutron. Deuterium, helium, lithium, beryllium, boron, and carbon are all light materials suitable from this standpoint. Of these, however, lithium and boron are strong neutron absorbers and unsuitable as moderators.

It is necessary to have a large number of moderator nuclei in a given volume (large density) so that a neutron does not have to travel long distances before encountering a moderator nucleus. Gases are therefore ineffective as moderators. Hydrogen and deuterium are effective as part of a heavier molecule in liquid or solid form, such as light and heavy water, hydrocarbons, zirconium hydride, polyethylene, and others.

Water, heavy water, graphite, and beryllium are the most practical moderators. Water is the most effective in terms of the path length necessary. It is plentiful and cheap but absorbs neutrons slightly. Heavy water absorbs no neutrons but is costly. Graphite is good but weak structurally. Beryllium is costly and is used only where cost is not of prime importance.

Reactors dependent primarily on thermal neutrons for fission are called *thermal reactors*. Reactors using highly enriched fuels and containing no moderator are called *fast reactors*.

1-8. CONVERSION AND BREEDING

The nonfission capture of neutrons by a fertile U^{238} nucleus results in a series of reactions culminating in fissionable Pu^{239} as follows:

$$\begin{aligned} _{0}n^1 + {}_{92}U^{238} &\longrightarrow {}_{92}U^{239} + \gamma \\ {}_{92}U^{239} &\longrightarrow {}_{93}Np^{239} + {}_{-1}e^0 \\ {}_{93}Np^{239} &\longrightarrow {}_{94}Pu^{239} + {}_{-1}e^0 \end{aligned} \tag{1-5}$$

In this process, called *conversion,* U^{239} is an unstable or *radioactive* nucleus. Within a very short time after its birth, it emits a β particle (electron) and is transmuted into Np^{239}, which likewise is transmuted into Pu^{239}, a long-lived stable isotope of plutonium.

Fertile Th^{232} can also be converted to fissionable U^{233}:

$$\begin{aligned} _{0}n^1 + {}_{90}Th^{232} &\longrightarrow {}_{90}Th^{233} + \gamma \\ {}_{90}Th^{233} &\longrightarrow {}_{91}Pa^{233} + {}_{-1}e^0 \\ {}_{91}Pa^{233} &\longrightarrow {}_{92}U^{233} + {}_{-1}e^0 \end{aligned} \tag{1-6}$$

A reactor in which fissionable nuclei, different from those in the original core loading, are produced is called a *converter*. If more fissionable nuclei are produced than are consumed by fission or if the same type of nuclei are produced as are consumed, the reactor is called a *breeder*. A reactor which is designed to produce power and fuel is called a *production* reactor.

1-9. FISSION POWER PLANTS

The energy produced in fission shows up mainly in the form of kinetic energy of fission fragments and, to a lesser extent, of emitted neutrons and other particles and radiations such as γ (gamma) rays. As these different particles slow down or are absorbed, their energies are converted into heat. This heat is removed by a *coolant* and then usually utilized in a thermodynamic cycle to produce power.

The process of *heat removal* is a very important one. The amount of heat produced in a reactor is not a sole function of its size but rather of its size and type. Heat removal is more difficult, the smaller the reactor, because of the difficulty of providing an ample heat-transfer surface. Thus the maximum power that can be obtained from a reactor depends upon heat transfer rather than nuclear considerations. This is the reason for the development of superior heat-transfer systems using coolant fluids and flow rates that are capable of high heat-transfer coefficients.

There are many reactor types. Reactors may be classified according to (1) general purpose or function, (2) type of moderator, (3) type of coolant, (4) neutron-energy classification, (5) type of fuel, (6) core internal design (homogeneous or heterogeneous), and others.

By general purpose is meant whether a reactor is used for generating power, for research, training, breeding, or a combination of these. In all cases, of course, the heat generated must be removed. Only in power reactors, however, do core temperatures become large. Such reactors may be further classified as to type of power plant associated with them.

Some of the above classifications determine others. For example, a water-cooled (and -moderated) reactor is necessarily a thermal one. While many combinations of design variables, and consequently numerous reactors and reactor concepts, are technically feasible, a few reactor power plants stand out. Some of these are now described, although more detail on each will be given later in this book.

Liquid-Cooled-Reactor Plants

Two schematics of these are shown in Figs. 1-3 and 1-4. The liquid

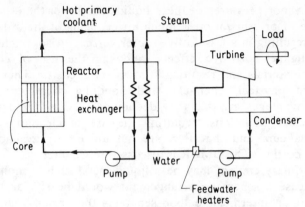

FIG. 1-3. Schematic arrangement of liquid-cooled-reactor
power plant.

coolant picks up reactor heat and leaves at a temperature high enough to generate steam in a heat exchanger.   The coolant pressure may be higher than the saturation pressure corresponding to the maximum coolant temperature in the reactor, and no coolant boiling occurs there (see Fig. 1-3).   The coolant loop is called the primary loop.   The water-steam loop is called the *secondary,* or *working-fluid,* loop.   The steam generated in the heat exchanger expands in an expander (turbine) where useful power is generated, the steam is condensed, and the condensate is pumped back to the heat exchanger.

The primary coolant may be water, which may double as moderator. In this case the reactor is called a *pressurized-water* reactor, abbreviated PWR.   The pressures in a PWR are of the order of 2,200 psia in the primary loop and 900 psia in the secondary.   The main advantages of this system are that water is cheap and plentiful (this does not apply to heavy water) and safe and easy to handle.   It has good heat-transfer characteristics and well-known physical and thermodynamic properties. Water does not suffer to a great extent from nuclear radiation damage (transformation or decomposition of molecules under reactor radiations). Also, induced radioactivity in pure water results in relatively low-level, short-lived radiations.   The main disadvantages are that reactor pressurization is necessary because of the high vapor pressure of water, if boiling is to be avoided, resulting in costly components and reactor vessel. Water is corrosive at high temperatures.   Light water absorbs some neutrons, necessitating the use of enriched fuels.   Heavy water absorbs no neutrons but is expensive.   The PWR is discussed in Chapters 5 and 6.

The primary coolant may be an organic liquid (such as terphenyl) which also doubles as moderator. Such a plant is called the *organic-moderated* and *-cooled-* reactor power plant, abbreviated OMCR.   Because the vapor pressures of these liquids are lower than that of water, the degree of pressurization is much lower, being of the order of 100 to 400 psia, resulting in a less costly reactor vessel. Also, organic coolants have negligible corrosive effects, allowing the use of conventional structural materials and resulting in low capital costs. They have fairly well-known physical, chemical, and handling characteristics. Their main disadvantages are that they decompose at high temperatures and under nuclear radiations, requiring the use of cleanup systems and coolant makeup, and that they are not universally available and are somewhat costly. The OMCR is discussed in Chapter 12.

The primary coolant may be a liquid metal such as molten sodium, in which case a schematic arrangement would be similar to Fig. 1-3, except that an intermediate loop separates the primary and secondary loops and that two heat exchangers are used. This is called a *liquid-metal-cooled*-reactor (LMCR) power plant. The costly intermediate loop is necessary to isolate the working-fluid loop from the high radioactivity induced in the primary sodium. A moderator such as graphite or heavy water may or may not be used, depending on whether the reactor is thermal or fast. Again, because of the very low vapor pressures of liquid metals, the degree of pressurization is low, and reactor pressure, mainly determined by fluid frictional losses, is of the order of 100 psia or less. The working fluid may be H_2O, as above, or a gas driving a gas turbine.

Other characteristics of liquid-metal coolants are as follows: They have excellent heat-transfer characteristics. They have wide ranges of temperatures in which they remain in the liquid state (sodium melts at 208°F, boils at 1621°F). They can thus operate at high temperatures, resulting in good power-plant thermal efficiencies. The relatively high freezing point of sodium, however, necessitates the use of electric or other heaters to keep the coolant from freezing during low-power operation or extended shutdown. Liquid metals are practically the only coolants suitable for fast reactors. They have good resistance to nuclear radiation damage. Their high thermal conductivities and low specific heats cause the temperature gradients in the coolant system to be low. Coupled with high boiling temperatures, local hot spots, and conduit warping are minimized. On the debit side, liquid metals are chemically active and corrosive, necessitating the use of costly structural materials and handling techniques. Oxygen, present even in small quantities, oxidizes sodium to Na_2O which is highly soluble in Na. It later precipitates on cold walls and causes clogging problems. Liquid metals are not universally available and are costly. Liquid-metal-cooled fast breeder reactors (FBR) are discussed in Chapters 9 and 10.

If the primary-coolant pressure corresponds to saturation at reactor temperatures, boiling of the coolant occurs within the reactor. Vapor thus produced may be used directly as the working fluid, resulting in a relatively simple cycle (see Fig. 1-4). The coolant that has undergone a large degree of development in this type is water, in which case the plant is called a *boiling-water*-reactor power plant, abbreviated BWR. The

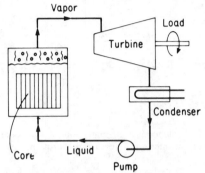

FIG. 1-4. Schematic arrangement
of boiling-reactor power plant.

reactor pressure in such a case is usually between 600 and 1,000 psia. In addition to the advantages and disadvantages of PWR, BWR enjoys higher heat-transfer rates per pound of coolant and a simpler, one-loop cycle. However, there exist large moderator-density changes within the core of a BWR, giving rise to nuclear and hydrodynamic problems. The BWR is discussed in Chapters 3 and 4.

A combination PWR and BWR cycle, called the *dual cycle,* is discussed in Chapter 4.

Gas-Cooled Reactor Power Plants

Gas coolants have negligible moderating capabilities at reactor pressures, necessitating a separate moderator such as graphite or heavy water. The *gas-cooled* reactor power plant, GCR, is mainly of three types.

The first operates on a *closed indirect* cycle, in which a primary coolant, such as CO_2 or He, is continuously circulated in a primary loop, giving off heat to a working fluid such as H_2O or another gas. The schematic for this is similar to Fig. 1-3. The primary-coolant pressure is low, mainly determined by the mass-flow rate desired and the frictional losses. It is of the order of 200 to 700 psia.

The second is the *direct open* cycle, shown in Fig. 1-5, the components of which are those of the Brayton cycle used in gas-turbine work but where the reactor takes the place of the combustion chamber. Since the cycle is open, the atmosphere (air) is the only possible coolant.

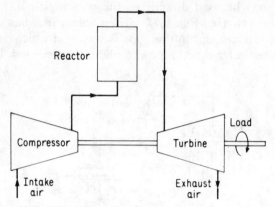

FIG. 1-5. Schematic arrangement of gas-cooled
direct-open-cycle reactor power plant.

Cycle pressures are determined from the thermodynamics of the system (pressure ratios for maximum thermal efficiency or power).

The third is the *direct closed* cycle (see Fig. 1-6), in which the gas is continuously recirculated throughout the power plant. The necessary heat rejection (to complete a thermodynamic cycle) is to cooling water, air, or other available media or by thermal radiation, where such media are not available (as in outer space). The primary coolant here may be one with better heat-transfer and thermodynamic behavior than air, such as He.

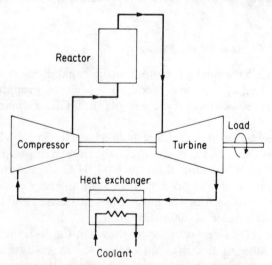

FIG. 1-6. Schematic arrangement of gas-cooled
direct-closed-cycle reactor power plant.

Gaseous coolants are generally available, cheap, safe, and easy to handle. They can operate at high reactor outlet temperatures, resulting in high plant thermal efficiencies. When clean, they do not present a serious problem because of radioactivity. Because of their low densities, they do not present great nuclear problems. For example, they do not absorb neutrons to any great extent. Gases, however, have poor heat-transfer characteristics and low volumetric heat capacities, and they require greater pumping powers and larger ducts than do liquid coolants. Pressurizing is necessary to reduce pumping requirements. Leak-proof systems are needed, especially for low-molecular-mass gases such as He. Because of the poor heat transfer, high fuel temperatures are required if high heat-removal rates from the reactor are to be achieved. Finned or specially designed fuel elements may be necessary, adding to the costs of the system. The GCR is discussed in Chapters 7 and 8. Gas-cooled fast-breeder reactors are discussed in Chapter 10.

Fluid-Fueled Reactors

In this category, the fuel is fluidized, eliminating the problem of solid-element fabrication. ·This is done by one of several techniques. It may be in the form of a salt, such as UO_2SO_4, dissolved in light or heavy water, acting as both fuel carrier and moderator. This is called the *aqueous-fuel* system. It may be a metallic fuel dissolved in a liquid metal such as molten bismuth, in which case the reactor may be thermalized by a solid moderator such as beryllium. This is the *liquid-metal-fueled* reactor, abbreviated LMFR. Fuel compounds such as uranium and Thorium fluorides may be dissolved in fused salts such as LiF and BeF_2 (also acting as moderator). The fuel may be in a finely ground state and suspended in a moving liquid, called a *slurry*, or it may be in the form of dust carried by a gas such as helium.

Figure 1-7 shows a schematic of a circulating or fluid-fueled system.

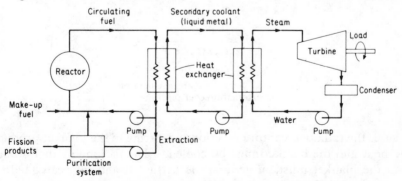

FIG. 1-7.   Schematic arrangement of circulating or fluid-fueled system.

The reactor core is essentially an expansion or bulge in the fuel piping. In this core the amount of fuel and moderator, if any, corresponds to criticality, and energy is produced there. The circulating fuel then carries this energy as sensible heat and transfers it to an intermediate loop of liquid metal. The intermediate loop is necessitated by the highly radioactive primary fluid. Fission products and other poisons can be continually removed and fresh fuel continually added, without the necessity of reactor shutdown, by a chemical purification system as shown. In this system there is no temperature limitation, as would be imposed on solid fuels, and higher reactor outlet temperatures (1000°F and higher) as well as higher steam temperatures and pressures are possible than with the PWR or BWR. Since there is a minimum of structural members within a fluid-fueled core, parasitic capture of neutrons is minimized. Corrosion problems are severe, however, especially at high rates of circulation. Excessive vapor formation (especially with aqueous systems) may cause serious fluctuations in reactivity, as well as explosion hazards necessitating precautionary measures, such as recombination cells for O_2 and H_2. Also, reactor design for very high pressures is made necessary. Pumping, leakage, and other problems must be solved. Fluid-fueled reactors are discussed in Chapter 11.

Any of these reactor categories can be a breeder, and all are, of course, converters to a certain degree (except when no fertile material is present in the fuel). A reactor can be an *internal* breeder, if the new fuel is produced within the active core, or it may be an *external* breeder, if the core is surrounded by a *blanket* of fertile fuel (see Fig. 1-8). In

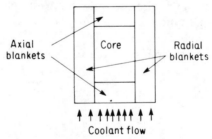

FIG. 1-8. Schematic arrangement of
an external-breeder reactor.

this case, the radiative-capture reactions occurring in the blanket produce some heat and the blanket must be cooled. The fissionable fuel produced in the blanket must, in general, be extracted and fabricated before it can be used in the core.

1-10. DIRECT CONVERSION OF NUCLEAR ENERGY

Direct energy conversion usually implies the elimination of the mechanical rotary machinery (turbines) used in the previous examples. Direct conversion of nuclear energy to electricity may be done in a single step (direct collection devices, Chapter 15) or in two steps involving intermediate thermal energy (thermionics, Chapter 13, and thermoelectricity, Chapter 14).

In the *direct collection devices,* nuclear radiations such as β or positively charged fission particles are emitted by a nuclear *cathode* and collected on an *anode* across a gap and voltage difference. This is the most direct way of nuclear energy conversion. It is characterized by very low currents, high voltages, and very low efficiencies.

Thermionics is the direct conversion of heat to electricity. In its simplest form, a thermionic converter consists of two closely spaced metallic plates, Fig. 1-9. One (heated) is the cathode, the other is the anode. Heat drives or boils off electrons from the surface of the cathode into free space towards the anode. The minimum amount of energy required is equivalent to the work that must be done against electric fields imposed by the atoms at the surface of the cathode. This is called the

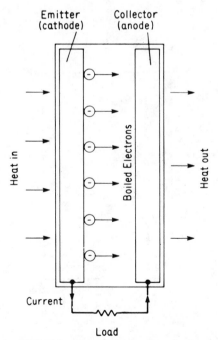

FIG. 1-9. Elements of a thermionic converter.

work function. In the case of tungsten it is about 4.5 ev. The cathode work function must be higher than the anode work function. A major problem with a thermionic *diode* is due to the boiled off electrons forming a cloud of negative charges that will repel electrons emitted later back toward the cathode. This *space-charge* effect is counteracted by very close spacings, the *vacuum diode,* or more effectively, by the introduction of a gas, such as cesium vapor, containing positively charged particles, the *plasma diode.* (The mixture of positive and negative charged particles is called a *plasma.*)

In *thermoelectricity* use is made of the Seebeck effect, which states that two dissimilar metals joined at a hot junction will develop an electromotive force (voltage) across it. The effect, discovered in 1821, was used only in thermocouples to measure and control temperatures. Only recently have efficient materials, the semiconductors, made possible the production of useful amounts of electric power. The semiconductors used in a thermoelectric converter, Fig. 1-10, are *p*-type and *n*-type semiconductors. They are made by deliberately introducing impurities to metals. In the *p*-type, the impurity atoms do not have enough valence electrons necessary to satisfy the valence-bond requirements of the main atoms and the resulting atomic lattice fills with positive *holes* which move

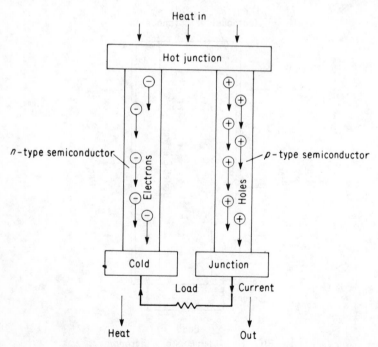

FIG. 1-10. Elements of a thermoelectric converter.

through the semiconductor like positive charges. In the *n*-type, the impurity atoms have more than enough electrons and the lattice fills with negative electrons. When *p*- and *n*-type materials are joined and the junction heated electrons will flow from the *p*- to *n*-type material. This is equivalent to current flowing from the *n*- to the *p*-type material. The reverse is true at the colder junction. A practical thermoelectric converter produces low currents and voltages and, compared to a Rankine cycle, low efficiency.

chapter **2**

Some Thermodynamic Aspects of Nuclear Power

2-1. INTRODUCTION

This chapter deals primarily with heat-power cycles and their adaptation to the conversion of nuclear energy to useful power. It is pressumed that the reader possesses a knowledge of basic thermodynamics. Nevertheless, we shall first review some fundamental concepts of thermodynamics necessary to the discussion that follows. In particular, the concept of cycle efficiency, the parameters that affect it, and the peculiar role that it plays in the case of nuclear power will be discussed. The most common nuclear-power-plant types and cycles and some of their characteristics have been discussed in Sec. 1-9. Not all thermodynamic aspects can be discussed in one chapter, however, and much will be left to the remaining chapters when individual nuclear-power-plant types will be discussed in more detail.

Table 2-1 lists property relationships and energy changes for ideal or perfect gases undergoing various processes. These are gases which obey the law $pv = RT$.

For real gases, the expressions of Table 2-1 are sufficiently accurate for most engineering calculations. However, for highly cooled or compressed gases, a correction in the form of a *compressibility factor* may be applied to the relationships between the properties p, v, and T [3].

The relationships for vapors are usually quite complex and are given in tables or charts. Appendix B contains abstracts of the Keenan and Keyes steam tables [4] as well as tables for organic [5] and metallic [6] fluids.

Table 2-2 lists the symbols and their units used in Table 2-1, in the tables and charts of Appendix B, and in the discussions that follow. For energy crossing the boundary of a system containing a fluid, the convention is that heat added is positive and heat rejected is negative. Work done by the fluid is positive, while work done on it is negative.

TABLE 2-1

Perfect-gas Relationships (constant specific heats)

Process	p, v, T relationships	$u_2 - u_1$	$h_2 - h_1$	$s_2 - s_1$	W (nonflow)	Q
Isothermal	$T = \text{const}$ $p_1/p_2 = v_2/v_1$	0	0	$(R/J)\ln(v_2/v_1)$	$(p_1v_1/J)\ln(v_2/v_1)$	$(p_1v_1/J)\ln(v_2/v_1)$
Constant pressure	$p = \text{const}$ $T_2/T_1 = v_2/v_1$	$c_v(T_2 - T_1)$	$c_p(T_2 - T_1)$	$c_p\ln(T_2/T_1)$	$p(v_2 - v_1)/J$	$c_p(T_2 - T_1)$
Constant volume	$v = \text{const}$ $T_2/T_1 = p_2/p_1$	$c_v(T_2 - T_1)$	$c_p(T_2 - T_1)$	$c_v\ln(T_2/T_1)$	0	$c_v(T_2 - T_1)$
Isentropic (adiabatic reversible)	$s = \text{const}$ $p_1v_1^\gamma = p_2v_2^\gamma$ $T_2/T_1 = (v_1/v_2)^{\gamma-1}$ $T_2/T_1 = (p_2/p_1)^{(\gamma-1)/\gamma}$	$c_v(T_2 - T_1)$	$c_p(T_2 - T_1)$	0	$\dfrac{p_2v_2 - p_1v_1}{J(1-\gamma)}$	0
Throttling	$h = \text{const}$ $T = \text{const}$ $p_1/p_2 = v_2/v_1$	0	0	$(R/J)\ln(v_2/v_1)$	0	0
Polytropic	$p_1v_1^n = p_2v_2^n$ $T_2/T_1 = (v_1/v_2)^{n-1}$ $T_2/T_1 = (p_2/p_1)^{(n-1)/n}$	$c_v(T_2 - T_1)$	$c_p(T_2 - T_1)$	$c_v\ln(p_2/p_1)$ $+\, c_p\ln(v_2/v_1)$	$\dfrac{p_2v_2 - p_1v_1}{J(1-n)}$	$c_v\left(\dfrac{\gamma - n}{1 - n}\right)(T_2 - T_1)$

TABLE 2-2

Some Common Thermodynamic Symbols

c_p = specific heat at constant pressure, Btu/lb$_m$ °F
c_v = specific heat at constant volume, Btu/lb$_m$ °F
h = specific enthalpy, Btu/lb$_m$
H = total enthalpy, Btu
J = energy conversion factor = 778.16 ft-lb$_f$/Btu
M = molecular mass, lb$_m$/lb-mole
n = polytropic exponent, dimensionless
p = absolute pressure (gauge pressure + barometric pressure),
     lb$_f$/ft^2; unit may be lb$_f$/in.^{2}, commonly written psia
ΔQ = heat transferred to or from system, Btu/lb$_m$ or Btu/cycle
R = gas constant, lb$_f$-ft/lb$_m$ °R = $\bar{R}/M$
$\bar{R}$ = universal gas constant = 1,545.33, lb$_f$-ft/lb-mole °R
s = specific entropy, Btu/lb$_m$ °R
S = total entropy, Btu/°R
t = temperature, °F
T = temperature on absolute scale, °R
u = specific internal energy, Btu/lb$_m$
U = total internal energy, Btu
v = specific volume, ft^3/lb$_m$
V = total volume, ft^3
ΔW = work done by or on system, lb$_f$-ft/lb$_m$ or **Btu/cycle**
x = quality of a two-phase mixture = mass of vapor
     divided by total mass, dimensionless
γ = ratio of specific heats, c_p/c_v, dimensionless
η = efficiency, as dimensionless fraction or percent
Subscripts used in vapor tables:
f refers to saturated liquid.
g refers to saturated vapor.
fg refers to change in property because of change from
     saturated liquid to saturated vapor.

2-2. THE CARNOT CYCLE AND EFFICIENCY

In 1824 Sadi Carnot [7] introduced his now famous *Carnot cycle* along with two new important concepts.   These were (1) the concept of reversibility and (2) the principle that no engine can produce more work for the same amount of heat added and the same heat source and sink temperatures than one operating on a reversible cycle.   This latter concept is the now well-known *Carnot principle.*

The Carnot cycle is composed of four reversible processes; two are isothermal (constant temperature) and two are reversible adiabatic and therefore isentropic.   On the *TS* diagram given in Fig. 2-1, the cycle is

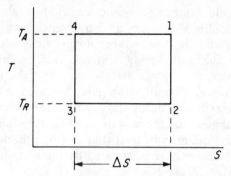

FIG. 2-1. *TS* diagram of Carnot cycle.

therefore a rectangular one. Heat is added during the isothermal process 4-1 and rejected during the isothermal process 2-3. Processes 1-2 and 3-4 represent isentropic expansion and compression, respectively. By definition, the heat added or rejected in a reversible process is given by $dQ = T\, dS$. It follows, then, that for the Carnot cycle

$$\Delta Q_A = T_A(S_1 - S_4) = T_A\, \Delta S \qquad (2\text{-}1)$$

and
$$\Delta Q_R = T_R\, \Delta S \qquad (2\text{-}2)$$

where ΔQ_A, ΔQ_R = amounts of heat added and rejected, Btu/cycle
$\quad\quad T_A$, T_R = absolute temperatures at which heat is added and rejected, °R
$\quad\quad\quad S$ = entropy of fluid undergoing cycle, Btu/°R
The thermal efficiency η_{th} of any cycle is given by

$$\eta_{\text{th}} = \frac{\Delta W_n}{\Delta Q_A} = \frac{\Delta Q_A - \Delta Q_R}{\Delta Q_A} \qquad (2\text{-}3)$$

where ΔW_n is the net work, in Btu per cycle. Applied to the Carnot cycle, the Carnot efficiency η_c is therefore given by

$$\eta_c = \frac{T_A - T_R}{T_A} \qquad (2\text{-}4)$$

The Carnot principle states that this efficiency is the maximum that can be attained between the temperature limits T_A and T_R. This efficiency is therefore a goal that we should strive to approach when designing actual cycles. We shall first examine this efficiency and then examine some of the reasons why actual cycles fail to achieve or attain such an efficiency, the methods that may be adopted to improve actual cycle efficiencies, and the practical limitations to those methods.

High efficiency (in both ideal and nonideal cycles) is attained, Eq. 2-4, with high T_A, low T_R, or both. In other words, heat should be added at the highest possible temperature and rejected at the lowest possible temperature. (In the Carnot cycle no temperature difference exists between the fluid and the heat source during heat addition or between the fluid and the heat sink during heat rejection.) The highest possible temperature is governed by the heat source and the metallurgy of the system. For example, if combustion gases are used in a cycle, a temperature of about 5000°F can be attained with near-stoichiometric air-fuel ratios. In nuclear reactors there is no limit to the neutron-flux level and consequently to the quantity of heat that can be produced. However, a practical limit on temperature is dictated by the metallurgy of the fuel or cladding materials.

The sink temperature is governed by whatever coolant is available at low temperatures and in sufficient quantities. This may be the earth's atmosphere or a large body of water. In the former case, T_R may vary from 450 to 580°R (-10 to 120°F), and the range is narrower for the latter. In outer space and very-high-altitude applications, where no such fluids are available, heat rejection would necessarily have to be by thermal radiation to the skies. While the effective temperature of the sky (not directly looking at bright stars) is quite low, the temperature of the heat radiator may have to be quite high, 1000 to 1500°F, in order to reject large quantities of heat per unit surface area of radiator. Remember that size and mass are of prime importance in such applications. Figure 2-2 is a plot of Carnot-cycle efficiency versus t_A, for different

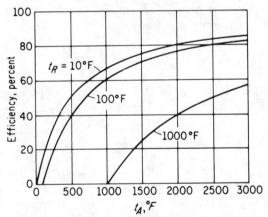

FIG. 2-2.   Efficiency of Carnot cycle for various
values of t_A and t_R.

values of t_R. It shows that, for $t_R = 1000°F$ (space applications), heat addition at very high temperatures, i.e., high t_A, is necessary if reasonable values of efficiency are to be obtained. Note finally that the efficiency of the Carnot cycle is independent of the type or phase of the working fluid.

In general, the thermal efficiency of a power plant is of prime importance. In fossil-fueled plants, fuel and operating costs represent a substantial portion of the cost of power produced. A high thermal efficiency cuts down the fuel and operating costs and thus has a major effect on power costs. In nuclear power plants, the fuel costs in general are small compared with the capital cost of the plant, and thermal efficiency, while important, is not of such prime importance as in fossil-fueled plants. Maximum attainable thermal efficiencies of different-type nuclear power plants vary. A high thermal efficiency of one type does not necessarily mean lower power costs than those of another type, however. This is because such things as capital costs and permissible fuel burnup vary appreciably from one type to another. For a single-type nuclear power plant, on the other hand, improvement in plant thermal efficiency means that the capital cost per unit kilowatt installed is low. It also lengthens the time between fuel processing and thus reduces the cost of the fuel cycle (removal, shipment, fabrication, etc.). A high thermal efficiency, however, means lower heat rejection and, therefore, less environmental problems.

2-3. ACTUAL CYCLES AND THE CONCEPT OF REVERSIBILITY

The Carnot cycle and other ideal reversible cycles with constant-temperature heat addition and rejection, such as the Stirling and Ericsson cycles (having regeneration at constant volume and pressure, respectively), all have the same efficiency when operating between the same temperature limits. Such cycles are reversible for two reasons:

1. Heat transfer from the heat source to the working fluid occurs with zero temperature difference. Thus if the temperature of the working fluid during heat addition were T_1 (see Fig. 2-3), the cycle is reversible only if $T_1 = T_A$. Similarly, if the temperature of the fluid during heat rejection were T_2, then, for reversibility, $T_2 = T_R$. This is called *external* reversibility.

2. No fluid friction occurs in the working fluid as it expands and contracts throughout the cycle. This, primarily, is *internal* reversibility.

In an actual cycle, heat is added and rejected in heat exchangers with finite temperature differences $\Delta T_A = T_A - T_1$ and $\Delta T_R = T_2 - T_R$. The degree of external irreversibility inherent here is a function of ΔT_A and ΔT_R and can be reduced by reducing the magnitudes of ΔT_A and ΔT_R.

It can be shown that finite temperature differences are wasteful, i.e., mean a loss of work (for the same heat added, $T_A \Delta S$). Without going

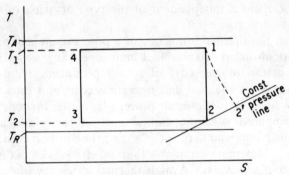

FIG. 2-3. Carnot-type cycle with temperature difference during heat addition and rejection.

into great detail, let us imagine an engine placed between T_A and T_1.
This engine would produce work not realized in the actual case. The
same could be said for the low-temperature end of the cycle. Heat
added and rejected in the cycle of Fig. 2-3 may be written in the forms

$$\Delta Q_A = U_A A_A \Delta T_A \tag{2-5a}$$

and
$$\Delta Q_R = U_R A_R \Delta T_R \tag{2-5b}$$

where U is an overall heat-transfer coefficient (not to be confused with
internal energy), A is the heat-transfer surface area of each heat exchanger, and the subscripts A and R refer to heat addition and rejection.
Thus for the same values of ΔQ and U, irreversibility due to temperature
difference can be minimized by increasing the heat-transfer surface area
of the heat exchangers. There is however, an optimum heat-exchanger
size beyond which the increased capital cost of the heat exchanger more
than obviates the operative gain due to increased efficiency of the cycle.

The other main source of irreversibility, internal fluid friction, is also
wasteful of work since, for example, part of the energy of the working
fluid at point 1 in Fig. 2-3 is converted to internal energy of the fluid
instead of useful work, ending at a higher enthalpy, point 2′ instead of
point 2. Points 2 and 2′ are on the same pressure line. Internal fluid
friction can be minimized by reducing the velocity of the working fluid
even to laminar-flow conditions. This reduces pressure losses as well as
increases cycle reversibility. However, the overall heat-transfer coefficients U decrease to such low values that here again the size of the heat
exchangers would have to be further increased, with consequences similar
to those stated above.

In actual power plants, individual machines are additional sources
of irreversibility such as mechanical friction, electrical losses (windage,
magnetic, etc.). In general, these irreversibilities have to be reduced

as much as possible by engineering skill, resulting in improved efficiency of the plant as well as reduced sizes and costs of the machines. In the case of mechanical friction, the use of special materials may be helpful. Of course, mechanical friction may be reduced by increasing the size of the machines or reducing their speeds, resulting also in reduced internal fluid friction but again in increased size and cost.

Again the reversibility of an actual cycle may be increased if ΔT_A and ΔT_R are as small as possible during the *entire* processes of heat addition and rejection. The temperatures of the heat source, heat sink, and working fluid in many actual cycles change, however, during heat addition and rejection. Another approach to reversibility can be achieved if these processes are made to take place at constant or near-constant temperatures, i.e., like the Carnot cycle.

Many thermodynamic cycles in nature have been conceived as ideals for actual engines or cycles, such as the Otto, Diesel, Brayton, and Rankine cycles. Of these only the Rankine cycle usually receives most of its heat and rejects all at constant temperature. This cycle, which is the ideal for most fossil- and nuclear-fueled power plants, will now be discussed in some detail.

2-4. THE RANKINE CYCLE

An internally reversible Rankine cycle is shown as 1-2-3-B-4 on the Ts plane in Fig. 2-4. The elements of a power plant operating on this cycle are the working-fluid loop of Fig. 2-5. Saturated vapor at temperature T_1 expands from point 1 adiabatically and reversibly in an expander or turbine to point 2. At point 2, the vapor, slightly wet at temperature

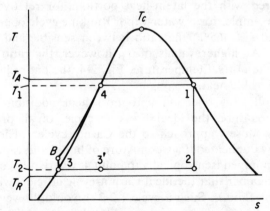

FIG. 2-4. Internally reversible Rankine cycle
with saturated vapor.

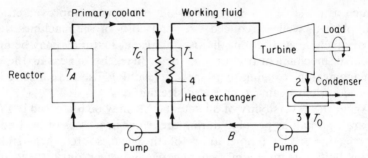

FIG. 2-5. Schematic of two-loop nuclear power plant.

T_2, is condensed at constant temperature and pressure to point 3. There, the fluid is a saturated liquid, also at temperature T_2. At point 3, the liquid is pumped to point B which is at the same pressure as point 1 (or in the actual case, slightly higher). For small pressure differences, B is so close to point 3 that, on a Ts diagram drawn to scale, the two points seem to coincide. The compressed liquid at B is then fed to a heat exchanger or boiler where it is heated at constant pressure along line B-4-1, receiving sensible heat between points B and 4 and latent heat of vaporization between points 4 and 1.

In the above cycle, heat is rejected at constant temperature, T_2. Heat added is composed of the sensible heat, added between B and 4, during which the temperature of the compressed liquid rises, and the latent heat of vaporization, added at constant temperature between points 4 and 1. The difference between a near-constant-temperature heat source and the working fluid in the cycle is therefore finite and varying, thus contributing to irreversibility. The sensible-heat portion, however, is small compared with the latent-heat portion, for relatively low values of T_1. For example, for a water-vapor Rankine cycle operating between $t_2 = 100°F$ and $t_1 = 465°F$, the ratio between sensible and latent heats is roughly 1:2. At higher values of T_1, however, the ratio quickly rises. (At $T_1 = T_c$, the critical temperature, Fig. 2-4, the entire heat added is in the form of sensible heat.)

Because all heat rejection and most heat addition take place at constant temperature, the ideal Rankine cycle, of all practical cycles, represents the closest approach to the Carnot cycle. The Carnot cycle may be exactly duplicated if a wet mixture of liquid and vapor represented by point 3' is pumped isentropically to point 4, where it becomes saturated liquid. (Remember that the ideal Carnot-cycle efficiency is independent of the working fluid.) Such a cycle 1-2-3'-4 would, however, show little advantage in efficiency over the normal Rankine cycle 1-2-3-B-4, because the pumping work of the liquid-vapor mixture (a function of specific volume) would be materially increased over the pumping work of the liquid only.

η_R, the efficiency of the basic Rankine cycle 1-2-3-B-4, is given by

$$\eta_R = \frac{(\text{work of expander}) - (\text{work of pump})}{\text{heat added}} = \frac{(h_1 - h_2) - (h_B - h_3)}{h_1 - h_B}$$

(2-6)

where h is the specific enthalpy Btu/lb$_m$ of the fluid used. At low maximum cycle pressure, i.e., low values of T_1, the pumping work is negligible compared with either the work of the expander or the heat added. In other words, $h_B \cong h_3$. Thus

$$\eta_R \cong \frac{h_1 - h_2}{h_1 - h_3}$$

(2-7)

2-5. THE EFFICIENCY AND WORK OF A NUCLEAR PLANT OPERATING ON AN IRREVERSIBLE CYCLE

A general treatment of efficiency and work output of an irreversible cycle, irrespective of the type of coolant, will now be attempted [8]. We shall assume that we have a two-coolant-loop nuclear power plant, such as that shown in Fig. 2-5. The primary loop is composed of a heat source (reactor) and a circulating pump. The working-fluid loop is composed of a turbine, condenser, and pump. The two loops are coupled with a heat exchanger. For simplicity we shall assume that the working-fluid loop operates on the cycle represented on the TS plane in Fig. 2-6.

It will be further assumed for simplicity that heat is produced in the reactor at a constant temperature T_A, that the temperature of the primary coolant is a constant T_C, and that the working fluid receives heat only along line 4-1 at constant temperature T_1. Heat is rejected at a constant temperature T_2.

We shall evaluate the effects of irreversibilities due to fluid friction during expansion of the working fluid in the turbine and due to the temperature differences during heat addition. The former irreversibility manifests itself in an increased entropy of the fluid after expansion. Irreversibilities during heat rejection (which cannot be easily controlled because we are limited by whatever heat sink is available and by condenser size) and during compression (which are usually small) will be ignored.

Cycle 1-2-3-4 in Fig. 2-6 is an idealized internally reversible representation of the working-fluid circuit. Cycle 1-2′-3-4, however, shows the actual irreversible expansion in the turbine. The maximum theoret-

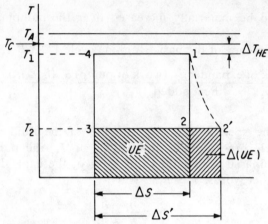

FIG. 2-6. *TS* diagram of cycle with irreversible
expansion and irreversible constant-temperature heat
addition.

ical thermal efficiency of the reversible cycle 1-2-3-4 is equal to the
efficiency of the Carnot cycle operating between T_1 and T_2 and thus is
given by

$$\eta_{rev} = 1 - \frac{T_2}{T_1} \tag{2-8}$$

The efficiency of the irreversible cycle is given by

$$\eta_{act} = 1 - \frac{\Delta Q_R}{\Delta Q_A} = 1 - \frac{T_2 \Delta S'}{T_1 \Delta S} \tag{2-9}$$

where η_{rev}, η_{act} = reversible and actual cycle efficiencies

ΔS, $\Delta S'$ = entropy change during heat rejection for reversible and
irreversible cycles, Btu/°R

ΔS and $\Delta S'$ are related by

$$T_2 \Delta S = UE$$

and
$$T_2 \Delta S' = UE + \Delta(UE)$$

Thus
$$\frac{\Delta S'}{\Delta S} = 1 + \frac{\Delta(UE)}{UE} = 1 + i \tag{2-10}$$

where *UE* is the *unavailable energy* in the reversible cycle, represented by
the area below line 2-3, and *i* is the fraction increase in unavailable energy
that occurs with irreversibility. (Unavailable energy is that portion of
ΔQ_A that cannot be converted to useful work since it is equal to the heat
rejected by the reversible cycle which has the highest possible efficiency
when operating between T_1 and T_2.) Combining Eqs. 2-9 and 2-10 gives

$$\eta_{act} = 1 - \frac{T_2}{T_1}(1 + i) \tag{2-11}$$

i increases only slightly as T_1 increases. This can be ascertained from the TS diagram of Fig. 2-6 where, for fixed T_2, point 2′ will shift further to the right with an increase in T_1. The expansion efficiency of turbines (actual enthalpy drop divided by isentropic enthalpy drop between the same inlet and exhaust pressures) increases, however, with an increase in turbine inlet temperature T_1. Thus point 2′ shifts only slightly to the right, and i may be assumed to be independent of T_1. Also i is small compared with unity (it approaches zero as the system becomes more reversible). It can thus be seen from Eq. 2-11 that η_{act} continuously increases with T_1.

ΔQ_A is proportional to $T_A - T_C$. $T_C - T_1$ largely depends on heat transfer and construction costs of the heat exchanger and is optimized by them. Thus ΔQ_A may be assumed to vary directly with the quantity $T_A - T_1 - \Delta T_{HE}$,

$$\Delta Q_A = C(T_A - T_1 - \Delta T_{HE}) \tag{2-12}$$

where $\Delta T_{HE} = T_C - T_1$, and C is a proportionality constant which includes the heat-transfer coefficient and heat-transfer area in the reactor core. Thus if T_A is fixed by metallurgical considerations, (and if it is assumed that ΔT_{HE} is a fixed fraction of $T_A - T_1$), it can be seen that ΔQ_A decreases as T_1 increases.

Neglecting heat losses to the outside, the actual work of the cycle ΔW_{act} therefore is given by

$$\Delta W_{act} = \Delta Q_A \eta_{act} \tag{2-13a}$$

or

$$\Delta W_{act} = C(T_A - T_1 - \Delta T_{HE}) \left[1 - \frac{T_2}{T_1}(1 + i) \right] \tag{2-13b}$$

Now ΔW_{act} (Eqs. 2-13) is equal to the product of ΔQ_A, which decreases with T_1, and η_{act}, which increases with it. We can thus see that there is an optimum value of T_1 which gives the maximum value of ΔW_{act}. Note that this optimum value of T_1 corresponds to an actual efficiency lower than the maximum attainable since the maximum attainable efficiency occurs when T_1 is a maximum, i.e., equal to T_A and ΔT_{HE} is zero. In this case, however, the work is zero since ΔQ_A is also zero.

The optimum value of T_1, $(T_1)_{opt}$, may be obtained by determining the derivative of ΔW_{act} with respect to T_1 from Eq. 2-13b (holding T_A, ΔT_{HE}, T_2 and i constant), equating it to zero, and solving for T_1. This gives

$$(T_1)_{opt} = \sqrt{(T_A - \Delta T_{HE})T_2(1 + i)} \tag{2-14}$$

For the completely reversible cycle where $\Delta T_{HE} = 0$ and $i = 0$,

$$(T_1)_{opt} = \sqrt{T_A T_2} \tag{2-15}$$

The efficiency of the power cycle operating with an optimum temperature of the working fluid is obtained by substituting $(T_1)_{opt}$ from Eq. 2-14 into Eq. 2-11 to give

$$\eta_{act} = 1 - \sqrt{\frac{T_2(1 + i)}{T_A - \Delta T_{HE}}} \tag{2-16}$$

For a completely reversible cycle the efficiency at maximum power conditions is $1 - \sqrt{T_2/T_A}$. Thus, to increase the efficiency of the plant at maximum power, the operating reactor-core temperature T_A must be increased and/or the temperature at which heat is rejected, T_2, decreased.

If the heat losses from the power cycle are neglected, the efficiency of the entire nuclear power plant is equal to the above efficiency multiplied by η_R, the *reactor thermal efficiency*. η_R is that portion of the *reactor thermal output* showing up as heat transported to the primary coolant. Reactor thermal output is defined as the total heat produced *within* the reactor, including all energy produced as a result of the fission process and by the radioactive decay or interaction of fission products; it includes heat lost to shields, pressure vessel, etc., but *not* the energy carried away by the neutrinos. The latter portion accounts for about 5 percent of the total thermal output but is nonrecoverable [2]. Because of losses, mostly due to long-range γ energy, only about 90 to 94 percent of the reactor thermal output is picked up by the coolant.

In discussing the efficiency of a nuclear power plant, our thinking is influenced by the fact that, in contrast to fossil-fueled power plants, capital costs contribute the largest share of nuclear-power costs whereas the contribution of fuel costs is relatively minor. This will probably be true for some time to come—that is, until enough nuclear plants have been built so that those types that are most feasible for particular applications have been built on a sufficiently large scale to reduce materially their capital costs. Even then, it is expected that fuel and fuel-fabrication costs will also decrease, so that capital costs will probably continue to contribute the larger share of nuclear-power costs.

Improving the efficiency of a nuclear power plant, however, results in an improvement in fuel burnup and therefore lower fuel costs if the plant is operated at the same power output, or in improvement in capital costs if the plant operates with the same reactor output, since the plant output will increase with efficiency. It also reduces heat rejection to the environment. With these thoughts in mind, we see that the most profitable operation of a particular power plant may be somewhere between maximum power and maximum efficiency (dictated by allowable temperature levels), though possibly closer to the former. The final decision is reached only after careful analysis of all the variables.

It is to be noted also that a nuclear power plant of one type may not necessarily produce cheaper power than a plant of another type and of lower efficiency. This is because capital, operational fuel costs, and allowable fuel burnup vary materially from one type of nuclear power plant to another.

2-6.   REGENERATION

The irreversibility due to the sensible-heat-addition portion of the Rankine cycle can be reduced by *regeneration*. In the ideal regeneration treatment, the compressed liquid between B and point 4 in Fig. 2-4 receives heat continuously at the same temperature that it is at. This can be accomplished, hypothetically, by using the vapor in the turbine, expanding between points 1 and 2, as a continuous heating agent for the compressed liquid. This may be done, say, by pumping the compressed liquid through an annular space in the casing of the turbine and in oppo-·site direction to the flow of steam. It will thus flow in a counterflow-heat-exchanger fashion, and if its velocity is low enough and if the heat-transfer area is large enough, it will always be facing vapor at the same temperature on the other side of the casing. It will thus receive heat reversibly from the vapor at all points in its path. The resulting cycle will be that shown on the TS diagram of Fig. 2-7 as 1-2-3-B-4. Heat added to the working fluid from an outside heat source will be only along line 4-1.

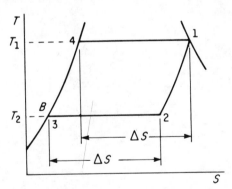

FIG. 2-7.   Rankine cycle with ideal regeneration.

It will now be shown that the efficiency of such a cycle would be equal to that of a Carnot cycle operating between the same temperature limits T_1 and T_2. The heat exchange between the compressed liquid and expanding vapor is done internally and does not enter into the evaluation for efficiency. The heat transferred from the expanding vapor is equal to the area under line 1-2. The heat added to the compressed liquid is equal to the area under line 3-4. Both these areas must be equal (barring heat losses to the surroundings, an assumption always made for an ideal cycle); therefore $S_1 - S_4 = S_2 - S_3 = \Delta S$, and the efficiency of the ideal regenerative Rankine cycle η_{RR} is

$$\eta_{RR} = \frac{\text{(heat added)} - \text{(heat rejected)}}{\text{heat added}} = \frac{T_1 \Delta S - T_2 \Delta S}{T_1 \Delta S} = \frac{T_1 - T_2}{T_1}$$

$$(2\text{-}17)$$

which is the same as the efficiency of a Carnot cycle operating between the same temperature limits of T_1 and T_2.

This ideal regeneration cycle is, of course, difficult to build. For one thing, the casing would be complicated, having an inner wall that must be relatively thin and that must have high thermal conductivity for maximum heat transfer and temperature equilibrium. However, temperature equilibrium would require low flow rates, themselves conducive to low heat-transfer coefficients. Furthermore, the vapor in the exhaust stage of the turbine, approaching point 2 in Fig. 2-7, becomes so wet that it would damage the blades of the turbine in the last few stages.

In practice, regeneration is accomplished in a finite number of steps. It is done by *bleeding* or extracting a small fraction of the expanding vapor at one or more points, such as a and b in the expansion line 1-2 of Fig. 2-8. The bled vapor is then used to heat the compressed liquid.

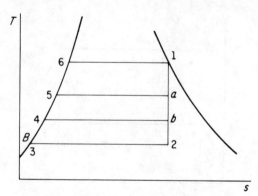

FIG. 2-8.   Rankine cycle with extraction feedwater heating.

Vapor extracted at point b heats liquid from point B to point 4. Similarly, vapor extracted at a heats up liquid from 4 to 5. The number of extraction or regenerative stations and regenerative feedwater heaters is made greater the higher the maximum cycle pressure. Feedwater heaters may be of the open type, where the extracted vapor mixes with the compressed liquid. They may be of the closed type, i.e., in the form of a shell-and-tube heat exchanger, and the condensate resulting from the extracted vapor is then fed back to the compressed-water line at a lower pressure stage. Open-type feedwater heaters are simpler and more direct in themselves but require one feed pump per heater. Figure 2-9 shows a

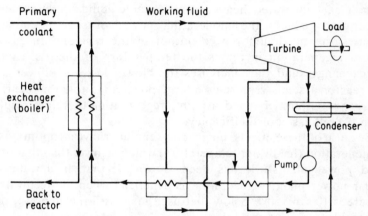

FIG. 2-9. Schematic of Rankine cycle with two closed-type feedwater heaters.

diagrammatic layout of the components of a power plant operating on a Rankine-type cycle, with two closed feedwater heaters. The method of analysis of a Rankine cycle with regenerative feedwater heating can be found in many standard books on thermodynamics.

Regeneration results in reducing the amount of heat not added at constant temperature. In Fig. 2-8 the only sensible heat added is from points 5 to 6. The higher the system pressure, the higher the sensible-heat portion of heat added, and the more important it becomes to add feedwater heaters. Note that, in Fig. 2-8, points 1, a, b, and 2 do not represent the same quantity of fluid. Neither do points B, 4, 5, and 6. However, Fig. 2-8 still represents the state (specific or intensive properties) of the fluid present at these points.

Regenerative heating has the effect of increasing the efficiency of the working-fluid cycle by as much as 20 percent (although this results in a lowering of cycle power output because the maximum mass-flow rate of the working fluid is not maintained in the various stages of the turbine). In a fossil-fueled power plant, regeneration has a decided effect on thermal efficiency, with maximum benefit occurring if the coolant enters the boiler at or near the saturation temperature. The degree of regeneration here is limited only by the increase in capital costs with the increase in number of feedwater heaters. In a nuclear power plant, however, the effect of regeneration is not so clear-cut. This is because regeneration increases the temperature of the working fluid entering the heat exchanger and therefore increases the temperature of the primary coolant leaving the heat exchanger and entering the reactor (see Fig. 2-9). Initially, this has a beneficial effect in that it equalizes the temperature of the primary coolant going through the reactor core. If this coolant doubles as moderator (e.g., light water, heavy water, organic liquids), a large degree of regeneration reduces the moderating power of such a coolant and decreases the maximum power output of the reactor. In gas-cooled reactors regeneration increases the temperature of the gas leaving the heat exchanger and thus increases the blower losses. Blower work in such reactor systems represents a large portion (15 to 25 percent) of the plant gross output. Beyond a point, regeneration therefore results in a loss in overall gas-cooled efficiency.

Because of these effects, nuclear power plants have an optimum degree of regeneration (feedwater temperature) which is lower than that of fossil-fueled power plants. It has been shown [9] that, in a water-cooled nuclear power plant, regeneration is beneficial when the feedwater temperature is 100 to 150°F below that used in a fossil-fueled power plant of comparable operating conditions.

The optimum degree of regeneration must be determined after careful consideration of its effects on capital, fuel, and operating costs. The relationship between feedwater heating and plant efficiency is a complex one, depending, among other things, upon the number of feedwater heaters and of steam extraction pressures. An expression for the efficiency can be obtained by careful analysis of the thermodynamic cycle.

2-7. HEAT ADDITION WITH A VARIABLE-TEMPERATURE HEAT SOURCE

One main source of irreversibility, as has been pointed out, is due to the temperature differences between the fuel and the primary coolant and between the primary coolant and the working fluid. These differences are not constant during the heat-addition process.

The temperature difference between the fuel and primary coolant, under normal conditions, is maximum at the center of the core. It then decreases both radially and axially. Except for deviations that develop within the core, the difference follows smooth cosine or Bessel functions, becoming minimum at the edges of the core [2].

The general level of temperature difference in this case may be reduced by increasing the heat-transfer coefficient, or decreasing the power generated by the fuel. A reduction may also be accomplished by changing the fuel enrichment or cross section of each fuel element in an axial direction (and of the fuel elements as a whole in a radial direction) so that the fuel temperature increases with increased coolant temperature as it travels through the core. The practical difficulties encountered in such an attempt are many and obvious and are not warranted by the thermodynamic gain that may be achieved. In cases where a change in phase of the coolant takes place (such as in a boiling-water reactor) the bulk temperature of the coolant remains fairly constant over the upper portions of the core, further increasing the irreversibility effects.

The second source of irreversibility in heat addition, that due to the temperature differences between the primary coolant and the working fluid, will now be discussed, with the help of a Rankine cycle producing saturated vapor, such as that shown in Figs. 2-4 and 2-8.

If the primary coolant (combustion gases in fossil-fueled plants; water, organic liquids, gases, or liquid metals in nuclear plants) does not undergo a change in phase in the primary loop, the processes taking place in the heat exchanger, producing saturated vapor, can be represented on the temperature-enthalpy diagram by Figs. 2-10 or 2-11. Line ab represents the primary coolant. This line is straight, or nearly straight, if the specific heat of that coolant is constant or varies only slowly and smoothly over the temperature range T_a to T_b. Line B-4-1 (or a line corresponding to 5-6-1 in the regenerative cycle in Fig. 2-8) represents the working fluid.

Figure 2-10 is for a counterflow heat exchanger, while Fig. 2-11 is for a parallel-flow heat exchanger. Note that in the parallel-flow case the temperature differences between the primary coolant and the working fluid are quite large, and the irreversibility is therefore correspondingly large. In the counterflow case (Fig. 2-10) the temperature differences between the primary coolant and the working fluid are smaller.

In the counterflow case, the two lines approach each other at points d and 4. This is called the *pinch point*. The smaller the temperature difference at the pinch point, the smaller the overall temperature differences between the two lines, and the less irreversible the process of heat addition becomes. Practical considerations, however, limit the degree of approach between the two lines. A very small temperature difference there would necessitate the use of a steam generator with a large heat-transfer area, size, mass, and capital cost. Pinch-point temperature differences that result in optimum compromises between the two conflicting requirements depend upon the temperature level of the primary coolant. They, for example, have the range 20 to 30°F for low-temperature gas-cooled reactors, and 100 to 150°F for high-temperature gas-cooled and liquid-metal-cooled reactors.

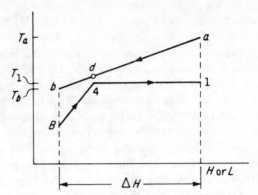

FIG. 2-10.   Heat addition to vaporizing fluid
with a variable-temperature source; counterflow heat exchanger.

Another advantage of the counterflow system over the parallel-flow
system is the ability to attain higher working-fluid pressures for the same
primary-coolant temperatures.   In the parallel-flow system (Fig. 2-11)
the saturation temperature represented by point 1 must be lower than the
coolant temperature at point b.   (The two lines never cross, since heat
transfer can be in only one direction.)   In the counterflow system
(Fig. 2-10) point 1 can be higher than point b, resulting in higher saturation temperatures and consequently higher working-fluid pressures and
higher cycle efficiencies.

We shall now concentrate on the counterflow system used in most
heat-exchange applications.   In Fig. 2-10, the abscissa is enthalpy.   It
may be replaced by length along the path of the fluids.   Note then that
points on the same vertical line (same enthalpy) correspond to adjacent

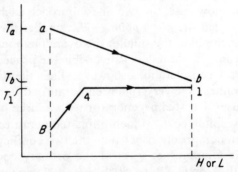

FIG. 2-11.   Heat addition to vaporizing fluid
with a variable-temperature source; parallel-
flow heat exchanger.

fluids in the heat exchanger. This is necessarily true since between any two adjacent points, and for negligible losses to the outside,

$$-\Delta H_c = \Delta H_f \tag{2-18}$$

where H is total enthalpy in Btu and the subscripts c and f refer to the primary coolant and the working fluid, respectively. For the entire system

$$H_a - H_b = H_1 - H_B \tag{2-19}$$

Equation (2-19) may be written in the form

$$m_c(h_a - h_b) = m_c c_{p_c}(T_a - T_b) = h_1 - h_B$$
$$= h_{fg} + h_s \tag{2-20}$$

where m_c = mass-flow rate of coolant per pound mass flow rate of working fluid per cycle.

c_{p_c} = average specific heat of coolant for temperature range $t_a - t_b$, Btu/lb$_m$ °F

h_{fg} = latent heat of vaporization of working fluid, Btu/lb$_m$

h_s = sensible heat added to working fluid when in liquid form
$= h_4 - h_B$, Btu/lb$_m$

The slopes of the lines in Fig. 2-10 can be given by

$$\left.\frac{dT}{dH}\right]_{a-b} = \frac{1}{m_c c_{p_c}}$$

$$\left.\frac{dT}{dH}\right]_{B-4} = \frac{1}{c_{p_f}}\Big]_{\text{liquid}} \tag{2-21}$$

$$\left.\frac{dT}{dH}\right]_{4-1} = 0$$

where c_{p_f} is the specific heat of the working fluid when in liquid form (between points B and 4) in Btu/lb$_m$ °F.

The ratio of the slopes of lines ab and $B - 4$ is therefore given by $c_{p_f}/m_c c_{p_c}$. Because the working fluid changes phase and the primary coolant does not, the specific enthalpy change of the working fluid is much greater than that of the coolant. It follows then that in order to satisfy Eq. 2-20 the value of m_c must be much greater than unity. If allowance is made for a smaller difference between c_{p_f} and c_{p_c}, the slope of the coolant line is much smaller than that of the working fluid when in liquid form, resulting in the pinch point at points d and 4, as mentioned previously.

If the total enthalpy change of the working fluid is fixed and if the temperature of the coolant entering the heat exchanger, T_a, is fixed, it follows then that T_b is also fixed. Note that for fixed boundary conditions

$m_c c_{p_c}$ is fixed and the ratio of slopes is dependent only on the type of working fluid but is independent of the type of coolant used. The following example illustrates some heat-exchanger calculations.

Example 2-1. A pressurized-water reactor produces water at 2,000 psia and 520°F. This is supposed to produce saturated steam at 500 psia in a once-through steam generator. Feed water is at 300°F. Calculate (a) the number of pounds mass of reactor coolant required per pound mass of working-fluid water for a pinch temperature difference of 20°F and (b) the temperature of the water returning to the reactor. Neglect pressure losses in the steam generator.

Solution. With reference to Fig. 2-10,

$$h_a = \text{enthalpy of compressed liquid entering steam generator}$$

This enthalpy is very nearly equal to the enthalpy of the saturated liquid at 520°F, obtained from the Keenan and Keyes steam tables (Table B-2); i.e., it equals 511.90 Btu/lb$_m$. Because the liquid is highly compressed, a small correction factor may be applied from Table 4 of the steam tables [4] if more accuracy is desired. Thus (by interpolation) $h_a = 511.9 - 0.8 = 511.1$ Btu/lb$_m$. Similarly, $h_B = 269.59 + 0.79 = 270.38$ Btu/lb$_m$.

Also
$$h_4 = 449.40 \text{ Btu/lb}_m$$
$$h_1 = 1204.40 \text{ Btu/lb}_m$$
$$t_4 = 467.01°F$$

Thus
$$t_d = 467.01 + 20 = 487.01°F$$
and
$$h_d = 472.53 \text{ Btu/lb}_m \qquad \text{By interpolation}$$

(a) The number of pounds mass of reactor coolant water per pound mass of steam produced is

$$m_c = \frac{h_1 - h_4}{h_a - h_d}$$
$$= \frac{1204.4 - 449.4}{511.1 - 472.53}$$
$$= \frac{755}{38.57} = 19.57$$

(b)
$$h_4 - h_B = m_c(h_d - h_b)$$
$$449.4 - 270.4 = 19.57(472.53 - h_b)$$

from which
$$h_b = 463.39 \text{ Btu/lb}_m$$
Thus (by interpolation)
$$t_b = 479°F$$

The temperature variation of the primary-coolant water through the reactor is therefore 479 to 520°F. This information is important from the point of view of reactor physics, since it determines the water density and consequently the neutron macroscopic cross section, mean free paths, and moderating qualities. If a smaller change is desired, the conditions of the plant should be changed to allow a greater primary-coolant mass-flow rate per pound mass of steam produced.

2-8.   SUPERHEAT AND REHEAT CYCLES

The average temperature difference in a counterflow heat exchanger may be further reduced by superheating.   This case, illustrated in Fig. 2-12, shows the saturated vapor at point 1 superheated to point 1′.   The specific heat of the vapor is usually less than that of the liquid, and on a *TH* diagram, line 1-1′ is steeper than line *B*-4.   It can be seen that there now are two approach points, the pinch point *d*-4 and the superheat point *a*-1′.   For reasons similar to those stated for the pinch point, the

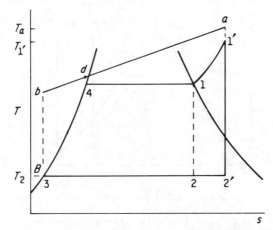

FIG. 2-12.   Internally reversible Rankine cycle with superheat and a variable-temperature heat source.

temperature difference at the latter point is optimized between operating and capital costs.   The new superheat cycle 1′-2′-3-*B*-4-1, shown in Fig. 2-12, gives higher thermal efficiency than its saturated counterpart, i.e., the one that has the same saturation pressure and temperature.   Furthermore, the superheat cycle allows drier vapor in the exhaust stages of the turbine (point 2′) than its saturated counterpart (point 2).   This reduces the problem of blade erosion and increases the turbine efficiency in an actual cycle.

The superheat Rankine cycle, however, deviates further from a Carnot cycle operating between its own temperature extremes, i.e., between $T_1′$ and the condenser temperature T_2.   Note that, when a primary coolant of fairly low temperature is used, superheat is not advisable since it lowers the *average* temperature at which most of the heat is received by the working fluid (line 4-1) and thus results in lower cycle efficiencies.   Such is the case in pressurized-water reactors where the reactor outlet temperature is theoretically limited to the water critical temperature of 705°F but actually to some 600°F because of pressure and other limitations.

If a high-temperature primary coolant (gas or liquid metal) but a low-critical-temperature working fluid (water) are to be used, the average temperature at which heat is received by the latter may be increased by the use of reheat. A cycle with one stage of reheat is shown by the solid line on the *Ts* diagram in Fig. 2-13. In this cycle the saturated

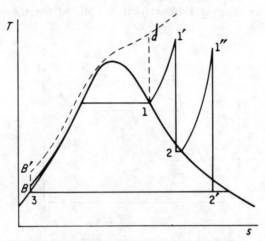

FIG. 2-13. *Ts* diagram of internally reversible supercritical and reheat cycles.

vapor at point 1 is superheated to point 1′. It is then partially expanded to point 2, after which it is led back to the heat exchanger where it is reheated to point 1″. It is then finally expanded to the condenser pressure at point 2′. Note that reheat also results in drier vapor in the latter stages of the turbine.

A modification of the above is the use of supercritical pressure heating, shown by the dashed line in Fig. 2-13. In this case the liquid is pressurized to supercritical pressures at point *B′* and heated to point *a* (with only gradual change in density but not in phase). Note that expansion from point *a* to the condenser pressure would result in very wet vapor in the later stages of the turbine. Thus supercritical heating must be accompanied by reheat.

Power plants operating on such cycles have been developed for fossil-fuel heating, where a high-temperature heat source is easily available. With nuclear heating the availability of a high-temperature heat source is limited to a few reactor types. Also, the complexity of such power plants may prove uneconomical even for these types.

2-9. THE CHOICE OF WORKING FLUID

Almost all Rankine-type power plants on earth use water-steam as the working fluid. The reasons are many: Water is available and plentiful in most places. It is cheap and, thanks to extended use, has well-known physical, thermodynamic, and handling properties. Water, however, has many disadvantages with which we have to contend. The most outstanding of these is the high pressures and low temperatures allowable in a water-steam cycle. The critical temperature and pressure of light water are 705°F and 3,206 psia, respectively. Unless superheating is used, with consequent reduction in efficiency ratio (efficiency of the proposed cycle divided by the efficiency of the Carnot cycle operating between the same temperature limits), the maximum theoretical temperature allowable in a Rankine cycle, using water as the working fluid, is limited to 705°F (actually to about 600°F, as indicated previously for the case of a presurized-water reactor). These are relatively low temperatures, considering that there are now available turbine materials that can withstand stresses at 1400°F and higher.

Associated with these low temperatures are relatively high saturation pressures. For example, at 600°F, the saturation pressure is 1,543 psia*. This necessitates component design for high pressures with consequent high capital costs. Water partly makes up for this deficiency at the low-temperature end of the cycle. With heat rejection usually in the neighborhood of room temperatures, the saturation pressures are of the order of 1 psia. These pressures correspond to partial vacuums that can easily be maintained within the condenser shell.

In order to increase the temperature at which heat is added and to take advantage of newly developed turbine materials (without resort to cycles that deviate appreciably from the Carnot cycle), other fluids having higher critical temperatures, and preferably lower saturation pressures, should be used. An example of the mandatory use of such a fluid is the case where the temperature at which heat is rejected is high, as, for example, in power plants operating outside the earth's atmosphere or at very high altitudes where radiant cooling, at 1000 to 1500°F, is the only method of heat rejection. To obtain positive work from such systems, i.e., positive efficiencies, heat must be added at much higher temperatures than those of the radiator (see Fig. 2-2).

Figure 2-14 shows vapor-pressure–temperature lines for water, mercury, and sodium. It can be seen that liquid metals in general are more suitable for high-temperature work. A conceptual design for a 20,000-kw nuclear turboelectric power supply for a manned space vehicle [10]

---

* The pressure necessary to suppress boiling at the hot spot [2] is over 2,000 psia.

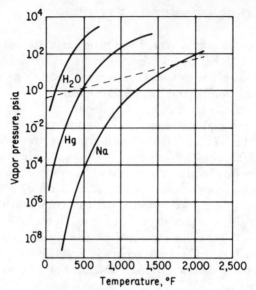

FIG. 2-14. Vapor pressure versus temperature
of some fluids.

specifies the use of sodium as both primary coolant and working fluid. The primary coolant leaves the reactor at 2340°F and 200 psi. The working fluid enters the turbine at 2040°F and 69 psi. It rejects heat at 1340°F and 2.7 psia in the radiator. Organic liquids, which have lower vapor pressures than water, do not have sufficiently high critical temperatures to justify their use in the above case. They also have the disadvantage of decomposition at high temperatures and under nuclear irradiation but have enough favorable qualities to justify use in other cases.

There are available a number of liquid metals with a wide range of pressure-temperature relationships, so that an appropriate one may be chosen to suit a particular situation. The difficulty with liquid metals is at low temperatures. At 100°F the saturation pressure is about 10^{-5} psia for mercury and below 10^{-10} psia for sodium. Such pressures, of course, are extremely difficult to maintain.

One other interesting characteristic of fluids is the shape of their Ts diagrams. Figure 2-15 shows three types. In Fig. 2-15a the diagram has an open dome. Water and liquid metals belong to this type. Expansion of saturated vapor between p_1 and p_2 here results in wet vapor in the exhaust stages of the turbine. In Fig. 2-15b, expansion results in superheated vapor but reduced work between p_1 and p_2, since the constant-pressure lines slant upward in the superheat region. Most organic compounds belong to this type. In Fig. 2-15c the saturated-vapor line

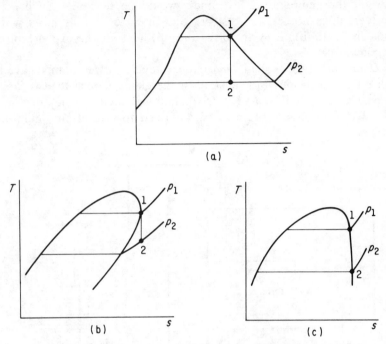

FIG. 2-15. Three *Ts* diagrams with isentropic expansions of saturated
vapor between constant-pressure lines p_1 and p_2.

is almost vertical, resulting in little change in the quality of the expanding
vapor. Phosphorus, carbon tetrachloride, chlorobenzene, and other
materials belong to this group.

Appendix B contains thermodynamic tables and charts for some
fluids of interest.

2-10. MULTIFLUID VAPOR CYCLES

It has been shown in the previous section that some fluids have good
vapor-pressure–temperature characteristics at high temperatures and
some have them at low temperatures. These characteristics are represented by lines which are roughly parallel, as shown in Fig. 2-14. Where
both a high-temperature heat source and a low-temperature heat sink
are available, an ideal fluid would be one represented by the dashed line
of Fig. 2-14. This hypothetical fluid has a relatively low vapor pressure
at the high-temperature end and a relatively high vapor pressure at the
low-temperature end. In this case no superheat or reheat is necessary,
and a cycle efficiency approaching that of a Carnot cycle, operating

between the temperature extremes, would be feasible. Such a fluid, however, does not exist. Instead, a teaming of two or more working fluids in a multifluid cycle solves the problem, at the expense of plant complication.

Dual-fluid- (also called *binary-vapor*) cycle power plants have been built on a large scale in the past, using mercury-steam as the two working fluids [11, 12]. A flow and a *TS* diagram of such a cycle are shown in Figs. 2-16 and 2-17. Mercury is vaporized to saturation conditions in

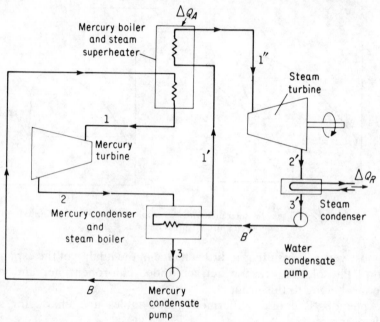

FIG. 2-16. Schematic of a mercury-steam binary-vapor power plant.

a mercury boiler. Saturated mercury vapor at point 1, say at 400 psia and 1148°F, now expands in a mercury turbine to point 2, where the pressure may be 2 psia, corresponding to 505°F. It is then condensed to saturated liquid mercury (point 3). Liquid mercury is then pumped back to the boiler in the usual fashion, and the mercury cycle is repeated.

Because the temperature of mercury during condensation is high, the cooling medium in the mercury condenser can be water at a low pressure, such as 422.6 psia. Water at point B' is now heated in the mercury condenser, to the corresponding saturation temperature of 450°F and then evaporated to point 1′. The resulting saturated steam may then be superheated at constant pressure to point 1″, utilizing the gases leaving the mercury boiler which must be at a temperature not much lower than that

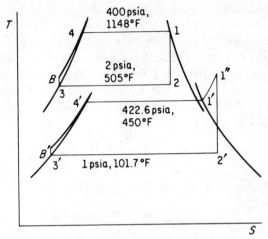

FIG. 2-17. *TS* diagram of internally reversible mercury-steam cycle.

of point 4 (see Fig. 2-10). Superheated steam now expands in a steam turbine to point 2'. It is condensed in the usual manner (e.g., using cooling water), say at 1 psia and 101.7°F, to point 3'. It is then pumped back to the mercury condenser, completing the steam cycle.

Note that heat, ΔQ_A, is mostly added to the composite cycle in the mercury boiler. This is done mainly at the constant temperature of 1148°F. Heat rejection, ΔQ_R, also at the constant temperature of 101.7°F, takes place only in the steam condenser. The heat exchanged in the mercury condenser is, barring heat losses to the outside, internal and does not figure in the heat balance or efficiency of the composite cycle. If water alone were used, it could receive heat at constant temperature only if this temperature were sufficiently below its critical temperature of 705.4°F and also at relatively high pressures. The maximum pressures of the composite cycle are in the neighborhood of 400 psia. Note also that the larger the temperature differences between lines 2-3 and 4'-1', the smaller and more compact will be the heat exchanger (mercury condenser) and the smaller the work done per pound mass of fluid in each cycle. Since for the same heat added and rejected the same work is obtained from the cycle, a large temperature difference there, however, necessitates the use of a larger mass-flow rate of the fluids in turbines, heat exchangers, etc. This may mean larger components and/or pressure losses.

Some disadvantages of the mercury-steam cycle are that mercury is expensive and toxic and that the cycle is relatively more complicated than a single-fluid one. Thus, despite the many advantages of the dual-fluid cycle, its use was discontinued many years ago, especially since techniques of utilizing and constructing materials for high-pressure work were developed.

Lately, however, more experience and information regarding the handling of liquid metals have been gained, and renewed interest in multifluid cycles has appeared. Multifluid cycles present a solution to the problem of the efficient use of high-temperature sources of heat. The efficiency of the cycle of the type shown in Figs. 2-16 and 2-17 may be given as usual by

$$\eta = \frac{\Delta Q_A - \Delta Q_R}{\Delta Q_A}$$

or, in terms of work done (pump work neglected),

$$\eta = \frac{m(h_1 - h_2) + m'(h_{1''} - h_{2'})}{\Delta Q_A} \tag{2-22}$$

where ΔQ_A, ΔQ_R = heat added and rejected, Btu/cycle
     m, m' = lb$_m$/cycle of high- and low-temperature fluids
     h = specific enthalpy, Btu/lb$_m$, of fluids as denoted by various subscripts

The ratio m/m' depends upon the fluid types and their temperatures and pressures when in the mercury condenser. Thus, where no feedwater heating (by regeneration or an economizer) is attempted in the steam cycle,

$$\frac{m}{m'} = \frac{h_{1'} - h_{B'}}{h_2 - h_3} = \frac{h_{1'} - h_{3'}}{h_2 - h_3} \tag{2-23}$$

For the conditions of Fig. 2-17, it can be shown, using the tables in Appendix B, that 8.97 lb$_m$ of mercury must be circulated in the cycle for every pound mass of water. The TS diagrams of the two cycles, as drawn in Fig. 2-17, are for corresponding masses.

Figure 2-18 shows Ts diagrams of water, mercury, sodium, and diphenyl, each drawn for 1 lb$_m$ of the respective substances. It can be seen that a sodium-water dual-fluid cycle is not feasible since it would involve extremely low sodium pressures in the condenser-boiler. A sodium-mercury dual cycle, where both heat addition and rejection must be at a high temperature, is, however, feasible. When a heat source at extremely high temperature and a heat sink at extremely low temperature are available, a *trifluid* cycle, such as one using sodium, mercury, and water as working fluids, is feasible and may in fact be necessary. Many other combinations utilizing other fluids (such as other liquid metals, organics, the freons, etc.) present good thermodynamic solutions of specific problems.

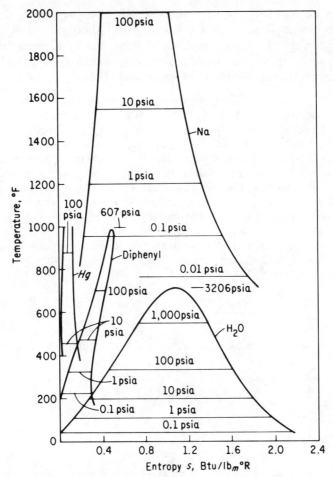

FIG. 2-18. *Ts* saturated-liquid and vapor lines for four fluids.

When regeneration is added to multifluid cycles, the combined efficiency approaches the Carnot-cycle efficiency between the temperature limits. For example, a sodium-mercury-steam cycle operating between 2000°F (when materials that withstand such temperatures become available) and 101.7°F gives an efficiency of 71 percent as compared with a Carnot efficiency, for the same temperature limits, of 77.2 percent. A single-fluid-steam supercritical cycle operating between these same temperature limits would yield an efficiency of approximately 50 percent or less.

It should be noted here that many problems remain to be solved before multifluid plants generating large quantities of power become a practical reality. Such problems include leakage; heat-transfer and fluid-flow behavior of the various fluids when in the liquid, vapor, and changing phases (boiling and condensation); thermodynamic behavior (such as during expansion and compression); availability of compatible materials (erosion, corrosion, etc.); the complication, cost, and mass of the power-plant components to be used; and others.

Example 2-2. A sodium-mercury-steam cycle operates between 1840 and 101.7°F. Sodium rejects heat at 1240°F to mercury. Mercury boils at 350 psia and rejects heat at 2 psia. Both the sodium and mercury cycles are saturated. Steam is formed at 420 psia and is superheated in the sodium boiler to 640°F. It rejects heat at 1 psia. Assume isentropic expansions, no heat losses, and no regeneration and neglect pumping work. Find (a) the number of pounds mass of sodium and mercury used per pound mass of steam; (b) the heat added and rejected in the composite cycle per pound mass of steam; (c) the total work done per pound mass of steam; (d) the efficiency of the composite cycle; (e) the efficiency of the corresponding Carnot cycle; and (f) the work, heat added, and efficiency of a supercritical pressure steam (single-fluid) cycle operating at 3,500 psia and between the same temperature limits.

Solution. Refer to Fig. 2-19.

(a)
$$\frac{m'}{m''} = \frac{h_E - h_C}{h_b - h_c} \qquad \text{and} \qquad \frac{m}{m'} = \frac{h_a - h_c}{h_2 - h_3}$$

where m, m', and m'' are the mass rates of flow of sodium, mercury, and steam, respectively.

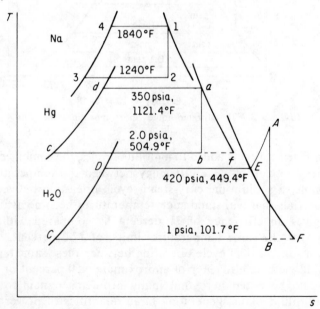

FIG. 2-19. *TS* diagrams for trifluid cycle of Example 2-2.

For steam, from Table B-1,

$$h_E = 1204.6 \text{ Btu/lb}_m$$
$$h_C = 69.7 \text{ Btu/lb}_m$$

For mercury, from Table B-7, Appendix B, by interpolation,

$$h_a = 184.59$$
$$h_c = 37.37 \text{ Btu/lb}_m$$

To obtain h_b: Since $a - b$ represents isentropic expansion,

$$s_b = s_a = 0.2048 \text{ Btu/lb}_m \text{ °R}$$

and $$x_b = \frac{s_b - s_c}{s_f - s_c} = \frac{0.2048 - 0.1098}{0.2435 - 0.1098} = 0.7105$$

Thus $$h_b = h_c + x_b(h_f - h_c)$$
$$= 37.37 + 0.7105(163.93 - 37.37) = 127.29 \text{Btu/lb}_m$$

For sodium, from Table B-4, following a procedure as above, $h_2 = 1881.2$ Btu/lb_m. At point 2 the quality is approximately 77 percent.

h_2 and h_b may also be obtained from Figs. B-2 and B-5 following constant entropy lines from points 1 and a on the 100 percent quality lines through the wet region until the intersection with a constant pressure (and temperature) line representing heat rejection.

$h_3 = 528.5 \text{ Btu/lb}_m$. This is the saturated liquid enthalpy at 1240°F (1700°R), Table B-4. Based on $m'' = 1 \text{ lb}_m$ of steam,

$$m' = \frac{1204.6 - 69.7}{127.29 - 37.37} = 12.62 \text{ lb}_m \text{ Hg/lb}_m \text{ H}_2\text{O}$$

(b) $$m = m' \frac{184.59 - 37.37}{1881.2 - 528.5} = 1.374 \text{ lb}_m \text{ Na/lb}_m \text{ H}_2\text{O}$$

$$\Delta Q_A = m(h_1 - h_3) + (h_A - h_E)$$
$$= 1.374(2330.7 - 528.5) + (1328.3 - 1204.6) = 2599.9 \text{ Btu/lb}_m \text{ H}_2\text{O}$$
$$\Delta Q_R = h_B - h_C = 898 - 69.7 = 828.3 \text{ Btu/lb}_m \text{ H}_2\text{O}$$

h_A and h_B are obtained from Tables B-3 and B-1 respectively.

(c) Total work done $\Delta W_t = \Delta Q_A - \Delta Q_R = 1771.6 \text{ Btu/lb}_m \text{ H}_2\text{O}$

ΔW_t may also be obtained by adding the work done in the individual cycles. Thus

$$\Delta W_t = m(h_1 - h_2) + m'(h_a - h_b) + (h_A - h_B)$$
$$= 1.374(2330.7 - 1881.2) + 12.62(184.59 - 127.29) + (1328.3 - 898)$$
$$= 617.6 + 723.1 + 430.3 = 1771.0 \text{ Btu/lb}_m \text{ H}_2\text{O}$$

(d) Efficiency of composite cycle $= \dfrac{\Delta W_t}{\Delta Q_A} = \dfrac{1771.6}{2599.9} = 68.1$ percent

(e) Efficiency of corresponding Carnot cycle $= \dfrac{1840 - 101.7}{(1840 + 460)} = 75.6$ percent

Thus this cycle attains an efficiency ratio of 90 percent, i.e., an efficiency which is 90 percent that of a Carnot cycle operating between the same temperature limits. If regeneration were incorporated, as it would be in an actual plant, the efficiency ratio would improve. The above calculations, however, yield efficiencies higher than those of an actual power plant, since the effects of irreversibilities (nonisentropic expansions, heat losses to the atmosphere, etc.) were not taken into account.

(f) For a supercritical-pressure steam cycle operating between the same temperature limits (refer to Fig. 2-20),

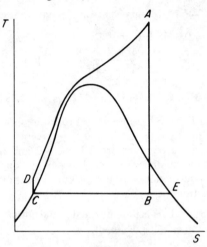

FIG. 2-20.   Supercritical-pressure cycle.

$$h_A = 1970 \text{ Btu/lb}_m \quad\Big\}\quad \text{at 3,500 psia and 1840°F (from Table B-3)}$$
$$s_A = 1.765 \text{ Btu/lb}_m \text{ °R} \quad\quad \text{by extrapolation}$$
$$s_B = s_A = s_C + x_B(s_E - s_C)$$
$$1.765 = 0.1326 + x_B(1.8456) \quad \text{from which } x_B = 0.884$$

Thus $h_B = h_C + x_B(h_E - h_C) = 69.7 + 0.884(1036.3) = 985.8 \text{ Btu/lb}_m$

Because of the high pressure in this cycle, the work of compression of the liquid W_{C-D} cannot be neglected.   Thus

$$W_{C-D} = h_D - h_C = \frac{v_c(p_D - p_C)}{J}$$

where v = specific volume
p = pressure
J = mechanical equivalent of heat

$$W_{C-D} = \frac{0.01614(3,500 - 1)144}{778} = 10.45 \text{ Btu/lb}_m$$

Net work of cycle $= (h_A - h_B) - W_{C-D}$
$$= (1970 - 985.8) - 10.45 = 973.75 \text{ Btu/lb}_m$$

Heat added $= h_A - h_D = (h_A - h_C) - W_{C-D}$
$$= (1970 - 69.7) - 10.45 = 1889.85 \text{ Btu/lb}_m$$

$$\text{Efficiency} = \frac{973.75}{1889.85} = 51.5 \text{ percent}$$

Compare this with the multifluid-cycle efficiency of 68.1 percent and Carnot-cycle efficiency of 75.6 percent.

PROBLEMS

2-1. Three 1-ft³ constant-volume chambers contain two-phase mixtures of mercury, sodium, and H_2O, respectively. The temperature and quality in all cases are 1100°R and 90 percent. Determine for each case (a) the mass of mixture in lb_m, (b) the chamber pressure in psia, and (c) the specific enthalpy in Btu/lb_m.

2-2. Three Carnot cycles using mercury, sodium, and H_2O as working fluids receive heat causing these fluids to vaporize at 100 psia. Determine for each cycle the pressure and temperature at which heat must be rejected if the efficiency in all cases is 35 percent.

2-3. If in Prob. 2-2 the three fluids change from saturated liquid to saturated vapor during the heat-addition processes, determine in each case (a) the quality at the end of expansion; (b) the quality at the beginning of compression; and (c) the heat added, heat rejected, the work of the expander, and the work of compression, in Btu/lb_m.

2-4. A reactor power plant represented by Fig. 2-5 operates on the Ts diagram of Fig. 2-6. For metallurgical reasons, the average surface temperature of the reactor fuel is limited to 700°F. The temperature drop in the heat exchanger is 30°F. The adiabatic efficiency of the turbine (actual enthalpy drop divided by isentropic enthalpy drop) is 90 percent. Steam is the working fluid. Heat is rejected at 1 psia condenser pressure. Find the optimum power of the cycle, in Btu/hr, and the corresponding efficiency if the overall heat-transfer coefficient of the heat exchanger is 500 Btu/hr ft² °F and its heat-transfer surface is 4,000 ft².

2-5. A gas (CO_2)-cooled reactor power plant uses steam as the working fluid.

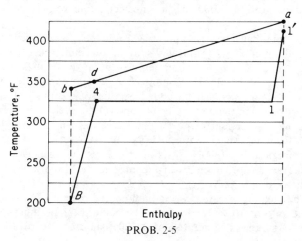

PROB. 2-5

The temperature-enthalpy diagram for the heat exchanger is shown. No regeneration is used. Find (a) the number of pounds mass of CO_2 per pound mass of steam used and (b) the working-cycle efficiency, assuming isentropic expansion and no heat losses.

2-6. A gas (CO_2)-cooled reactor power plant operates on the heat-exchanger conditions shown in the diagram.

(a) Find the maximum steam pressure if, by regeneration, the feed water enters the heat exchanger at 250°F instead of 200°F. The temperature difference at the pinch point and the gas temperatures remain the same.

(b) Find the new cycle efficiency, assuming isentropic expansion and no heat losses.

2-7. The turbine inlet temperature in a pressurized-water-reactor power plant is limited to 550°F. Two nonregenerative Rankine cycles are proposed, one using saturated steam and the other steam that is 46.9°F superheated. The condensate temperature in both cases is 105°F. Assuming both cycles to be internally reversible and neglecting the pump work, find for each case (a) the cycle thermal efficiency, and (b) the quality of the turbine exhaust.

2-8. A gas-cooled reactor uses helium as primary coolant. Helium then gives off its heat to feed water coming from the steam cycle at 250°F. The gas leaves the heat exchanger at 400°F. The maximum steam pressure is 900 psia. Calculate (a) the number of pounds mass of He per pound mass of steam if the temperature difference at the pinch point is 23°F and (b) the temperature of the gas leaving the reactor if the steam is superheated to 1000°F. Assume that helium has constant $c_p = 1.25$.

2-9. A high-temperature binary-vapor cycle uses lithium and sodium as working fluids. Heat is added to the lithium, producing saturated vapor at 3400°R. In the lithium condenser–sodium boiler, the lithium is at 2300°R and the sodium is at 2200°R. The sodium is then superheated in the lithium boiler to 2500°R. Heat is rejected from the cycle at 1600°R. Assuming isentropic expansions and no heat losses, calculate (a) the number of pounds mass of lithium per pound mass of sodium required, (b) the cycle power in Btu/hr if the total sodium flow is 1,000 lb_m/hr (neglect pump work), and (c) the cycle efficiency.

2-10. Assume that a space-boiling reactor channel is designed to completely vaporize its sodium coolant. Saturated sodium thus emanating from the reactor is used to drive a turbine. Sodium enters the turbine at 2600°R, and the condenser-radiator at 1600°R. Assuming an ideal nonregenerative Rankine cycle, find (a) the amount of heat added per lb_m of sodium in the nonboiling region of the core, (b) in the boiling region, and (c) the cycle efficiency.

2-11. A SNAP-type reactor power plant uses sodium as a working fluid in a direct-boiling Rankine cycle. Saturated Na enters the turbine at 9.15 psia. The condenser temperature is 1300°R. Assuming isentropic expansion, find (a) the Na mass flow rate in lb_m/hr necessary to produce 1 Mw of power, and (b) the necessary reactor output in Mw(t).

2-12. The steam generator in a sodium-cooled reactor power plant receives sodium at the rate of 10×10^6 lb_m/hr. The sodium temperatures are 600 and 1000°F. 800°F steam is generated at 1,000 psia from feedwater at 340°F.

Assuming no heat losses, find the temperature difference at the pinch point in the generator. (c_p for Na = 0.3.)

2-13. Derive a relationship between the thermal pollution (heat rejected) from a power plant of a given output, and the thermal efficiency of the plant. Show that the thermal pollution of a 30 percent efficient plant (current water reactors) is some 55 percent greater than from a 40 percent efficient plant (current gas-cooled reactors and fossil-fueled plants), 133 percent greater than from a 50 percent efficient plant, and 250 percent greater than from a 60 percent efficient plant (projected fusion plants).

2-14. A power plant is designed to generate power for a life support system on a space station. The plant should be nuclear because the length of the mission prohibits the lifting off of sufficient fossil-fuel supplies. The upper temperature of the power plant cycle is fixed by metallurgical considerations. Because the plant operates in space, it can only reject heat by thermal radiation to space which is at $0°R$, according to $\Delta Q_R = \sigma \epsilon A_R T_R^4$ where σ is the Stefan-Boltzmann constant, ϵ is an emissivity factor of the radiator, assumed constant, and A_R and T_R are the surface area and temperature of the radiator surface respectively. Noting that the radiator represents by far the largest mass in the power plant, it was found that, from a mass point of view, it is preferable to operate the plant at relatively low efficiency. Derive a relationship between mass and cycle efficiency, assuming for simplicity that the plant operates on a Carnot cycle. Also find the optimum efficiency at which the mass is minimum.

chapter **3**

The Boiling Reactor

3-1. INTRODUCTION

Except for direct-energy-conversion techniques, reactors in which the coolant boils within the core are probably the most direct means of converting nuclear energy into useful power. The boiling reactor has a function closely resembling that of the boiler in a conventional fossil-fuel steam power plant and is basically simpler than it. In the boiler, heat is transmitted from the furnace to the water indirectly—partly by radiation, partly by convection, and partly by conduction, with combustion gases used as an intermediate agent or coolant. In the boiling-water reactor, the coolant is in direct contact with the heat-producing nuclear fuel and boils in the same compartment in which the fuel is located.

There are many possible coolants for this type of reactor. Ordinary water has been used most extensively, however, because of its availability and the advanced state of knowledge concerning it. Heavy water, organic liquids, and liquid metals are the other possible coolants. Because of the moderating action, boiling reactors using water or organic liquids are necessarily of the thermal type. The coolant thus serves the triple function of coolant, moderator, and working fluid. If a liquid metal is used in a thermal reactor, a separate, usually solid, moderator is necessary.

In its simplest form (Fig. 3-1), a boiling-reactor power plant consists of a reactor, a turbine generator, a condenser and associated equipment (such as air ejector, cooling system, etc.), and a feed pump. Slightly subcooled liquid enters the reactor core at the bottom where it receives sensible heat to saturation plus some latent heat of vaporization. When it reaches the top of the core, it has been converted into a very wet mixture of liquid and vapor. The vapor separates from the liquid, flows to the turbine, does work, and is condensed by the condenser, and the condensate is then pumped back to the reactor by the feed pump.

The saturated liquid that separates from the vapor at the top of the reactor or in a steam separator flows downward via *downcomers* within or outside the reactor and mixes with the return condensate. This recircul-

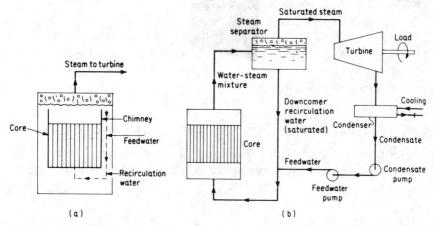

FIG. 3-1. Simple boiling-water-reactor systems. (a) Internal recirculation; (b) external recirculation.

ating coolant flows either naturally, by the density differential between the liquid in the downcomer and the two-phase mixture in the core, or by a recirculating pump in the downcomer line, not shown.   The recirculation pump usually is used with external circulation but can be incorporated with internal circulation by placing it on the side of the reactor vessel.   Internal recirculation has the advantage of simplicity (less piping, no vapor separator drum).   External recirculation has the advantage of more efficient vapor separation and more natural driving head, because of the added height. Good vapor separation in internal recirculation can be accomplished only by low downcomer velocity or by incorporating separating devices inside the reactor.   It can be seen that, in either system, we have a one-loop arrangement with inherent high plant thermal efficiency.

The ratio of the recirculation liquid to the saturated vapor produced is called the *recirculation ratio*.   It is a function of the core average exit quality (Eq. 3-4, below).   Boiling-water core exit qualities are low, of the order of a few percent, so that recirculation ratios in the range 10:1 to 25:1 are common.   This is necessary to avoid large void fractions in the core which would materially lower the moderating powers of the coolant, and possibly result in low heat transfer coefficients, or vapor blanketing and burnout [2].

Unstable and erratic behavior (chugging oscillations, etc.) due to the boiling process may occur with this type of reactor, at low pressures, because of the unsteady nature of low-pressure voids.   Such oscillations place a limit on the specific power attainable in a low-pressure boiling reactor.   Boiling-water reactor operating pressures of 600 to 1100 psia are used to overcome this and to take advantage of the favorable water burnout heat fluxes [2] at such pressures.

This chapter will cover the thermodynamic and hydrodynamic characteristics* and the void and pressure induced kinetics of a boiling core. The latter have a bearing on the cycle design of the different types of boiling reactors. These types and the power plants associated with them are the subject of the next chapter.

3-2. BOILING REACTOR MASS AND HEAT BALANCE

As shown in Fig. 3-2 for both internal and the external recirculation boiling reactors, a slightly subcooled liquid enters the core bottom at a

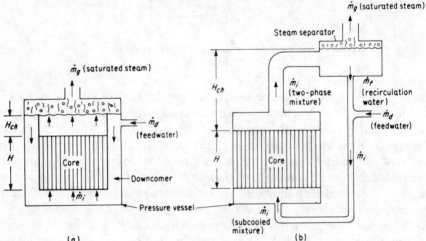

FIG. 3-2. Mass-flow rates in simple boiling-water reactor with (a) internal and (b) external recirculation.

rate of $\dot{m}_i$ lb_m/hr. This liquid rises through the core and *chimney,* if any. The chimney is an unheated section above the core which helps to increase the driving pressure in natural circulation, Sec. 3-5. The resulting vapor separates and proceeds to the power plant at a rate of $\dot{m}_g$ lb_m/hr. The saturated recirculation liquid flows via the downcomer at the rate of $\dot{m}_f$ lb_m/hr. There it mixes with the relatively cold return feed liquid $\dot{m}_d$ from the power plant to form the slighty subcooled inlet liquid $\dot{m}_i$.

An *overall* mass balance in the reactor core is given by

$$\dot{m}_d = \dot{m}_g \tag{3-1}$$

$$\dot{m}_g + \dot{m}_f = \dot{m}_i \tag{3-2}$$

---

* Sections 3-2 through 3-6 in this book are repeated from Chapter 14 of the author's companion book *Nuclear Heat Transport* [2].

The *average* exit quality of the entire core $\bar{x}_e$, that is, the quality of all the vapor-liquid mixture at the core exit, is given by

$$\bar{x}_e = \frac{\dot{m}_g}{\dot{m}_g + \dot{m}_f} = \frac{\dot{m}_d}{\dot{m}_d + \dot{m}_f} = \frac{\dot{m}_d}{\dot{m}_i} \tag{3-3}$$

The *recirculation ratio* is the ratio of recirculation liquid to vapor produced. It is given by modifying Eq. 3-3 as follows:

$$\frac{\dot{m}_f}{\dot{m}_g} = \frac{1 - \bar{x}_e}{\bar{x}_e} \tag{3-4}$$

Now if the incoming feed liquid has a specific enthalpy h_d Btu/lb$_m$ and the saturated recirculated liquid has a specific enthalpy h_f (at the system pressure), a heat balance is obtained, if we assume no heat losses to the outside (a good assumption) and neglect changes in kinetic and potential energies, as follows:

$$\dot{m}_i h_i = \dot{m}_f h_f + \dot{m}_d h_d \tag{3-5}$$

where h_i is the specific enthalpy of the liquid at the reactor-core inlet. Equation 3-5 can be modified to

$$h_i = (1 - \bar{x}_e)h_f + \bar{x}_e h_d \tag{3-6}$$

Rearranging gives the following expression for $\bar{x}_e$:

$$\bar{x}_e = \frac{h_f - h_i}{h_f - h_d} \tag{3-7}$$

The condition of the liquid entering the bottom of the core is given by the *enthalpy of subcooling,*

$$\Delta h_{\text{sub}} = h_f - h_i \tag{3-8a}$$

or

$$\Delta h_{\text{sub}} = \bar{x}_e(h_f - h_d) \tag{3-8b}$$

or by the *degree of·subcooling,*

$$\Delta t_{\text{sub}} = t_f - t_i \tag{3-9}$$

where t_i is the core inlet liquid temperature, corresponding to h_i.

The *total heat generated, Q_t,* can be obtained from a heat balance on the core as a system, or on the reactor as a system. Assuming negligible heat losses, the two relationships, which should yield identical results, are respectively

$$Q_t = \dot{m}_i[(h_f + \bar{x}_e h_{fg}) - h_i] \tag{3-10}$$

$$= \dot{m}_g(h_g - h_d) \tag{3-11}$$

Example 3-1. Find the overall heat and mass balance parameters of a boiling-water reactor operating at 1,000 psia and producing 10^7 lb$_m$/hr of saturated steam. The average exit.void fraction has been set by physics considerations at 40 percent. The slip ratio, S, assumed uniform throughout the core is 1.95. The feedwater is at 300°F.

Solution. Using the steam tables, Appendix B:

At 1,000 psia, $t_f = 544.6$°F, $h_f = 542.4$ Btu/lb$_m$, $h_g = 1191.8$ Btu/lb$_m$, $v_f = 0.0216$ ft^3/lb$_m$, $v_g = 0.4456$ ft^3/lb$_m$

At 300°F, $h_d = 269.6$ Btu/lb$_m$

Thus
$$\phi = \frac{v_f}{v_g} S = \frac{0.0216}{0.4456} \times 1.95 = 0.0945$$

Average core exit quality $\bar{x}_e = \dfrac{1}{1 + [(1 - \alpha_e)/\alpha_e]1/\phi}$, from [2]

$$= \frac{1}{1 + (0.6/0.4)/0.0945} = 0.0593$$

Recirculation ratio $= \dfrac{1 - \bar{x}_e}{\bar{x}_e} = \dfrac{1 - 0.0593}{0.0593} = 15.86$

Recirculation rate $= \dot{m}_f = 15.86 \times 10^7 = 1.586 \times 10^8$ lb$_m$/hr

Total core flow $= \dot{m}_i = (15.86 + 1)10^7 = 1.686 \times 10^8$ lb$_m$/hr

$\Delta h_{\text{sub}} = x_e(h_f - h_d) = 0.0593(542.4 - 269.6) = 16.2$ Btu/lb$_m$

$h_i = h_f - \Delta h_{\text{sub}} = 542.4 - 16.2 = 526.2$ Btu/lb$_m$

Core inlet temperature $= t_i =$ saturation temperature corresponding to a saturated liquid enthalpy of 526.2 Btu/lb$_m$ = 531.7°F (by interpolation).

$$\Delta t_{\text{sub}} = t_f - t_i = 544.6 - 531.7 = 12.9°\text{F}$$

Total heat generated in the core, Eq. 3-11,

$$Q_t = 10^7(1191.8 - 269.6) = 9.222 \times 10^9 \text{ Btu/hr}$$
$$= 9.222 \times 10^9 \times 2.931 \times 10^{-7} = 2703 \text{ Mw(t)}$$

The same result could be obtained from Eq. 3-10.

3-3. THE DRIVING PRESSURE IN A BOILING CHANNEL

A boiling core is usually composed of many flow channels. The individual channels do not generate the same heat, have different flows, have different pressure drops* and have different driving pressures. In this section we shall evaluate the driving pressure in a single channel. The case of a multichannel core will be taken up later.

The liquid in the downcomer of a boiling reactor is either saturated or slightly subcooled, depending upon whether the feed water is added

---

* The total pressure drop across each channel must, however, be the same for all channels in a core.

near the bottom or top of the core. In either case, the density in the downcomer, ρ_{dc}, is greater than the channel average density (in the core and chimney, if any) $\bar{\rho}_c$ where voids are found. Because of this density differential, a driving pressure, Δp_d is established. It is given, in the general case, by

Δp_d = (hydrostatic pressure drop in downcomer) − (hydrostatic pressure drop in channel) + (pressure rise, due to pumping, Δp_p, if any)

$$= [\rho_{dc}(H + H_{ch}) - \bar{\rho}_c(H + H_{ch})] \frac{g}{g_c} + \Delta p_p \qquad (3\text{-}12)$$

where H = height of core, ft

H_{ch} = height of chimney, ft

g = gravitational acceleration, ft/hr²

g_c = conversion factor = 4.17×10^8 lb$_m$ft/lb$_f$hr²

In the case of natural recirculation, and the case of forced recirculation during a power loss to the pump, $\Delta p_p = 0$. In the case of forced recirculation Δp_p may be much greater than the natural recirculation caused by the difference in the hydrostatic pressure terms, and the latter may be ignored. In the case of a reactor core without chimney, $H_{ch} = 0$.

The driving pressure, Δp_d, must equal the total system pressure losses, $\Sigma \Delta p$, at the desired rate of flow. The latter are given by

$$\Sigma \Delta p = \Sigma \Delta p_f + \Sigma \Delta p_a + \Sigma \Delta p_{c,e} \qquad (3\text{-}13)$$

where $\Sigma \Delta p$ = total pressure losses, lb$_f$/ft²

$\Sigma \Delta p_f$ = sum of frictional pressure losses in core, chimney (if any), and downcomer all computed in direction of flow (arrows in Fig. 3-3)

$\Sigma \Delta p_a$ = sum of acceleration pressure losses (these may be ignored in chimney and downcomer, since no large changes in density are encountered there)

$\Sigma \Delta p_{c,e}$ = sum of pressure losses due to area contractions and expansions such as at core entrance and exit and due to restrictions and submerged bodies such as at spacers, support plates, feedwater distribution ring, fuel-element handles, etc., again all added up in direction of flow indicated

If the driving pressure is less than the losses at a given flow rate, a new equilibrium condition will be established at a reduced flow rate (recall that all single and two-phase flow losses are proportional to the square of the flow rate) and consequently a reduced reactor power output.

Methods of obtaining the various losses have been outlined in Chapter 12 of [2]. The driving pressure will be evaluated below.

The density in the downcomer, ρ_{dc}, is equal to the density of the

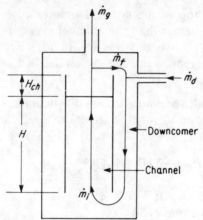

FIG. 3-3. Flow in a natural-recirculation reactor composed of a single channel core, downcomer, and
chimney.

subcooled water at core inlet, ρ_i, if feedwater is added near the top of
the core. In this case,

$$\rho_{dc} = \rho_i = \frac{1}{v_i} \tag{3-14}$$

where v_i is very nearly equal to the specific volume of the saturated liquid
obtained at temperature t_i. If feedwater is added near the bottom of the
core, then

$$\rho_{dc} = \rho_f = \frac{1}{v_f} \tag{3-15}$$

where v_f is the specific volume of saturated liquid at the pressure of the
system. It can be seen that since ρ_i is greater than ρ_f, it is advantageous
to add the feedwater near the top. Also the addition there of the
relatively cool feedwater helps prevent vapor bubble carry-under to the
downcomer where it would have an adverse effect on the driving pressure
(because of reduced downcomer density), and to the core where it would
have an adverse effect on moderation.

The average channel density in core and chimney is given by

$$\bar{\rho}_c = \frac{\bar{\rho}_0 H_0 + \bar{\rho}_B H_B + \bar{\rho}_{ch} H_{ch}}{H_0 + H_B + H_{ch}} \tag{3-16}$$

where H_0 = core nonboiling height, ft.
$\quad H_B$ = core boiling height, ft.
$\quad \bar{\rho}_0$ = average density in the nonboiling height H_0
$\quad \bar{\rho}_B$ = average density in the boiling height H_B
$\quad \bar{\rho}_{ch}$ = average density in the chimney, of height H_{ch}

We shall now evaluate these three densities reserving the most complex, $\bar{\rho}_B$, to last.

$\bar{\rho}_0$ is that constant density that, if existed in H_0, would give the same hydrostatic pressure as the actual varying density in H_0. Thus

$$(\Delta p_h)_{H_0} = \int_0^{H_0} \rho_0(z)\, dz\, \frac{g}{g_c} = \bar{\rho}_0 H_0 \frac{g}{g_c}$$

or

$$\bar{\rho}_0 = \frac{1}{H_0} \int_0^{H_0} \rho_0(z)\, dz \qquad (3\text{-}17)$$

where $\rho_0(z)$ is the density at axial position z within H_0. Evaluation of the above integral depends upon the axial heat-flux distribution in H_0 and may be a complex matter. Since, however, the degree of subcooling is usually small, the density of the liquid can be considered to vary linearly with position z, and it is therefore sufficiently accurate to use

$$\bar{\rho}_0 = \frac{1}{2}\,(\rho_i + \rho_f) = \frac{1}{2}\left(\frac{1}{v_i} + \frac{1}{v_f}\right) \qquad (3\text{-}18)$$

The chimney contains a two-phase mixture. The density there, ρ_{ch}, therefore depends upon the void fraction. Since no heat is added in the chimney, the quality in it will be equal to the channel exit quality. If the slip ratio remains constant (it changes at the sudden entrance to a chimney but returns to the original value), then the void fraction in the chimney will be equal to the channel exit void fraction, and

$$\bar{\rho}_{ch} = \rho_e \qquad (3\text{-}19)$$

Note therefore that the density in the chimney is the lowest in the reactor. An addition of a chimney therefore increases the driving pressure materially, by the quantity $(\rho_{dc} - \rho_e)\, H_{ch}\, (g/g_c)$, and a chimney is therefore often resorted to in natural circulation systems. Also the chimney contains no fuel rods and contains less obstruction to flow so that the added pressure losses due to it are less, per foot, than in the core. The driving pressure is now given by

$$\Delta p_d = \left[\rho_{dc}(H + H_{ch}) - (\bar{\rho}_0 H_0 + \bar{\rho}_B H_B + \rho_e H_{ch}) \right] \frac{g}{g_c} + \Delta p_p$$

$$(3\text{-}20)$$

There now only remains the average density in the boiling height $\bar{\rho}_B$ to be evaluated to complete the evaluation of driving pressure. This is done in the following section.

3-4. THE AVERAGE DENSITY IN A BOILING CHANNEL

The hydrostatic head in the boiling height is given by

$$(\Delta p_h)_{H_B} = \int_{H_0}^{H} \rho_B(z)\, dz\, \frac{g}{g_c} = \bar{\rho}_B H_B \frac{g}{g_c} \tag{3-21}$$

where $\rho_B(z)$ is the density of the two-phase mixture at height z within the boiling height. $\bar{\rho}_B$ may now be given by

$$\bar{\rho}_B = \frac{1}{H_B} \int_{H_0}^{H} \rho_B(z)\, dz \tag{3-22}$$

It is now necessary to obtain a relationship between $\rho_B(z)$ and z. With reference to Fig. 3-4, the derivation will involve obtaining $\rho_B(z)$ in terms

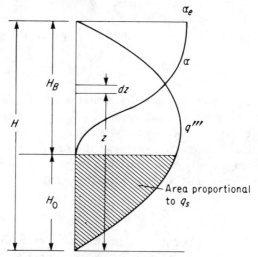

FIG. 3-4. Sinusoidal heat addition in a boiling
channel.

of void fraction at z, α_z, which in turn will be obtained in terms of quality at z, x_z. Finally an expression involving x_z and z will be obtained and integrated. $\rho_B(z)$ is given in terms of α_z as

$$\rho_B(z) = (1 - \alpha_z)\rho_f + \alpha_z \rho_g = \rho_f - \alpha_z(\rho_f - \rho_g) \tag{3-23}$$

where ρ_f and ρ_g are constants at the system pressure. Combining the above two equations gives

$$\bar{\rho}_B = \frac{1}{H_B} \int_{H_0}^{H} [\rho_f - \alpha_z(\rho_f - \rho_g)]\, dz \tag{3-24}$$

The relationship between α_z and x_z is given by [2]

$$\alpha_z = \frac{x_z}{\psi + (1 - \psi)x_z} \tag{3-25}$$

where $\psi = (v_f/v_g)\,S$, and S is the slip ratio. Thus

$$\bar{\rho}_B = \frac{1}{H_B} \int_{H_0}^{H} \left[\rho_f - \frac{x_z}{\psi + (1 - \psi)x_z} (\rho_f - \rho_g) \right] dz \tag{3-26}$$

Next a relationship between x_z and z is obtained. This is done by relating z to the heat addition as follows:

$$q_z = (h_f + x_z h_{fg}) - h_i \tag{3-27}$$

$$q_t = (h_f + x_e h_{fg}) - h_i \tag{3-28}$$

where q_z is the heat generated (Btu/lb$_m$) and transferred up to the height z, and q_t is the total heat generated and transferred in the entire channel (all the fuel element), per pound mass of coolant entering the channel.

The above equations apply to all modes of heat addition. An expression between x_z and z, however, depends upon the axial heat-flux distribution. An expression will be derived for the case of *sinusoidal* axial heat flux.

In the case of sinusoidal heat addition, q_z and q_t are related by [2]

$$\frac{q_z}{q_t} = \frac{1}{2} \left(1 - \cos \frac{\pi z}{H} \right) \tag{3-29}$$

Equations 3-27 to 3-29 are now combined to give

$$x_z = c_1 + c_2 \cos \frac{\pi z}{H} \tag{3-30}$$

where

$$c_1 = \frac{q_t}{2h_{fg}} - \frac{h_f - h_i}{h_{fg}} \tag{3-31a}$$

and

$$c_2 = -\frac{q_t}{2h_{fg}} \tag{3-31b}$$

Equations 3-25 and 3-30 may now be combined to give the desired relation between α_z and z. Substituting Eq. 3-30 into Eq. 3-26 and rearranging give

$$\bar{\rho}_B = \frac{1}{H_B} \left[\int_{H_0}^{H} \rho_f\, dz - (\rho_f - \rho_g) \int_{H_0}^{H} \frac{c_1 + c_2 \cos (\pi z/H)}{c_3 + c_4 \cos (\pi z/H)} \right] dz$$

$$= \rho_f - \frac{\rho_f - \rho_g}{H_B} \int_{H_0}^{H} \frac{c_1 + c_2 \cos (\pi z/H)}{c_3 + c_4 \cos (\pi z/H)}\, dz \tag{3-32}$$

where
$$c_3 = \psi + (1 - \psi)c_1 \tag{3-31c}$$

and
$$c_4 = (1 - \psi)c_2 \tag{3-31d}$$

Equation 3-32 can be integrated by putting

$$y = \frac{\pi z}{H} \tag{3-33a}$$

and
$$dz = \frac{H}{\pi} dy \tag{3-33b}$$

Thus

$$\bar{\rho}_B = \rho_f - \frac{\rho_f - \rho_g}{H_B} \frac{H}{\pi} \int \frac{c_1 + c_2 \cos y}{c_3 + c_4 \cos y} \, dy \tag{3-34a}$$

which can be broken down into

$$\bar{\rho}_B = \rho_f - \frac{\rho_f - \rho_g}{H_B} \frac{H}{\pi} \left(c_1 \int \frac{dy}{c_3 + c_4 \cos y} + c_2 \int \frac{\cos y \, dy}{c_3 + c_4 \cos y} \right)$$

This reduces to

$$\bar{\rho}_B = \rho_f - \frac{\rho_f - \rho_g}{H_B} \frac{H}{\pi} \left(c_1 \int \frac{dy}{c_3 + c_4 \cos y} + \frac{c_2}{c_4} y - \frac{c_2 c_3}{c_4} \int \frac{dy}{c_3 + c_4 \cos y} \right)$$

$$= \rho_f - \frac{\rho_f - \rho_g}{H_B} \frac{H}{\pi} \left[\frac{c_2}{c_4} y + \left(c_1 - \frac{c_2 c_3}{c_4} \right) \int \frac{dy}{c_3 + c_4 \cos y} \right] \tag{3-34b}$$

The limits of the integration are

$$z = H_0 \quad \text{to} \quad z = H \tag{3-35a}$$

or
$$y = \pi \frac{H_0}{H} \quad \text{to} \quad y = \pi \tag{3-35b}$$

The integral in Eq. 3-34b has two solutions [13].
   1. For $c_3^2 > c_4^2$.

$$\bar{\rho}_B = \rho_f - \frac{\rho_f - \rho_g}{H_B} \frac{H}{\pi} \left[\frac{c_2}{c_4} y + \left(c_1 - \frac{c_2 c_3}{c_4} \right) \frac{2}{\sqrt{c_3^2 - c_4^2}} \right.$$

$$\left. \tan^{-1} \frac{(c_3 - c_4) \tan (y/2)}{\sqrt{c_3^2 - c_4^2}} \right]_{\frac{\pi H_0}{H}}^{\pi}$$

Introducing the limits and rearranging give

$$\bar{\rho}_B = \rho_f - (\rho_f - \rho_g) \left\{ \frac{c_2}{c_4} + \frac{c_1 c_4 - c_2 c_3}{c_4 \sqrt{c_3^2 - c_4^2}} \frac{H}{H_B} \right.$$

$$\left. \left[1 - \frac{2}{\pi} \tan^{-1} \frac{(c_3 - c_4) \tan (\pi H_0/2H)}{\sqrt{c_3^2 - c_4^2}} \right] \right\} \tag{3-36}$$

2. For $c_4^2 > c_3^2$:

$$\bar{\rho}_B = \rho_f - \frac{\rho_f - \rho_g}{H_B} \frac{H}{\pi} \left[\frac{c_2}{c_4} y + \left(c_1 - \frac{c_2 c_3}{c_4} \right) \right.$$
$$\left. \frac{1}{\sqrt{c_4^2 - c_3^2}} \ln \frac{(c_4 - c_3) \tan(y/2) + \sqrt{c_4^2 - c_3^2}}{(c_4 - c_3) \tan(y/2) - \sqrt{c_4^2 - c_3^2}} \right]_{\frac{\pi H_0}{H}}^{\pi}$$

Again introducing the limits and rearranging give

$$\bar{\rho}_B = \rho_f - (\rho_f - \rho_g) \left[\frac{c_2}{c_4} - \frac{c_1 c_4 - c_2 c_3}{c_4 \sqrt{c_4^2 - c_3^2}} \frac{H}{H_B} \frac{1}{\pi} \right.$$
$$\left. \ln \frac{(c_4 - c_3) \tan(\pi H_0/2H) + \sqrt{c_4^2 - c_3^2}}{(c_4 - c_3) \tan(\pi H_0/2H) - \sqrt{c_4^2 - c_3^2}} \right] \quad (3\text{-}37)$$

For the case of *uniform axial heating,* a similar and simpler procedure than the one used above can be employed for evaluating $\bar{\rho}_B$. The result, which may be verified by the reader (Prob. 3-5), is

$$\bar{\rho}_B = \rho_f - (\rho_f - \rho_g) \frac{1}{1 - \psi} \left\{ 1 - \left[\frac{1}{\alpha_e(1 - \psi)} - 1 \right] \ln \frac{1}{1 - \alpha_e(1 - \psi)} \right\}$$
$$(3\text{-}38)$$

If the axial flux distribution is such that the heat addition is neither sinusoidal nor uniform, $\bar{\rho}_B$ may be evaluated by a similar analytical technique, provided that distribution is representable by a function that can be easily handled, or by a stepwise or graphical technique in which the channel is divided into several small segments in which the heat added may be considered uniform. Note, however, that Eq. 3-38 represents the average density in uniform heating of a channel portion starting with zero quality.

Example 3-2. Calculate the average density and the hydrostatic pressure for a 6-ft-high boiling-water-reactor channel (without chimney) in which heat is added sinusoidally at the rate of 50.05 Btu/lb$_m$ of incoming water. The sensible heat added is 28.1 Btu/lb$_m$. $t_i = 522°F$; $S = 1.9$. Channel average pressure = 1,000 psia.

Solution. The average specific volume of the water in the nonboiling height is

$$\bar{\rho}_0 = \frac{1}{2} \left(\frac{1}{v_i} + \frac{1}{v_f} \right) = \frac{1}{2} \left(\frac{1}{0.02094} + \frac{1}{0.0216} \right)$$
$$= 47.026 \, \text{lb}_m/\text{ft}^3$$

where v_i is evaluated at 522°F, the inlet temperature; $\psi = 1.9 \times 0.0216/0.4456 = 0.0921$. Using Eq. 3-29, after putting $z = H_0$ and $q_z = q_s$:

$$\frac{H_0}{H} = 0.539 \quad \text{and} \quad \frac{H_B}{H} = 0.461$$

Thus

$$c_1 = -0.00473 \qquad \text{from Eq. 3-31a}$$
$$c_2 = -0.03854 \qquad \text{from Eq. 3-31b}$$
$$c_3 = +0.08781 \qquad \text{from Eq. 3-31c}$$
$$c_4 = -0.03499 \qquad \text{from Eq. 3-31d}$$

Since $c_3^2 > c_4^2$, Eq. 3-36 applies:

$$\bar{\rho}_B = 46.296 - (46.296 - 2.244)\left[1.1015 - 2.7323\left(1 - \frac{2}{\pi}\tan^{-1} 1.7236\right)\right]$$
$$= 46.296 - 44.052[1.1015 - 2.7323(1 - 0.6652)]$$
$$= 46.296 - 44.052 \times 0.1867 = 38.07 \text{ lb}_m/\text{ft}^3$$

Thus

$$\bar{\rho}_c = \frac{1}{H}(\bar{\rho}_0 H_0 + \bar{\rho}_B H_B)$$
$$= \frac{1}{6}(47.026 \times 3.234 + 38.07 \times 2.766)$$
$$= \frac{1}{6}(257.384) = 42.897 \text{ lb}_m/\text{ft}^3$$

and

$$\Delta p_H = \bar{\rho}_c H \frac{g}{g_c} = 257.384 \text{ lb}_f/\text{ft}^2 = 1.787 \text{ lb}_f/\text{in}^2$$

where g/g_c is numerically equal to 1.0 in standard gravity.

3-5. THE CHIMNEY EFFECT

As indicated previously, a chimney, Fig. 3-2, is an unheated extension of the core. It usually has fewer walls, dividers, etc., and, of course, no fuel elements so that there is less friction than in the core. The quality in the chimney is substantially the same as that at the channel exit, that is, $x_{ch} = x_e$. (If the chimney is compartmentalized, the quality in each compartment is the same as the average exit quality of the fuel channels feeding into that compartment.) Since no heat is generated in or added to the chimney, x_{ch} does not vary in the axial direction. If the slip ratio in the chimney is assumed to be the same as that in the core, the chimney void fraction will be constant and equal to that at the channel exit. The density along the chimney will therefore be also equal to that at channel exit ρ_e. The addition of a chimney of height H_{ch} increases the driving pressure by the quantity

$$(\rho_{dc} - \rho_e)H_{ch}(g/g_c)$$

which may be desirable or necessary in natural circulation systems.

In the case of the computed driving pressure being less than the total system losses at any designated coolant rate of flow and core heat generation, the coolant rate of flow decreases. When this happens, the system losses, which are mostly functions of the coolant velocity to the second power, rapidly decrease. Also the driving pressure increases, since the same quantity of heat generates more voids in the reduced coolant flow. An equilibrium is reached when Δp_d and $\Sigma \Delta p$ become equal. The reverse occurs, of course, if the computed Δp_d is greater than the system losses.

In the former case, however, if it is desired to maintain a certain level of steam generation, i.e., a certain coolant rate of flow, a chimney is added in natural-recirculation systems, a larger pump is used in forced-recirculation systems, or a combination chimney and pump are used.

Of course the existence of a chimney increases pressure-vessel height and consequently its mass and cost and may lengthen and complicate the fuel-loading mechanism and the control-rod drives (if they are actuated from the top).

In most cases the height of the chimney necessary to maintain a certain coolant rate of flow is calculated by first computing the total losses associated with the increased path length of the coolant in terms of $H + H_{ch}$, where H_{ch} is an unknown quantity. This is then equated to the necessary driving pressure which will also be a function of the unknown H_{ch}. The equation is then solved for H_{ch}. This procedure is illustrated in the following example.

Example 3-3. Calculate the chimney height necessary to maintain design flow in a 6-ft-high natural recirculation, boiling-water reactor channel if the combined pressure losses $\Sigma \Delta p_{c,e}$ in the channel, chimney, and downcomer are 0.201 lb$_f$/in.², the two-phase friction losses in the chimney are 0.01 lb$_f$/in.² ft, and the friction losses in the downcomer are 0.005 lb$_f$/in.² ft. The feedwater is added near the top of the chimney at 522°F. Exit void fraction = 0.499. Friction drop in channel = 0.0864 lb$_f$/in.² Acceleration drop in channel = 0.0877 lb$_f$/in.², average channel density in core only $\bar{\rho} = 43.370$ lb$_m$/ft³.

Solution

$$\text{Friction drop in chimney} = 0.01 H_{ch} \qquad \text{where } H_{ch} \text{ is in feet}$$

$$\begin{aligned}\text{Friction drop in downcomer} &= 0.005\,(H + H_{ch}) \\ &= (0.03 + 0.005 H_{ch}) \qquad \text{lb}_f/\text{in.}^2\end{aligned}$$

$$\text{Acceleration drop in chimney and downcomer} \simeq 0$$

Therefore,

$$\begin{aligned}\text{Total system losses} &= (0.0864 + 0.01 H_{ch} + 0.03 + 0.005 H_{ch}) + 0.0877 + 0.201 \\ &= 0.4051 + 0.015 H_{ch} \qquad \text{lb}_f/\text{in.}^2\end{aligned}$$

$$\rho_e = \alpha_e \rho_g + (1 - \alpha_e)\rho_f$$

$$= 0.499 \; \frac{1}{0.4456} + (1 - 0.499) \; \frac{1}{0.0216} = 24.314 \; \text{lb}_m/\text{ft}^3$$

$$\rho_{dc} = \frac{1}{v_i} = \frac{1}{0.02094} = 47.755 \; \text{lb}_m/\text{ft}^3$$

$$\Delta p_d = [(\rho_{dc} - \bar{p})H + (\rho_{dc} - \rho_e)H_{ch}] \; \frac{g}{g_c}$$

$$= (47.755 - 43.370)6 + (47.755 - 24.314)H_{ch}$$

$$= 26.310 + 23.441 H_{ch} \qquad \text{lb}_f/\text{ft}^2$$

$$= 0.1827 + 0.1628 H_{ch} \qquad \text{lb}_f/\text{in.}^2$$

$$0.4051 + 0.015 H_{ch} = 0.1827 + 0.1628 H_{ch}$$

from which

$$H_{ch} = 1.50 \; \text{ft}$$

In case the computed driving pressure, without chimney, is greater than the system losses, the desired coolant flow rate may be maintained by increasing flow resistance, i.e., adding obstructions to flow in the fuel channels. Otherwise, an equilibrium condition will be established in which the coolant flow rate will be higher than at design conditions. An alternative procedure is to reduce the heat generation in the core.

In a forced-circulation boiling reactor, the determination of system losses and natural driving pressure is required to determine the pumping power. The addition of a chimney above the core in this case is, however, beneficial in case of accidental loss of pumping power so that a sufficient natural driving pressure will be maintained to take care of the reactor decay heat.

3-6. THE MULTICHANNEL BOILING CORE

Because neutron-flux distributions in the radial direction are never really uniform, the heat generated by the fuel differs from one channel to another. Also, the amount of steam generated and consequently the exit quality vary from one channel to another. In a natural-recirculation reactor, or a forced recirculation reactor operating under natural-recirculation conditions (due to pump power failure), the driving pressure is therefore not equal in all channels, being greatest for those channels with the highest neutron flux (i.e., near the core center) and least near the core periphery. This also results in coolant flows in the different channels that vary in a similar order.

Complete calculation of the fluid-flow pattern in the usual multichannel boiling core involves a multiple iteration procedure, requiring lengthy and tedious calculations that are best performed on a digital computer. The core is first regionized, i.e., divided into a number of regions, each usually containing a number of fuel channels of approximately the same neutron-flux level. For rough calculations, the core shown in Fig. 3-5, which contains 72 fuel channels (or subassemblies),

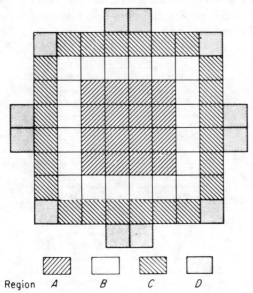

Region *A*   *B*   *C*   *D*

FIG. 3-5. Top view of 72-fuel subassembly
core, divided into four regions.

may be divided into four regions: *A*, *B*, *C*, and *D*. Note that these regions do not necessarily contain the same number of fuel channels, nor do they necessarily contain adjacent fuel channels. These regions contain the following number of fuel channels.

Region	No. of fuel channels
A	16
B	20
C	24
D	/12

For more accuracy, and much more work, the core may be divided into 11 regions, as numbered in Fig. 3-6. Each of these 11 regions contains those fuel channels that are symmetrically situated with respect to the axis of the core. The four regions of Fig. 3-5 may allow a solution using only a desk calculator. Even this is possible only if the axial flux distribution is simple enough so that calculations of the driving pressure do not require a stepwise solution. On the other hand, a large number of regions or an axial flow distribution that requires a stepwise solution almost always require the use of digital computers.

We shall now outline a procedure by which a multiregion core may be analyzed as to chimney requirements, mass-flow rate, quality and void fraction distribution. Let us assume that the data given in a particular case include the power plant electric output, thermal efficiency, and therefore core thermal power, the core operating pressure, the feedwater enthalpy, (or temperature) and the general core design, i.e., fuel type and distribution, core shape (such as shown in Fig. 3-6), and general description,

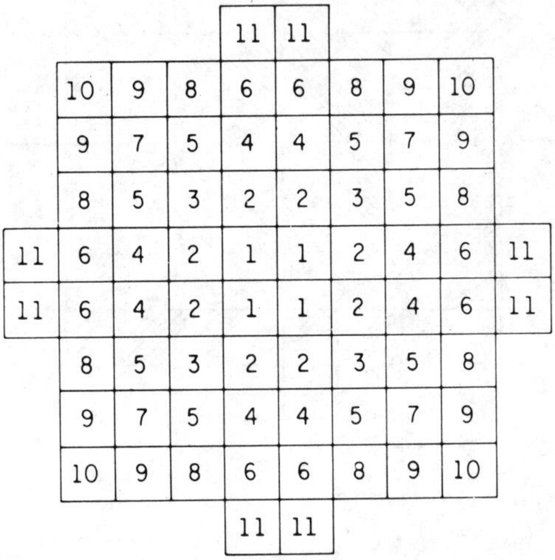

			11	11				
	10	9	8	6	6	8	9	10
	9	7	5	4	4	5	7	9
	8	5	3	2	2	3	5	8
11	6	4	2	1	1	2	4	6	11
11	6	4	2	1	1	2	4	6	11
	8	5	3	2	2	3	5	8
	9	7	5	4	4	5	7	9
	10	9	8	6	6	8	9	10
			11	11				

FIG. 3-6. A 72-fuel subassembly core divided
into 11 regions.

i.e., natural internal recirculation, etc. The value of the maximum permissible void fraction in the core is also usually set by nuclear considerations, such as its effects on moderation, etc. The steps may be as follows:

1. The heat generated in each of the regions selected is calculated from the total core thermal power and known (or assumed) neutron flux distribution throughout the core.

2. A value for the core average exit quality $\bar{x}_e$ is first assumed. With this, the core recirculation ratio can be calculated and consequently the total flow in the downcomer. If the system pressure (and thus h_f and h_g) and the feedwater enthalpy h_d are known, the properties of the inlet water at the reactor bottom, such as h_i, the degree of subcooling, t_i, etc., may be found. (Alternatively, h_i may be assumed and the other quantities calculated.)

3. Starting with the region of maximum heat generation, for example A in Fig. 3-5, here called the hot region, its exit void fraction, α_{e_A},

is set equal to the maximum permissible in the core. From available experimental or theoretical correlations, an appropriate value for the slip ratio S is selected. The exit quality for the hot region, x_{e_A}, is now calculated. This should naturally be greater than the assumed value for the core average exit quality $\bar{x}_e$. (The difference between the two depends upon the degree to which the radial flux in the core is flat.) If not, a lower value for $\bar{x}_e$ is chosen and steps 2 and 3 are repeated.

4. The total flow in the hot region is now calculated. From h_i and the heat generated in that region and the axial flux distribution, the nonboiling and boiling heights in that region are also calculated. (These will not be the same in all regions.) Also, all the pressure losses in that region are calculated

5. A downcomer velocity is selected. This selection is usually predicated on two criteria. First, it is desired that no vapor carry-under into the downcomer should take place, since it reduces the downcomer density (and consequently the driving pressure) and reduces the core power by reducing the moderating power in the lower parts of the core. In this respect, a low downcomer velocity is preferred. Second, too low a downcomer velocity means a large downcomer flow area and therefore a large and consequently costly pressure vessel. Usually a downcomer velocity of no more than 1.5 ft/sec is a good compromise in natural recirculation with no artificial means of bubble separation. If larger velocities are preferred, some form of steam separator should be incorporated in the reactor design.

6. After selection of the downcomer velocity, the downcomer area, based on total downcomer flow calculated in step 2, is determined. The pressure losses in the downcomer are then calculated.

7. The driving pressure in the hot region is now computed in terms of an unknown chimney height. The chimney height is then calculated by a method similar to that given in Example 3-3.

8. The next hottest region, B, is now tackled. An exit quality for that region, x_{e_B}, is assumed. It should naturally be lower than x_{e_A}. The flow and the corresponding pressure losses in that region,.including the already known chimney height, are also calculated. From x_{e_B}, the heat generation in that region and the axial flux distribution, the exit void, α_{e_B}, the boiling and nonboiling heights, the average density and the driving pressure for that region are calculated. This driving pressure is compared with the losses in the region plus those in the downcomer (calculated in step 6). If the two values are not equal, it is necessary to iterate by making another assumption for x_{e_B}, recalculating and repeating until agreement is attained. This fixes x_{e_B} and the corresponding α_{e_B}.

9. Step 8 is repeated for each of the other regions. This fixes values of flow, exit quality and exit void fraction for each of the regions selected.

10. With these values, the total core flow and the average core exit quality $\bar{x}_e$ are calculated. These are compared with the values assumed in step 2. Steps 2 to 10 are repeated until all the assumed and calculated values are equal.

It is important to remember here that neutron-flux distributions in the core are functions of void fractions. Note that the void fractions assumed and calculated in the sequence above were based on an assumed flux distribution (step 1). The new calculations yield new void fractions and, therefore, new flux distributions are computed from the physics of the reactor. These are used again and the entire process is repeated until a final solution which couples the hydrodynamic and nuclear parameters of the reactor is obtained.

It can be seen that the final solution yields flux and void distributions that are interdependent. Since voids abound in the upper part of the core, the moderating power is highest in the nonboiling portion of the core. This causes the peak of the neutron flux, and consequently the power-density in a boiling core to shift from the center position, as encountered in sinusoidal flux distributions, toward the bottom of the core. This is also accompanied by an increase in the peak-to-average ratio of both flux and power density. The effect is more serious in the case of large-size boiling reactors, where the dimensions are large compared with the neutron migration length. Typical flux distributions for both small and large boiling reactors are shown in Fig. 3-7.

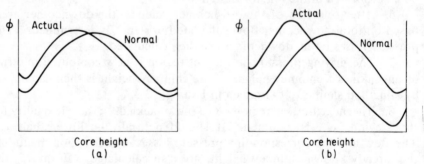

FIG. 3-7. Deviation of actual flux from normal in boiling-water reactors.
(a) Small reactors; (b) large reactors.

In a multichannel core, the effect of low flux and power density in the peripheral channels causes the nonboiling heights to be predominant there. Thus moderating power in such channels is greatest, a situation which tends to partially flatten the radial power and void distribution. Further radial flattening may be accomplished by orificing, control-rod and fuel enrichment programming. Figure 3-8 shows calculated relative power-density and void-fraction distributions in a large boiling core under

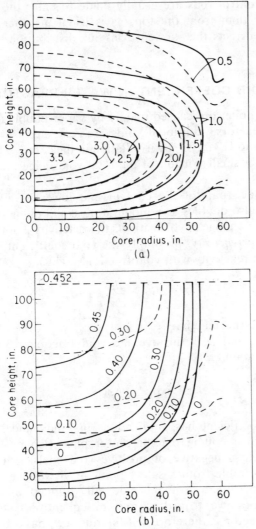

FIG. 3-8. Two-dimensional plots of (a) ratios of local to average power density, and (b) void fractions for large boiling-water reactors. (Dotted lines indicate inlet orificing). (Ref. 14.)

normal and orificing conditions [14]. Note that orificing shifts the maximum power density and the nonboiling heights nearer the center of the core and has a flattening effect on the nonboiling heights and void fractions in general.

Because the power density is greatest near the core bottom in boiling reactors, the control rods are usually made to enter the core from the bottom, rather than from the top, as is usual in other reactor types. This partially corrects the skewed axial flux distribution in such reactor cores.

3-7.   THE VOID COEFFICIENT IN WATER REACTORS

Voids in a boiler are caused mainly by vapor-bubble formation. In water-cooled reactors, additional voids are generated by the decomposition of water into H_2 and O_2 by irradiation in the core. These, however, constitute only a small portion of the total voids generated in a boiling-water reactor.

Voids displace part of the coolant and lower its average density. If this coolant also doubles as a moderator, as in boiling or pressurized-water reactors, a change in voids affects the reactivity ρ of the core. The *void coefficient of reactivity*, or simply the void coefficient, V_D, is defined as the change in reactivity with void fraction. Thus

$$V_D = \frac{d\rho}{d\alpha} \tag{3-39}$$

where ρ here is the reactivity $= (k_{eff}-1)/k_{eff}$, k_{eff} is the effective multiplication factor, and α is the average void fraction. Since k_{eff} varies only slightly from unity,

$$V_D = \frac{dk_{eff}}{d\alpha} \tag{3-40}$$

The void coefficient has implications in safety and load following characteristics in a boiling reactor power plant (Chap. 4). A boiling reactor may have a negative or positive void coefficient. A negative coefficient is desirable from a safety point of view since an increase in steam formation, due to some accidentally applied excess reactivity, causes the reactor power to go down and consequently less steam to be generated. The reactor is therefore self-limiting, i.e., has a built-in protection against runaway accidents. A positive void coefficient on the other hand if used in a carefully designed system has some utility in plant load following characteristics. These aspects will become more apparent in later sections.

In this section we shall examine the parameters that affect the sign and magnitude of the void coefficient. In order to do this k_{eff} is formulated in terms of these parameters and the effect of α on these parameters is examined. Recall that

$$k_{eff} = k_\infty P \tag{3-41}$$

where k_∞ = the infinite multiplication factor, given by the four-factor formula

$$k_\infty = \epsilon p f \eta \qquad (3\text{-}42)$$

and ϵ = fast-fission factor = the factor by which neutrons increase due to the fission of U^{238} by high-energy neutrons

p = resonance escape probability = fraction of neutrons that escape nonfission resonance absorption as they slow down to thermal energies

f = thermal utilization factor = fraction of all thermal neutrons absorbed that are absorbed in the fuel

η = thermal fission factor = average number of neutrons produced per thermal neutron absorbed in fuel

P = nonleakage probability = fraction of fast and thermal neutrons that do not leak out of the core boundaries

The above two equations are combined and differentiated with respect to α to give.

$$\frac{1}{k_{\text{eff}}} \frac{dk_{\text{eff}}}{d\alpha} = \frac{1}{\epsilon} \frac{d\epsilon}{d\alpha} + \frac{1}{p} \frac{dp}{d\alpha} + \frac{1}{f} \frac{df}{d\alpha} + \frac{1}{\eta} \frac{d\eta}{d\alpha} + \frac{1}{P} \frac{dP}{d\alpha} \qquad (3\text{-}43)$$

For water reactors η (a fuel constant) is fairly independent of voids. Contrasting this is the case of sodium-cooled fast reactors where the small moderating effect of sodium causes some hardening of the fast-neutron flux (neutron spectrum shifts towards higher energies), causing η to increase with neutron energy, and resulting in a positive $d\eta/d\alpha$. This is discussed further in Chapter 9 on fast reactors. This effect is, however, small, and can be ignored in thermal water reactors. For these reactors, therefore, the above equation reduces to

$$\frac{1}{k_{\text{eff}}} \frac{dk_{\text{eff}}}{d\alpha} = \frac{1}{\epsilon} \frac{d\epsilon}{d\alpha} + \frac{1}{p} \frac{dp}{d\alpha} + \frac{1}{f} \frac{df}{d\alpha} + \frac{1}{P} \frac{dP}{d\alpha} \qquad (3\text{-}44)$$

The exact evaluation of the different terms in this equation is a complex matter. For our purposes here we shall make some reasonable approximations for some important cases of interest.

3-8. THE CASE OF ORDINARY-WATER REACTORS WITH HIGHLY ENRICHED FUELS

Highly enriched fuels contain little nonfissionable isotopes that contribute to fast fission and resonance absorption.

The first two terms, $d\epsilon/d\alpha$ and $dp/d\alpha$, may therefore be considered negligible. Equation 3-44 now reduces to

$$\frac{1}{k_{\text{eff}}} \frac{dk_{\text{eff}}}{d\alpha} = \frac{1}{f} \frac{df}{d\alpha} + \frac{1}{P} \frac{dP}{d\alpha} \tag{3-45}$$

The two terms on the right-hand side of this equation will now be evaluated. Assuming for simplicity a homogeneous reactor, the thermal utilization factor is given by [1]

$$f = \frac{(\Sigma_a)_U}{(\Sigma_a)_U + (\Sigma_a)_{H_2O} + (\Sigma_a)_{\text{cl}} + \cdots} = \frac{(\Sigma_a)_U}{(\Sigma_a)_{\text{tot}}} \tag{3-46}$$

where Σ_a is the macroscopic neutron absorption cross section and the subscripts U, H_2O, cl, and tot refer to fuel, water, cladding, and total. Since voids affect $(\Sigma_a)_{H_2O}$ only, the above equation differentiates with respect to α as

$$\frac{df}{d\alpha} = -\frac{(\Sigma_a)_U}{(\Sigma_a)_{\text{tot}}^2} \frac{d(\Sigma_a)_{H_2O}}{d\alpha} \tag{3-47}$$

Combining Eqs. 3-46 and 3-47 gives

$$\frac{df}{d\alpha} = -f \frac{1}{(\Sigma_a)_{\text{tot}}} \frac{d(\Sigma_a)_{H_2O}}{d\alpha} \tag{3-48}$$

Introducing a *water utilization factor*, f_{H_2O}, equal to the fraction of all neutrons absorbed that are absorbed in water

$$f_{H_2O} = \frac{(\Sigma_a)_{H_2O}}{(\Sigma_a)_{\text{tot}}} \tag{3-49}$$

and combining with Eq. 3-48, the first term in Eq. 3-45 becomes

$$\frac{1}{f} \frac{df}{d\alpha} = -f_{H_2O} \frac{1}{(\Sigma_a)_{H_2O}} \frac{d(\Sigma_a)_{H_2O}}{d\alpha} \tag{3-50a}$$

and since $(\Sigma_a)_{H_2O}$ is proportional to the density of the water two-phase mixture, ρ, the above equation becomes

$$\frac{1}{f} \frac{df}{d\alpha} = -f_{H_2O} \frac{1}{\rho} \frac{d\rho}{d\alpha} \tag{3-50b}$$

(Note that ρ here and in Eq. 3-39 are not the same.)

The second term in Eq. 3-45 is evaluated with the help of the two-group approximation. Thus

$$P = \frac{1}{1 + M^2 B^2} \tag{3-51}$$

where M^2 is the neutron migration area and B^2 the core geometric buckling.

$$\frac{dP}{d\alpha} = -\frac{M^2 B^2}{(1 + M^2 B^2)^2}\left(\frac{1}{M^2}\frac{dM^2}{d\alpha} + \frac{1}{B^2}\frac{dB^2}{d\alpha}\right) \tag{3-52}$$

$$\frac{1}{P}\frac{dP}{d\alpha} = -\frac{M^2 B^2}{(1 + M^2 B^2)}\left(\frac{1}{M^2}\frac{dM^2}{d\alpha} + \frac{1}{B^2}\frac{dB^2}{d\alpha}\right) \tag{3-53}$$

The migration area M^2 is inversely proportional to macroscopic fast and thermal neutron scattering as well as the thermal absorption cross sections of the water and other materials in the reactor core. It is therefore not inversely proportional to the square of the water density (as would have been the case if only water existed in the core) but to the water density to some power n, less than 2, which depends upon the core composition. Thus M^2 is proportional to ρ^{-n} and

$$\frac{1}{M^2}\frac{dM^2}{d\alpha} = -\frac{n}{\rho}\frac{d\rho}{d\alpha} \tag{3-54}$$

Combining with Eq. 3-53 gives

$$\frac{1}{P}\frac{dP}{d\alpha} = \frac{M^2 B^2}{(1 + M^2 B^2)}\left(\frac{n}{\rho}\frac{d\rho}{d\alpha} - \frac{1}{B^2}\frac{dB^2}{d\alpha}\right) \tag{3-55}$$

Combining Eqs. 3-50 and 3-53 with 3-45 gives

$$\frac{1}{k_{\text{eff}}}\frac{dk_{\text{eff}}}{d\alpha} = -f_{\text{H}_2\text{O}}\frac{1}{\rho}\frac{d\rho}{d\alpha} + \frac{M^2 B^2}{(1 + M^2 B^2)}\left(\frac{n}{\rho}\frac{d\rho}{d\alpha} - \frac{1}{B^2}\frac{dB^2}{d\alpha}\right) \tag{3-56}$$

The derivative $d\rho/d\alpha$ will now be evaluated. The density of the two-phase mixture ρ is a function of the void fraction, given by

$$\rho = (1 - \alpha)\rho_f + \alpha\rho_g$$

At pressures far removed from critical ρ_g is much smaller than ρ_f and can be ignored, so that

$$\frac{d\rho}{d\alpha} = -\rho_f$$

and

$$\frac{1}{\rho}\frac{d\rho}{d\alpha} = -\frac{1}{1 - \alpha} \tag{3-57}$$

Equation 3-56 now becomes

$$\frac{1}{k_{\text{eff}}}\frac{dk_{\text{eff}}}{d\alpha} = \left(\frac{1}{1 - \alpha}\right)f_{\text{H}_2\text{O}} - \frac{M^2 B^2}{(1 + M^2 B^2)}\left(\frac{n}{1 - \alpha} - \frac{1}{B^2}\frac{dB^2}{d\alpha}\right) \tag{3-58}$$

The value of f_{H_2O} has been found in Borax type reactors, which use 90 percent enriched aluminum-alloy fuel plates, to be approximately 0.3. If all neutron absorptions are assumed to occur only in fuel and water, the value of f_{H_2O} would simply be equal to $(1 - f)$. The value of n has been found in the Borax III reactor to be equal to 1.7.

In the case of a *bare* reactor, the effect of the change in α on buckling is negligible, so that $dB^2/d\alpha$ may be dropped and Eq. 3-58 simplifies to

$$\frac{1}{k_{eff}} \frac{dk_{eff}}{d\alpha} = \frac{1}{(1-\alpha)} \left[f_{H_2O} - \frac{nM^2B^2}{(1 + M^2B^2)} \right] \tag{3-59}$$

from which the void coefficient may be calculated. Eq. 3-59 shows that, for reactors using highly enriched fuels, the void coefficient may either be positive or negative depending upon the relative magnitudes of f_{H_2O} and $nM^2B^2/(1 + M^2B^2)$. For the void coefficient to be negative,

$$\frac{nM^2B^2}{1 + M^2B^2} > f_{H_2O}$$

or

$$B^2 > \frac{f_{H_2O}}{M^2(n - f_{H_2O})} \tag{3-60}$$

Thus the buckling must be greater than a limiting value, or the core size must be *less* than a limiting size if the void coefficient of a highly enriched reactor is to be negative. In general, then, a large highly enriched water reactor is likely to have a postitive void coefficient, while a small one is likely to have a negative void coefficient.

The effect of a *reflector* will now be discussed. A water reactor is usually reflected by water outside the core, or occasionally by graphite or other reflector. The effect of the reflector is usually expressed in terms of its savings [1, 2]. A reflector normally reduces the size of a core, so that the limiting size of a reflected core is smaller than that of a bare core (above). An additional effect, however arises.

Since the reflector usually remains in single phase, i.e., generates no voids, it reduces the magnitude of the change of the nonleakage probability with voids, $dP/d\alpha$, and the reflector savings do not therefore, remain constant, but increase as the void fraction increases. This means an increase in buckling, and, unlike the bare core, the quantity $dB^2/d\alpha$ is not affected by the presence of a reflector, $dP/d\alpha$ can only be reduced if a positive $dB^2/d\alpha$ is introduced. From Eq. 3-58, therefore, in order to keep $dk_{eff}/d\alpha$ negative

$$B^2 > \frac{f_{H_2O}}{M^2(n - f_{H_2O})} + \frac{1 - \alpha}{n - f_{H_2O}} \frac{dB^2}{d\alpha} \tag{3-61}$$

and since $dB^2/d\alpha$ is positive, and $n > f_{\text{H}_2\text{O}}$, the limiting value for the buckling of a reflected core must be greater than that given by Eq. 3-60 for the bare core.

Thus the limiting size that guarantees a negative void coefficient is smaller for a reflected core than for a bare core because (a) the size of reflected core is normally smaller than a bare core, and (b) the change in reflector savings with void fraction results in a positive contribution to the buckling.

3-9. THE CASE OF ORDINARY-WATER REACTORS WITH LOW-ENRICHED FUELS

For highly enriched reactors, all terms in Eq. 3-43 were dropped except those due to the thermal utilization factor f and the nonleakage probability P. For reactors, using low-enriched fuels, the effect of the change in voids on η is the same as in highly enriched fuels, but the effects on the fast fission factor and the resonance escape probability are not negligible because of the abundance of nonfissionable fuel. All terms (except η) must therefore be considered, and Eq. 3-44 applies.

$d\epsilon/d\alpha$ is positive, since an increase in voids causes an increase in fast fission. The value, however, is usually small, but increases with increased fuel-to-moderator ratio and with reduced fuel-element diameter. No simple expression for $d\epsilon/d\alpha$ is available.

The resonance escape probability p depends upon the fuel composition, core lattice arrangement, the fuel-to-moderator (water) ratio, and the water density. For small changes in density p may be given by

$$p = e^{-A/\rho} \tag{3-62}$$

where A is a constant depending upon fuel composition and lattice arrangement, and ρ is the water density. Equation 3-62 is defferentiated with respect to α to give

$$\frac{1}{p}\frac{dp}{d\alpha} = -\ln p \frac{1}{\rho}\frac{d\rho}{d\alpha} \tag{3-63}$$

Combining with Eq. 3-57 gives

$$\frac{1}{p}\frac{dp}{d\alpha} = \left(\frac{1}{1-\alpha}\right)\ln p \tag{3-64}$$

which is negative, since p is less than unity.

The last two terms in Eq. 3-44 have already been evaluated in Eqs. 3-50 and 3-53. Thus for a low-enriched fuel,

$$\frac{1}{k_{eff}} \frac{dk_{eff}}{d\alpha} = \frac{1}{\epsilon} \frac{d\epsilon}{d\alpha} + \left(\frac{1}{1-\alpha}\right) \ln p + \left(\frac{1}{1-\alpha}\right) f_{H_2O}$$

$$- \frac{M^2 B^2}{1 + M^2 B^2} \left(\frac{1}{M^2} \frac{dM^2}{d\alpha} + \frac{1}{B^2} \frac{dB^2}{d\alpha}\right) \tag{3-65}$$

In the last term $dM^2/d\alpha = dL^2/d\alpha + dL_f^2/d\alpha$, where L is the thermal diffusion length and L_f the slowing-down length [1, 2]. However, as with the first term on the right-hand side of the above equation, there is no simple expression to evaluate these terms in this case.

In Eq. 3-65 the first term on the right-hand side is small but positive, the second is negative, the third is positive and the fourth is positive, but its contribution is negative because of the negative sign. For EBWR, a low-enriched boiling-water reactor, Sec. 4-3, at the operating temperature of 486°F, and between $\bar{\alpha} = 0$ and $\bar{\alpha} = 0.15$, the value of $(1/k_{eff}) (\Delta k_{eff}/\Delta\alpha)$ is -0.108, and the terms corresponding to those on the right-hand side of Eq. 3-65 are $+0.019$, -0.169, $+0.122$ and -0.080 respectively. EBWR is a small reactor (4-ft-high × 4-ft-diam core) where the nonleakage probability P, at a core average void fraction $\bar{\alpha} = 0.15$, is only 0.91. B^2 at the same $\bar{\alpha}$ is 15.08×10^{-4} cm^{-2}, a relatively large number. It can be seen that, for small reactors, the negative terms dominate and the void coefficient is usually negative. For modern large-sized boiling-water reactors, such as Browns Ferry, Sec. 4-7, with a 12 ft × 16.55 ft core, P is expected to be much higher, B^2 much lower, so that the last term would be numerically small, as is the first term. These two terms may therefore, be ignored (they may just cancel each other out). Approximately, then,

$$\frac{1}{k_{eff}} \frac{dk_{eff}}{d\alpha} = \left(\frac{1}{1-\alpha}\right) \ln p + \left(\frac{1}{1-\alpha}\right) f_{H_2O} \tag{3-66}$$

In the above equation the p-term is negative and f-term is positive. The void coefficient of large low-enriched reactors can be made positive or negative depending upon the relative magnitudes of these two terms. It can, for example, be made strongly negative by designing the core lattice to give a small value of p, resulting in a large negative $\ln p$. This is usually done by increasing the fuel-moderator ratio. This also results in decreasing f_{H_2O}, adding to the desired effect. An overmoderated reactor, on the other hand, would be expected to have a positive void coefficient. This last mode of operation is made use of to obtain the desired load-following characteristics in one case, Sec. 4-6.

Void coefficients vary with the void fraction α, with temperature, and with core life (burnup). Figure 3-9 shows this variation at operating temperatures, at the beginning of core life, and after 10,000 Mw-day/ton burnup for the Browns Ferry boiling-water reactors, Sec. 4-7. Note that

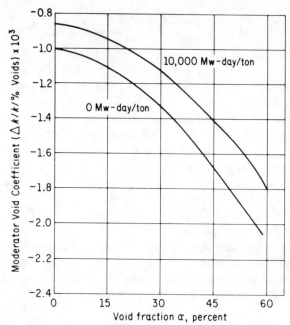

FIG. 3-9.   Void coefficient vs. void fraction at beginning of life and after 10,000 Mw-day/ton burnup for
the Browns Ferry BWR.   (Ref. 15.)

the void coefficient becomes more negative as the voids increase but less
negative with burnup.

3-10.   THE CASE OF HEAVY-WATER REACTORS

In heavy-water reactors (such as the Canadian PWR's and Halden
BWR in Norway and the Marviken BWR in Sweden), the fraction of
neutrons that gets absorbed in water, given by f_{H_2O}, is negligible.   Also
the last term in Eq. 3-65 is relatively large, even for large reactors with
small B^2, because M^2 for D_2O is large.   The void coefficient is therefore
almost invariably negative.   A value of p larger than in ordinary water
reactors may therefore be tolerated (though this may be limited by
considerations of conversion ratio.)   Finally, the variations of void coefficient with voids and temperature in heavy-water reactors are much
smaller than those in ordinary water reactors such as shown in Fig. 3-9.

3-11.   PRESSURE COEFFICIENTS IN BOILING REACTORS

The pressure coefficient of reactivity, analogous to the void and
temperature coefficients of reactivity, is defined as

$$P_D = \frac{d\rho}{dp} = \frac{dk_{\text{eff}}}{dp} \tag{3-67}$$

where p here is the system pressure.

The pressure coefficient is usually positive. This can be explained by the following: A small increase in reactivity in a boiling-water reactor causes an increase in pressure because of the production of more steam that must be pushed out of the core against the resistance of the existing steam passages. The increase in pressure causes a reduction in quality and void fraction [1, 2], which increases the moderating power of the water-steam mixture, which in turn causes a further increase in reactor power. The positive pressure coefficient causes the reactor to be unstable, much as a positive temperature coefficient would. It will be shown below that the value of the pressure coefficient is rather large at low pressures and thus is believed to be the major cause of the reactor instability observed in the early low-pressure boiling-reactor systems. At high steam pressures, however, the pressure coefficient becomes very small in value, because of the larger steam enthalpies and lower specific volumes. It is believed that pressure effects cause little oscillation at high system pressures.

Pressure coefficients in boiling reactors can be calculated from the following relationship:

$$\frac{dk_{\text{eff}}}{dp} = \frac{dk_{\text{eff}}}{d\alpha} \frac{d\alpha}{dx} \frac{dx}{dp} \tag{3-68}$$

The three components on the right-hand side of this equation will now be evaluated.

1. $dk_{\text{eff}}/d\alpha$: This can be estimated from the known system characteristics, as described in Secs. 3-7 to 3-10.

2. $d\alpha/dx$: The basic relationship between α and x can be used to derive $d\alpha/dx$.

$$\alpha = \frac{x}{\psi + (1 - \psi)x} \tag{3-69}$$

where

$$\psi = \frac{v_f}{v_g} S \tag{3-70}$$

α in Eq. 3-69 can now be differentiated with respect to x as follows:

$$\frac{d\alpha}{dx} = \frac{[\psi + (1 - \psi)x] - x\{d[\psi + (1 - \psi)x]/dx\}}{[\psi + (1 - \psi)x]^2}$$

$$= \frac{\psi}{[\psi + (1 - \psi)x]^2} = \left(\frac{\alpha}{x}\right)^2 \psi \tag{3-71a}$$

This relationship may be changed to a more convenient form by combining with Eq. 3-69 and rearranging to give

$$\frac{d\alpha}{dx} = \frac{\alpha(1-\alpha)}{x(1-x)} \tag{3-71b}$$

In using this relationship, if α is known, the corresponding value of x is evaluated from the slip ratio and other system constants.

3. dx/dp: This is the rate of change of steam quality x with respect to pressure. It can best be obtained as a function of dh_f/dp and dh_g/dp. These last two quantities are sole functions of pressure (and saturation temperature) in any liquid-vapor system. The enthalpy of 1 lb_m of liquid-vapor mixture, h, is given by

$$h = h_f + x h_{fg}$$

Rearranging, $$x = \frac{h - h_f}{h_{fg}} \tag{3-72}$$

Differentiating with respect to p,

$$\frac{dx}{dp} = d(h/h_{fg})/dp - d(h_f/h_{fg})/dp$$

$$= \frac{1}{(h_{fg})^2} \left[\left(h_{fg} \frac{dh}{dp} - h \frac{dh_{fg}}{dp} \right) - \left(h_{fg} \frac{dh_f}{dp} - h_f \frac{dh_{fg}}{dp} \right) \right]$$

Assuming that the system heat balance is such that the enthalpy of the water-steam mixture will not change with pressure, that is, $dh/dp = 0$, and using $h_{fg} = h_g - h_f$, so that $dh_{fg} = dh_g - dh_f$, we can rewrite the above equation in the form

$$\frac{dx}{dp} = - \left(\frac{1-x}{h_{fg}} \frac{dh_f}{dp} \right) - \left(\frac{x}{h_{fg}} \frac{dh_g}{dp} \right) \tag{3-73}$$

As indicated above, the derivatives dh_f/dp and dh_g/dp are sole functions of the pressure. They are given in Fig. 3-10 for the water–steam system.

Example 3-4. Compute the pressure coefficient of three boiling-water reactors operating at atmospheric, 1,000-psia, and 2,000-psia pressure, respectively. Assume that all these reactors have the following identical characteristics:

$$\text{Average void fraction in core } \bar{\alpha}_c = 0.10$$

$$\text{Void coefficient } \frac{dk_{eff}}{d\alpha} = -0.10$$

$$\text{Slip ratio } S = 2.0$$

Solution. The necessary components of Eqs. 3-71b and 3-73 are obtained from the steam tables, and Fig. 3-10. These, the data given, and the answers are given in Table 3-1.

TABLE 3-1

	Pressure		
	Atmospheric	1,000 psia	2,000 psia
$dk_{eff}/d\alpha$ (given)	-0.10	-0.10	-0.10
α (given)	0.10	0.10	0.10
S (given)	2.00	2.00	2.00
x (from Eq. 3-69)	1.386×10^{-4}	1.066×10^{-2}	2.951×10^{-2}
$d\alpha/dx$ (from Eq. 3-71b)	6.5416×10^{2}	8.634	3.371
h_{fg}, Btu / lb$_m$ (steam tables)	970.3	649.4	463.4
dh_f/dp, Btu / lb$_m$ psia (Fig. 3-10)	$+3.500$	$+0.152$	$+0.119$
dh_g/dp, Btu / lb$_m$ psia (Fig. 3-10)	$+1.300$	-0.040	-0.073
dx/dp, psia^{-1} (Eq. 3-73)	-3.607×10^{-3}	-2.322×10^{-4}	-2.539×10^{-4}
dk_{eff}/dp, psia^{-1} (Eq. 3-68)	$+0.236$	$+1.98 \times 10^{-4}$	$+8.556 \times 10^{-5}$
Percent / psia	$+23.6$	$+0.0198$	$+0.008556$

It can be seen that the positive pressure coefficient decreases in value with pressure and becomes negligible at high pressures, such as those used in power plants. The value of the pressure coefficient can also be shown to increase, though not directly, with α.

3-12. BOILING-WATER-REACTOR STABILITY

Early work on boiling-water reactors was made up of the much discussed Borax and Spert experiments and, later, in the first power-producing Experimental Boiling Water Reactor (EBWR). A summary of operational characteristics of these is given in Table 3-2.

TABLE 3-2

Experiment	Operator	Fuel	Maximum Pressure, psig
Borax I	Argonne	U^{235} + Al plates, 0.06 in. thick	130
Borax II	Argonne	U^{235} + Al plates, 0.06 in. thick	300
Borax III	Argonne	U^{235} + Al plates 0.06 in. thick	300
Borax IV	Argonne	$U^{235}O_2$ + ThO$_2$ rods, 0.23 in. diam	300
Spert I	Phillips Petroleum Company	U^{235} + Al plates 0.06 in. thick	0
EBWR	Argonne	U (1.44% enriched) + Zr-Nb plates, 0.279 in thick	600

In most of these initial experiments, power oscillations were experienced, especially at high power densities. They were thought to exert limiting effects on the performance of all boiling reactors. These

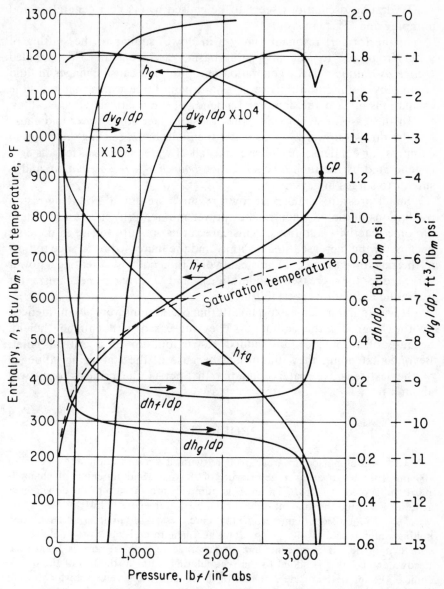

FIG. 3-10. Some thermodynamic properties of saturated water and steam.

oscillations frequently had amplitudes equivalent to 100 percent of full power and durations varying from a fraction of a second to a few seconds. Experiments with EBWR and later experiments conducted by General Electric Company indicated, however, that boiling-water reactors operating at high pressures are quite stable and pose no unsurmountable control problems.

Hydraulic oscillations seem to be caused by (1) heat generation and pumping and (2) steam voids.

In the former, resonant changes in flow, "similar to the shifting of water from side to side in a large basin," are created. This is called *U-tube oscillation.* Heat generation and pumping cause changes in fluid volume by boiling and thus serve as a driving force which maintains the oscillations against the damping tendency of the relatively viscous water.

In the case of steam voids, the voids displace some water and cause it to be thrown upward. This water then falls back, causing the steam to compress or condense. The rise and fall of water repeats as long as heat is generated by the fuel. This is called *bouncing-piston oscillation,* and is similar to surges in geysers.

Small pressure changes at low pressures are due to bouncing-piston oscillations, initiated when reactor power momentarily increases because of some reactivity addition. This causes the pressure to rise in the core because of the increased steam supply and the increased resistance to flow in the reactor passages. The steam voids then collapse, creating a power surge (due to increased moderation), and a driving mechanism for hydraulic oscillations is established.

It was found in some experiments that oscillations were also influenced by the height of the chimney. They increased with chimney height, because of turbulence and added restriction, up to a critical point. Beyond that point, they diminished because of the stabilizing influence of the less-sensitive water column. In Spert I the critical height was about 4 ft.

PROBLEMS

3-1. A boiling-sodium reactor core with a cross sectional flow area of 100 in.^{2} operates at an average pressure of 38 psia. Sodium enters all channels 100°F-subcooled at 3 fps. The feed is pumped into the top of the downcomer at 1,300°R. Find the amount of heat generated in the core in kw(t).

3-2. The outside diameter of the core vessel of an internal-circulation boiling-water reactor is 4 ft. The total heat generated is 60 Mw(t), the average core exit quality is 6 percent, and the average reactor pressure is 1,000 psia. Feedwater at 250°F is added to the recirculating water at the top of the downcomer. Estimate the inside diameter of the reactor pressure vessel which would result in little or no bubble carryunder into the downcomer.

3-3. A natural-internal-recirculation boiling-water-reactor core is simulated by a single 4-ft-high 4.5 × 0.5-in. rectangular channel. Heat is added in the channel uniformly at the rate of 10^5 Btu/hr ft^2 of the wide sides only. The channel operates at an average pressure of 1,200 psia and with an inlet velocity of 2 fps. Feedwater is at 200°F. Determine (a) the exit quality, (b) the recirculation ratio, and (c) the degree of subcooling in °F.

3-4. If the channel of the previous problem has a 1.5-ft-high chimney (of the same cross-sectional dimensions as the channel itself), determine the driving pressure of the system in psi. Take $S = 2.0$.

3-5. Derive the expression given by Eq. 3-38 for the average boiling density in a uniformly heated channel.

3-6. The flow into a forced-recirculation boiling-water-reactor channel is 200,000 lb_m/hr of 22°F subcooled water. The inlet pressure is 900 psia. The channel generates 5 Mw(t). Other pertinent data: channel height = 5 ft, equivalent diameter = 0.5 in., slip ratio = 1.8. Calculate the hydrostatic pressure drop in the channel.

3-7. The core of a forced-external-recirculation boiling-water reactor is 6 ft high. The top of the core is 30 ft below the steam-separation drum. The reactor produces 1,000,000 lb_m/hr of steam at 1,000 psia. The recirculation ratio is 19:1, feedwater is at 230°F, the average boiling density is 41 lb_m/ft^3, and the slip ratio is 1.8. The axial flux distribution changes linearly from 0 at core bottom to maximum in 2 ft, remains at maximum for 1 more foot, then drops linearly, becoming 0 again at core top. Neglecting the variations in the radial flux, determine (a) the nonboiling height, (b) the pressure rise across the pump, and (c) the corresponding pump horsepower necessary to maintain flow if the contraction, expansion, submerged body, and other similar pressure losses in the loop are 3.26 psi, and the total frictional and acceleration losses are 4.53 psi.

3-8. A boiling-water reactor operates with an average exit void fraction of 0.50, an average inlet velocity of 3 ft/sec, and a slip ratio of 1.9. The average reactor pressure is 1,000 psia. One channel in the reactor, operating at these conditions, is 6 ft high and square, 4.85 in. on each side, and contains 36 fuel rods each 0.55 in. in diameter. Feedwater is at 400°F. For uniform heat flux (axial and radial) calculate the height of the chimney necessary for natural recirculation in that channel. The entrance, exit, contraction, expansion, spacer, and other similar losses are 0.301 psi in the channel and chimney and 0.182 psi in the downcomer. The frictional losses in the downcomer are 0.008 psi/ft. The fuel-rod cladding corresponds to smooth-drawn tubing.

3-9. An internal-natural-recirculation boiling-sodium reactor is composed of a single channel 4 ft high with a 2-ft chimney of the same cross-sectional dimensions. The heat, generated uniformly in the reactor, is adjusted at all conditions so that the exit quality is always 11.5 percent. Liquid sodium returns from the power plant at 1800°R and is fed to the reactor near the top of the downcomer. The reactor average pressure is 150.5 psia, and the slip ratio is 4.0. The reactor was first operated on earth and then on the lunar surface where the gravitational field is one-sixth of that on earth. Calculate (a) the change in driving pressure, psi, and (b) the percent decrease in reactor power necessary to maintain exit quality as mentioned, if the pressure losses in the core and downcomer varied as the (rate of flow)².

3-10. An external-natural-recirculation boiling-water reactor was designed to produce 150,000 lb_m/hr of steam. The following data apply: $H_0 = 1.23$ ft, $H_B = 4.77$ ft, $H_{ch} = 12.26$ ft; $\rho_i = 46.3$, $\rho_0 = 46.2$, $\rho_B = 33.1$, and $\rho_e = 24.3$ (all in lb_m/ft^3). After an initial period of operation, a decision was made to reduce reactor power so that only 100,000 lb_m/hr of steam are produced. The same pressure, exit quality, and feedwater temperatures were maintained. Because the pressure losses decreased below the available driving pressure, it was decided that the simplest way to correct the situation was to add an orifice in the down-comer. Calculate the orifice diameter (coefficient of discharge = 0.85). Assume all pressure losses to be proportional to the square of the flow speed and that the downcomer diameter is much larger than the orifice diameter.

3-11. A natural internal recirculation BWR operates at an average pressure of 1,000 psia. The average densities in the nonboiling and boiling heights are 47 and 38 lb_m/ft^3 respectively. These heights are 2 and 3 ft respectively. The radial flux is uniform. Downcomer temperature is 520°F. The core exit quality is 8 percent. The slip ratio is 2:1. The total pressure losses are 0.527 psi (including chimney). Calculate the necessary chimney height in feet.

3-12. A boiling-internal-natural-recirculation reactor has a 5-ft high core and a 2-ft chimney. The density in the downcomer is 46.5 lb_m/ft^3. The average density in the core is 40.0 lb_m/ft^3. The core exit density is 30.0 lb_m/ft^3. The pressure losses in core, chimney and downcomer may be given by $\Delta p/foot = kG^2$ where k is a constant and G is the mass velocity of coolant. Assuming densities do not change, find the percent change in mass flow rate if the chimney height is increased to 3 ft.

3-13. A forced-recirculation BWR operates at 1,000 psia with a core exit quality of 7.75 percent. The inlet water to the core is at 500°F. The average core density is 43 lb_m/ft^3, $S = 2$. The core height is 8 ft *plus* chimney. The total core, chimney, and downcomer losses are 2.8 psi. A pump produces 2 psi. Calculate the height of the chimney.

3-14. A forced-recirculation BWR generates 600 Mw(t) at 1,000 psia. A total of 2×10^7 lb_m/hr of 44.6°F subcooled water enters the core. The average density in the core only is 43 lb_m/ft^3. The slip ratio is 2. The core is 8 ft. high. There is a chimney which is 3 ft high. The total core and downcomer losses are 2.8 psi. What should the pump pressure be in psi?

3-15. A large boiling-water reactor using low-enriched fuels operates with an average quality of 5 percent. The reactor pressure is 1,000 psia. The slip ratio is 2.06. Estimate the resonance escape probability that must be designed into the core if it were to have a pressure coefficient of $10^{-4}/psia$. Take $f_{H_2O} = 0.3$.

3-16. A bare boiling-water reactor is in the form of a cylinder of equal height and diameter. It uses Borax-type fuel. The migration area is 40 cm^2. Find the maximum diameter of the core that will result in a negative void coefficient. Ignore the extrapolation lengths.

3-17. A boiling-water reactor using highly-enriched fuel has a buckling of 0.002 cm^{-2}, a water utilization factor of 0.3 and a migration area of 40 cm^2. It operates at a steady state with a void fraction of 0.6. A sudden increase in reactivity caused the void fraction to increase by 5 percent. Calculate the new effective multiplication factor.

3-18. Water enters a boiling channel saturated at 1,000 psia. In the channel it receives 32.47 Btu/lb_m and exits with a void fraction of 0.60. A sudden increase in reactivity caused the heat added to increase by 10 percent. Calculate the new exit void fraction.

3-19. A large boiling-water reactor uses low enriched fuel. The thermal utilization factor is 0.55. The reactor has a zero void coefficient at low temperatures. Assuming no neutron absorptions except in fuel and water, find the resonance escape probability.

3-20. Calculate the pressure coefficient of a boiling-water reactor that operates at 1,500 psia with an average quality of 5 percent, a slip ratio of 2, and a void coefficient of -0.01.

chapter **4**

Boiling-Reactor
Power Plants

4-1. INTRODUCTION

In the preceding chapter we discussed the thermodynamic and fluid-flow characteristics of the boiling reactor and the effects of voids and pressure on the behavior of the reactor. In this chapter we shall discuss the effect of this behavior on the power plant to which it is coupled. We shall then describe the various types of power plants associated with the boiling-type reactor and describe some representative plants that have been or are in the process of being built.

Since in boiling-water reactors the coolant is allowed to boil directly in the reactor, the reactor pressure does not have to be high in order to avoid bulk boiling of the coolant. Thus, whereas the pressurized-water reactor operates at about 2,200 psia, the boiling-water reactor may operate at one-half (or less) that pressure and still produce comparable steam temperatures and pressures. Lower coolant temperatures within the core result in lower fuel and cladding temperatures.

Because of boiling within the core, only low power densities are possible. Also, since the steam is collected in the compartment (reactor) where boiling takes place, saturated steam at best is admitted to the turbine. Thus, unless special provision is made to superheat the steam or to extract moisture from it as it expands through the turbine, it becomes exceedingly wet in the last stages of the turbine, with detrimental effects on the blades there. There will be more discussion of this later. The thermal efficiency of a boiling-water reactor power plant is about 30-32 percent because of the use of relatively low-pressure saturated steam. This compares with 40 percent for a modern fossil-fueled plant operating with superheated high-pressure steam. The *cost* of power, however, depends upon capital and operating (including fuel cycle) costs (Chapter 17), and could be competitive.

Of importance in the coupling of a reactor to the power plant cycle is the load-following characteristics of the reactor. A boiling-water reactor can attain criticality at various control-rod positions. Withdrawing a control rod increases core reactivity and power. This causes boiling and voids to increase until a point is reached where the negative change in reactivity due to increased voids balances the positive change in reactivity caused by rod withdrawal, and new criticality is attained. In boiling-water reactors load following is accomplished by one of four means: (a) *control by bypass,* (b) *control by subcooling,* (c) *recirculation control,* and (d) *control by positive void coefficient.* The examples that are given in this chapter cover these four methods.

4-2. RADIOACTIVITY OF THE STEAM SYSTEM

Before proceeding with the discussion of the boiling-water reactor, a brief discussion of a question that inevitably comes to mind is in order. Since the same fluid that passes through the core also passes through the power units (turbine, condenser, piping, pumps, etc.), it is of interest to evaluate the extent of induced radioactivity in the fluid.

If the mineral content of the coolant in water-cooled reactor systems is kept low enough (below about 1 ppm), the main radioactivity will be that due to the neutron capture by the oxygen in the water. Neutron capture by hydrogen converts it into nonradioactive deuterium. The most important of the oxygen reactions is the $O^{16}(n,p)N^{16}$ reaction given by

$$_8O^{16} + _0n^1 \longrightarrow _7N^{16} + _1H^1 \tag{4-1}$$

The microscopic cross section of this reaction is 1.4×10^{-5} barn. Nitrogen 16 is a radioactive β emitter (reverting to O^{16}) with a half-life of 7.2 sec. The β rays are mainly of 3.8, 4.3, and 10.5 Mev energy. Gamma rays accompanying this β emission are mainly of 6.13 and 7.10 Mev energy. This half-life of N^{16} is short enough so that only a small fraction of radioactivity remains when the steam reaches the turbine. Thus moisture remaining in contact with turbines and other equipment after shutdown is not dangerously radioactive, and maintenance work can usually be undertaken a short time after shutdown.

Another reaction of somewhat less importance than the above is that due to the neutron capture by O^{17} present to the extent of 0.037 percent of all oxygen. This also is an (n,p) reaction, resulting in N^{17}. Nitrogen 17 is another β emitter of 4.16-sec half-life. The microscopic cross section of this reaction is higher than the one mentioned above by a factor of about 10^3. This is outweighed, of course, by the small concentration of O^{17} in water.

A third reaction of some importance is the $O^{18}(n,\gamma)O^{19}$ reaction.

Oxygen 19 is also a β emitter converting to stable fluorine 19, with a half-life of 29.0 sec. This and other reactions that are not very important produce only a few weak radiations.

If the mineral content of the coolant is high, long-lived and strong radiations result. The radioactive particles may embed themselves in component parts, making maintenance difficult. Coolant treatment is thus an important feature in nuclear plants. Also, care should be taken in designing components to avoid pockets and crevices that may collect and retain radioactive particles.

4-3. THE DIRECT CYCLE POWER PLANT

This system is shown diagramatically in its simplest form in Fig. 4-1. The reactor supplies saturated (or near saturated) vapor directly to the turbine. After expansion through the turbine, the exhaust wet vapor is condensed, and the condensate is pumped back to the reactor. The

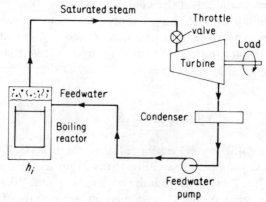

FIG. 4-1.   Direct-cycle boiling-reactor plant.

reactor may be of the internal (shown) or external-recirculation type, of either natural or forced recirculation. Not shown are auxiliaries and accessories, such as for feedwater heating, water treating, etc.

The direct cycle has the advantages of simplicity and of relatively low capital costs. Because of the direct loop arrangement, there are·no heat exchangers.

A major drawback of the direct cycle is that the reactor is not load-following (although, usually with a negative void coefficient, it may be self-controlling). To illustrate this, let us assume that the reactor is operating at some power level determined by its flow and control-rod setting. Let us also assume that a larger load is applied to the turbine, causing the turbine governor to increase the throttle valve opening, i.e., call for more steam. This reduces the flow resistance in the steam passages between the core and turbine. This in turn reduces the reactor pressure. If the reactor has a negative void coefficient, and a negative dx/dp, as is usually the case, it will then have a positive pressure coefficient (Sec. 3-11). A reduction in pressure thus decreases reactor power resulting in less steam generation, opposite to the desired effect. A similar argument applies to a demand for reduced load.

That a reduction in pressure increases reactor voids, i.e., dx/dp is negative, can be shown by a simple heat balance in which the reactor is considered to be an adiabatic steady-flow system, that is $dh/dp = 0$, where h is enthalpy and p is pressure. Equation 3-73, here repeated, therefore applies.

$$\frac{dx}{dp} = - \left(\frac{1 - x}{h_{fg}} \frac{dh_f}{dp} \right) - \left(\frac{x}{h_{fg}} \frac{dh_g}{dp} \right) \qquad [3\text{-}73]$$

Noting that equation and Fig. 3-10, it can be seen that the first term on the right-hand side is always negative (because of the negative sign). The second term could either be positive or negative depending upon the pressure. Below 500 psia, dh_g/dp is always positive, so that dx/dp is always negative below 500 psia. Above 500 psia, the second term becomes positive and dx/dp is either positive or negative depending upon the relative contribution of the two terms. This is a function of quality x. It can be shown that dx/dp becomes positive only at high values of x, greater than those realized in reactors. For example, at 1,000 psia dx/dp is positive if x is greater than 79.3 percent, and the quality decreases with a decrease in pressure. Boiling-water reactors operate at much lower qualities, and therefore dx/dp is almost always negative. Figure 4-2 shows this state of affairs. The pressure coefficient is therefore almost always positive if the void coefficient is negative.

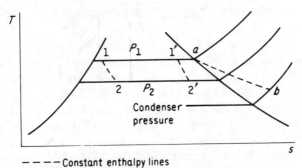

$----$ Constant enthalpy lines

FIG. 4-2.   Constant-enthalpy processes for wet and dry steam. $x_2 > x_1$; $x_2' < x_1'$; $p_1 > p_2$.

The fact that a demand for more steam in the direct cycle results in the opposite-effect plant response must be corrected. *Bypass control* is employed here. This is done by operating the reactor itself continuously at full load, and bypassing excess steam at part load around the turbine. This method was employed in the EBWR, described in Sec. 4-4.

The bypass method is shown diagrammatically in Fig. 4-3. A fraction of the steam produced in the reactor is bypassed around the turbine,

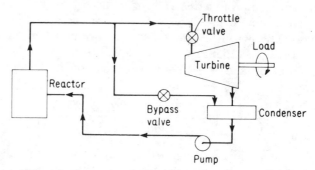

FIG. 4-3. Bypass controlled direct-cycle boiling-reactor
plant.

via a bypass valve, to the main condenser. A demand for more steam momentarily reduces the reactor pressure (as before). However, a pressure-sensitive device on the reactor rapidly reduces the bypass-valve opening, thus maintaining the reactor pressure and diverting the necessary amount of steam from the bypass line to the turbine. The reverse is true. The reactor maintains a steady output. It is obvious that this system is very inefficient at part loads because of the degradation of useful energy by throttling to the condenser pressure. Even at full loads, a certain amount of steam is bypassed to accommodate overload conditions.

4-4. THE EXPERIMENTAL BOILING-WATER REACTOR (EBWR)

Although several boiling-water reactors were built prior to it, EBWR [16] was the first boiling-water-reactor power plant to be built. Despite the disadvantages of the direct cycle, EBWR was built on its principle, since it was desired at the time to demonstrate the feasibility and stability of the boiling-water reactor as a power producer rather than to solve all the other problems of control, etc. The EBWR was designed, built, and operated by the Argonne National Laboratory. It has since been used for various studies including plutonium-fueled thermal reactor operation, and has subsequently been mothballed.

Figure 4-4 shows a cutaway of the EBWR reactor and associated

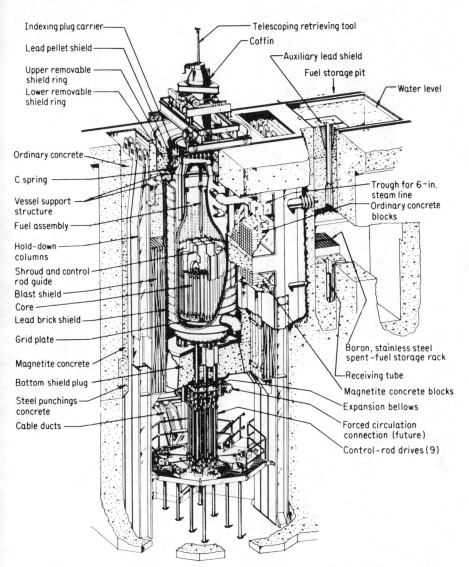

Indexing plug carrier —
Lead pellet shield —
Upper removable shield ring
Lower removable shield ring
Telescoping retrieving tool
Coffin
Auxiliary lead shield
Fuel storage pit
Water level

Ordinary concrete

C spring

Vessel support structure
Fuel assembly

Hold-down columns
Shroud and control rod guide
Blast shield
Core
Lead brick shield

Grid plate

Magnetite concrete

Bottom shield plug

Steel punchings concrete
Cable ducts

Trough for 6-in. steam line
Ordinary concrete blocks

Boron, stainless steel spent-fuel storage rack
Receiving tube
Magnetite concrete blocks
Expansion bellows
Forced circulation connection (future)
Control-rod drives (9)

FIG. 4-4. Cutaway of EBWR and components. (Courtesy Argonne National Laboratory.)

equipment. It is a natural-internal-recirculation reactor with chimney. The main design and operational data of the plant as initially operated at 4,500 kw(e) and 20,000 kw(t) are given in Table 4-1. (Operation at much higher output was subsequently achieved.)

Figure 4-5 shows a flow diagram of the EBWR power plant. Steam from the reactor is passed through a steam dryer-cooler before it goes to the turbine. This unit is a horizontally mounted stainless-steel drum

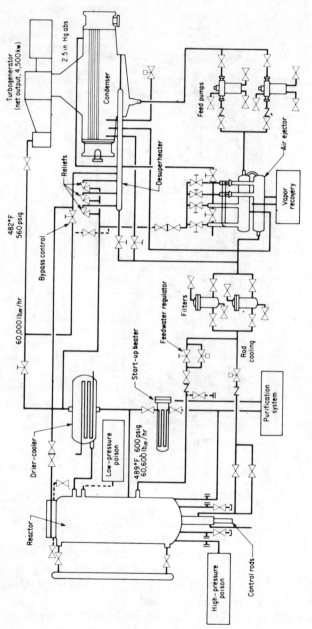

FIG. 4-5.   Flow diagram of EBWR.   (Ref. 16.)

TABLE 4-1
Some Design and Operational Data for EBWR

Reactor heat output	20,000 kw(t)
Plant net power	4,500 kw(e) (5,000 kw gross)
Coolant-moderator	Light water
Reactor outlet pressure	600 psig
Reactor outlet temperature	498°F
Steam flow	60,600 lb_m/hr
Fuel-element configuration	Plate type, six each in 112 subassemblies in 4-in. center-to-center lattice
Feedwater temperature	130°F
Coolant speed in core	2.5 ft/sec (0.9 ft/sec in downcomer)
Average core heat flux	45,500 Btu/hr ft² (155,000 max.)
Estimated burnout flux	750,000 Btu/hr ft²
Core dimensions	Cylindrical, 4-ft-high, 4-ft-diameter, assemblies arranged 4 in. center-to-center lattice
Fuel type	U metal 93.5%, Zr 5%, Nb 1.5% by mass
Fuel enrichment	76 central subassemblies, 1.44% and 36 subassemblies, natural U
Total core heat-transfer surface . . .	1,500 ft²
Maximum fuel center-line temperature.	660°F
Average core nonboiling height . . .	1.6 ft
Average core boiling height	2.4 ft
Average void fraction at core exit . .	28.5%
Core inlet subcooling	5.2°F
Slip ratio	1.5:1
Recirculation ratio	63:1
Control rods	Nine total, X-shaped (five, hafnium and Zircaloy 2; four, 2% boron, Type 304 stainless steel and Zircaloy 2)

which contains a large number of scrubbers (in the form of parallel corrugated plates of 820-ft² surface area). These scrubbers entrain moisture and other particles, leaving the steam with less than 0.1 mass percent moisture and less than 0.001 ppm solids. The condensate from the dryer-cooler leaves at bottom center and goes to a steam trap (not shown) and then to the main plant condenser.

The lower part of the dryer-cooler shell contains stainless-steel tubes designed to remove 1,000 kw of heat in an emergency. It also provides shutdown cooling of the reactor decay heat. Water for the cooler is supplied from a 15,000-gal overhead storage tank (not shown) via a solenoid-operated air-diaphragm valve. The condensate in such a case passes back through the reactor. The unit is normally radioactive from $N^{16}\gamma$ radiation (Sec. 4-2), so that normal access to it is restricted and periodic flushing of it is recommended.

Dry steam leaves the dryer-cooled and goes to the turbine, now at 560 psig and 482°F. The turbine is a 5,000-kw, 3,600-rpm, multistage impulse turbine (one stage Curtis, eleven stages Rateau).

In the EBWR, load control is by steam bypass. The reactor operates at a constant power level determined by the control-rod setting. (Note that in a boiling-water reactor criticality can be attained at several control-rod settings.) The pressure in the reactor is the main control parameter of the plant. An increase in turbine load, resulting in demand for more steam, increases the turbine valve opening and reduces reactor pressure. The difference between the new pressure and a set (reference) pressure acts as a signal actuating a bypass valve in the right direction (closing it in this case) to restore reactor pressure to the set value.

The bypass valve is actuated either automatically, manually, or electrically. Automatic operation, accomplished hydraulically, works in the following cases:

1. An increase in turbine load results in closing the bypass valve, thus putting more steam through the turbine. (Control rods come into play only when an increase in turbine load beyond that which can be handled by the bypass valve takes place.)

2. An increase in reactor power over that demanded by the turbine opens up the bypass valve through a pressure-sensitive element in the automatic regulator that senses the increased system pressure.

3. A turbine overspeeding results in closing the turbine trip throttle valve by an overspeed governor which automatically opens the bypass valve and dumps the entire steam load into the desuperheater and condenser.

The three relief valves shown open successively to provide protection against accidental failure of the bypass valve to open in the last case and in the case of system overpressure.

Bypass steam is throttled and therefore superheated. It therefore passes through a desuperheater prior to the condenser. The direct use of superheated steam in a condenser lowers the condensing heat-transfer coefficient and should be avoided. The desuperheater is a 14-in.-diameter pipe that extends through the condenser just above the hot well water level. Steam coming into the desuperheater after throttling is substantially at the same enthalpy as the high-pressure reactor steam (line *ab*, Fig. 4-2). At condenser pressure (2.5. in. Hg abs) this steam has a 110°F superheat. Desuperheating in normal operation is accomplished by the condensate water shower coming down in the condenser main shell. The bypassed steam may, however, be subjected to two water sprays, via two nozzles, supplied from the condensate line (beyond the air ejector and prior to the filters). The desuperheater is designed to handle the entire reactor steam load if necessary.

The main condenser is a single-pass, surface-type horizontal unit. Except for a rupture disk, all joints of the condenser are welded to prevent contamination through atmospheric air leaking into the condenser. Provision also is made to avoid any mixing of untreated condenser cooling water with reactor water. This is done by securing the ends of the cooling-water tubes in double walls, with the space between them evacuated to a drain tank. The condenser is also longitudinally divided to allow isolation of one-half in case of a tube leak. The tubes are aluminum-clad on the inside, instead of the usual copper alloys, to reduce possible radioactivity in the system.

Drains are incorporated everywhere in the system to collect leaks of the highly treated water. Makeup water from a 200-gal demineralized water tank (supplied from two mixed-bed ionexchange units at pH of 7.0 and below 0.2 ppm total solids) is fed directly into the condenser. Deaerating takes place in the condenser itself. The makeup-water tank is smaller than in conventional steam-plant practice because of the leak-tightness of the system.

One of two 180-gpm multistage centrifugal vertical feedwater pumps (the other is kept as a spare) returns the condensate to the reactor. These units thus serve the dual role of condensate and feedwater pumps, with the first stage of each acting as the condensate pump. This design reduces the number of pump-shaft seals.

The condensate leaves the pumps at approximately 900 psig and flows through a two-pass intercooler and a single-pass aftercooler which takes care of the steam operating the air ejectors. These air-ejector coolers require a certain minimum flow rate through them (70 gpm). At low reactor powers and low feedwater rates, this minimum requirement is maintained by recirculating some of the feed water back to the condenser.

The air ejectors evacuate the air which leaks to the condenser through the glands, joints, etc. (estimated at 1 lb_m/hr or 0.25 cfm) and the oxygen and hydrogen formed by reactor-water decomposition (estimated at 4 lb_m/hr). These noncondensable gases reduce the heat-transfer coefficient in the condenser. They are also corrosive because of the presence of free oxygen. Except during plant starting, when a large quantity of air is entrained (about 2,000 ft³), only one of the two ejectors is normally in operation.

The feedwater now enters two disposable cotton-fiber filters (one a standby) capable of removing particulate matter down to 2 to 5 μ. There also are provisions for sampling the water, before and after the filter, as well as for continuous monitoring of the activity there.

Water is then led into the reactor vessel via feedwater regulators. Some water, at the rate of 10 gpm, is continuously removed from the bottom of the reactor, cooled in a regenerative heat exchanger, and delivered to a water purification system (not shown) which contains two mixed-bed ion-exchanger units (one normally a standby) with accompanying secondary coolers, pre- and after-filters, pumps, etc. The treated water returns via the regenerative heat exchanger where it is reheated, after which it is pumped back to the feedwater line. The purification system reaches levels of radioactivity up to 30 to 40 r/hr in normal use (and much higher in case of fuel-element failure). The prefilters and ion-exchanger columns are therefore lead-shielded.

The startup heater, shown next to the reactor vessel in Fig. 4-5, heats reactor water up to 325°F (corresponding to a saturation pressure of 100 psig) with auxiliary steam at 378°F, prior to the withdrawal of the control rods. The flow of the reactor water through the startup heater is by natural circulation. The heater is also connected to the purification system, which can be put into operation prior to starting.

The EBWR plant uses corrosion-resistant materials in a large number of places, the choice of which was dictated by need and economy. Type 304 stainless steel is used for the reactor-vessel lining, the dryer-cooler, and the piping in between. Piping and valves leading to the turbine, as well as the major portion of the turbine surfaces, are nickel-plated. The main condenser shell, inner plates, and piping to the feedwater pumps are constructed of carbon steel. The condenser tubes are aluminum-clad on the inside, the feedwater pumps are made of 12 percent chrome steel, and the remainder of the system is of either stainless or nickel-plated carbon steel.

4-5. THE DUAL-CYCLE POWER PLANT

The *dual-cycle* boiling-water-reactor power plant has been devised to obviate the two main drawbacks of the direct cycle: low power density (an unattractive feature, especially in large central power stations) and poor load-control response (which necessitates steam bypassing). Load control is accomplished by varying the degree of subcooling of the reactor inlet water. Higher power density is accomplished by increasing the degree of subcooling at higher loads.

The basic elements of a dual cycle are shown in Fig. 4-6. The wet mixture of steam and water produced by the reactor leaves at point 1 and goes to a steam separator. Saturated steam separates at slightly less than reactor pressure (because of losses) and goes to the turbine via 2. Saturated water leaves the steam separator and goes down via 3 to a secondary steam generator, where it gives up some of its sensible heat to low-

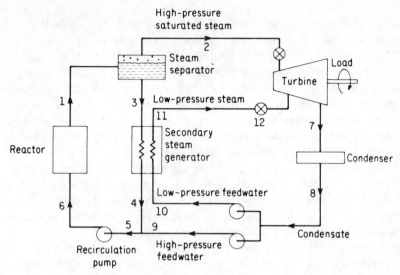

FIG. 4-6. Basic elements of dual-cycle boiling-reactor power plant.

pressure feedwater, and leaves, subcooled, at 4. There it mixes with high-pressure feedwater (5) and is pumped back to the reactor vessel. (6)

Low-pressure steam generated in the secondary steam generator leaves at 11 and enters the turbine at a low-pressure stage. High- and low-pressure steam expand through the turbine and enter the condenser at 7. The total condensate leaves at 8. The condensate is then split into two paths. In one, the feedwater is pumped by a high-pressure pump, leaves at 9 and mixes with the secondary-steam-generator return water at 5. In the other path, the remaining feedwater is pumped to a lower pressure and is lead at 10 into the secondary steam generator. Because it is at lower pressure than the reactor recirculation water, its saturation temperature is lower than the temperature of the recirculation water. Also, because of the recirculation ratio is high, the rate of flow of the recirculation water is many times that of the low-pressure water and thus carries with it sufficient energy to raise the water at point 10 to its saturation temperature and to boil it completely into steam.

The reason for the higher power density in the dual cycle is now apparent. In the direct-cycle boiling-reactor, the reactor-core inlet water is only slightly subcooled, because of the addition of a small percentage (1/recirculation ratio) of relatively cool feedwater to the saturated recirculation water. In the dual cycle, the degree of subcooling is much greater because the recirculation water loses a portion of its sensible heat in the secondary steam generator. For the same core exit quality, the dual-cycle core has a greater percentage of its height as a nonboiling height than does the direct-cycle. This means that a smaller portion of the core is subjected to the same void fraction in the former and consequently a greater power density can be attained.

Note, however, that the same reasoning shows that the dual cycle would have a lower natural driving pressure which (coupled with the increased flow resistances) necessitates the use of more chimney heights, such as a high riser, or forced circulation, or both. (A certain natural driving pressure necessary to take care of decay-heat cooling in the case of pump failure is necessary.) Because of the secondary steam generator, the dual cycle is normally of the external-recirculation type.

Control of the dual cycle is accomplished in the following manner: A change in steam demand results in the turbine governor actuating mainly the *low-pressure* inlet valve, 12. In the case of demand for more steam this valve is opened wider by the turbine governor and more low-pressure coolant is passed through the secondary steam generator. More heat is consequently extracted from the high-pressure recirculation line, the temperature at 4 is reduced, and the degree of subcooling of the reactor return water at 6 is increased. A smaller portion of the core is subjected to voids, and more heat is generated in the core, resulting in the desired effect. Both the void and temperature coefficients of reactivity, therefore, take effect here.

Note that while the reactor is a boiling-water reactor, the dual cycle is essentially a combination boiling-water and pressurized-water cycle. The pressurized-water portion is in the water-to-water secondary steam generator, although this occurs at lower pressures than in a full pressurized-water plant.

An alternative method of generating low-pressure steam is shown in

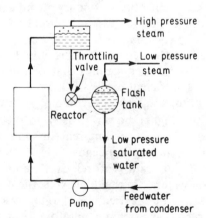

FIG. 4-7. Alternative method of
secondary steam generation in dual-
cycle boiling-reactor plant.

Fig. 4-7. The recirculation water is throttled down to a lower pressure in a flash tank. Part of its sensible heat is thus converted into latent heat of vaporization and part of that water flashes into steam which is admitted to the turbine at a low-pressure stage. The remaining water is pumped back to the reactor, together with the main condensate from the plant.

Several boiling-water-reactor dual cycles were designed and built by General Electric Company. One is the Dresden I nuclear power station, described in the next section.

4-6. THE DRESDEN I NUCLEAR POWER PLANT

The Dresden 1 nuclear power station [17] is a 180,000-kw(e), 627,300-kw(t) plant built by General Electric Company for Commonwealth Edison Company. It is situated on the Kankakee and Des Plaines rivers, some 50 miles southwest of Chicago, Ill. Dresden II and III, later built on the same site, are of the type described in Sec. 4-8. The main design and operational data of the Dresden I plant at rated load are given in Table 4-2.

Figure 4-8 shows a flow diagram of the plant with the properties of the fluids given at rated load (h is enthalpy, Btu/lb_m). The dashed lines indicate steam while the solid lines represent water.

The reactor, shown to the left of the diagram, is of the forced-external-recirculation type. The steam-water mixture leaves the reactor at 990 psig and 4.92 percent quality and flows up risers to a 2.5-ft-diameter and 60-ft-long horizontal steam separator, called the primary steam drum, 85 ft above the reactor vessel. In this drum 1,407,400 lb_m/hr of dry steam is produced after a pressure drop to 965 psig, and 24,295,000 lb_m/hr of saturated recirculation water flows via downcomers at a saturation temperature of 543°F, corresponding to 990 psig. This water is pumped back to the reactor bottom via four secondary steam generators, where 1,200,620 lb_m/hr of low-pressure feedwater coming in at 405°F, is converted into 1,187,660 lb_m/hr of saturated steam at 475 psig and 461°F. The balance of 12,960 lb_m/hr goes into blowdown and leakage.

In this process, the recirculation water suffers a reduction in enthalpy from 540.4 to 500.1 Btu/lb_m. It then mixes with return water coming from the high-pressure feedwater circuit before returning to the reactor, both at 504.8°F and 493.6 Btu/lb_m enthalpy. At the reactor pressure of about 1,015 psig, this corresponds to a degree of subcooling of about 43°F—much higher than in the case of direct-cycle boiling-water reactors.

TABLE 4-2
Main Design Data of Dresden I Plant

Reactor heat output	627,300 kw(t)
Plant net power	180,000 kw(e)
Coolant-moderator	Light water
Reactor outlet pressure	990 psig (1,015 psig in reactor)
Reactor outlet temperature	542.5°F
Steam flow	1,407,000 lb_m/hr (high pressure); 1,187,660 lb_m/hr (low pressure)
Reactor inlet temperature	504.8°F
Coolant velocity in core	4.5 ft/sec at inlets, 4.75 ft/sec at saturation (4 ft up), and 9 ft/sec average at exit
Average core heat flux	100,000 Btu/hr ft² (300,000 max.)
Estimated burnout heat flux	2.6 times actual heat flux at worst combination of heat flux and quality
Core dimensions	Cylindrical, 10 ft high, 10½ ft in diameter
Fuel-element configuration	Rods approximately ½ in. diameter, 104 in. active length; 25 each in square array in normal subassemblies, '16 in subassemblies with control rods; total, 700 subassemblies and 17,000 fuel rods
Fuel type	UO_2 sintered solid cylindrical pellets
Fuel enrichment	1.5%
Maximum fuel center-line temperature.	2800°F
Average core nonboiling height	About 6 ft
Average core boiling height	About 4 ft
Average void fraction at core exit	About 47%
Core inlet subcooling	About 43°F
Slip ratio	3:1
Recirculation ratio	17:1
Control rods	80 X-shaped, boron carbide-stainless steel alloy, clad in Type 347 stainless steel
Net thermal efficiency	28.7%

The physical arrangement of the reactor, primary steam drum, secondary steam generators, and piping and pumps is depicted in Fig. 4-9. The risers are shown in solid black. The risers and downcomers are bent sufficiently to allow room for fuel handling. The downcomers feed into two horizontal headers which in turn feed into the secondary steam generators. Each of the secondary steam generators has separate shield-

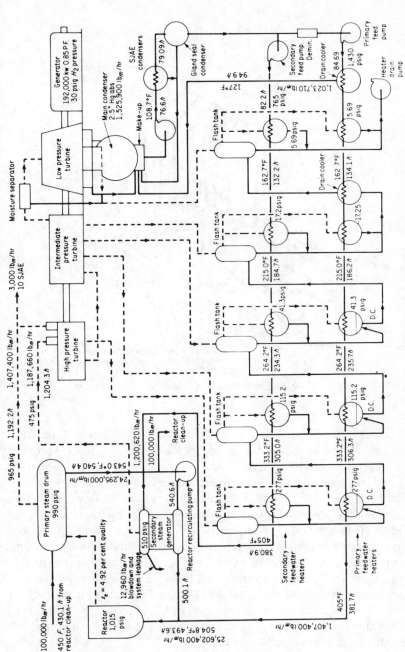

FIG. 4-8. Heat balance diagram of Dresden I power plant. (Ref. 17.)

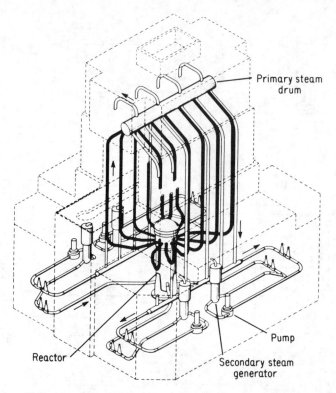

FIG. 4-9.   Perspective view of Dresden I reactor package.
(Ref. 17.)

ing and can be isolated from the others for maintenance purposes.
The plant is designed to be able to operate with up to two secondary
steam generators out of service.

Load control in the Dresden plant is accomplished as previously
outlined for the dual cycle.   The turbine governor actuates the secondary steam-admissions valves, and load control is by reactor inlet subcooling.   This method is completely automatic, i.e., without resort to
steam bypassing or control-rod actuation, over a range from 40 to 100 percent of rated load.   A steam bypass system (not shown) is provided for
emergency cases, such as a turbine trip, however.   Control-rod actuation
is necessary in the following cases: (1) to start up and shut down, (2) to
scram, (3) to flatten the neutron flux distribution in the core, (4) to
regulate the ratio of primary to secondary steam, and (5) to compensate
for reactivity changes due to fuel burnup.

Control rods are made of a boron carbide and stainless steel alloy, clad in Type 347 stainless steel. They extend into the core from the bottom and are approximately 9.5 ft long, cruciform-shaped, 3 in. wide on each side, and 3/8 in. thick. Under normal operation, control-rod travel at 1/2 ft/sec is accomplished by means of a motor-driven screw actuator, which properly positions the end stop of a piston, normally pressurized on the opposite side to counteract the movable stop. In case of a scram, hydraulic pressure is applied to the other side of the piston, causing the rods to travel at 10 ft/sec.

The primary steam (at about 965 psig) is fed into the high-pressure turbine, Fig. 4-8. The secondary steam (at 475 psig) is fed into a lower stage of the same turbine. The exhaust steam from the high-pressure turbine goes to the intermediate-pressure turbine via a moisture separator. The steam in the turbine becomes quite wet on expansion because inlet steam is at high pressure and is only saturated. Besides the moisture separator moisture extraction buckets are built into the turbine, Sec. 4-12. The high-, intermediate-, and low-pressure turbines are built on one shaft, rotating at 1,800 rpm. The turbine is provided with the necessary seals which allow it to be filled with decontaminating fluid if necessary.

The turbine exhaust goes to the main condenser at 2.5 in. Hg abs. From there the condensate is pumped via a steam-jet air ejector (SJAE), where it acts as cooling water for the steam jet. The water formed in the SJAE is returned to the main condenser makeup compartment. The primary steam line, shown just above the high-pressure turbine, supplies 3,000 lb_m/hr of steam for the SJAE. The SJAE removes air, oxygen, and hydrogen (due to water decomposition) from the system. These non-condensable-gases are retained at least 3 min before ejection to the atmosphere, allowing sufficient time for any radioactivity to decay to safe limits. The condensate increases in enthalpy across the SJAE. It then passes through a gland-seal condenser, where it again acts as cooling water to condenser steam from the gland seals of the turbine (line not shown). The resulting water is also returned to the main condenser makeup.

The condensate then divides into two paths. One, represented by the upper feedwater line in Fig. 4-8, goes through a secondary feed pump which pressurizes it to 765 psig, then into five regenerative surface-type feedwater heaters. Regeneration steam is bled from the turbines as shown. Feedwater leaves the last regeneration stage at 405°F and an enthalpy of 380.9 Btu/lb_m and enters the four secondary steam generators.

In the primary feedwater line (the lower one in the diagram), the condensate first goes through two of three demineralizers (only one is shown in the diagram), and is then pumped to a pressure of 1,430 psi, sufficient to overcome succeeding pressure losses and yet to enter the reactor at the reactor pressure of 1,015 psi. It then goes through a drain cooler, in which various drains are collected, cooled, and fed back to the main condenser. In this process the primary water receives some sensible heat. It then goes through five stages of regeneration, emerging at 405°F and 381.7 Btu/lb$_m$ enthalpy. It then mixes with the water coming out of the secondary steam generator, and the mixture enters the reactor at 504.8°F and 493.6 Btu/lb$_m$ enthalpy.

In the process of regeneration, the extracted steam at each stage is first drawn into a flash tank in which drier steam is taken at the top, divided into two paths, and fed to the two surface feedwater heaters (the primary and the secondary) for that stage. After this steam is condensed in the two heaters, it is mixed with the condensate from the flash tank of the same stage, and the mixture is admitted to a point near the bottom of the flash tank of the next lower-pressure stage. There it undergoes a throttling process and is partially converted into steam, which is taken out at the top of the flash tank along with the steam resulting from the extraction line for that stage. In the lowest-pressure stage, the water mixture is pumped by the heater drain pump, shown near the right bottom, back into the condenser makeup via the drain cooler mentioned above.

Not shown are the following auxiliary systems: (1) a reactor cleanup system, (2) a condensate demineralizer system, (3) a waste-disposal system, (4) a fuel-handling system, and (5) various safety systems.

4-7. PLANT CONTROL BY RECIRCULATION FLOW

It has been shown that in a boiling-water reactor plant, the void coefficient of reactivity is usually negative and the pressure coefficient is usually positive and that the reactor is therefore not by itself load following. In the two previous examples load-following control was attained by steam bypass and by inlet-subcooling control in the direct and dual cycles respectively. Bypass control results in poor part-load efficiencies. The dual cycle is fairly complicated and costly.

A third method of load following control uses a direct cycle, but with variable recirculation flow. If the total heat generated in a boiling reactor is Q_t, the core inlet enthalpy h_i and the average core exit quality is $\bar{x}_e$, then by Eqs. 3-10, 3-7, and 3-3,

$$Q_t = \dot{m}_i(h_f + \bar{x}_e h_{fg} - h_i)$$

$$= \dot{m}_i(h_f - h_i) + \dot{m}_i \bar{x}_e h_{fg} \qquad (4-2)$$

and

$$h_i = h_f - \bar{x}_e(h_f - h_d)$$
$$= h_f - \frac{\dot{m}_d}{\dot{m}_i}(h_f - h_d) \qquad (4\text{-}3)$$

Combining Eqs. 4-2 and 4-3 gives

$$Q_t = \dot{m}_i \left[h_f - h_f + \frac{\dot{m}_d}{\dot{m}_i}(h_f - h_d) \right] + m_i\bar{x}_e h_{fg}$$
$$= \dot{m}_d(h_f - h_d) + \dot{m}_i\bar{x}_e h_{fg} \qquad (4\text{-}4)$$

where $\dot{m}_d$ and $\dot{m}_i$ are the feedwater (and steam), and core inlet mass flow rates respectively. A reactor operating at a steady state will have constant Q_t, $\dot{m}_d$ and h_d. h_f and h_{fg} are fixed at the system pressure. The product $\dot{m}_i\bar{x}_e$ therefore is constant.

If the reactor power is to be increased the value of $\dot{m}_i$ is increased. This momentarily lowers $\bar{x}_e$ (since Q_t has not yet changed). The normally negative void coefficient then acts to increase k_{eff} above unity. k_{eff} returns to unity when increased power increases the core void fraction and quality to their original values.

Except at low loads h_d varies only slightly with load and h_i is therefore alomost independent of load, if $\bar{x}_e$ remains the same (see Eq. 3-7). Equation 4-2 then indicates that Q_t changes linearly with $\dot{m}_i$, except at low loads. The range of change in Q_t depends upon the control rod positions. This will be discussed further in Sec. 4-8 with the help of Fig. 4-17.

A change in the core inlet flow is accomplished by changing the recirculation flow in a forced-recirculation system. This is accomplished by varying the speeds of the recirculation pumps by using, for example, variable frequency motor-generator set drives for these pumps.

The following two examples of boiling-water reactors, the Browns Ferry and the Pathfinder power plants use recirculation control in direct cycle arrangements. The Browns Ferry reactors use a combination of external centrifugal and internal jet pumps. The Pathfinder uses external centrifugal pumps only.

4-8. THE BROWNS FERRY NUCLEAR POWER STATION

The Browns Ferry Nuclear power station [15] is a three-unit boiling-water power plant located at Browns Ferry on the north shore of Wheeler reservoir, Alabama. The plant was designed, built, and is owned by the Tennessee Valley Authority. The nuclear reactors, steam supply system, turbogenerator, and auxiliary equipment are furnished by General Electric Company. The plant represents a nuclear breakthrough from an economics standpoint since it proved more economical in open bidding than competitive bids based on fossil fuel plants.

The boiling-water reactors are of the direct-cycle, forced-internal-recirculation type. They are similar, except for size, to all large boiling-water reactors designed and constructed in the late 1960's and early 1970's. The three reactors produce 3,293 Mw(t) and, 1,098 Mw(e) gross each. The turbogenerators are designed for a maximum output of 1,152 Mw(e) gross each with all feedwater heaters in, or 1,188 with the top heaters out (and thus more steam flow in the turbine). The main design data of Browns Ferry are given in Table 4-3.

TABLE 4-3
Main Design Data of Browns Ferry

Heat output per reactor	3,293 Mw(t)
Electrical output per reactor	1098 Mw(e) gross
Coolant-moderator	Ordinary water
Core operating pressure	1,000 psia
Feedwater temperature	376.1°F
Steam flow per reactor, $\dot{m}_g$	13.38 × 10⁶ lb$_m$/hr
Total core flow per reactor, $\dot{m}_i$	99.6 × 10⁶ lb$_m$/hr
Average exit quality, $\bar{x}_e$	13.6%
Core void fractions, $\bar{\alpha}$ and $\bar{\alpha}_e$	37.3% (average), 75.4% (maximum)
Enthalpy of subcooling, Δh_{sub}	25.5 Btu/lb$_m$
Heat flux, q''	163,200 (average), 425,000 (maximum) Btu/hr ft²
Minimum critical heat flux ratio	1.9
Power density	50.8 kw/liter
Total peaking factor	2.6 (1.4 radial, 1.5 axial, 1.24 local channel)
Pressure vessel	20 ft, 11 in. inside diam., 72 ft, 7⅝ in. inside length
Fuel type and enrichment	UO₂, 2.19% initial enrichment
Fuel configuration	0.488 in. pellet diam., clad in Zircaloy-2, 0.562 in. OD, 0.032 in. thick; 12 ft active length with 16 in. gas plenum
Fuel assembly	764 fuel channels containing 7 × 7 fuel rods on 0.738 in. pitch in Zircaloy-4 channels
Fuel temperature	1100°F (average), 4380°F (maximum)
Control rods	185 cruciform, 9.75 in. width on 12.0 pitch. Each with 840.188 in. OD stainless steel tubes per rod containing B₄C granules, 143 in. active length, 144 in. stroke

TABLE 4-3 *(continued)*

Temporary control curtains	356 8.5-in. wide, 0.0625-in. thick flat sheets between fuel channels; stainless steel with 5,700 ppm boron, 143 in. active length
Core reactivities	0.25 excess, cold clean core, 0.96 k_{eff} all control rods in; < 0.99 k_{eff} with strongest rod out
Control-rods worth	-0.17
Control-curtains worth	-0.12
Moderator temperature coefficient .	-5×10^{-5} cold, -39×10^{-5} hot (no voids)
Moderator void coefficient	-1×10^{-3} hot (no voids), -1.6×10^{-3} (operating)
Fuel temperature (Doppler) coefficient.	-1.3×10^{-5} cold, -1.2×10^{-5} hot (no voids), $<1.3 \times 10^{-5}$ (operating)
Recirculation loops per reactor . . .	2, contained in dry well, 28 in. pipe diam.
Pumps per reactor	2 external centrifugal pumps, 20 internal jet pumps
Primary containment	Pressure suppression, 62 psig design internal pressure
Secondary containment	Reinforced concrete and steel superstructure with metal siding, 0.25 psig design internal pressure

The reactor (fig. 4-10) is fueled with UO_2 pellets clad in Zircaloy-2 tubing. The rods contain helium and have a plenum at the top to prevent excessive internal pressure due to the helium, the volatile content of UO_2 and the gaseous fission products that are not retained within the fuel pellets. (A quantity of 1.35×10^{-3} g-moles of fission gases are produced per Mwd. Experimental data show that, up to 15,000 Mwd/ton, 0.5 percent are released below 3000°F, 20 percent between 3000 and 3450°F. At higher than 15,000 Mwd/ton, 4 percent are released below 3000°F and 100 percent above 3000°F.) The fuel rods are 12 ft active length (cold), except for the central rod in each subassembly which has intermediate plugs which serve to position the spacers. Each subassembly, Fig. 4-11, contains a 7 × 7 array of rods of which 19 have reduced U^{235} enrichment to reduce the local power peaking factor, Fig. 4-12. The fuel rods are spaced and supported by upper and lower tie plates. Coil springs at the top of each rod allow rod axial expansion by sliding within the holes in the upper tie plate. There are seven Zircaloy-4 spacers along the length of the rods to keep them properly aligned and separated. The fuel subassembly is enclosed in a Zircaloy-4 channel.

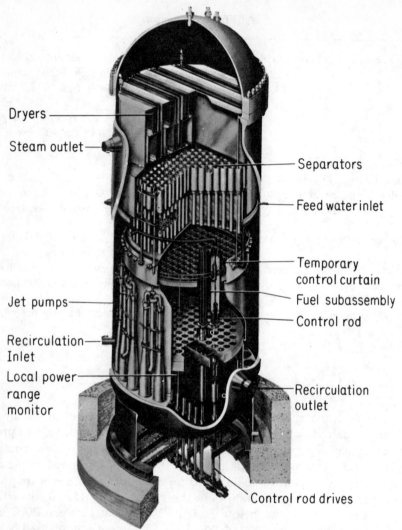

Dryers

Steam outlet

Separators

Feed water inlet

Temporary
control curtain

Fuel subassembly

Jet pumps

Control rod

Recirculation
Inlet

Local power
range
monitor

Recirculation
outlet

Control rod drives

FIG. 4-10.   Browns Ferry reactor   (Courtesy General Electric Company.)

Some coolant flows outside these channels to cool the control rods,
temporary control curtains, neutron sources and in-core instrumentation.
The lower part of the channel makes a sliding seal fit on the lower tie
plate and contains an orifice which establishes the relative flow in each
channel.   The upper part is attached to the upper tie plate by a capscrew.

Control rods are used to start the reactor, shut it down, and to shape
the core power distribution.   Operating adjustment of reactor power and
load following are accomplished by varying recirculation flow, above.
The control rods are cruciform-shaped, Fig. 4-13, and are of the bottom-

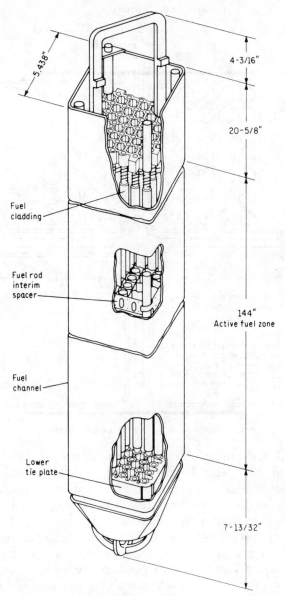

FIG. 4-11. Browns Ferry reactor fuel subassembly.
(Ref. 15.)

entry type. Each has its own rod drive and scram device. The drives are hydraulically operated and are of the locking-piston type. In the event of a loss of power, stored energy available from gas-charged accumulators and from reactor pressure provides hydraulic power for

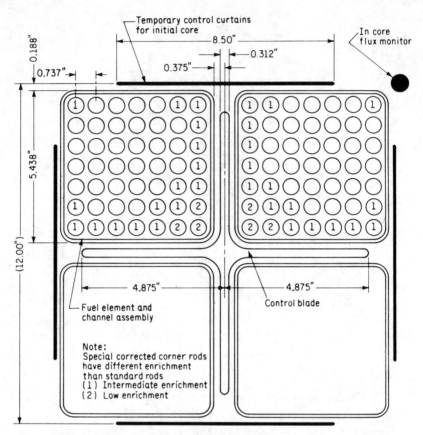

FIG. 4-12.   Browns Ferry reactor core lattice arrangement.   (Ref. 15.)

scramming all rods simultaneously. The blades of the control rods contain 3/16-in.-diam. sealed stainless-steel tubes filled with compacted boron carbide powder. The reactor also contains temporary control curtains made of boron stainless steel. These supplement the control rods for reactivity control of the initial core.

In the reactor (Fig. 4-10) the water-steam mixture leaves at the top of the core and passes through steam separators and dryers located within the pressure vessel. The separators are an array of standpipes welded to a common base, each containing a centrifugal-type steam separator assembly with no moving parts located at the top. In each separator, the mixture impinges on vanes which impact a spin to establish a vortex which separates steam from water. Steam, still slightly wet, then leaves at the top and passes into a wet steam plenum below the dryers. The separated water enters a pool that surrounds the stand-pipes, from which it enters the downcomer annulus. Moisture is re-

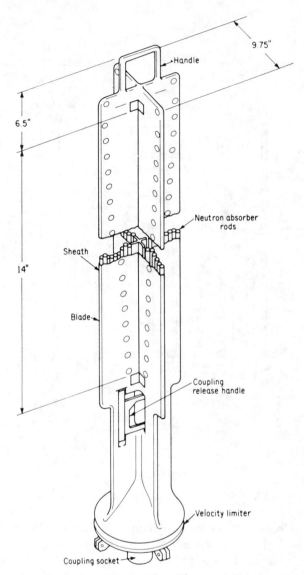

FIG. 4-13. Browns Ferry reactor control rod. (Ref. 15.)

moved from steam flowing upward and outward through driers mounted above the separators and flows through troughs and tubes to a pool surrounding the dryers. A shroud separates the dry and wet steam plenums. Dry steam finally leaves the vessel through outlet nozzles to the turbine.

The recirculation system for each reactor consists of two external

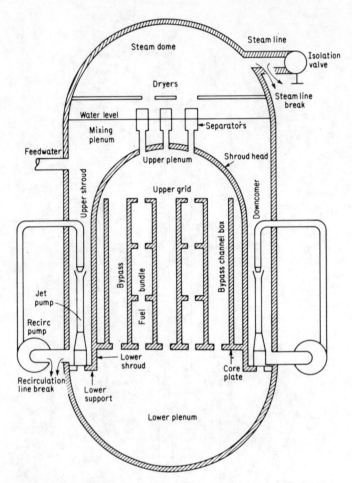

FIG. 4-14.   Browns Ferry reactor flow arrangement.   (Ref. 15.)

recirculation pump loops and 20 internal jet pumps located inside the reactor vessel, Figs. 4-14 and 4-15. The two external recirculation pumps are rated at 45,000 gallons per minute each. They are single-stage vertical centrifugal units with mechanical shaft seals, driven by variable-frequency motor-generator sets. Changing the pump speed changes recirculation flow, which in turn changes reactor power, Sec. 4-7. The pumped water provides the driving force for the 20 internal jet pumps pumping the main internal recirculation flow of 99.6×10^6 lb_m/hr.

The jet pumps are made of stainless steel and consist of a driving nozzle, suction inlet, throat or mixing section, and diffuser, Fig. 4-15.

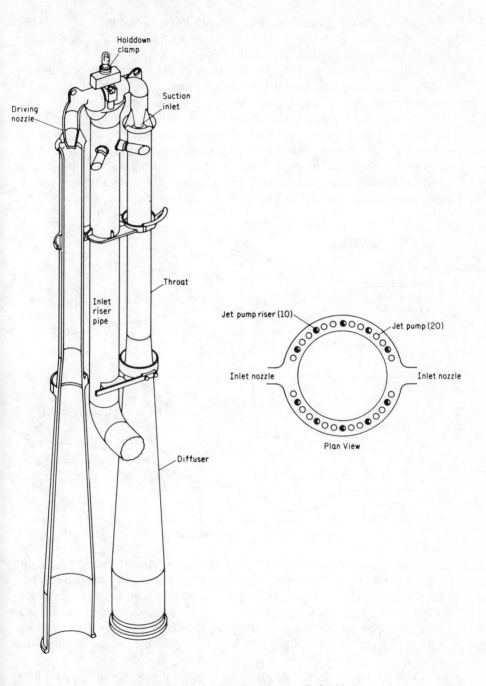

FIG. 4-15. Jet pump arrangement.   (Ref. 15.)

The jet pumps are grouped in the outer periphery around the core. They are arranged in pairs, each having a common inlet riser. The diffusers are welded at the bottom to a baffle plate.

The above external-internal pumping arrangement allows a considerable reduction in external flow. Only one third of the total reactor recirculation water is pumped through the two external pumps. This flow leaves the reactor vessel through two outlet nozzles, Fig. 4-10, and returns to the jet pump inlet riser pipes to become the *driving flow* for the jet pumps. The pressure to which the driving fluid is raised in the external pumps is, however, higher than required in an all-pumped recirculation system, Fig. 4-16. This driving flow goes through the jet pump nozzles and acquires high velocity and momentum. By a process

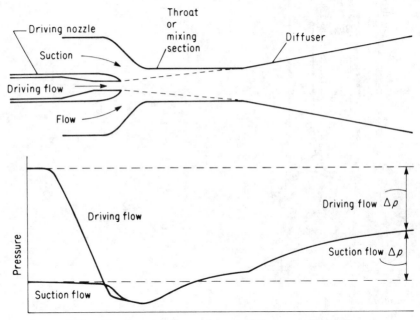

FIG. 4-16.   Jet pump flow and pressure diagram.   (Ref. 15.)

of momentum exchange, it entrains the remainder of the recirculation flow, called the *suction flow,* which is at a lower inlet pressure. The combined flow then enters the throat or mixing section of the jet pump where the momentum decreases and the pressure rises. Additional pressure recovery to the exit pressure occurs in the diffuser. The resultant suction flow Δp is sufficient to overcome losses through the reactor.

The jet pump design has the advantages of less moving parts, lower probabilities of major line ruptures, capability to reflood the vessel in case of such rupture (loss of coolant accident), improved natural recirculation (lower pressure losses) in the event of power loss to the recirculation pumps, and relative freedom from cavitation—a problem that arises in pumps when the liquid is near saturation conditions.

The operation map of Fig. 4-17 shows the effect of varying external

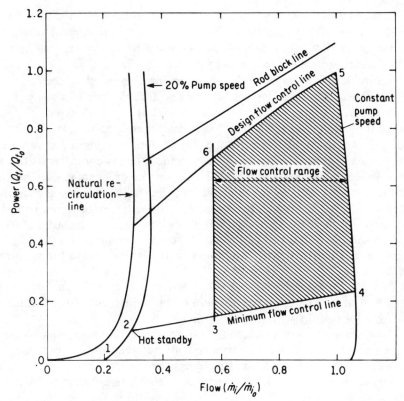

FIG. 4-17.   Operating map of BWR recirculation control. $P_0 = 1,670$ Mw(t), $\dot{m}_0 = 57.6 \times 10^6$ lb$_m$/hr.   (Ref. 15.)

pump speed on flow $(\dot{m}_i/\dot{m}_{i_0})$ and power generation (Q_t/Q_{t_0}) where $\dot{m}_{i_0}$ and Q_{t_0} are the rated flow and power respectively.   Zero pump speed results in the natural-recirculation line shown at extreme left.   The 20 percent pump speed line is next to it.   The vertical lines 3-6 and 4-5 show two constant pump speed lines that bracket the operation range of the recirculation control system.   The lines 3-4 and 6-5 indicate control rod positions for minimum flow control, and design flow control, respectively. Recall that boiling-water reactors can attain criticality at various control rod positions.   Both lines are approximately linear, as was predicted in Sec. 4-7, and show a control range of approximately 25 percent without control rod motion. The flow control range shown is determined by no fuel damage if power were lost at flows above that indicated by point 3.

The feedwater flow rate is controlled by varying feedwater pump speed to maintain reactor water level between specified limits. A feedwater controller accomplishes this by utilizing water level and steam and feedwater flow measurements.

In normal operation, when reactor power is changed, the initial pressure regulator (IPR) adjusts the turbine main steam control valves to change steam flow and maintain constant reactor pressure, causing the turbine output to follow reactor output.

Recirculation flow is adjusted by varying the speed of the recirculation pumps. Motor-generator sets with adjustable speed couplings vary the frequency supply to the pump motors to vary their speed. To change the reactor power, a demand signal from the operator, or a load-frequency error signal from the speed-governing mechanism is supplied to the master controller. A signal from the latter is compared with the actual speed of the generator. The resulting error signal causes adjustment of the set points of the controller of each adjustable speed coupling and therefore of the generator speed to reduce the error signal to zero. The speed of the recirculation pump motor is adjusted in accordance with the new output frequency of the generator. Recirculation flow adjusts reactor power and steam flow and the IPR repositions the turbine control valves.

The reactor operational control system provides sufficient reactivity compensation to make the reactor subcritical if the highest-worth control rod is stuck in a fully withdrawn position.

Overpressure transients resulting from sudden turbine valve closure are restricted by opening relief valves on the steam line and a turbine bypass line, similar to that used in bypass control. The bypass system also causes rapid partial load rejection without the necessity of reactor scram.

Startup of the reactor is performed manually. The recirculation pumps are started at a reduced speed while the reactor water is cold. Control rods are then withdrawn according to a predetermined schedule to achieve reactor criticality. Power is raised to a level that will give a desired rate of temperature increase (specified so that certain temperature differentials are not exceeded between certain parts of the reactor vessel) until operating pressure and temperature are reached. Neutron flux and power monitors are used during subcritical and power operation. While the operating pressure is not reached, the turbine is on turning gear and condenser vacuum is not required. When reactor steam becomes available the turbine shaft seal system is placed in service, the mechanical vacuum pump is started and partial vacuum is attained in the condenser.

Steam is allowed to flow to the condenser through the bypass valves. This is then gradually transferred to the turbine until rated speed and temperatures are achieved. The unit is then synchronized with the system. When the reactor reaches operating temperature, recirculation flow is increased to its maximum value before control rods are withdrawn further to increase reactor power. The rods are withdrawn one at a time in a symmetrical manner by quadrants.

4-9. THE PATHFINDER BOILING REACTOR

Another boiling-water-reactor system in which load control is accomplished by varying recirculation flow has been designed by the Allis-Chalmers Manufacturing Company for the Northern States Power Company [18]. It has now been shut down for economic reasons. It is interesting, however, because it has integral superheat, i.e., boiling and superheat within one reactor vessel. Research and development for that reactor have been carried on in association with the Central Utilities Atomic Power Associates (CUAPA), a group of Midwestern utilities. Originally called the Controlled-Recirculation Boiling Reactor (CRBR), and more commonly known as the *Pathfinder*, the small 62-Mw(e) power plant is built near Sioux Falls, S.D. Some design and operational data of the plant are given in Table 4-4.

A cross section of the CRBR is shown in Fig. 4-18. Feedwater comes in at 360°F and 650 psia at lower right, goes to a feedwater distribution ring from which it mixes with saturated recirculation water flowing downward. The mixture, now at 486°F and 620 psia, is pumped out by three 20,000-gpm recirculation pumps via three outlets (one shown at lower left). The pumps are housed in separate shielded compartments to permit repairs without plant shutdown.

The pumped water comes back via three inlets (one shown) at 486°F and 642 psia and flows upward through the boiler region converting to a mixture of 2.558 percent quality at 489°F. Most of the steam separates from the water at the free-liquid level shown. 48 steam separators, each 10 in. in diameter and 14 ft long, attached to the reactor-vessel wall, are used to separate the remaining entrained steam. Water containing entrained steam enters these separators tangentially via longitudinal inlet nozzles extending to about 5 ft below the water level. A vortex is imparted to the water, causing the lighter entrained steam to accumulate in the center and to discharge upward through mist eliminators (which remove entrained liquid from the steam to less than 0.1 percent); this steam joins the main steam in the reactor-vessel dome. The recirculation water, now virtually free from steam, discharges through nozzles at the

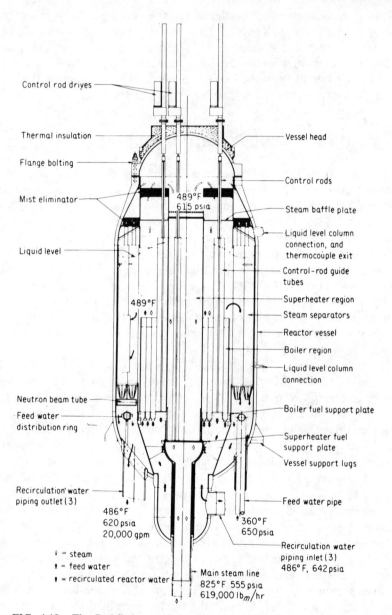

Control rod drives

Thermal insulation

Flange bolting

Mist eliminator

Liquid level

489°F
615 psia

489°F

Neutron beam tube

Feed water
distribution ring

Recirculation water
piping outlet (3)

486°F
620 psia
20,000 gpm

Vessel head

Control rods

Steam baffle plate

Liquid level column
connection, and
thermocouple exit

Control-rod guide
tubes

Superheater region

Steam separators

Reactor vessel

Boiler region

Liquid level column
connection

Boiler fuel support plate

Superheater fuel
support plate

Vessel support lugs

Feed water pipe

360°F
650 psia

Recirculation water
piping inlet (3)
486°F, 642 psia

$\dot{v}$ = steam
$\dagger$ = feed water
$\dagger$ = recirculated reactor water

Main steam line
825°F 555 psia
619,000 lb_m/hr

FIG. 4-18.   The Pathfinder reactor.   (Courtesy Allis-Chalmers Manufacturing Co., Atomic Energy Division.)

TABLE 4-4
Main Design Data of Pathfinder

Reactor heat output (boiler region)	164,000 kw(t)
Reactor heat output (superheater region)	39,700 kw(t)
Plant net power	62,000 kw(e) (66,000 gross)
Net thermal efficiency	30.5%
Coolant-moderator	Light water
Reactor outlet pressure	540 psi
Reactor outlet temperature	825°F
Reactor operation pressure	600 psi
Steam generation	600,000 lb_m/hr
Coolant inlet velocity (boiler region) .	14.2 fps
Coolant inlet velocity (superheater region)	99 fps (116 fps outlet)
Average heat flux (boiler)	131,000 Btu/hr ft²
Average heat flux (superheater) . . .	85,000 Btu/hr ft²
Estimated burnout heat flux (boiler) .	1,000,000 Btu/hr ft²
Maximum heat flux (superheater) . .	208,000 Btu/hr ft²
Core dimensions	Cylindrical, 72-in. diameter; central superheater 30 in. in diameter; both 72 in. high
Fuel-element configuration	Boiler, 92 subassemblies of square lattice bundles of fuel rods
Fuel type	Boiler, 1.8% enriched UO_2 pellets with proposed X-8001 aluminum alloy cladding; superheater 93% enriched UO_2 clad in Type 304L stainless steel
Maximum fuel temperature (boiler)	4800°F
Maximum surface temperature (superheater)	1300°F
Average exit void fraction (boiler) . .	42.4%
Average exit quality (boiler)	2.558%
Recirculation ratio (forced)	39:1
Feedwater temperature	360°F
Control rods (boiler)	16 rods, 10-in. cruciform-shaped, made of 1/4-in.-thick 2% stainless steel
Control rods (superheater)	Four rods, 9-in, cruciform-shaped
Pressure drop (boiler)	18.1 psi
Pressure drop (superheater)	55 psi

bottom of the separators and mixes with incoming feedwater, completing the recirculation loop.

The saturated steam from the boiler region flows toward the center of the reactor and downward through the superheater region. It emerges at bottom center through an 18-in. line at 555 psia and 825°F, corresponding to about 347 degrees of superheat. From there it goes directly to the turbine.

The fuel in the boiler region is made up of 1.8 percent enriched UO_2 pellets 0.338 in. in diameter, clad in Zircaloy. The fuel subassemblies contain square arrays of 9×9 tubes at the bottom and 8×8 or 7×7 at the top. This arrangement helps flatten the axial thermal-neutron flux.

The superheater is 30 in. in diameter and 6 ft high. It contains 429 fuel elements and four control rods. It is steam-cooled and water-moderated. Each of the 429 fuel elements is made of two concentric cylinders, of 0.015-in.-thick 93 percent enriched UO_2-stainless-steel cermets clad in Type 304L stainless steel. The outer fuel cylinder is 0.726 in. OD and 0.672 in. ID, and the inner cylinder is 0.512 in. OD and 0.458 in. ID (including cladding). The fuel elements surround a center burnable poison* pin, 0.298 in. in diameter. Concentricity is maintained by spiral wire spacers wound on a 12 in. pitch. The fuel shells are surrounded by a double-walled stainless-steel cylinder, enclosed at the top so that its annular space (0.057 in. wide) contains stagnant steam to serve as thermal insulation between the live steam and the superheater moderator water. This stagnant steam reduces thermal losses to the moderator water by about 1.7 percent of superheater power. Gamma and neutron heating amount to 7.3 percent. The superheater section is orificed at the top.

The superheater moderator water enters the superheater region through an orifice at the bottom and fills the space outside the double-walled insulation tubes. It is circulated at the rate of 4,000 gpm by the main circulation system of the reactor. This circulation keeps it from excessive boiling and limits its exit void fraction to 10 percent.

A centrally located superheater, rather than one surrounding a central boiler, was chosen in order to use fewer fuel elements for the same superheater power output, to obtain a more uniform radial flux (an outside superheater would cause a large radial flux gradient and a depression in the boiler region flux at the boundary, Fig. 4-19), and to design a good steam-flow path into and out of the superheater.

Special attention had to be paid to reactivity change due to accidental or intentional (after shutdown) flooding of the superheater steam passages with water. This flooding corresponds to a change in k_∞ of about 0.2 percent and was expected to decrease with fuel burnup.

---

* A *burnable poison* is a material with nuclei of high neutron absorption cross sections which, on capturing neutrons, convert to nuclei of low absorption cross sections. The material is, therefore, depleted, or burned. By proper choice of location and quantity, it is made to partially compensate for the loss of reactivity due to fuel depletion and fission product buildup. A suitable material is boron-10 (3837 barn for thermal neutrons), converting to lithium-7 (0.037 b).

Other reactivity effects in the superheater, such as those caused by changes in fuel, moderator, and coolant temperatures, are so small that the superheater itself does not adversely affect the physics of the nuclearly coupled system. The boiler is the controlling region of the two, and the superheater follows flux variations in it.

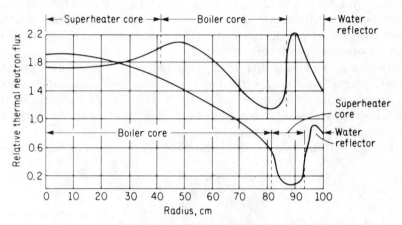

FIG. 4-19. Comparison between neutron flux distribution of central and peripheral superheater designs.

Pathfinder was designed to operate as a base-load station. Besides load following control by varying recirculation flow, it is interesting to note that the reactor is partly self-regulating as far as steam conditions at exit are concerned. If the reactor power is increased, the boiler generates more steam, and the superheater, which is nuclearly coupled to the boiler, maintains exit steam temperature. Some temperature transients occur, however, due to the time lag between the neutron and heat fluxes in the boiler, to the long time constant of the UO_2 fuel, and to the flow resistance in the steam dome and the superheater. However, because of forced recirculation, the plant was found not to be susceptible to high-frequency instabilities that may accompany rapid changes in heat flux.

A possible problem with this type of reactor is the loss of steam coolant to the superheater fuel elements that may occur in a sudden shutdown. An automatically controlled reduction in reactor pressure is planned for such cases. This will cause water in the boiler region to flash into steam, maintaining coolant flow into the superheater for a sufficient period of time.

Nuclear superheat results in good plant thermal efficiencies (Table 4-4), as well as drier steam in the last stages of the steam turbine, but increased plant complexity and capital costs.

4-10.   A GRAPHITE-WATER BOILING SUPERHEAT REACTOR

Another interesting design of an integral superheat boiling-water reactor is that of a 100-Mw(e) high-pressure forced-external-recirculation power plant built in the USSR [19].   In this reactor an array of vertical individual boiling and superheating fuel subassemblies are encased in a graphite cylindrical core 23.6 ft in diameter and 19.7 ft high (plus graphite radial and axial reflectors).   In the core 730 boiler subassemblies produce a wet mixture of steam and water, and 268 superheater subassemblies produce superheated steam

The water coolant is divided into two independent circuits–a primary circuit which passes through the boiler elements, and a secondary circuit which passes through the superheater elements.   The fuel elements in the subassemblies themselves are embedded in graphite and therefore are constructed with pressure tubes.

Figure 4-20 shows a cross section of either a boiling or superheating

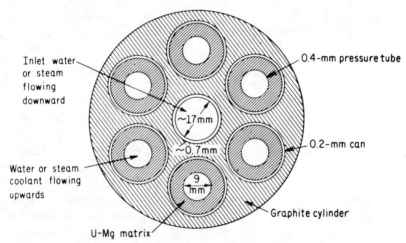

Inlet water or steam flowing downward

0.4-mm pressure tube

~17mm

~0.7mm

0.2-mm can

9 mm

Water or steam coolant flowing upwards

Graphite cylinder

U-Mg matrix

FIG. 4-20.   Approximate cross section of fuel subassembly for graphite-water superheating reactor.   (Ref. 19.)

fuel subassembly consisting of a vertical graphite cylinder of the same height as the core and having seven vertical holes in it.   The center hole has a simple stainless-steel pressure tube.   The six surrounding holes have hollow cylindrical fuel elements made of a 1.3 percent enriched uranium-magnesium mixture.   The fuel elements are clad on the inside in stainless-steel pressure tubes and are surrounded on the outside by a thin stainless-steel can.   There is no metallic bonding between the pressure tubes and the fuel, and good contact for heat transfer is apparently ensured by mechanical contact under pressure.

In the boiler subassemblies, the water comes in at the top and flows downward through the center tube. At the bottom, the water is then distributed to the six surrounding fuel tubes, flowing upward through them, where it emerges as a mixture of steam and water. Similarly, in the superheater, incoming saturated steam flows downward through the center tube and upward through the surrounding fuel elements, emerging in a superheated condition. The superheater elements are identical to the boiler elements except for the grade of stainless steel used for the pressure tubes.

The neutrons are thus moderated mainly by the graphite but also to a lesser extent by the water in the boiler subassemblies. The lattice chosen makes the reactivity not very sensitive to water conditions. A decrease in the void fraction in the boiler subassemblies causes a decrease in resonance capture and an increase in neutron absorption, compensating each other. The size of the core and the amount of moderator-reflector used are such that neutron leakage is negligible. The close spacing of the lattice also reduces resonance capture. A high exit quality of about 33 percent in the boiler subassemblies is therefore tolerated.

Also of interest is the fact that the lattice and fuel-element design minimizes the amount of stainless steel (which has a high cross section for neutrons) and places it, and the water, in the region of minimum thermal-neutron flux in the subassembly. Because of these effects, a fuel enrichment of only 1.3 percent was made possible.

Because the superheater elements are flooded at starting and because of the possibility of accidental flooding during operation, the superheater elements are situated in the core at positions where reactivity is least sensitive to changes in water content. These positions are in an annulus roughly intermediate between the center and periphery of the core.

The power-plant flow diagram (Fig. 4-21) shows primary water entering the boiler assemblies at 572°F and 2,278 psi. It emerges at 2 with 33 percent quality and goes to the steam separator 3. The saturated water from that separator goes into the preheater 4 where it is cooled from a saturation temperature of 644°F (at the separator pressure of 2,204 psi) to 582°F, transferring its heat to the feedwater of the working fluid in the secondary circuit.

Feedwater of the secondary circuit enters the preheater at 419°F where it is heated to the saturation temperature at 1,616 psi. It then goes to the steam generator 5 where it is converted to steam of 20 percent quality by the saturated steam from the primary steam separator 3. The primary-steam condensate then mixes with the recirculation water from the steam separator. Cooled primary water then reenters the boiler subassemblies via the recirculation pumps 6.

Secondary steam, produced from the steam generator at 1,616 psi, goes to the superheating fuel subassemblies 8 where it is superheated to 950°F. It then enters the turbine 7 at 1,323 psi and 932°F. It can

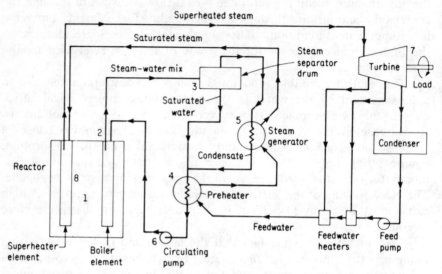

FIG. 4-21.   Flow diagram of superheating-reactor power plant.   (Ref. 19.)

be seen that the fluids in the boiler and superheater subassemblies do not mix and are only coupled by heat transfer in the preheater and steam generator.

TABLE 4-5
Some Performance Characteristics of the
Graphite-Water Superheat Reactor

	Boiler	Superheater
Maximum heat output per subassembly, kw(t)	405	363
Heat flux, Btu/ft² hr	194,000	177,000
Maximum temperature of pressure tube, °F	671	986
Maximum temperature of fuel, °F	752	1022
Maximum temperature of graphite, °F	1022	1337

Because of steam superheat, a thermal efficiency of 35 to 37 percent is expected.   Some performance characteristics of the fuel subassemblies are given in Table 4-5.

4-11. THE VARIABLE MODERATOR BOILING REACTOR (VMR)

The fourth method of load-following control of boiling-water reactors, that of a positive void coefficient of reactivity, is used in conjunction with a direct cycle in VMR. As its name implies, the variable-moderator reactor is one in which neutron moderation is not only dependent on the coolant but also on a separate moderator that can be controlled as to level and density. This is done by dividing the water in the core into two regions, one for coolant and the other for moderator. The fuel channels are sealed between upper and lower tie plates as shown schematcally in Fig. 4-22. The water passing through these channels is the coolant, undergoes boiling and generates steam in the usual manner but also contributes to moderation.

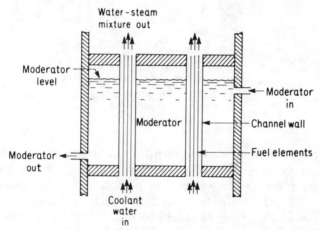

FIG. 4-22. Schematic of variable-moderator boiling reactor.

The portion of the core volume outside the channels is filled with water which is the main moderator and is not allowed to boil. Since 4 to 8 percent of the total reactor heat is produced in the moderator (owing to the acquisition of energy from the scattered neutrons, absorption of γ and other radiations, and heat transfer from the coolant), the moderator is kept below its saturation temperature by continually recirculating it through a heat exchanger outside the core. The level of the moderator is changed to take care of large load changes and fuel burnup and can be dumped for scram without endangering fuel-element cooling, since the coolant continues to flow in the fuel channels.

The evaluation of the effect of voids on multiplication in this core design must take into account the effects of both core regions on neutron energy. Remember (Sec. 3-9) that in the boiling region the resonance escape probability p and the nonleakage probability P decrease with void fraction while the thermal utilization factor f increases with it. In the moderator region, these are not affected.

The change of k_{eff} with α is dependent upon the volume ratio of coolant water to moderator water and of each to the fuel. Under proper design conditions, the net effect is one in which k_{eff} can be made to increase initially with α, reach a maximum at some optimum value of α, and then continually decrease with it. Such an effect is shown in Fig. 4-23. (The same is generally true for k_∞, but because of the change of P with α, k_∞ peaks at a point to the right of that of k_{eff}.)

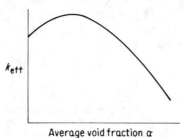

FIG. 4-23. Effective multiplication factor versus average void fraction in VMR.

It can be seen that the void coefficient of reactivity $dk_{eff}/d\alpha$, which is the slope of the line in Fig. 4-23, is positive for values of α below optimum and negative for higher values. A positive void coefficient means a negative pressure coefficient. A direct cycle, such as that shown in Fig. 4-1, using this type of reactor, would therefore be load following provided that the operating void fraction is kept below optimum. If the turbine calls for more steam, the throttle-valve opening is increased and the pressure in the reactor drops. This increases voids, which increases reactivity, i.e., reactor thermal output and consequently steam generation and a new equilibrium is attained. The reverse is also true.

This type of reactor is also capable of larger power densities than the simple direct cycle due to greater moderation by the nonboiling moderator.

A power plant has been designed (but never built) to operate on the above principles by the Atomic Energy Division of the American Standard Corporation [20]. The reactor, shown in Fig. 4-24, has fuel subassemblies in mostly hexagonal channels, fabricated from stainless-steel plate 0.050 in. thick, which are seal-welded to lower and upper grid plates and which split the core volume into two approximately equal regions for coolant and moderator. Pressure-equalizing pipes are used to equalize pressure in the two regions and avoid channel design for

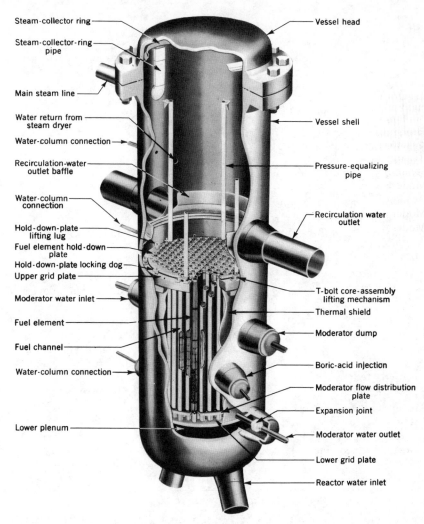

Steam-collector ring

Steam-collector-ring pipe

Main steam line

Water return from steam dryer

Water-column connection

Recirculation-water outlet baffle

Water-column connection

Hold-down-plate lifting lug

Fuel element hold-down plate

Hold-down-plate locking dog

Upper grid plate

Moderator water inlet

Fuel element

Fuel channel

Water-column connection

Lower plenum

Vessel head

Vessel shell

Pressure-equalizing pipe

Recirculation water outlet

T-bolt core-assembly lifting mechanism

Thermal shield

Moderator dump

Boric-acid injection

Moderator flow distribution plate

Expansion joint

Moderator water outlet

Lower grid plate

Reactor water inlet

FIG. 4-24. Cutaway of the VMR. (Courtesy Advanced Technology Laboratories, a Division of American Standard Corporation.)

pressure.   The reactor contains no control rods.   The main design data of the plant are given in Table 4-6 and Fig. 4-25 shows a flow diagram of the plant.

Steam is separated from the water near the top of the reactor and is collected by main steam pipe 9 at the reactor dome.   This steam then flows to steam separator 10 (that drains back to the reactor) and into the turbine.   Reactor recirculation water goes to two pumps (only one is shown, at 4) after mixing with the feedwater coming from the main

TABLE 4-6
Main Design Data of VMR

Reactor heat output	75,000 kw(t)
Plant power (net)	20,000 kw(e), 22,329 kw(e) gross
Coolant	Ordinary water
Moderator	Ordinary water
Reactor outlet pressure	600 psia
Reactor outlet temperature	586°F
Steam flow	286,560 lb_m/hr
Moderator flow	600 gpm
Reactor inlet temperature	374°F
Moderator inlet temperature	423°F
Moderator outlet temperature	466°F
Average core heat flux	75,700 Btu/hr ft² (331,000 max.)
Core dimensions	Approximately cylindrical, 85-in. diameter at upper grid plate, 75-in. diameter at lower grid plate; active height, 6 ft; height of moderator, 40 in. above lower grid plate
Fuel-element configuration	163 hexagonal subassemblies each containing 61 vertical fuel pins 0.39 in. in diameter, each containing 0.32-in.-diamater fuel pellets in Zircaloy 2 cladding
Fuel type	UO_2 pellets
Fuel enrichment	1.78% initial (1.5% final)
Maximum fuel center-line temperature	2018°F
Average void fraction at core exit	48%
Core inlet subcooling	About 110°F
Slip ratio	2:1 (estimated)
Recirculation ratio	41:1

condenser. The mixture is then pumped to the bottom of the reactor via two recirculating-water flow control valves 5 and then evenly distributed to the fuel channels by orifices in the lower grid plate.

The moderator water is circulated, at 600 gpm, in the space between the fuel channels, by one of the two moderator pumps 3, from top to bottom, counterflow to the coolant water. Higher moderator density is thus provided near the core top, where it is needed most. Moderator flow is distributed equally around the fuel channels by an orificed distribution plate. The moderator picks up an estimated 4.3 percent of the total heat generated in the core. The maximum moderator temperature is kept at 20°F below saturation by circulating it through a moderator cooler 17 where it loses about 43°F to the feedwater. Also, temperature

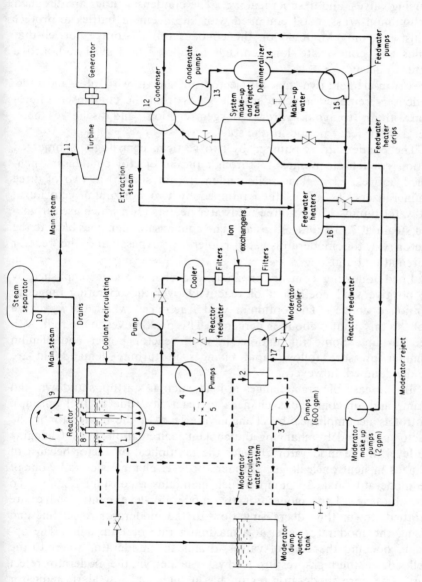

FIG. 4-25. Simplified VMR power-plant flow diagram. (Ref. 20.)

controllers actuate a feedwater bypass valve around the moderator cooler.

The main condenser is designed to handle the entire reactor steam load, in case of turbine trip, via a bypass line (not shown), pressure-reducing valves, and desuperheaters. The condenser tubes are designed of aluminum-brass metal equipped with impingement baffles to protect against erosion. Mounted on the condenser is a storage hotwell that retains the condensate long enough to allow short-lived radioactive matter to decay.

A twin-element two-stage air ejector (not shown) with inter- and after-condensers removes noncondensable gases from the condenser, dilutes them with air (to avoid hydrogen-oxygen reaction), and discharges them, after a short decay period, to the stack.

The condensate is pumped by two vertical, centrifugal, submerged suction condensate pumps 13 through the air-ejector condensers mentioned above and a gland-seal condenser (not shown) to two of three demineralizers, 14. It is then pumped by two horizontal, centrifugal feedwater pumps 15 into three feedwater heaters (16), which use turbine-bled steam at 7, 35, and 118 psia. The condensate then goes back to the reactor via the moderator-water cooler (or bypass) and the reactor recirculating pump.

Load following control for small load changes is accomplished by the positive void coefficient of reactivity over the operating range, as mentioned above. The optimum void fraction at which the k_{eff} curve peaks in this plant is about 25 percent. This peak moves to the right with fuel burnup. Some fluctuations in both reactor power and plenum chamber pressure follow rapid changes in turbine demand but are rapidly damped, however.

In the case of large changes in load, such as startup, shutdown, and scram, and of continuous changes in reactivity (due to fuel burnup), control is accomplished by changing the level of the moderator in the reactor core and by changing the coolant recirculation rate. Changing the level of the moderator changes the multiplication factor because of changes in neutron leakage. Moderator-level change is done by one of the moderator makeup pumps which maintains a constant 2 gpm. To raise the level of the moderator, two moderator-level control valves are throttled down, thus decreasing flow to the moderator reject line and raising the moderator level at a maximum rate of 1 in./min. The opposite, opening the control valves, drains the moderator water to the shell side of the high-pressure feedwater heater via the moderator reject line. The core moderator region has a high and low alarm indicator (connected to scram). A provision for dumping the moderator out of the reactor is also provided.

4-12. BOILING-WATER-REACTOR TURBINES

The design of turbines for use in nuclear boiling-water, nonsuperheat, reactor plants should, in addition to normal design conditions, meet two important requirements. These are (1) to withstand wet, oxygenated steam, especially in the last stages of the turbine, and (2) to cope with radioactivity associated with the reactor-produced steam. (Note that turbines for pressurized-water reactors are also required to meet the first of the above requirements since these plants also generate saturated or near-saturated steam as a working fluid.)

When saturated steam expands through a turbine from the inlet pressure to the condenser pressure, a large quantity of moisture is formed. For example, if expansion is between 1,000 psia and 1 or 2 in. Hg abs, the percent moisture in the last stages reaches 20 to 25 percent by mass (80 to 75 percent quality). This condition is worse, i.e., the percent moisture is higher, the higher the initial steam pressure. This can be ascertained from the temperature-entropy and enthalpy-entropy (Mollier) diagrams of Fig. 4-26, where points a and b represent the conditions of the steam at the last stage of the turbine for high and low inlet pressures, respectively.

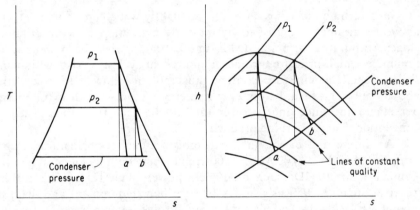

FIG. 4-26. Effect of pressure on exhaust-steam quality $(p_1 > p_2)$.

Steam produced in a reactor also has higher concentrations of oxygen (of the order of 100 ppm, because of decomposition by irradiation) than is customary in nonnuclear steam power plants. The effects of high moisture content and high oxygen content on turbine materials are related to steam velocity, bucket velocity, and density. However, other things being equal, the effects can be summarized as follows:

1. Erosion
   (a) Due to the impingement of water droplets against the surfaces

of the last stage buckets which are moving at high velocity. (Turbine buckets in the last stage have large diameters and consequently move at high tip velocities. When the tip velocity exceeds 1,000 ft/sec and the moisture exceeds 4 percent, the buckets are normally protected with erosion shields.)

(b) Due to surface washing of a metallic surface by high-velocity water; can be minimized by the use of special materials such as Cr steels.

(c) Due to wire drawing, defined as the undercutting of metal caused by wet steam leaking through narrow passages at sufficiently high velocity.

2. Corrosion due to concentration of oxygen in pockets in the turbine, causing attack and pitting.

Because of the effects of wet steam on turbines, it has been the practice in nonnuclear applications to limit the moisture content of the steam to a maximum value of 12 percent by mass in the last steam-turbine stage. This was accomplished by the use of superheated steam and, in the case of very high steam pressures, by the additional use of reheat (Sec. 2-8).

In a turbine receiving saturated or slightly wet steam at its inlet, the above requirement is satisfied by moisture extraction between different stages within the turbine. (Moisture extraction has been applied to nonnuclear high-pressure turbines in nonreheat cycles.) Also, within any single stage, the moisture content must not be allowed to exceed the above limit. This can be done by reducing the pressure drop (or energy per stage) to values much lower (about one-half) than those used in conventional steam-turbine practice.

An interesting design in which moisture-extraction features are incorporated is the 180,000-kw, 1,800-rpm turbine built by General Electric Company for the Dresden dual-cycle power plant [21]. This turbine receives steam at 950 psig and 540°F (primary admission) and 460 psig and 462°F (secondary admission). The turbine is composed of a 15-stage high-pressure section, operating between 950 and 180 psig; a seven-stage intermediate-pressure section, operating between 180 psig and 11.5 psia; and a three-stage, double-flow low-pressure section (38-in. buckets in the last stage) exhausting to the condenser at 2.5 in. Hg abs. This design makes the stage diameters of this 1,800-rpm turbine relatively small (about equivalent to those of a turbine run at twice its rpm). This results in low fluid and tip velocities and thus reduces the possibility of erosion caused by high steam and water velocities.

Moisture extraction in the Dresden turbine is accomplished by the use of specially designed buckets, shown in Fig. 4-27. The stationary buckets (nozzles) are of conventional design, but the moving buckets have grooves milled on the back and leading edges. Water which strikes the back surface at high bucket velocity and normally bounces off as droplets adheres instead to the grooves and forms a water film, owing to surface tension. This film of water then flows out of the bucket tips by centrifugal force into a peripheral chamber in the turbine casing, from which it is drained away.

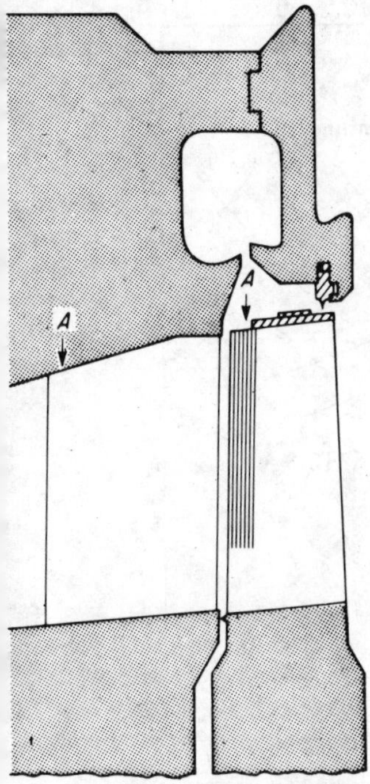

Turbine stage number	Type of moisture extraction		Moisture extracted	
	MEB*	FWH†	lb_m/hr	Total moisture in stage (before removal), %
11	X	X	12,600	8.2
15	X		25,500	13.1
16	X	X	18,300	9.2
19	X	X	46,400	18.7
21	X	X	52,600	21.4
22	X		42,400	19.9
XO	Crossover separator		79,300	47.0
23	X	X	20,200	16.6
24	X		23,500	16.8
		Total	320,800	

* Moisture extracting bucket

† Extraction for feedwater heating

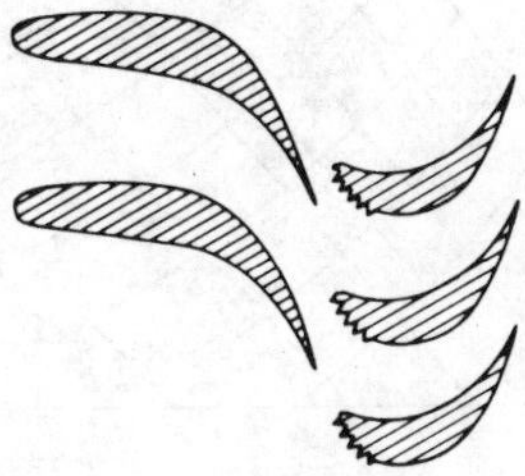

View A

FIG. 4-27. Typical moisture-extracting stage in Dresden turbine. (Ref. 21.)

Eight such moisture-extraction stations are provided at the stages shown in the table accompanying Fig. 4-27. Five of these double as extraction stations for the feedwater heaters (previously shown in Fig. 4-8). In addition, moisture extraction is also accomplished by the separator in the large rectangular crossover between the intermediate- and low-pressure turbines (after the 22d stage). This separator contains corrugated stainless-steel plates. The water droplets strike the vertical corrugations and flow downward, in the form of a water film, to a drain.

The state of the wet steam in the turbine is shown by the Mollier diagram of Fig. 4-28. The dotted line represents the state of the fluid with no moisture extraction, showing a possible moisture content of about 24 percent in the last stage of the turbine. The solid line represents the expected state of the fluid with moisture extraction. The numbers shown identify the stages at which extraction is undertaken. It can be seen that the moisture content is made to remain largely within the 12 percent limit used in conventional steam-turbine practice.

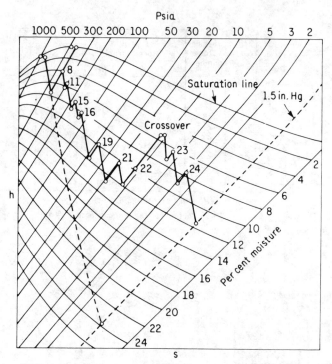

FIG. 4-28.   Expansion line indicating steam state in various stages of Dresden turbine.   (Ref. 21.)

The EBWR turbine uses moisture-deflecting lips on the diaphragm faces of the last two-stage rows. The lips direct the water down to a number of interstage drains as well as help to protect the turbine casing against erosion.

The problem of corrosion of turbine materials by the high oxygen concentrations encountered in boiling-reactor plants has been experimentally investigated [22]. It was found that, except for cupronickels and BTH-leaded bronze materials, oxygen corrosion was not a very serious problem. When these two materials are avoided, the choice of turbine materials should be based primarily on the need for erosion resistance. Erosion-resisting materials include the stainless steels, Monel, the Stellites, chromium steels, nickel steels, and Ni-Resist. (The leading edges of the last-stage buckets in the EBWR turbine, where the moisture content is allowed to rise to 18 percent, are all protected with a layer of Stellite.)

The second major problem in boiling-water-reactor turbines is that of withstanding the effects of radioactivity in the steam. It has been shown (Sec. 4-2) that when the solids content of steam is kept below 1 ppm no serious effect is encountered, because of the short half-lives of the radioactive species. Special provisions, however, are usually incorporated in turbine design to guard against the effects of such mishaps as fuel-cladding rupture. These are (1) the elimination, as far as possible, of pockets and crevices in which radioactive materials may tend to accumulate, (2) the provision of drainage for those pockets that are necessitated by design requirements, and (3) provision for filling (and draining) the turbine with decontaminating fluids when necessary and before dismantling for maintenance.

PROBLEMS

4-1. A boiling-water reactor core is 8 ft in diameter and 12 ft high. It is fueled with 93 percent enriched uranium-zirconium alloy. The thermal utilization factor $= 0.71$. The migration area $= 40$ cm^2 but varies as the water density to the power -1.70. Assume that neutron absorptions occur in fuel and water only. Determine whether the reactor is load following when used in a direct cycle.

4-2. A direct-cycle boiling-water-reactor plant generates 10^6 lb$_m$/hr of steam at part load. The main steam pipe is 24 in. in diameter and 100 ft long. The turbine inlet valve causes a pressure drop which is usually given in terms of an equivalent length of straight pipe of the same diameter as follows:

Valve	Equivalent Length, Diameters
Fully open	7
Three-quarter open	40
Half open	200
One-quarter open	800

At the above flow, the valve was half open. A demand for full power caused the valve to fully open. For the case of no bypass or rod control, estimate the percent change in k_{eff} immediately following valve opening.

Reactor data: Pressure $= 1,200$ psia, $p = 0.819$, $f = 0.828$, $\bar{x}_e = 0.10$, $t_d = 300°$F, flux flat in radial and sinusoidal in axial direction, enrichment low, slip ratio $= 2$, neutron absorption ignored in all but fuel and water. Take $\bar{\alpha} = 0.12$.

4-3. A boiling-water reactor generates 10^7 lb$_m$/hr of steam at 1000 psia.

The exit quality is 0.10. $f = 0.84$. $p = 0.811$. Fuel is low enriched. The slip ratio is 2. Load following is by bypass control. If the mass flow rate through the core is increased by 11.1 percent, find the percent change in k_{eff} immediately following flow increase. Ignore neutron absorptions in all but fuel and water.

4-4. A dual-cycle boiling-water reactor power plant operating at full load generates 1.8×10^6 lb_m/hr of steam at 1,000 psia and 1.5×10^6 lb_m/hr of steam at 500 psia. The reactor exit quality is 6 percent. Both high and low pressure feedwater leave the last stage feedwater heaters at 400°F Calculate the reactor inlet degree of subcooling, °F, (a) at the above full-load conditions, and (b) at 80 percent of total full-load steam flow if only low-pressure steam flow is affected.

4-5. A boiling-water reactor of the recirculation-control type generates 2,488 Mw(t) at full load. It operates at 1,000 psia with feedwater and core inlet temperatures of 370 and 520°F respectively. Calculate the steam mass flow rate, total core flow rate and exit quality (a) at full load, (b) immediately after recirculation control to change to 80 percent of full load, and (c) after load change is affected.

4-6. In VMR, k_∞ is less sensitive to moderador level than k_{eff}. Discuss and show that for a fixed $k_\infty - \alpha$ curve, a series of $k_{eff} - \alpha$ curves may be obtained for various moderator levels, with the peaks becoming higher and moving to the right as the moderator level is increased.

4-7. (a) Calculate the work done by the steam in Btu/hr and Mw, and the isentropic efficiency (actual work divided by isentropic work between inlet and outlet pressures) of the Dresden I turbine. (b) What would have these values been if there were no moisture extraction?

chapter **5**

Pressurized-Water
Reactors

5-1. INTRODUCTION

In pressurized-water reactors water has the multiple function of coolant, moderator, and reflector. Pressurized-water reactors are therefore thermal. Water is an excellent heat-transfer agent; it is safe, and it has well-known thermodynamic and physical properties. Unlike the boiling-water reactor, the coolant in a pressurized-water reactor remains in the liquid phase over the entire core path, although some degree of subcooled boiling is usually allowed under high-power conditions. The coolant therefore exerts strong neutron moderation all over the core volume. Because of the strong moderating power of light water, a large number of control rods is required. The relatively high neutron absorption of light water, on the other hand, necessitates the use of slightly enriched fuels. The degree of enrichment varies from about 1.5 percent to highly enriched, depending upon the nuclear design and the presence of other nuclei in the fuel elements, through alloying or the use of ceramic fuels. Water, of course, has a strong corrosive effect on materials, especially at high temperatures. The problem of corrosion is discussed in Sec. 5-2.

Because of the absence of bulk boiling within the core, natural circulation in pressurized-water reactors can be made possible only if the core is sufficiently high to take advantage of reduced water density. This rules out natural circulation for many applications (such as submarine).

The operating reactor pressure is rather high (around 2,200 psia), necessitating a thick, heavy, and costly pressure vessel. The high pressures encountered make the design of the pressure-vessel head (top cover), which must allow for fuel loading and unloading, a bit of a problem. Either an offset handling mechanism or complete head removal for loading may be necessary.

Unlike boiling-water reactors, pressurized-water reactors need steam generators, external to the reactor, to produce steam, resulting in the

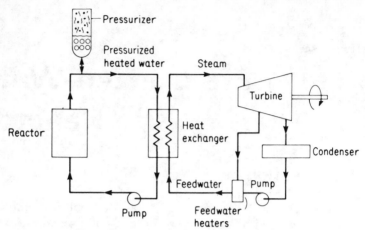

FIG. 5-1. Simple flow diagram of a pressurized-water reactor power
plant.

separation of primary coolant and working fluid, Fig. 5-1. Because of
this, no radioactive steam enters the turbine.

Pressurized-heavy-water reactors can use uranium fuel of natural
enrichment. Heavy water has lower moderating power [1] so that these
reactors tend to be large in diameter, further increasing the cost of the
pressure vessel. This is avoided by surrounding the fuel elements by
sealed tubes that are capable of easily withstanding the system pressure.
This is called a *pressure-tube* design. Heavy water has physical and thermodynamic properties approximately the same as those for ordinary
water, but is not as readily available and is costly. Heavy water reactors
have received greatest attention in Canada, Secs. 5-6 and 6-5.

In·general, the PWR concept results in relatively compact cores that
use available, safe and easy-to-handle fluids. Its extensive use in marine,
submarine, and land-based power plants accounts for the fact that the
largest single group of reactors built to date is of the PWR type.

This chapter will discuss some material and components of pressurized-water reactors, the method of control known as *chemical shim*
commonly used with many pressurized-water reactors, and the concept
of the pressurized-heavy-water reactor. Representative PWR power
plants will be presented in the next chapter.

5-2. MATERIALS FOR WATER REACTORS

In addition to nuclear and heat-transfer characteristics, the selection
of reactor coolants depends to a large extent on the corrosion characteristics of reactor and other component materials in these coolants. Cor-

rosion will result not only in the destruction of the material in question but also in the increase of mineral content of the coolant, with the accompanying problems of increased activity, erosion, deposition on heat-transfer surfaces, and increased load on the plant demineralizing system.

In the case of water (also aqueous homogeneous fuels; see Sec. 11-4), the problem of corrosion is aggravated by the production of oxygen because of the dissociation of water under reactor irradiation.

Corrosion rates usually increase logarithmically with temperature. The exact corrosion-rate–temperature relations depend upon the specific material used and other effects, themselves dependent on temperature, such as the solubility of oxygen and other impurities in water. Many metals are strongly susceptible to dissolved oxygen. Type 300 stainless steel, however, shows good resistance to high oxygen concentrations in high-temperature water. Dissolved gases should, however, be removed from reactor water. One method is mechanical degasification which consists of partial vaporization; the noncondensable gases have higher vapor pressures than water vapor and are readily vented with some water vapor. Also, it has been found that the addition of hydrogen usually depresses the oxygen concentration through recombination.

Salts that are soluble in water have also been found to accelerate corrosion, even when in very low concentrations of a few parts per million (ppm). Particularly troublesome in this respect [23] are the chloride ions, the heavy metals, and the salts of Cu, Cd, Co, Au, Pb, and Ag. The former attack the stainless steels, Be, Al, and Al alloys. The latter deposit in metallic form and thus provide bimetallic electrochemical reactions. Soluble salts can be reduced materially by bypassing a portion of the coolant to a demineralizer.

The pH* condition of the water is an important factor in controlling corrosion in water. Different metals and metal combinations, however, react differently to acidic or alkaline conditions, and the pH most suitable for use must be determined for each set of reactor materials. For example, slightly acidic conditions are suitable for aluminum and aluminum alloys, while slightly alkaline conditions are preferred for ferrous alloys. High velocities, such as are used in reactors, also tend to increase corrosion and erosion rates.

Table 5-1 shows the general corrosion resistance of various metals in pure water at different temperatures. The table shows no effects due to water impurities, radiation, or velocity and therefore should be taken only

---

* The pH is a number between 0 and 14 which indicates the degree of acidity or alkalinity of water at 60°F. pH = 7.0 indicates pure, neutral water. pH less than 7.0 indicates acidic and greater than 7.0 indicates alkaline water. pH is the logarithm (base 10) of the reciprocal of the hydrogen-ion concentration. Neutral water at 60°F has a concentration of 10^{-7} g_m/liter.

TABLE 5-1

General Corrosion Resistance of Metals in Pure Water[†]

Material	Corrosion resistance [‡] at temperature, °F					
	200	300	400	500	600	650
Aluminum	G	G	D	P	P	P
Beryllium	G	D	P	P	P	P
Chromium plate	G	G	G	P	P	P
Cobalt alloys	G	G	G	G	G	G[§]
Copper	G	G	D	P	P	P
70-30 Cu-Ni	G	G	G	G	P	P
Bronze	G	G	D	P	P	P
Gold	G	G	G	G	G	G[§]
Magnesium	D	P	P	P	P	P
Nickel	G	G	G	D	P	P
Nickel alloys:						
Monel	G	G	G	G	D	P
Inconel	G	G	G	G	D	P
Hastelloy	G	D	P	P	P	P
Platinum	G	G	G	G	G	G[§]
Steels:						
Carbon	G	G	G	G	G	G
Stainless:						
Austenitic	G	G	G	G	G	G[§]
Ferritic	G	G	G	G	G	G
Heat-resisting	G	G	G	G	G	G[§]
Martensitic	G	G	G	G	G	G
Precipitation hardening	G	G	G	G	G	G
Silver	G	D	D	P	P	P
Titanium	G	G	G	G	G	G[§]
Thorium	D	P	P	P	P	P
Uranium	P	P	P	P	P	P
Zirconium (crystal bar)	G	G	G	G	G	G
Zircaloy 2 ($1\frac{1}{2}$% Sn + Fe, Ni, Cr)	G	G	G	G	G	G[§]
Zircaloy 3 ($1\frac{1}{4}$% Sn, $\frac{1}{4}$% Fe)	G	G	G	G	G	G[§]

[†] From Ref. 23.

[‡] G = good: adherent film, no loose corrosion products, metal loss less than 50 mg/cm²mo. D = doubtful: adherent film from loose corrosion products, metal loss greater than 50 mg/cm²mo. P = poor: loose film, metal loss greater than 100 mg/cm²mo.; intergranular or other preferential attack.

[§] Recent tests show that these materials have satisfactory corrosion resistance in 750°F superheated steam.

as a general rather than definite guide to corrosion resistance. Tests simulating reactor conditions should be used to determine the behavior of a particular material. In general, it can be said that aluminum and its alloys are suitable in low-temperature water-reactor systems (such as swimming-pool, tank-type, and other research and training reactors). High-temperature operation (power reactors) requires the use of stainless steel and zirconium and its alloys. Other materials shown to be suitable in Table 5-1 are either too costly or have poor nuclear and thermal characteristics.

Of importance is the behavior of bare fuels in water. Uranium and throium metals have low corrosion resistance in water at all temperatures and therefore must be clad when used in water. Alloys of uranium and thorium have higher resistance to corrosion but not sufficiently high to rule out cladding. The higher resistance to corrosion in this case is, however, beneficial in case of cladding failure. Examples of these alloys are the zirconium-uranium alloys and uranium dioxide dispersed in stainless steel. They require the use of relatively high enrichment because of fissionable-material dilution. Other uranium alloys such as the so-called distorted alpha-phase (used in EBWR) have high uranium content, low neutron absorption, and good corrosion resistance.

Uranium dioxide has excellent corrosion resistance in water, good resistance to radiation damage, and low neutron absorption. Its main disadvantage, however, is in its poor structural strength. Cladding is therefore required in order to provide structural support in a heterogeneous reactor. Uranium dioxide is used in present day boiling- and pressurized-water reactors, clad in Zircaloy 2 or 4. In special cases where structural support is not needed, such as in the fluidized-bed (Sec. 6-7) or the pebble-bed (Sec. 8-10) reactor, the fuel may be used unclad.

Highly enriched fuels cause fewer problems than low-enriched ones, since much more alloying material can be added to the basic fuel, and therefore greater freedom of choice and use of these additives can be exercised.

Cladding materials suitable for use in water are aluminum (up to about 300°F), aluminum-base alloys containing nickel (up to about 570°F), low-hafnium zirconium and zirconium alloys (up to about 660°F), and the stainless steels at all temperature ranges and exceeding 660°F. The latter have the disadvantage of higher neutron cross sections.

5-3. STEAM GENERATORS

While boiling-water reactors generate their own steam internally, pressurized-water reactor power plants need steam generators, external to the reactor, to generate steam. The generators use pressurized water, heated in the reactor (the primary coolant) as the hot fluid which imparts its heat to lower-pressure feedwater (the working fluid) thus converting it to steam for use in the turbogenerator.

Nuclear-plant steam generators must be fabricated to high quality standards, especially in the welding and cladding operations. Heat exchange must be accomplished without transferring radioactive contamination from the reactor.

A typical pressurized-water steam generator has a capacity of over 250,000 kw(t), is about 67 ft high, 14 ft in diameter, and weighs about 330 tons [24]. Figure 5-2 shows a cross-section of such a steam generator manufactured by Westinghouse Electric Corporation. It is of the

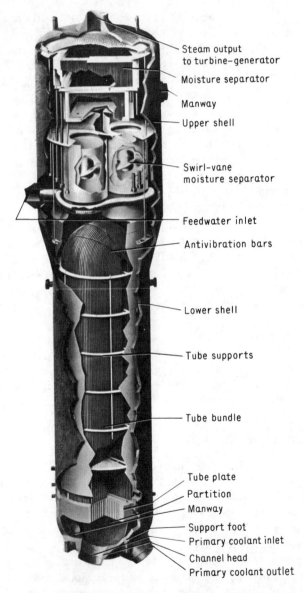

Steam output to turbine–generator

Moisture separator

Manway

Upper shell

Swirl-vane moisture separator

Feedwater inlet

Antivibration bars

Lower shell

Tube supports

Tube bundle

Tube plate

Partition

Manway

Support foot

Primary coolant inlet

Channel head

Primary coolant outlet

FIG. 5-2. PWR steam generator. (Courtesy Westinghouse Electric Corporation.)

vertical U-tube recirculation type and consists basically of an evaporator and an upper shell.

The evaporator contains the U-tube bundle. The primary coolant enters the steam generator through the primary coolant inlet at the bottom, flows through these tubes and exits from the primary coolant outlet, also at the bottom. The working fluid (feedwater) enters the generator through the feedwater inlet nozzle above the U-tube bundle and mixes with water which is recirculated from the moisture separators located near the top of the generator. The mixture then flows downward to the bottom of the shell via an annular downcomer between the lower shell and the tube bundle wrapper. It then flows upwards by natural convection through the tube bundle where it partially boils into a steam-water mixture.

The steam-water mixture is separated in the upper shell, first by swirl vane separators, and finally by vane-type separators. Dry saturated steam discharges through the steam outlet nozzle at the top. The saturated water leaves the separators and mixes with feedwater before entering the downcomer. The steam generator is therefore not unlike a natural-internal-circulation boiling-water reactor (Sec. 3-1) except that heat is supplied from reactor heated water instead of directly from the fuel.

One of the most critical areas in PWR steam generators is the weld between the tubes and the tube plate. A layer of cladding, carefully deposited on the tube plate surface (either by explosive bonding of a wrought sheet to the tube plate surface or by overlay weld deposit cladding utilizing a metal-inert-gas method) is welded to the tubes.

5-4. REACTOR-SYSTEM PRESSURIZERS

It has been shown that in reactor systems where the coolants are made to remain in the liquid phase, such as pressurized-water reactors, organic-cooled and -moderated reactors, liquid-metal-cooled reactors, and aqueous homogeneous-fueled reactors, the system pressure must be maintained at a value greater than the saturation pressure corresponding to the maximum system-operating temperature, to avoid bulk boiling of the coolant. In pressurized-water reactors, this pressure is of the order of 2,200 psia. Because liquids are practically incompressible, small changes of volume, due to changes in coolant temperatures, because of normal load changes or accidental reactivity insertions, or due to unforeseen expansions or contractions in the loop components, cause severe or oscillatory pressure changes. These may be quite unsafe when the changes are positive. They cause flashing into steam and consequent disruption of the reactor nuclear characteristics and possible burnout of reactor elements and coolant-loop caviation when they are negative. For this reason, it is necessary to provide a surge chamber which will accommodate volume changes in the primary coolant and maintain system pressure at or within prescribed limits. Such a chamber is called a *pressurizer*.

The pressurizer maintains the required coolant pressure during steady-state operation, limits pressure changes caused by thermal expansion and contraction during normal load transients, and prevents primary coolant pressure from exceeding safe limits. There are two main types of pressurizers in common use. These are (1) vapor pressurizers and (2) gas pressurizers.

1. *Vapor pressurizers* are essentially small boilers (Fig. 5-3) in which liquid, the same as the primary coolant, is maintained by controlled heating at a constant temperature and consequently a constant vapor pressure above its free surface. This pressure is the same as that of the primary coolant at the junction between the pressurizer and the primary loop. This temperature in the pressurizer is slightly higher than the temperature of the primary coolant, since the latter is normally subcooled.

Heating is usually done with electric immersion heaters located in the

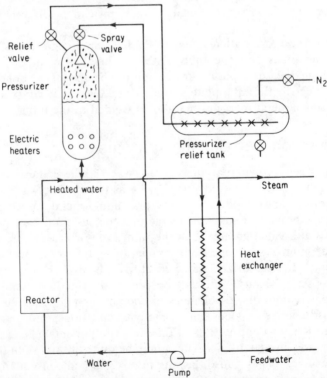

FIG. 5-3. Simple flow diagram of a pressurized-water reactor power plant with a vapor-type pressurizer system.

lower section of the pressurizer vessel. These heaters are also used to heat the pressurizer and its contents at the desired rate during plant startup.

The bottom of the pressurizer is usually connected to the hot leg of the primary coolant system, Fig. 5-3. A spray nozzle located at the top of the pressurizer is connected, via control valves, to the cold leg of the primary coolant system, after the pump. Under normal full-power operation, the pressurizer is about half full of water. The top half is full of vapor.

During a positive surge, the volume of the primary coolant increases and the vapor in the top half is compressed. Entry of the cooler primary coolant into the pressurizer condenses some of the vapor, limiting the pressure rise. In addition, the spray valves are power actuated, and a cool spray (under pump pressure) enters the top, helping condense vapor at a rapid rate and limiting pressure rise. (The spray valves may also be manually operated.) A small continuous spray is usually provided to prevent excessive cooling of the spray piping and maintain equal boron concentrations in the primary coolant and pressurizer water. Boron is used for chemical shim (Sec. 5-5).

A negative surge decreases the primary coolant volume and expands the vapor in the pressurizer, causing a momentary reduction in pressurizer pressure. The liquid in the pressurizer then partially flashes into vapor, and, assisted by further steam generation because of the automatic actuation of the electric heaters, the pressure is maintained above a minimum allowable limit.

A power-operated relief valve is attached to the top of the pressurizer to protect against pressure surges that are beyond the capacity of the pressurizer. The relief valve, in such a case, discharges steam into a pressurizer relief tank which is partly filled with water at near room temperature, and where the vapor condenses. The condensate then goes to the waste-disposal system.

In a typical design of a 500 Mw(e) PWR, both the pressurizer and pressurizer relief tank had volumes of 800 ft³ each, compared to 3,700 ft³ inside volume of the reactor pressure vessel. Figure 5-4 shows a photograph of a pressurizer for a PWR plant built by Westinghouse Electric Corporation.

The following is a simplified treatment of the pressure change in an insurge (when a liquid expansion in the primary circuit causes a rush of primary coolant into the pressurizer) or, conversely, an outsurge. The situation is treated as a nonsteady flow case. The general energy equation in that case is written as

$$PE_1 + KE_1 + \Delta m_i h_i + m_{f_1} u_{f_1} + m_{g_1} u_{g_1} + \Delta Q$$
$$= PE_2 + KE_2 + \Delta m_e h_e + m_{f_2} u_{f_2} + m_{g_2} u_{g_2} + \Delta W \qquad (5\text{-}1)$$

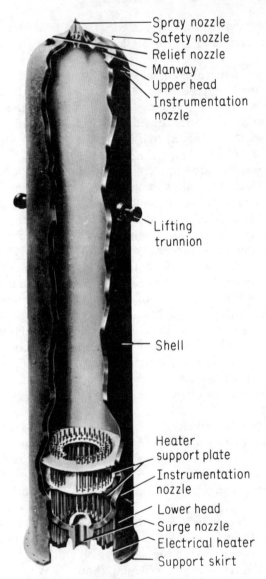

Spray nozzle
Safety nozzle
Relief nozzle
Manway
Upper head
Instrumentation nozzle

Lifting trunnion

Shell

Heater support plate
Instrumentation nozzle
Lower head
Surge nozzle
Electrical heater
Support skirt

FIG. 5-4. PWR plant pressurizer.
(Courtesy Westinghouse Electric Corporation.)

where the subscripts 1 and 2 refer to conditions within the pressurizer
before and after the insurge (or outsurge), and i and e refer to incoming
and outgoing fluids. The changes in potential energy PE and kinetic
energy KE as well as the work ΔW are zero, and the heat added ΔQ is
negligible for a pressurizer. Equation 5-1 then becomes.

$$\Delta m_i h_i + m_{f_1} u_{f_1} + m_{g_1} u_{g_1} = \Delta m_e h_e + m_{f_2} u_{f_2} + m_{g_2} u_{g_2} \qquad (5\text{-}2)$$

The above is an energy balance. A mass balance gives

$$\Delta m_i + m_{f_1} + m_{g_1} = \Delta m_e + m_{f_2} + m_{g_2} \qquad (5\text{-}3)$$

A volume balance gives

$$m_{f_1} v_{f_1} + m_{g_1} v_{g_1} = m_{f_2} v_{f_2} + m_{g_2} v_{g_2} = V \qquad (5\text{-}4)$$

where Δm = mass of water entering or leaving pressurizer; m_f, m_g = masses of liquid and steam within pressurizer; h = enthalpy of water entering or leaving pressurizer; u_f, u_g = internal energies of water and steam within pressurizer; v_f, v_g = specific volumes of water and steam within pressurizer; V = total volume of pressurizer.

Noting that the pressurizer is large enough so that there will always be both water and steam in it, the values of u and v above are for the saturated fluids at the corresponding pressures.

The above three equations are solved for the case of an insurge, where

$$\Delta m_e = 0 \qquad (5\text{-}5)$$

or an outsurge, where

$$\Delta m_i = 0 \qquad (5\text{-}6)$$

Solutions, for given primary system conditions and permissible pressure fluctuations result in required total pressurizer volume. For given pressurizer volume and primary system temperature fluctuations, the solutions (usually by trial and error) give resulting pressure fluctuations. As expected the larger the pressurizer, the smaller the pressure surges.

Basic equations of vapor-type pressurizers predicting system transient, pressure-time relationships during changes in reactor power level have been formulated [25] from energy, mass, and volume balances.

2. *Gas pressurizers* are those in which a volume of noncondensable gas is maintained above the primary coolant surface. Since there is no volume exchange due to condensing or flashing, the volume of gas required is rather large. Because of pressure-vessel-size limitations, this type is more suited to low-pressure systems. For example, in sodium-cooled reactors where low pressures (of the order of 100 psia) are needed, volume relief is provided by an inert-gas blanket on top of the sodium pool in the reactor. This blanket also isolates the liquid metal from air and water vapor and therefore prevents chemical reactions between them [2].

In boiling-water reactors, relief is naturally taken care of by the large volume of steam produced. However, because even small fluctuations in pressure affect the nuclear behavior of the reactor, an increase in the normal steam volume is provided by the use of a drum through which steam from the reactor is passed. Pressure transients are then dampened by the combined effects of drum capacity and flow resistance in the connecting piping. This drum can be a steam dryer, as in the EBWR (Sec. 4-4), or a steam separator (primary steam drum), as in the Dresden I plant (Sec. 4-6).

5-5. CHEMICAL-SHIM CONTROL

The term *chemical shim* refers to the use of a soluble neutron absorber, such as boric acid, which is dissolved in the primary reactor coolant. Control is then accomplished by varying the concentration of this absorber in the coolant. This, of course, is a slow process and is used only to control slowly varying reactivity effects, in addition to conventional control rods.

Boric acid has good water solubility and has been used experimentally in both pressurized [26] and boiling-water [27] reactors. Its use in commercial power reactors, however, is now restricted to the former. Since boron has no radioactive isotopes, no coolant radioactivity problems arise from it. (Boric acid has also been used as a shutdown device in many reactors). The concentration of boric acid in the coolant is changed at startup and during the lifetime of a core to compensate for (1) changes in reactivity resulting from changes in moderator temperature from a cold shutdown condition to a hot operating, zero power condition, (2) changes in reactivity due to the buildup of xenon-135 and samarium-149 concentrations in the core, and (3) reactivity losses due to fuel depletion and the buildup of long-lived fission products other than xenon and samarium.

Rapid reactor transients are handled by usual control rods. The requirements here include (1) transients due to the power and Doppler coefficient of reactivity, (2) safety and shutdown margins, (3) rapid changes in the average moderator temperature, and (4) void formation (boiling).

Boron worth, W_B, is defined as the fractional change in the effective multiplication factor, k_{eff}, per unit concentration of boron, C_B, in parts per million, ppm, or

$$W_B = \frac{1}{k_{eff}} \cdot \frac{dk_{eff}}{dC_B} \tag{5-7}$$

Boron concentration in the coolant may be adjusted by the *feed-and-bleed* method. In this method boron content is increased by feeding into the core some of a more concentrated boron solution than is in the core, and decreased by feeding in some pure water or a less concentrated solution. Some coolant must necessarily be bled off to make room for the feed. This may be discarded, or processed by distillation and the resulting concentrate and distillate reused for subsequent adjustments.

It has been found to date that the dissolved boron does not increase corrosion rates and that most materials suitable in neutral or high pH water at high temperature are also suitable in a boric acid solution. Long-term tests, however, need to be made. It has also been shown that boron does not increase the precipitation of corrosion products which would have posed problems of heat-transfer surface fouling and radioactivity levels in deposition locations.

Since chemical shim permits a reduction in the amount of reactivity controlled by control rods, the number and/or size of rods may be reduced resulting in simplified design and reduced costs. Also, the rod blackness may be reduced–i.e., rod materials of lower neutron cross sections, such as stainless steel instead of Ag-In-Cd, Hf or boron may be used. Note that because of strong water moderation, a pressurized-water reactor normally needs a large number of control rods.

A chemical absorber does not by itself materially affect the spatial power distribution, since it is uniformly distributed throughout the core, except for a minor effect due to increased density in the lower half of the core because of cooler water temperatures there. The use of chemical shim results, however, in improvements in spatial power distribution and therefore increased average-to-maximum power density in the core because of reduced blackness, size or the degree of insertion of control rods.

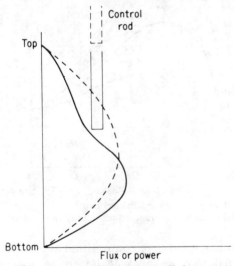

FIG. 5-5. Axial flux in a controlled reactor channel.

The control rods that remain inserted in a chemically shimmed core will, as in any core, distort the axial power distribution so that peak power density is pushed from the core center to a point farther from the rod entrance, Fig. 5-5. The axial maximum-to-average power ratio is low when the rod is fully withdrawn because of the uniformity of the channel, then increases with insertion up to a maximum, then decreases as the rod approaches full insertion. Near full insertion the ratio increases again to the original value since the rod channel approaches uniform composition again. A hypothetical rod of zero reactivity worth would not perturb the channel at all. The degree of skewing is of course severer the higher the reactivity worth of the rod. These effects, which apply whether or not chemical shim is used, are shown by Fig. 5-6 which is representative of large pressurized-water reactors. Specific values will

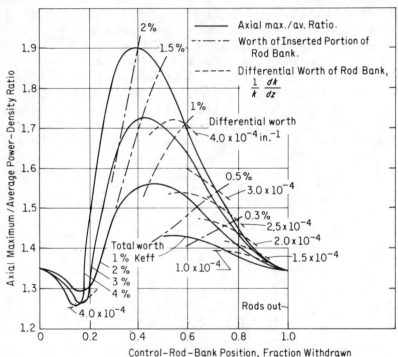

FIG. 5-6. Effect of reactivity worth and depth of insertion of control bank on axial maximum average power-density ratio. (Ref. 30.)

vary greatly depending upon core composition and height, with the effects of partially inserted rods on peaking becoming severer the larger the worth. The solid lines in Fig. 5-6 are for rods of different reactivity worths. Superimposed on these lines are lines of constant inserted worth, giving the depth of insertion of various control rods to compensate for a given reactivity. Also shown are lines of constant differential worth of rods. These show the amount of reactivity change per inch of insertion. Figure 5-6 shows that for a given inserted worth value, a more favorable axial distribution (lower axial maximum-to-average ratio) is obtained by the use of relatively low worths (and deeper insertion). For a given differential worth, on the other hand, a more favorable distribution is obtained by the use of relatively high rod worths (and not much insertion).

It can be seen that rods of lower worths, obtainable when chemical shim supplements rods, are conducive to a more favorable axial power distribution.

Another effect on axial power distribution in large reactors is power operation. The Doppler effect in oxide fuels causes a large reactivity loss in fuel at high temperatures, and, since peaking results in higher fuel temperature, the Doppler effect tends to reduce this peaking effect. The solid lines in Fig. 5-6 are for full power operation. At zero power conditions the ratios were found to increase for the conditions of Fig. 5-6 by about 15 percent for 1 percent worth rods to about 25 percent for 4 percent worth rods. Other effects of lesser magnitude are due to equilibrium xenon poisoning at high power and due to increased water temperature from bottom to top of core.

At the beginning of core life, control rods are usually used to flatten the radial power distribution. Because of burnup, these rods are later moved out and the power distribution becomes less flat. When a chemical absorber is used, however, power can be flattened at the beginning of core life by spatial variations in fuel enrichment or core composition. This favorable power distribution is then maintained to a great degree by simply varying the concentration of the chemical absorber, while the rod positions are maintained.

With favorable power distributions, the use of chemical shim results in increased fuel burnup for a given number of control rods. Note that without a chemical absorber, fuel burnup is limited by the number of control rods.

On the debit side, while chemical shim results in a decrease in number, size or blackness of control rods, the effect is not linear. There actually is a slight decrease in rod worth with increasing concentration of chemical absorber.

Another important effect is that the moderator temperature coefficient, in a chemically shimmed reactor, is less negative than in a fully rodded one and may even be positive. This is because the rods contribute to the negative coefficient and in a chemically shimmed core, they are fewer or less black or partially inserted; and because the boron concentration decreases, together with the water density, as the latter's temperature increases. This results in a positive contribution to the moderator temperature coefficient. Note, however, that the moderator temperature coefficient becomes more negative with time due to the decreased boron concentration with burnup.

The moderator temperature coefficient in a chemically shimmed pressurized-water reactor [28] is estimated at $+1 \times 10^{-4}$ to -3×10^{-4}

($^\circ$F)$^{-1}$. The first-core average burnup in that reactor is estimated at 21,800 Mwd/ton. The boron concentration necessary to insure a minimum shutdown margin at room temperature is estimated at 3200 ppm. Figure 5-7 shows the boron worth as a function of boron concentration for two moderator-fuel ratios, two fuel enrichments, and two temper-

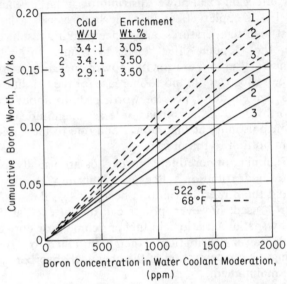

FIG. 5-7. Calculated reactivity worth of dissolved natural boron in three pressurized-water reactors. (Ref. 31.)

atures. Note that all the hot curves lie below the cold curves. The effects of the positive coefficient, resulting from the use of chemical shim together with high burnups in a pressurized-water reactor, have been investigated and found that they do not present a safety problem in the operation of such a reactor.

Boron worth decreases slightly with fuel enrichment. This effect, also evident from Fig. 5-7, is to be expected since the richer fuel competes more effectively against boron for neutrons. It can also be seen that a higher moderator-to-fuel ratio core is affected more by boron, an expected effect.

When boric acid is used, means for maintaining rigid pH control and purity of coolant must be provided. If the feed-and-bleed method of concentration control is used, and coolant bled from the reactor is discarded, greater capacity of the reactor plant waste-disposal system would be required. The water purification system (ion exchange) must use resins which will not remove boron, or the small amount (2 ppm) of LiOH or KOH used to reduce deposition of corrosion products, while removing other impurities.

One problem with chemical shim is referred to as *hideout* and *plateout*. Hideout is defined as the precipitation of boron from solution onto solid surfaces or in deposits of corrosion products adhering to surfaces of the core and coolant system. It may later reenter the system as a result of changes in operation or water conditions. This is plateout. Hideout and plateout are of course undesirable since they would result in positive and negative reactivity drifts. They represent the only significant safety problem resulting from the use of chemical shim. Tests have shown that boron deposition occurs to a limited extent onto corrosion products adhering to surfaces rather than on clean surfaces. Indications are also that it occurs on fouled surfaces when boiling occurs when the thickness of fouling is at least 0.3–0.4 mil. In this case deposition was proportional to the rate of evaporation. No rapid release of boron from a surface into the coolant has been observed. However, reactivity insertion from such rapid release is assumed in order to safely estimate the control rod requirements in a chemically shimmed reactor.

Typical reactivity requirements in a pressurized-water, chemically shimmed reactor are given in Table 5-2.

TABLE 5-2

Typical Reactivity Requirements in a Chemically Shimmed PWR*

Reactivity requirement	Reactivity, Percent	
	Rods	Boron
Safety shutdown, cold to operating temperature change	3.0	—
Doppler effect	2.2	2.0
Samarium poisoning	—	0.8
Xenon poisoning	—	2.2
Operating control	0.8	—
Core lifetime (fuel depletion)	—	9.0

* Ref. 29.

When overall power-plant control problems are considered, the one important difference between an all-rodded pressurized-water reactor and one using chemical shim is in the reduction in magnitude of the negative moderator temperature coefficient of reactivity when chemical shim is used. A large negative coefficient is relied upon in the design of automatic load following between reactor and turbine-generator into the overall control system. Note that an increase in load increases steam flow and therefore reduces reactor inlet temperature which in turn increases reactivity and a new equilibrium is reached. The degree of reduction in the temperature coefficient with chemical shim is not constant with time but decreases with burnup as the amount of dissolved absorber is reduced. These considerations must therefore be taken into account in designing control systems of a chemically shimmed pressurized-water reactor power plant.

5-6.   PRESSURIZED HEAVY-WATER REACTORS

Heavy water has almost the same physical, thermodynamic and chemical characteristics as ordinary water (214.6°F vs. 212°F normal boiling point, 38.9°F vs. 32°F melting point, 1.10 vs. 1.0 g_m/cm³ density at room temperature).   Power reactors using heavy water at high temperature, must as those using ordinary water operate at high pressure.   The pressures used vary between 900 and 1800 psia.   Heavy water, on the other hand, has nuclear characteristics markedly different from those of ordinary water.   Its slowing-down length is about twice that of ordinary water (10.9 vs. 5.7 cm at room temperature) but its moderating ratio is about 90 times greater (5,600 vs. 60).   Heavy water therefore is an excellent moderator and reflector.   Its cost however, is much greater, being of the order of \$30/lb_m (1970 prices).

The greatest advantage of heavy-water moderated reactors is their ability to use natural uranium fuels or fuels of near natural enrichment with high neutron economy, long reactivity duration, and therefore high fuel utilization.   This is done by utilizing the fissionable isotope effectively, i.e., they extract a large amount of energy per fissionable isotope burned, and by minimizing the fissionable isotope inventory.   In other reactor types, the tie-up of large quantities of fissionable isotope in inventory represents a large economic drain.   Burnups in the neighborhood of 10,000 Mwd/ton of natural uranium fuel are possible in heavy-water reactors (higher burnups have been achieved in other reactors, but with enriched fuels.   Also natural uranium is used in graphite-moderated reactors, but the burnups there are considerably lower because of physics and metallurgical reasons).   There is a great incentive, therefore, for the development of heavy water particularly for those countries with no or limited fuel-enriching facilities, though the development is not limited to these countries.

Another advantage, an indirect one, is in the reduction in fissile isotope losses in the isotope separation process of spent fuel elements. This is made possible by the ability to achieve long exposures with low-enriched fuels in heavy-water reactors.   Also, low enrichment results in reactors with flux distribution that are low in the epithermal energy range because of the abundance of the resonance absorbing U^{238}.   The additional effect of heavy water makes these reactors have a lower capture-to-fission cross-section ratio α than ordinary water thermal reactors of higher enrichment. Heavy-water reactors, therefore, have higher $1/(1 + \alpha)$ which means improved conversion, Sec. 9-3. Thus this type of reactor is an excellent vehicle, short of breeding, of conserving and extending present uranium resources.

The above advantage of high-potential neutron economy and fuel utilization is accentuated if plutonium is recycled and if the moderator temperature is kept low. Low moderator temperature softens the neutron spectrum further, adding to the above effect, lowers resonance absorption in reactor materials, and minimizes the effect of the low value of the thermal fission factor η (and therefore high α) of Pu^{239} in the region of its 0.3 ev resonance, and therefore allows more effective utilization of plutonium as fuel (plutonium recycle). Keeping the moderator cool in a power reactor is achieved in *pressure-tube* design in which the main moderator can be kept cooler than the direct coolant. Such a design has been adopted in the United States for the Carolinas-Virginia Tube Reactor (CVTR), and in Canada in various reactors, Sec. 6-5.

The price of uranium has a small influence on fuel-cycle costs in natural-uranium, heavy-water-moderated reactors. If the process of plutonium recycle is developed and becomes economical, this influence is decreased further and uranium ores that are relatively more expensive to mine could be utilized. Such expensive ores constitute the greatest uranium reserve in the world today.

Other economic advantages claimed for natural-uranium heavy-water reactors are the low unit fabrication cost of natural fuels because of the absence of criticality hazards in fuel-fabrication plants built specifically to handle natural fuels. If natural fuels are fabricated in plants designed to handle enriched fuels, this advantage, of course, disappears.

The economic incentives to develop heavy-water reactors thus vary in different countries and the economics depend on different methods of estimations and projections. In Canada, for example, it has been shown that natural-uranium heavy-water-moderated reactors can produce power at a lower cost than enriched-fuel reactors and Canada has concentrated on developing this type of reactor. This view, plus other advantages already cited, has been shared elsewhere, and Canadian-type reactors are now being built in such countries as India and Pakistan. In the United States, however, most studies have indicated cheaper power from enriched-fuel light-water reactors (BWR, PWR) and organic-moderated reactors (OMR) and that, even if heavy water is used, power would be cheaper if enriched fuels are used. The overriding factors in this case are the higher capital costs and the large expensive heavy-water inventory, coupled with high fixed charge rates. Hence the United States effort in D_2O reactors is rather meager. Other countries developing such reactors are Czechoslovakia, France, the Soviet Union, the United Kingdom, and others. In general fuel costs, according to estimates of the late 1960's fall in the 1 to 1.5 mills/kwhr range for D_2O reactors depending on normal economic and technical factors (natural uranium can be bought on the open market) and influenced strongly by fabrication costs. H_2O reactors have fuel costs in the 1.5-2.5 mill/kwhr range and depend on a set of ground rules which determine such things as uranium price, fuel use charge, plutonium buy-back, chemical processing as well as fabrication (Chapter 17). The fuel cost advantage of the D_2O reactors is taken up by high capital costs and costs associated with the use of D_2O.

Heavy-water reactors can use either metallic or oxide natural fuels. The former are more desirable from the point of view of parasitic neutron absorption, the latter from the standpoint of resistance to radiation damage. The latter consideration favors the use of UO_2 in power reactors where high burnups are sought [32].

Cladding materials of low neutron absorption must be used in all natural uranium reactors. Zircaloy, beryllium, and beryllium magnesium alloys are suitable for high-temperature power reactors. Aluminum may be used in low-temperature reactors.

Heavy-water moderated reactors require large moderator-to-fuel volume ratios. Such reactors, therefore, require large-diameter pressure vessels. Large power reactors operating at high temperature and pressure, therefore, require larger, thicker, and costlier pressure vessels than ordinary-water reactors of comparable output. Both pressure vessel and pressure-tube designs have been used. The latter design [33] allows the use of a lower-pressure, less costly vessel but with the added expense of constructing a leaktight calandria vessel free of differential expansion. It also results in the separation of coolant and moderator, and, as already mentioned, a cool-moderator design can then be easily incorporated.

Some other problems associated with D_2O reactors is the loss by leakage of the expensive D_2O and the high activity associated with the decay of tritium formed in the reactor.

Section 6-5 presents a Canadian design reactor of the pressurized heavy-water moderated and cooled, pressure-tube variety. Other reactor types of interest that have seen various degrees of development are the D_2O-moderated, gas-cooled, organic-cooled, and ordinary-water-cooled reactors, all made possible by the pressure-tube concept. In the latter concept the ordinary water may be allowed to boil, resulting in the generation of steam within the pressure tubes [34]

Another reactor of some interest is the *spectral-shift* reactor [35] in which the heavy-water coolant is gradually replaced by ordinary water during reactor operation. The reactor has fewer control rods and the initial core loading is made to have little excess reactivity. As fuel burns up, the addition of ordinary water improves moderation and restores lost reactivity. The range of D_2O concentration is from 70 to 5 mole percent over the life of the core. Because more excess neutrons are absorbed in the fertile material instead of poison rods, more conversion, higher fuel burnups and lower-fuel cycle costs are made possible. Better flux distribution is also claimed for this type reactor by reducing control rods and proper fuel loading, resulting in high specific powers, contributing further to increased burnup.

PROBLEMS

5-1. Pressurized-water reactors are used exclusively on naval vessels. What feature of this type reactor makes it more suitable for this application than boiling-water reactors?

5-2. It is desired to assess the effectiveness of the steam generator of a pressurized-water reactor plant on plant design, in particular the reactor output, plant efficiency and the size of the steam generator itself. To simplify the problem assume that the primary coolant is at a constant uniform temperature T_A and the condenser operates at fixed temperature T_2. Further assume that the plant operates on a Carnot cycle between T_1 and T_2 where $T_A - T_1$ is the temperature drop in the steam generator. For a fixed plant output ΔW and fixed steam-generator overall heat-transfer coefficient U, show that reactor input ΔQ_A decreases and cycle efficiency η increases as T_1 increases, but that the steam generator size, as represented by the heat-transfer area A, reaches a minimum value at $T_1 = (T_A T_2)^{1/2}$

5-3. A pressurized-water reactor power plant operates with a constant uniform temperature $T_A = 600°F$ and constant condenser temperature $T_2 = 100°F$. The plant is assumed to operate on a Carnot cycle (as above). For a steam generator overall heat-transfer coefficient of 1000 Btu/hr ft² °F, and electric generator efficiency of 90 percent, and a plant output of 720 Mw(e), plot the reactor output in Mw(t), the cycle efficiency, and the steam generator heat-transfer area in ft² against the upper cycle temperature T_1 for all possible values of T_1.

5-4. A vapor-type PWR pressurizer is 1,000 ft³ in volume. Under normal conditions, the primary system is at 2,000 psia and 600°F, and the pressurizer is half full of water. The primary system has a total volume of 10,000 ft³ and is assumed to be mechanically rigid. Due to some reactivity insert, the primary water temperature suddenly rose to 605°F. What will be the final composition in the pressurizer and the final system pressure?

5-5. Repeat Prob. 5-4 but for the case where there was a reactivity loss, resulting in the primary water temperature suddenly dropping to 595°F.

5-6. It is desired to design a pressurizer for a PWR that operates normally at 2,200 psia and 610°F, and that has a total primary-circuit volume of 8,000 ft³. The pressurizer, to be normally half-filled with water should prevent a system pressure rise of more than 1 percent for an expected maximum primary-coolant temperature rise of 10°F.   Calculate the necessary pressurizer volume.   Assume all primary circuit components to be mechanically rigid.

5-7.   Show that the change in reactivity due to the presence of a burnable poison which depletes to a product of negligible neutron absorption cross section may be given by the expression

$$(\Delta \rho)_{bp} = -f \frac{(\sigma_a)_{bp}}{(\sigma_a)_{ff}} \frac{(N_0)_{bp}}{N_{ff}} e^{-(\sigma_a)_{bp}\bar{\varphi}\theta}$$

where $(\sigma_a)_{bp}$ and $(\sigma_a)_{ff}$ are the microscopic absorption cross sections for the burnable poison and fissionable fuel, respectively. $(N_0)_{pb}$ and N_{ff} are the number densities of the poison when initially added in the core (assumed uniformly distributed throughout) and fissionable-fuel nuclei, respectively. $\bar{\varphi}$ is the average reactor neutron flux. θ is the time elapsed after the addition of burnable poison.

5-8.   A PWR uses 3.05 percent enriched uranium fuel and has a water-to-fuel ratio of 3.4:1.   The core is chemically shimmed with boron.   A control rod bank is used for operational control of 1 percent $\Delta k/k$.   It is required to do this with a maximum insertion in the core of 60 percent of core height, and with a minimum insertion of 30 percent, without causing the ratio of maximum-to-average power density to exceed 1.72.   Calculate (a) the total rod worth, percent, and (b) the boron concentration that contributes the balance of a required 16 percent reactivity (hot conditions).

chapter **6**

Pressurized-Water
Reactor Power Plants

6-1. INTRODUCTION

The basic flow diagram of a pressurized-water reactor power plant was shown in Fig. 5-1. The primary coolant is maintained at a sufficiently high pressure so that bulk boiling is avoided, though some subcooled (or surface) boiling in the hot channels is permitted at high loads. The design reactor pressure is usually in the 2,000-2,500 psia range, permitting a reactor exit-water temperature of 550 to 620°F, below saturation at these pressures. Higher pressures result in complications in reactor pressure vessel design with little increase in temperature and, therefore, efficiency. The steam conditions are usually around 1,000 psia and 550°F. These and the resulting plant efficiency (around 30-32 percent) are not unlike those obtained in boiling-water reactor power plants.

The pressurized-water reactor concept originated from the U.S. Naval Submarine Reactor Program, which was intended to develop a class of PWR for use in submarine and surface vessels. The reactors were first developed on land. The first full-scale, land-based prototype, called the STR Mark I, achieved criticality in March 1953. The first atomic submarine, the U.S.S. *Nautilus,* started sea trials in January 1955, and was followed by a large number of military underwater and surface vessels using the same reactor type.

The first civilian pressurized-water reactor power plants to produce electricity were a small, 30 Mw(t), 5 Mw(e) station in the Soviet Union [36] which started operation on July 27, 1954, and the 231 Mw(t), 68 Mw(e), Shippingport Atomic Power Station in the United States which achieved criticality in December 1957 (Sec. 6-2). Shippingport was followed by Yankee and Indian Point I (Sec. 6-3). The PWR concept subsequently became one of the two mainstays (with BWR) of commercial nuclear power in the U.S. Merchant ships to use PWR for propulsion include the Lenin, an icebreaker in the Soviet Union (to be followed by other icebreakers), and the N.S. Savannah, a cargo-passenger carrier in the United States. The latter was decommissioned in 1969.

The above-mentioned reactors used ordinary water for the coolant-moderator. In Canada, development of the heavy-water cooled and moderated PWR is taking place. This type, unlike ordinary water, permits the use of naturally enriched fuels, an advantage that makes it economically feasible in certain countries, despite the high cost of heavy water.

Load following in a PWR power plant depends upon the negative temperature coefficient of the coolant-moderator. An increased load on

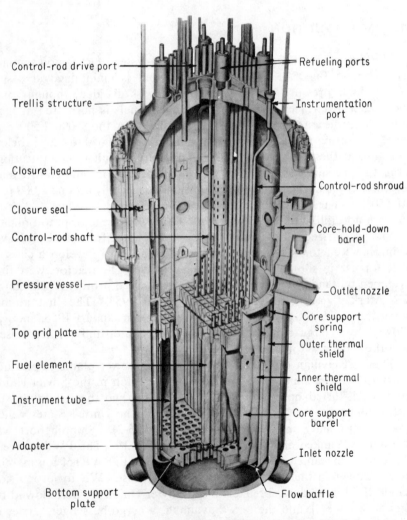

FIG. 6-1. Cutaway view of Shippingport PWR. (Courtesy Westinghouse Electric Corporation, Bettis Atomic Power Laboratory.)

the turbine increases reactor inlet water subcooling (as in a BWR dual cycle, Sec. 4-5) and increases reactor power.

In this chapter, some early and recent pressurized-water and power plants will be presented. Examples from both ordinary- and heavy-water reactors will be covered. The concept of the fluidized-bed reactor, in essence a PWR, will also be presented.

6-2. THE SHIPPINGPORT ATOMIC POWER STATION

Shippingport [37] was the first United States Central nuclear plant of any type designed and built for the purpose of generating electric power for civilian use. It is located at Shippingport, Pa., and has an output of 60 Mw(e). The reactor was designed and built by Westinghouse Electric Corporation for the U.S. Atomic Energy Commission and the Duquesne Light Company, which operates the plant. The primary objective of Shippingport was to gain information and experience to be used in the design of subsequent, more advanced, reactors.

A cutaway of the Shippingport reactor is shown in Fig. 6-1. The

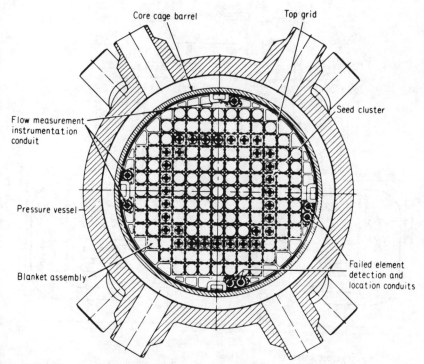

FIG. 6-2. Cross section of Shippingport pressure vessel above core. (Ref. 37.)

cylindrical core is a seed-blanket arrangement, containing a total of 145 fuel subassemblies of which 32 are in the seed. The seed assembly, which contains the active fuel, is in the form of a square annulus situated between inner and outer blanket assemblies, Fig. 6-2. The blankets contain pellets of natural uranium in UO_2.

The reactor is controlled by 32 cruciform-shaped rods with their lower absorbing sections made of hafnium, and the upper sections made of Zircaloy 2. The rods are cooled by allowing a small portion of the seed cooling water to flow up into the shrouds enclosing the rods and then out through slots. Shippingport has a relatively high negative temperature coefficient (approximately $-3.1 \times 10^{-4}/°F$), resulting in good load-following characteristics. The principal function of the control rods during normal operation is to compensate for long-range reactivity changes due to fission product poisoning and fuel burnup. At one time

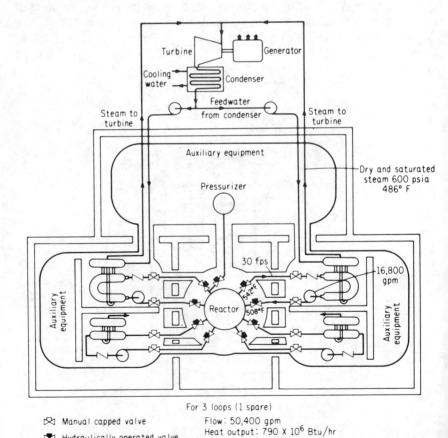

For 3 loops (1 spare)

⊠ Manual capped valve

▨ Hydraulically operated valve

Flow: 50,400 gpm
Heat output: 790 X 10^6 Btu/hr
Hot electrical output: 60,000 kw

FIG. 6-3.   Flow diagram of Shippingport pressurized-water nuclear power plant.

during the operation of Shippingport, a full-load trip occurred, owing to the loss of condenser vacuum. The reactor, not automatically scrammed when this happened, showed a slight increase in pressure from 2,000 to 2,090 psia, and leveled off with no operator action. Average system temperatures began drifting very slowly upward and were then corrected by control-rod adjustment.

A flow diagram of the Shippingport power plant is shown in Fig. 6-3. Table 6-1 lists some of its design and operational data.

TABLE 6-1
Some Design and Operational Data of Shippingport Power Plant

Reactor heat output	231,000 kw(t)
Electric output	68,000 kw(e) gross, 60,000 kw(e) net
Average thermal-neutron flux	6×10^{13} seed, 4×10^{13} blanket
Maximum thermal-neutron flux	2×10^{14} seed, 3×10^{14} blanket
Fuel material	Highly enriched U-Zr alloy in seed; natural uranium in UO_2 in blanket
Fuel configuration	32 subassemblies in seed, containing parallel-plate-type elements in seed; 113 subassemblies in blanket containing rod-type elements
Cladding material	Zircaloy 2, 0.015 in. thick in seed; 0.027 in. thick in blanket
Maximum fuel-surface temperature	636°F
Average fuel temperature	580°F in seed; 1000°F in blanket
Expected average burnup	3,000 Mw-day/ton
Average conversion ratio	0.85
Average specific power	360 kw/kg_m of U^{235} in seed; 2.57 kw/kg_m of natural U in blanket
Reactor core dimensions	6.8 ft average diameter; 6 ft high
Core configuration	Approximately cylindrical with annular squarish seed ring
Coolant-moderator-reflector	Ordinary water
Thermal shielding	Two stainless-steel concentric cylinders around core
Coolant velocity in core	20 fps in seed; 4.6–12.8 fps in blanket
Reactor coolant temperatures	508°F inlet, 538°F outlet
Reactor pressure	2,030 psig inlet, 45 psi drop
Primary-coolant flow rate	22.6×10^6 lb_m/hr
Number of primary loops	Four (three required for normal full power operation)
Pressurizer heat input	60 kw normal, 840 kw max.
Steam generators	Two U-tube plus two straight-tube
Steam generation	287,000 lb_m/hr per generator
Steam conditions	585 psig, 486°F
Net plant thermal efficiency	26%

6-3. THE INDIAN POINT I REACTOR POWER PLANT

Indian Point I [38] is a 255,000-kw(e)-net pressurized-water-reactor power plant built at Buchanan, Westchester County, N.Y. The reactor was designed by the Babcock & Wilcox Company for the Consolidated Edison Company of New York, which built and operates the plant. Indian Point I is interesting because it employs an oil-fired superheater in series with the reactor. Figure 6-4 shows a simplified flow diagram of the power plant.

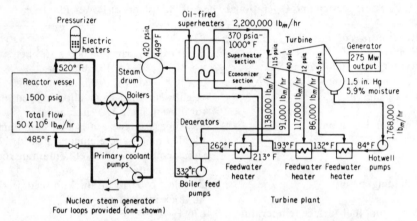

FIG. 6-4. Flow diagram of Indian Point power plant.

The reactor, a thermal heterogeneous internal thorium converter with no blanket, generates 585,000 kw(t). 53.9 × 10⁶ lb_m/hr of ordinary water flowing in four loops is heated in the reactor from 385.5 to 521°F. The reactor pressure is about 1,525 psig, corresponding to maximum fuel-surface temperatures around 600°F. Four horizontal U-shaped steam generators produce 550,000 lb_m/hr of steam at 405 psig and 449°F. This steam is fed into two oil-fired pressurized furnace superheaters operating in parallel. The superheaters are rated at 100,000 kw(t) each. They are designed to use either coal or oil firing.

The use of the superheaters results in steam conditions of 355 psig and 1000°F at the turbine inlet. The net thermal efficiency of the plant is 32 percent.

The reactor fuel is 93.5 percent enriched uranium in UO_2-ThO_2 mixture clad in stainless steel. The fuel elements are in the form of rods, 0.304 in. OD. There are 196 fuel rods per subassembly and a total of 120 subassemblies. The maximum design temperature of the ceramic fuel is 4352°F, and the expected average burnup is 18,000 Mw-day/ton of UO_2-ThO_2 mixture.

Reactor control is by one central regulating rod and 20 shim rods, all arranged on two concentric rings. The rods are cross-shaped and are fabricated from hafnium and 1 percent boron stainless steel.

6-4. THE POINT BEACH REACTOR POWER PLANT

Point Beach, located on the western shores of lake Michigan, is a recent design, two-power plant complex, each utilizing a 497 Mw(e) net,

TABLE 6-2
Some Design and Operational Data of Point Beach Power Plant

Reactor heat output	1,518 Mw(t)
Electric output	502.481 Mw(e) gross, 497.2 Mw(e) net
Fuel material	UO_2 enrichments: 3.40, 3.03 and 2.27% in 3 regions
Fuel configuration	14 × 14 array 121 subassemblies (33 containing 16-rod control clusters); fuel pellets 0.3669 in. OD, 0.600 in. long
Cladding	Zircaloy, 0.422 in. OD, 0.0243 in. thick
Reactor control	33 clusters of 16 rods (RCC) made of 80% Ag, 15% In and 5% Cd, clad in type 304-stainless steel
Chemical shim agent	Boric acid
Boron concentrations, ppm	
for reactor shutdown ($k_{eff} = 0.99$)	
with no rods, clean, cold/hot . . .	2,200/2,350
for control at power with no rods .	2,030 clean, 1,700 equilibrium Xe and Sm
Total rod worth	8.6%
Reactor vessel dimensions	37 ft 5 in. high, 11 ft ID, 0.156 in. stainless-steel clad
Reactor pressure	2,235 psig
Primary total coolant flow rate . . .	178,000 gpm
Number of primary loops	2.
Reactor coolant flow	66.7×10^6 lb_m/hr, 178,000 gpm
Reactor coolant temperatures . . .	551.8°F inlet, 605.4°F outlet, 650.7°F hot-channel outlet
Hot-channel factors	$F_q = 2.72$, $F_{\Delta H} = 1.58$
DNB ratios	1.84 nominal, 1.30 for design transients
Max fuel element temperatures . . .	657°F clad surface, 4050°F fuel center at 100% power, 4250°F overpower
Heat flux	175,800 Btu/hr ft² average, 491,000 maximum
Thermal output	5.7 kw/ft average, 16.0 maximum, 19.0 overpower
Steam generators	2 vertical shell and U-tube with integral moisture separators
Steam conditions (shell side, no load)	1,005 psig, 547°F

1,518 Mw(t), pressurized-water reactor built by Westinghouse for the
Wisconsin Michigan Power Company [39]. The main design and operational data of each plant are given in Table 6-2.

The reactor, shown in Fig. 6-5, contains a three-region core with a

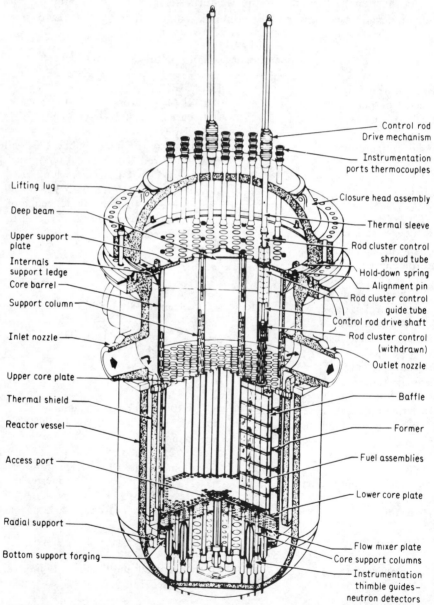

FIG. 6-5. Point Beach reactor vessel and internals. (Courtesy Wisconsin Electric
Power Company.)

total of 121 fuel subassemblies of which 33 contain rod cluster control (RCC) clusters, Fig. 6-6. In addition chemical shim (boric acid), Sec. 5-5, is used to control this reactor. The core is of the open type, i.e., the fuel subassemblies are not contained in individual channels. The fuel

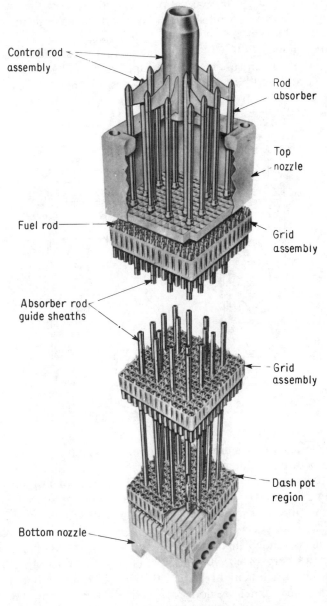

Control rod assembly

Rod absorber

Top nozzle

Fuel rod

Grid assembly

Absorber rod guide sheaths

Grid assembly

Dash pot region

Bottom nozzle

FIG. 6-6. Typical control cluster assembly. (Courtesy Wisconsin Electric Power Company.)

elements are Zircaloy clad rods, 0.422 in. OD, containing UO_2 pellets, each 0.3669 in. diam. and 0.600 in. long. The cladding tube is sealed at both ends by a plug welded to it. Sufficient void is left at the top to accommodate both gaseous fission products and fuel thermal expansion. A compression spring is placed within the void between the top plug and the top fuel pellet to prevent shifting the fuel during shipment.

All fuel subassemblies are about 13.5 ft long with a 12-ft. active fuel length, are composed of a 14×14 array of fuel rods, and are located on a square pitch. Each subassembly is supported axially by seven Inconel spring clip grids, and bottom and top nozzles, Fig. 6-7. Five of the grids

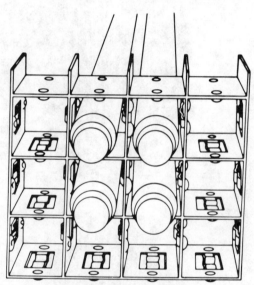

FIG. 6-7. Typical spring clip grid detail.
(Courtesy Wisconsin Electric Power
Company.)

are mixing grids which help intermix coolant within the core and thus reduce temperature gradients. Each fuel rod is supported in two perpendicular directions by spring clips whose forces (11–14 lb_f) are opposed by two rigid dimples. This provides rigid support and reduces flow-induced vibrations of the fuel rods. The rods are free to expand axially, reducing reactivity effects due to bowing.

The fuel is loaded in three approximately equal-volume concentric regions in the core of 40, 40 and 41 fuel subassemblies, Fig. 6-8, with first-core fuel enrichment of 3.40, 3.03 and 2.27 percent in the outer intermediate and inner regions respectively. Refueling takes place according to an inward loading schedule.

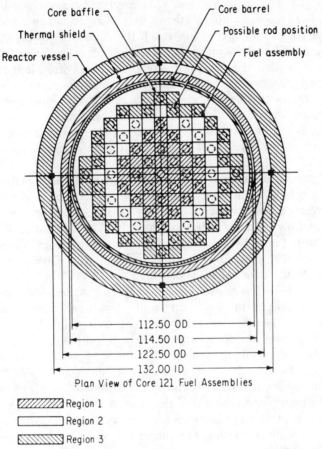

Core baffle

Thermal shield

Reactor vessel

Core barrel

Possible rod position

Fuel assembly

112.50 OD
114.50 ID
122.50 OD
132.00 ID

Plan View of Core 121 Fuel Assemblies

////// Region 1

Region 2

Region 3

FIG. 6-8. Point Beach reactor core cross section. (Ref. 39.)

Each control-rod cluster is composed of 16 control rods which are inserted directly into 16 guide thimbles welded to the grids and top and bottom nozzles of the fuel subassemblies, Fig. 6-6. This design is said to make it easy to insert the clusters and to make them conform easily to small misalignments without any binding. The control elements are fabricated of a silver (80 percent)-indium (15 percent)-cadmium (5 percent) alloy and are clad in stainless steel. Control-rod drives are of the magnetic-latch type. The latches are controlled by three magnetic coils which release the clusters upon loss of power, making them fall into the core by gravity to shut the reactor down. Some control-rod clusters, called the *control group,* are used to compensate for reactivity changes due to variations in reactor operating conditions such as power or temperature. At full power the control group moves within a prescribed band of travel, called the *operational maneuvering band,* to compensate for periodic changes in temperature or in xenon or boron concentration. When the clusters in this group reach the upper or lower limits of this band, the boron concentration is then changed. The rest of the control rod clusters, called the *shutdown group,* changes with changes in power due to the Doppler effect. The expected change in reactivity from zero to full power in this case is about 2 percent. The most severe control problems are at the end of core life when the moderator temperature coefficient is the most negative. Reactivity changes associated with void formation (boiling) are small in this pressurized-water reactor, being of the order of 0.1 percent at the end of core life.

The core is surrounded by a form-fitting baffle, Fig. 6-8, which restricts the bulk of upward coolant flow to the fuel. The baffle is in turn surrounded by the core barrel. A small amount of coolant is allowed to flow between baffle and barrel. The coolant is diffused uniformly into the core by a perforated flow-mixture plate situated between the core support plate and the lower core plate. One thermal shield, supported by the core barrel, is provided.

The effects of hydraulic lifting forces on the fuel rods are counteracted by the grid spring loads and bottom and top nozzles, and on the entire fuel subassembly by holddown leaf springs in the top nozzle assembly which bear against an upper core support plate.

The primary coolant system, Fig. 6-9, consists of two loops, each containing a steam generator, a pump, and necessary piping and instrumentation. One pressurizer is connected to one of the loops.

The primary coolant enters the reactor vessel via two nozzles and flows downward through the annulus between the core barrel and reactor wall, Fig. 6-8, thus cooling the thermal shield on both sides. It then enters a plenum at the bottom of the vessel, reverses direction, and passes through two exit nozzles at the same level as the inlet nozzles.

The steam generators are of the vertical shell and U-tube type, Fig. 5-2. The hot coolant enters the steam generator inlet channel head at the bottom, flows through the U-tubes, reversing direction to an outlet channel at the bottom. The inlet and outlet channels are separated by a partition. The tubes are made of Inconel.

The coolant is then pumped back to the reactor by two vertical single-stage centrifugal pumps of the controlled leakage type.

One primary coolant circuit contains a vapor-type pressurizer (Sec. 5-4) and auxiliary equipment, Fig. 6-9. A pressurizer relief tank contains water at or near ambient containment conditions which condenses any steam discharged from the pressurizer relief and safety valves. These come into operation during pressure surges which are beyond the pressure-limiting capacity of the pressurizer spray. The condensate is piped

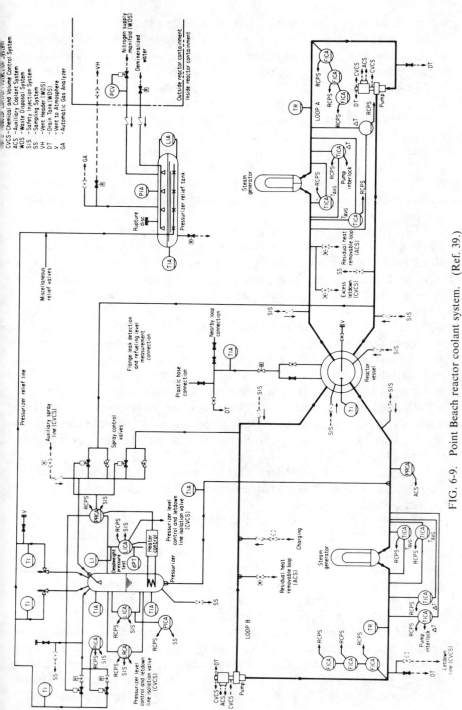

FIG. 6-9. Point Beach reactor coolant system. (Ref. 39.)

to the waste-disposal system. A nitrogen blanket covers the relief tank water. A spray and drain are used to reestablish normal conditions after a discharge. A rupture disk which discharges into the containment vessel guards against overpressure in the relief tank. The pressurizer is stainless steel clad. The 29 in. ID hot-leg pipes, the 27.5 in. ID cold-leg pipes, the 31 ID line between steam generators and pumps, and all valve surfaces in contact with coolant are made of stainless steel. The pump parts in contact with the reactor coolant and relief tank are constructed or coated with corrosion resistant materials.

The secondary or working fluid system is composed of the shell side of the steam generators, a turbine-generator, condenser, two condensate pumps, 5 stages of feedwater heating, two feedwater pumps and auxiliary equipment (Fig. 6-10).

The turbine-generator system is designed to produce a guaranteed maximum of 502.841 Mw(e) gross and an expected maximum of about 523.7 Mw(e) gross, should the primary system prove capable of such a capacity in the future. The 1,800-rpm steam turbine is composed of one double-flow, high-pressure element in tandem with two double-flow, low-pressure elements.

There are four combination moisture separator-reheater assemblies between the high- and low-pressure units, Fig. 6-10. Wet steam from the exhaust of the high-pressure elements enters each assembly at one end, is distributed by internal manifolds and rises through a wire mesh where the moisture is removed. It then flows over the tubes in the reheater where it is heated by high-pressure steam from the steam generators. This enters the other end of each assembly, passes through the tubes and leaves as condensate. The reheated steam flows back to the low-pressure turbines.

The ac generator, together with rotating rectifier-exciter, are mounted on the turbine shaft. The generator is rated at 582,000 kva and is inner-cooled by hydrogen.

Turbine exhaust steam from four manifolds enters a radial-flow type condenser. The condenser has a deaerating hotwell with sufficient storage for three minutes operation at full throttle and an equal free volume for surge flow. A steam-jet air ejector is provided. It has four first-stage elements and two second-stage elements mounted on shells of intermediate and after condensers. The ejector is driven with high-pressure steam from the steam-generator outlet.

Condensate from the condenser hotwell is pumped by two condensate pumps, then normally passed through hydrogen coolers, air ejectors, a gland steam condenser, then through the first four feedwater-heater stages. It is then pumped by two feedwater pumps through the fifth and last feedwater-heater stage back to the steam generators. An auxiliary

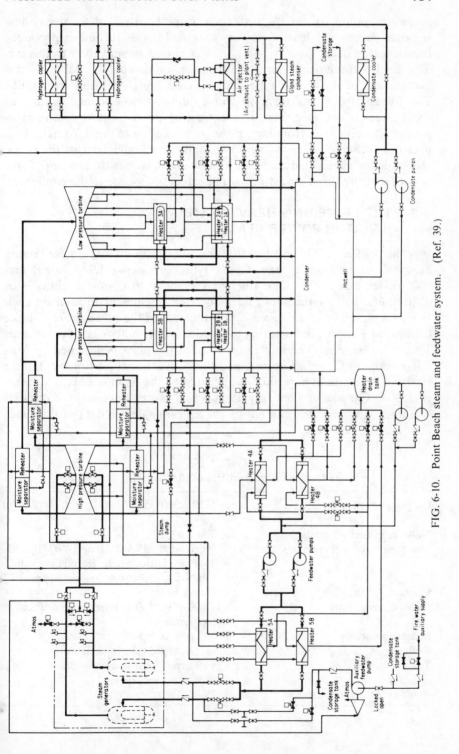

FIG. 6-10. Point Beach steam and feedwater system. (Ref. 39.)

feedwater pump driven by a steam turbine is provided for decay-heat removal in case of loss of power. Steam for this turbine is produced from reactor decay heat. The normally closed steam valves to this turbine open automatically on loss of power, or manually. All feedwater heaters are of the closed type and are twin units operating in parallel. The lowest two stages are joined in a duplex arrangement. Steam for the heaters is obtained from five extraction points, one from the high-pressure turbine, one from high-pressure exhaust, and three from the low-pressure turbines. Drains from the high-pressure heater cascade to the second-stage heater and then to a drain tank. Heater drain pumps force water from the three lowest-pressure heaters to cascade to the condenser.

6-5. THE PICKERING HEAVY-WATER REACTOR POWER PLANT

The Pickering Generating Station is a multiunit station containing heavy-water reactors of the *Candu* type, producing 1,743 Mw(t) and 500 Mw(e) each [40]. The station is planned to consist eventually of eight units, with 4 reactors and associated equipment located on each side of a common service area. It is located on the north shore of Lake Ontario at Fairport in Pickering township, twenty miles east of downtown Toronto, Canada. The nuclear portion of the plant is designed by Atomic Energy of Canada Limited, and the conventional portion by Ontario Hydro. Construction permit was granted by the Atomic Energy Control Board of Canada in February 1966. Full-power operation of the first two units is expected in the fall of 1971. Some design data of this plant are given in Table 6-3.

TABLE 6-3
Some Design Characteristics of the Pickering Station

Output per reactor	1718 Mw(t), 540 Mw(e) gross, 505 Mw(e) net
Number of fuel channels	390
Dimensions of fuel channels	Horizontal, 19.6 ft long, 4.07 in. ID, 11.25 in. lattice pitch, 28 rods per subassembly, 12 subassemblies per channel
Coolant	Heavy water
Coolant conditions	1,500 psig, 480°F inlet and 560°F outlet temperature
Steam conditions	593 psia, 485°F
Boilers	12, U-shaped tube-in-shell units
Turbine	1,800 rpm, tandem compound, with steam separators and reheat
Plant overall efficiency	29.4%

The reactors are of the so-called *Candu* type. The main features of this type are that the reactors have horizontal pressure-tube calandria cores and that they are heavy-water-moderated, pressurized heavy-water-cooled, and natural-uranium-dioxide fueled. The Pickering reactor (Fig. 6-11) consists of a core which is contained in a stainless-steel calandria shell (1) which has integral-steel end shield (7) cooled by ordinary water (6),

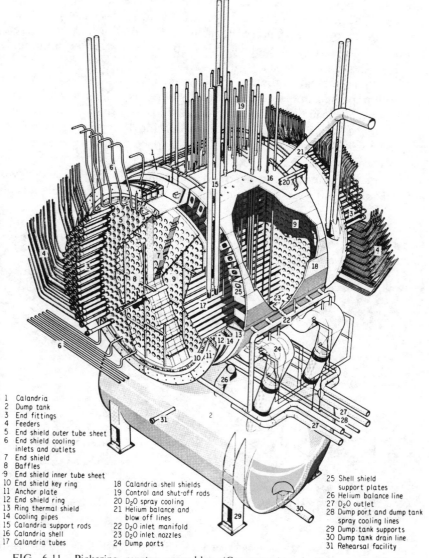

1 Calandria
2 Dump tank
3 End fittings
4 Feeders
5 End shield outer tube sheet
6 End shield cooling
   inlets and outlets
7 End shield
8 Baffles
9 End shield inner tube sheet
10 End shield key ring
11 Anchor plate
12 End shield ring
13 Ring thermal shield
14 Cooling pipes
15 Calandria support rods
16 Calandria shell
17 Calandria tubes

18 Calandria shell shields
19 Control and shut-off rods
20 D₂O spray cooling
21 Helium balance and
    blow off lines
22 D₂O inlet manifold
23 D₂O inlet nozzles
24 Dump ports

25 Shell shield
    support plates
26 Helium balance line
27 D₂O outlet
28 Dump port and dump tank
    spray cooling lines
29 Dump tank supports
30 Dump tank drain line
31 Rehearsal facility

FIG. 6-11. Pickering reactor assembly. (Courtesy Atomic Energy of Canada Limited.)

and stainless-steel circumferential shell thermal shields (13). The calandria contains 390 horizontal pressure tubes (17). Each pressure tube in turn contains a coolant tube which is supported in sliding bearings at the end shields of the calandria. The calandria and coolant tubes are separated by a sealed annulus containing CO_2 or nitrogen. Heavy water fills the calandria, outside the pressure tubes, and serves as moderator and reflector as well as coolant for the circumferential shell shields. There is a total of 276 tons of heavy water in the reactor.

Below the calandria is a dump tank (2) with a heavy aggregate concrete vault to provide shielding from reactor radiation. The concrete is cooled by one layer of cooling coils embedded in it. The dump tank contains a helium atmosphere.

The fuel elements are hermetically sealed Zircaloy-2 rods which contain compacted and sintered naturally enriched UO_2 pellets. 28 such elements are attached mechanically at their ends to form a cylindrical fuel subassembly (Fig. 6-12) 4.03 in. in diameter and 19.5 in. long. Spacers attached to the cladding maintains 0.05 in. space between the elements. The core power density is 5.73 $kw(e)/kg_m U$.

The heavy-water coolant enters the fuel channels at 480°F and leaves at 560°F and 1300 psig, via feeder pipes (4). 12 U-shaped tube-in-shell

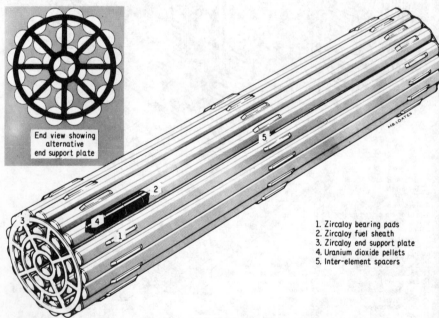

End view showing alternative end support plate

1. Zircaloy bearing pads
2. Zircaloy fuel sheath
3. Zircaloy end support plate
4. Uranium dioxide pellets
5. Inter-element spacers

FIG. 6-12. Pickering 28 element fuel bundle. (Courtesy Atomic Energy of Canada Limited.)

heat exchangers generate a total of 6.46×10^6 lb$_m$/hr of steam at 585 psia and 483.5°F from the ordinary-water working fluid.   A simplified plant flow diagram is shown in Fig. 6-13.

Two digital computer controllers (one a standby) control the reactor and turbogenerator, and operate the fueling machines and fuel transfer system.   Ordinary water in 14 chambers in the reactor will be used for the control of neutron-flux instabilities.   Flux shaping and xenon override will be provided by neutron absorber rods.

The reactor is protected by 11 shutoff rods and by the moderator dump.   The latter is resorted to only in case the shutoff rods fail to shut down the reactor.

The reactor building is part of a negative-pressure containment system which also includes a pressure relief duct and a vacuum building, and is designed for an internal pressure of 6 psig.

The fuel is loaded and removed in the reactor *on power* by two coordinated fueling machines located at opposite ends of the reactor, each on the underside of a bridge.   The bridges move vertically while the machines move horizontally.   In an on-power refueling operation, both machines locate, by remote control, on to an end fitting (3) of the selected channel.   The machines then remove and store end closures and shield plugs.   One machine then advances a fuel carrier containing the new fuel subassembly into the channel end fitting.   Simultaneously the other machine extends an empty fuel carrier at the other end of the channel.   The charge machine then rams the new fuel through a fuel latch into the channel, probing the spent fuel to enter the carrier in the discharge machine.   The carriers are then withdrawn, the shield plugs and end closures replaced, and the machines detached from the end fittings.   After the required number of bundles have thus been loaded, the discharge machine traverses to a spent-fuel port where it jettisons the spent fuel to an underwater storage canal via a transfer mechanism and a conveyor.

6-6. THE FLUIDIZED-BED CONCEPT

Fluidized beds have been extensively used in the chemical industry for mixing solids and fluids.   Examples are the chemical reduction of iron ores, the calcination of limestone. the catalytic cracking of petroleum, and many others.   A fluidized bed is one in which a fluid flows upward through a bed of solid particles which then become borne or fluidized but not transported or slurried.   The velocity of the fluid can be maintained over a wide range.   The lower limit is that necessary to just buoy the solids.   The upper limit is that which would just transport

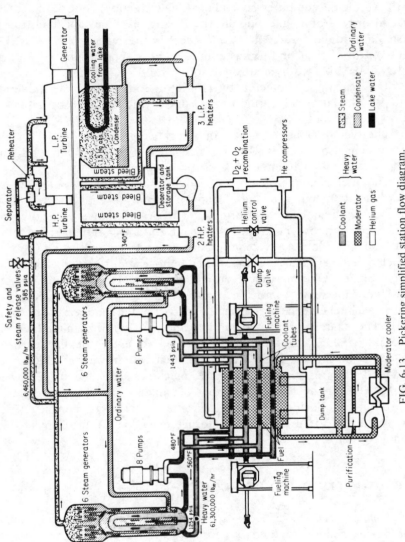

FIG. 6-13. Pickering simplified station flow diagram.

or carry the solids out of the bed. The bed is in a state of turbulence, and the solid particles are in constant motion, resulting in good mixing and high reaction and heat-transfer rates. The fluid may be either a liquid or a gas, although in the chemical industry the majority of the fluids used are gaseous.

Because of the intimate mixing and the resulting high rates of heat removal, and the simplicity of design, it was probably inevitable to think of the fluidized bed as a nuclear-reactor concept. In such a case the nuclear fuel in the form of small spheres or pellets takes the place of the solid particles and the coolant or coolant-moderator the place of the fluid in the fluidized bed (Fig. 6-14). In the collapsed condition, when the fuel elements are in the packed, unfluidized state, the physics of the core is such that the ratio of fuel to moderator is not critical.

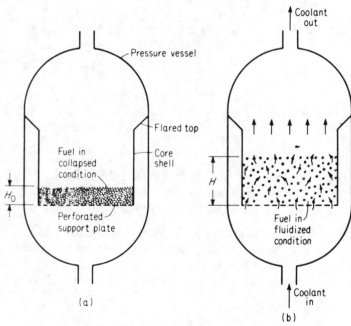

FIG. 6-14. Schematic of a fluidized-bed reactor. (a) Fuel in collapsed, and (b) in fluidized state.

A fluidized bed has many characteristics that are desirable from a reactor standpoint. These are discussed in the next section. In this section we shall derive some of the basic fluid-flow characteristics of fluidized beds.

The minimum fluid velocity that causes fluidization is obtained by equating the drag force on a pellet due to the motion of the fluid, to the weight of the pellet. Thus

$$C_D A_c \rho_f \frac{V^2}{2g_c} = V_s \rho_s \frac{g}{g_c} \tag{6-1}$$

where C_D = drag coefficient on pellet, a function of the Reynolds number, dimensionless

A_c = cross sectional area of pellet = πr^2 for a sphere, ft^2

r = radius of a spherical pellet, ft

ρ_f = density of fluid, lb$_m$/ft^3

V = velocity of fluid causing fluidization, ft/hr

V_s = volume of pellet = $\frac{4}{3}\pi r^3$ for a sphere, ft^3

ρ_s = density of solid-fuel pellet, lb$_m$/ft^3

g = gravitational acceleration, ft/hr^2

g_c = conversion factor = 4.17×10^8 lb$_m$ ft/lb$_f$ hr^2

Equation 6-1 can be solved for a spherical pellet to give

$$V = \sqrt{\frac{8}{3C_D} \frac{\rho_s}{\rho_f} rg} \tag{6-2}$$

(and the pellet becomes fluidized in stagnant fluid at zero gravity.)

The pressure drop Δp in a fluidized bed is composed of three components as follows:

$$\Delta p = \Delta p_w + \Delta p_s + \Delta p_f \tag{6-3}$$

where Δp_w = pressure drop due to frictional drag at walls

Δp_s = pressure drop due to static weight of solids in bed

Δp_f = pressure drop due to hydrostatic head of fluid

A fluidized-bed-reactor core would have a relatively large diameter, generally greater than its height, and a large mass of fuel pellets so that, in normal gravitational fields, Δp_w is negligibly small in comparison with Δp_s and Δp_f. Equation 6-3 can now be written for a fluidized bed as follows:

$$\Delta p = \Delta p_s + \Delta p_f = H \rho_B \frac{g}{g_c} \tag{6-4}$$

where Δp = pressure drop in fluidized bed, lb$_f$/ft^2

H = height of bed in fluidized state, ft

ρ_B = overall density of bed, including fluid and solid, lb$_m$/ft^3

Equation 6-4 can in turn be written in the form

$$\Delta p = \Delta p_s + \Delta p_f = H \left[(1 - \alpha)\rho_s + \alpha \rho_f \right] \frac{g}{g_c} \tag{6-5}$$

where α = average porosity or void fraction of bed = fraction of total bed volume not occupied by solids, dimensionless

Because the frictional drag at the walls is negligible and because the mass of the solids and the fluid does not change, the pressure drop in a fluidized bed is not a function of the coolant flow rate and remains constant so long as the bed is fluidized and maintains the same α. Δp_s can then be evaluated from the generally known characteristics of a packed bed at the *onset* of fluidization, since, for a packed bed, good experimental values of α are available, and H can be calculated.

As coolant flow is increased from zero through a packed bed, the flow is initially laminar, and Δp_s increases linearly with flow. At the onset of fluidization Δp_s reaches a value that just balances the weight of solids and fluid in the bed. Thus

$$\Delta p_s = H(1 - \alpha)\rho_s \frac{g}{g_c} = H_0(1 - \alpha_0)\rho_s \frac{g}{g_c} \qquad (6\text{-}6)$$

where H_0 and α_0 are the height and porosity of the bed in the collapsed or packed state (Fig. 6-14a). Van Heerden et al. [41], investigating the fluidization of powdered solids (carborundum, Fe_3O_4, coke) in ascending gases, found that, of four basic types of packing of spheres (rhombohedral, $\alpha_0 = 0.260$; tetragonal spheroidal, $\alpha_0 = 0.302$; orthorhombic, $\alpha_0 = 0.395$; and open cubical, $\alpha_0 = 0.476$), the experimental value for maximum porosity of a bed of spheres of equal diameter approached that of the orthorhombic. Their recommended value is 0.406. Mickley and Trilling [42] obtained values which ranged between 0.390 and 0.397.

In the fluidized state, when the height of the bed is H, Fig. 6-14b, the void fraction α is related to α_0, for constant bed area, by

$$\frac{1 - \alpha}{1 - \alpha_0} = \frac{H_0}{H} \qquad (6\text{-}7)$$

as was already evident in Eq. 6-6. The total pressure drop, Eq. 6-5, now becomes

$$\Delta p = [H_0(1 - \alpha_0)(\rho_s - \rho_f) + H\rho_f] \frac{g}{g_c} \qquad (6\text{-}8)$$

Example 6-1. A fluidized-bed-reactor core is 8 ft in diameter. It contains 18×10^6 spherical fuel pellets of $1/4$ in. diameter and density of 686 lb_m/ft^3. The fluid is pressurized water at 500°F. Calculate the pressure drop through the bed in the fluidized state when the height is 6 ft.

Solution. Equation 6-8 is used. H_0, the height of the bed in the collapsed state, must first be computed:

$$\text{Volume of fuel present} = 18 \times 10^6 \times \frac{\pi(0.25)^3}{6 \times 1,728}$$
$$= 85.222 \ \text{ft}^3$$

Choosing a value of 0.406 for α_0, volume occupied by fuel in collapsed state

$$= \frac{85.222}{1 - \alpha_0} = \frac{85.222}{0.594} = 143.47 \ \text{ft}^3$$

Therefore

$$H_0 = \frac{143.47}{(\pi/4)(8)^2} = 2.85 \ \text{ft}$$

Density of water at 500°F (from steam tables) = 49.02 lb_m/ft^3. Therefore

$$\Delta p = 2.85(1 - 0.406)(686 - 49.02) + 6 \times 49.02$$
$$= 1372.5 \ \text{lb}_f/\text{ft}^2 = 9.53 \ \text{lb}_f/\text{in}^2$$

The above pressure drop does not, of course, include that due to reactor entrances, exits, support plate, etc. Experiments on fluidized beds have frequently shown pressure drops greater than theoretical. This has been related [43] to the kinetic energy of the fluidized particles. This, called the *fluidization energy,* is in turn dissipated into the fluid in the form of thermal energy.

The nature of motion of fluidized beds is a complex one. Solids toss about at random and with varying velocities. Much attention has been devoted to fluid-flow and heat-transfer characteristics of fluidized beds [44]. Most of the work, however, is of interest to chemical engineering and was done on small solid particles (powders, sands, etc.) that are fluidized in gases. These particles are much smaller than the fuel pellets that might be used in nuclear reactors. Also, the flow conditions (velocities and passage equivalent diameters) encountered in the packed bed and at the onset of fluidization in most of the investigations were laminar. For such cases, good correlations for the velocity at the onset of fluidization (the critical velocity) were obtained by equating the pressure drop at the onset of fluidization (Eq. 6-6) to the pressure drop resulting from frictional drag between the fluid and the packed bed [41, 45] and solving for the velocity. This latter pressure drop is, of course, a function of the velocity and of a friction factor for which good experimental data are available [46]. Not many data are yet available on conditions applicable to nuclear reactors, which would use relatively large spherical pellets fluidized in a liquid.

Many of the above-mentioned investigations also resulted in various heat-transfer correlations. However, apparently because of the complex nature of fluidized-particle motion, a comparison [44] indicated that no coordinated correlation could result from them.

One type of reactor in which fairly large, heavy fuel elements remain in the packed state while a high-speed coolant, usually gas, passes through it is called the *pebble-bed reactor*. Fluid-flow and heat-transfer correlations for this type of reactor are known or can be calculated with more accuracy. The pebble-bed reactor and its characteristics are discussed in Secs. 8-10 and 8-11.

6-7. THE FLUIDIZED-BED REACTOR

The fluidized-bed concept has many characteristics very desirable from a reactor standpoint. Mentioned already are the high heat-transfer rates possible with this type. Also the intimate mixing and agitation result in a uniform temperature distribution throughout the bed, thus materially reducing the hot-spot factor. For the same reasons, there results an even, and consequently high, burnup of all the fuel, irrespective of the shape of the neutron-flux distribution curve.

A fluidized-bed reactor is simple in design. A typical design would have a cylindrical core with a perforated-plate bottom and a flared top (Fig. 6-14). The perforations at the bottom should be small enough to prevent the fuel from falling through but large enough to minimize coolant pressure drop across them. The perforations can be designed to produce any desired velocity distribution across the core. The plate itself should be designed to withstand the total weight of the fuel when in the collapsed condition, i.e., in case of insufficient or no coolant flow. The height of the active core–the height of the topmost fuel elements above the perforated plate–varies with coolant rate of flow. The core top is, however, flared to reduce the velocity of flow there and ensure against the escape of any fuel elements that may reach that height. No complicated system of fuel support or tie plates is necessary.

Fuel elements for a fluidized-bed reactor are simple small spherical pellets that may be clad by coating. UO_2 was suggested without cladding at all. Here, however, problems of abrasion of fuel and of erosion of primary-loop components due to UO_2 dust arise. Also, primary-loop design should ensure that abraded fuel dust is not allowed to settle anywhere where it would cause a serious radiation hazard.

Control of a fluidized-bed reactor can be achieved without control rods by varying the coolant flow rate within the wide range of fluidization. An increase in flow increases the height of the active core and therefore the distance between fuel pellets and thus the moderator-fuel ratio. Also, a fluidized bed has a large negative temperature coefficient, providing stability and safety against power surges.

The reactor must, of course, be designed to withstand the maximum system pressure which, in the case of a water coolant, is around 2,000 psia. The fluidized-bed concept can be used with lower-vapor-pressure coolants, such as the organics (Chapter 12) and liquid metals [2]. In the latter case, however, the problem of providing a suitable moderator for thermal reactors must be solved.

The simplicity of core loading in this type of reactor (all that is needed is the insertion of the necessary number of fuel pellets through a small opening to the core) makes feasible the continuous removal and addition of one or more pellets at a time. The high possible fuel burnup, together with this possibility of continuous loading and unloading, makes it unnecessary to retain much fuel for excess reactivity requirements or to use highly enriched fuels.

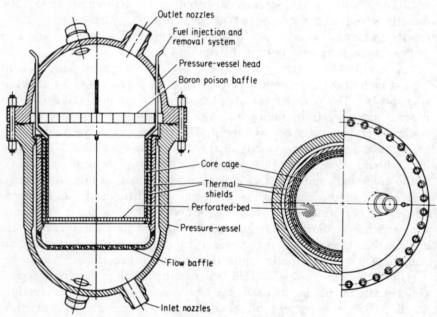

FIG. 6-15. LFBR pressure vessel. (Ref. 47.)

It can be seen that the fluidized-bed concept results in a simple reactor which can produce power economically on both capital, and fuel and operating charges. A major unknown in this type of reactor at this time, however, is the effect of the flow pattern on the criticality and neutron flux distribution.

A Liquid Fluidized Bed Reactor (LFBR) program has been carried on by the Martin Company of Baltimore, Md., for the AEC. Initial feasibility studies showed that the LFBR concept is economical and sound. The program [47] had as major objectives the conduct of critical experiments and the exploration of abrasion characteristics of both clad and unclad fuel pellets. Other objectives included studies of irradiation characteristics of fuel pellets, means of removing abrasion and fission products from the primary coolant, the experimental verification of thermodynamic heat-transfer and fluid-flow design calculations, and the study of stability and control of fluidized-bed-reactor power plants.

In the reference design [47], the reactor (Fig. 6-15) has an 8.5-ft-diameter, 6-ft-high core which contains 63,736 lb_m of 1.5 percent enriched UO_2 spherical pellets, each 1/4 in. in diameter. The core is flared at the top and is surrounded by two thermal shields and a pressure vessel. Approximately 15.21 × 10^6 lb_m/hr of ordinary water at 2,000 psia enters the core in three loops via three inlet nozzles, flows upward, and removes 1.366 × 10^9 Btu/hr of heat at full load. A small percentage of this water is used to cool the thermal shields and pressure vessel. The pressure vessel is to be fabricated out of SA-353, clad on the inside with stainless steel. In the initial design, the pressure-vessel head was flanged. This was done to facilitate removal of the core should erosion prove to be a serious problem. Otherwise, a less costly and troublesome welded design

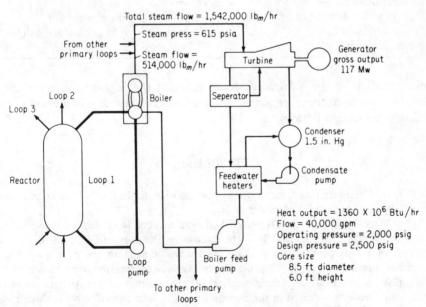

FIG. 6-16. Flow diagram of LFBR power plant. (Ref. 47.)

may be used in future vessels. The pressure vessel has provision for injecting and removing fuel pellets from the core without removing the head. Control is provided by varying the coolant flow through the core. For this, the primary-coolant pumps are rated 10 percent higher than the normal full-load flow.

The LFBR power plant reference design (Fig. 6-16) contains three steam generators (one per loop) and high- and low-pressure turbine sections in tandem. A moisture separator is interposed between the two,

TABLE 6-4
Some Design Data of LFBR

Reactor heat output	400 Mw(t)
Plant electrical output	117 Mw(e) gross
Number of primary-coolant loops . .	Three
Flow per loop	5.07 × 10⁶ lb$_m$/hr
Reactor coolant temperatures	491.5°F inlet, 565.5°F outlet
Reactor pressure	2,000 psia
Reactor pressure drop	75 psi
Fuel	Spherical pellets of UO_2, $\frac{1}{4}$-in.-diameter
Total fuel in core	63,736 lb$_m$
Core shape and dimensions	Cylindrical; $8\frac{1}{2}$ ft in diameter, 6 ft high, flared at top
Number of steam generators	Three
Rate of steam generation	455 × 10⁶ lb$_m$/hr per generator
Steam conditions	615 psia, 491.5°F at generators' exit
Plant efficiency	29.25% gross

necessitated by the saturated steam entering the high-pressure turbine. It also contains three regenerative feedwater heaters. Table 6-4 contains some design data of the LFBR.

PROBLEMS

6-1. Calculate the percent energy contribution to the total of the nonnuclear portion of the Indian Point power plant.

6-2. A 10-ft diameter fluidized-bed reactor core is cooled by water at an average temperature of 500°F. The fuel is in the form of 0.25-in. diameter pellets of 1.5 percent enriched UO_2. In the fluidized state, the core is 8-ft high and the pressure drop is 10 psi. The maximum thermal neutron flux is 10^{14}. Ignoring the extrapolation lengths, find (a) the number of fuel pellets, and (b) the total energy generated in the core in Mw(t). Take the effective fission cross section $\bar{\sigma}_f = 382.5$ barn.

6-3. A fluidized-bed reactor contains 1 in. diameter fuel pellets of 490 lb_m/ft^3 density. The water has a density of 49 lb_m/ft^3. At a water velocity of 2.5 fps, the bed was still in the collapsed state and the pressure drop was 4 psi. Find (a) the pressure drop at the onset of fluidization, (b) the height of the bed in the collapsed state, (c) the pressure drop in the fluidized state if the active core height is 10 ft, and (d) the water-to-fuel ratio by volume in normal operation.

chapter 7

Gas-Cooled Reactors

7-1. INTRODUCTION

Gas-cooled reactors have received great attention particularly in the United Kingdom but also in France, the United States, and the USSR. Both natural- and enriched-uranium fuels with CO_2 as coolant and graphite as moderator are used in the U.K. and France. In the United States, enriched fuels and helium coolant are used. Heavy-water-moderated, gas-cooled reactors are under consideration [48, 49].

The attractiveness of gas cooling lies in the facts that, in general, gases are safe, are relatively easy to handle, have low macroscopic neutron cross sections, are plentiful and cheap (except in the case of helium), and may be operated at high temperatures without pressurization.

The main disadvantages are in the lower heat-transfer and heat-transport characteristics of gases, requiring large contact surfaces and flow passages within the reactor and heat exchangers, and their high pumping requirements (between 8 to 20 percent of plant gross power), necessitating careful attention to the problems of fluid flow, pressure drops, etc.

To partially overcome the inherent disadvantages of gas coolants and at the same time to obtain attractive thermodynamic efficiencies, it is necessary to operate the fuel elements at as high temperatures as possible (commensurate with metallurgy) and permit a high gas-temperature rise in the reactor by reducing the gas mass-flow rate and pressurizing the gas. Because the fuel operates at high temperatures, fuel-element and cladding-material choice and fabrication in gas-cooled reactors present major problems, and the trend seems to be toward using ceramic rather than metallic fuels in such reactors. Also, because gas-cooled reactors are inherently large, they are particularly suited to large-capacity power plants. The reactor itself may impose structural and foundation problems. The size of the units can, of course, be reduced to a certain extent by increasing the fuel enrichment.

This chapter discusses some physical, thermodynamic, and flow characteristics of gas coolants. The next chapter presents gas-cooled reactor power plants of the thermal type. Gas-cooled fast-breeder reactors are presented in Chapter 10.

7-2. THERMODYNAMIC CYCLES

The hot gas emerging from a gas-cooled reactor can be used directly as the primary working fluid, i.e., expanding through a gas turbine or a nozzle, or it can be used indirectly by heating a secondary fluid acting as the working fluid. For each of these two cases, i.e., the direct or the indirect cycle, we may also have an open or a closed cycle. Following are possible combinations of these:

The Direct Open Cycle

This is shown diagrammatically and on a Ts diagram is Fig. 7-1.

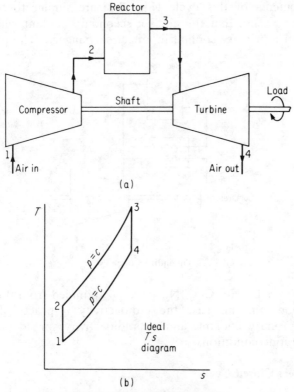

FIG. 7-1. The direct open cycle.
(a) Cycle diagram; (b) Ts diagram.

The coolant enters a compressor at point 1 where it is compressed to point 2. The gas then enters the reactor where it receives heat at constant pressure (ideally) and emerges hot at point 3. From there it expands through the turbine to point 4. The hot exhaust mixes with the atmosphere outside the cycle and a fresh cool supply is taken in at point 1. The turbine supplies the compressor power. Useful power may be supplied by the turbine or by the gas expanding further in a nozzle that supplies propulsion to the vehicle carrying the power plant. Because this is an open cycle, air is the only feasible coolant fluid (on earth). The above processes make up the familiar Brayton cycle used in gas-turbine or jet applications but where the reactor takes the place of the combustor. The cycle is more practical for turbojet than for stationary applications because of the danger of spreading radioactive gases associated with the air (Sec. 7-3) to the surrounding areas.

The Indirect Open Cycle

The elements of this cycle (Fig. 7-2) are similar to those in the previous one, except that the air is a secondary coolant which receives its heat from a primary coolant in a heat exchanger. The primary coo-

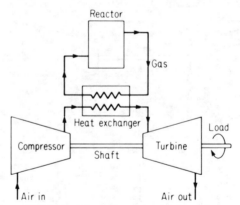

FIG. 7-2. The indirect open cycle.

lant, which may be He, CO_2, N_2, or air, is pumped from the reactor in a closed loop. In this case the radioactivity is practically confined within the primary coolant, and no radioactivity spread occurs under normal operating conditions.

The Direct Closed Cycle

In this cycle (Fig. 7-3), the gas coolant is heated in the reactor, expanded through the turbine, cooled in a heat exchanger, and then

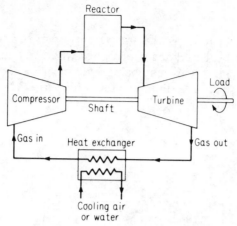

FIG. 7-3. The direct closed cycle.

compressed back to the reactor. In this cycle a gaseous working fluid other than air may be used. No effluent of radioactive gases passes into the atmosphere under normal operating conditions. Closed cycles permit pressurization of the working fluid with consequent reduction in the size of rotating machinery. A plant that operated on a direct closed cycle is described in Sec. 8-13.

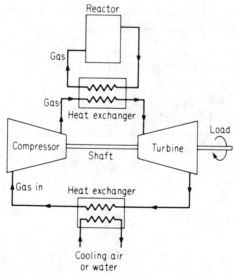

FIG. 7-4. The indirect closed cycle, gas to gas.

The Indirect Closed Cycle

The reactor heat is picked up by a primary gaseous coolant and transported to a heat exchanger or boiler where it transfers its heat to a secondary coolant. This may be another gas (Fig. 7-4) or, more commonly, a liquid such as ordinary water (Fig. 7-5). The steam produced

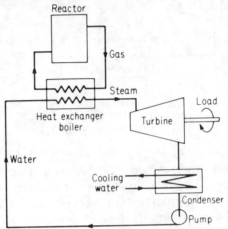

FIG. 7-5. The indirect closed cycle, gas to water.

in the latter case is then used in a steam plant in the conventional manner. An indirect closed-cycle gas-turbine plant is described in Sec. 8-14. Several gas-steam cycles are described in Chapter 8.

7-3. GAS-COOLANT RADIOACTIVITY

In designing open cycles, the extent of induced radioactivity in the coolant and its effect on plant and surroundings should be carefully evaluated. In closed cycles, the level of radioactivity may continually build up (if the rate of formation of the radioactive species is greater than its rate of decay) as the gas is recirculated through the reactor (e.g., the formation of C^{14} from C^{13}). In case of a gas leak to the surroundings, the level may be high and dangerous.

Following is a discussion of radioactivity induced in various gaseous coolants.

Air

One of the more serious problems is the radioactivity induced in some gaseous elements constituting air in open-cycle power plants, operating on or near the ground. When discharged to the atmosphere, these gases constitute a hazard to nearby life. In particular, two constituents of atmospheric air are troublesome. These are argon and carbon in carbon dioxide.

Of these two, the second presents the lesser danger. Ordinary carbon contains approximately 1.1 percent C^{13} which is converted to the radioactive isotope C^{14} which has a half-life of approximately 5,700 years. This would be quite a hazard, because of the absorption of carbon dioxide in biological processes, except for the fact that the concentration of $C^{13}O_2$ in the atmosphere is quite negligible. The problem due to C^{14} is, of course, more acute in reactors using CO_2 as coolant (see below).

Argon, on the other hand, is present as A^{40} to the extent of about 0.93 percent by volume in the atmosphere. Argon 40 has an absorption cross section for thermal neutrons of about 1.2 barns. By absorbing neutrons, A^{40} is converted into A^{41}, a radioactive product called "radio-argon." Radioargon is a β and γ emitter of 110-min. half-life (decaying into K^{41}). The amount of A^{41} formed in air-cooled reactors is some 10^4 times that of C^{14}. Being an inert gas, radioargon does not enter into biological processes, however, and thus is not assimilated by plant or animal life. The danger is in exposure to its β and γ radiation.

Because of the relatively short half-life of A^{41}, severe ground-level radioactivity is more or less localized. The extent of localization depends upon the extent of dispersion of the exhaust air. In ground installations this in turn is dependent upon many factors such as (1) the height of the reactor discharge above the ground, (2) the temperature and velocity of the exhaust air, (3) local wind velocity and turbulence, (4) local topography, buildings, etc. The possibility of the presence of particulate matter in the exhaust is also a complicating factor. Because of these usually ill-defined factors, no fixed or reliable rule can be specified in dealing with the problem. The best means of investigating the flow pattern in a particular terrain is by direct testing in a wind tunnel. However, there exist formulas, developed by Sutton [50], which give safe mathematical limits for the dispersion of effluent emanating from a point source under different atmospheric conditions.

In the case of aircraft and rockets, landing and launching fields should be located sufficiently far from inhabited areas and where the prevailing wind is in the opposite direction. The problem of radiation hazard on the ground may be partly solved by using conventional turbojets (or rockets) for landing and takeoff. In the aircraft or rocket, the reactor should be located far from the crew compartments (say near the wing tips in aircraft), and split shielding, i.e., around crew, their compartments, and the reactors, is recommended to reduce the overall mass of the vehicle.

Helium

In other than open cycles, the radioactivity that may be induced in the working fluid within the reactor is of interest only because of its effect on power-plant components that may come in contact with it.

Fortunately the helium-neutron reaction cross sections are quite low. Only the He^3 isotope has the relatively large cross section for thermal neutrons of about 1 barn. Its abundance in naturally occuring helium, however, is so low (0.00013 percent) that the induced radioactivity is negligible. The only serious radioactivity in a helium coolant is due to other gaseous impurities in it. Those that may exist in high-grade helium are hydrogen (about 5 ppm by mass), water vapor (about 50 ppm), and air (about 75 ppm) with its various constituents, including argon. Radioactivity due to the $O^{16}(n,p)N^{16}$ reaction in water vapor (Sec. 4-2) is possible but unimportant. Small but detectable quantities of radioargon may be produced from the air, as has been shown above.

It can thus be concluded that, except for the case of a fuel-element rupture or fission-gas leakage, the problem of induced radioactivity in high-grade helium is not of a serious nature.

Carbon Dioxide

Carbon dioxide, like helium, is used only in closed circuits. Radioactivity in carbon dioxide is induced within the reactor owing to the absorption of neutrons by the constituents O^{16} forming N^{16}, O^{18} forming O^{19} (as in water coolants), and C^{13}, present in natural carbon to the extent of 1.1 percent and having a cross section for thermal neutrons of about 0.9 millibarn. The reaction is $C^{13}(n,\gamma)C^{14}$. Carbon 14 has a half-life of about 5,700 years. Because of the long life of the C^{14} activity, the level continuously builds up as the coolant is recirculated through the reactor and power plant (Prob. 7-1).

Particulate Matter in Gases

Gas coolants can be carriers of foreign particulate matter that becomes radioactive in reactors. The particulate matter, such as dust, salt, and soot, can be introduced into poorly filtered open systems from the atmosphere. It can also be accumulated by abrasion of reactor materials, such as solid moderators (as graphite or beryllium oxide dust), cladding, etc., or from highly active fuel particles from ruptured fuel elements.

In the case of an open system, a good filter on the air-intake side that filters all but the smallest particles (0.5 to 1 μ) is sufficiently adequate, since these particles remain in suspension in the air after it is exhausted and are easily dispersed. To trap particles that originate from within the reactor, a good filter has to be installed in the exhaust side of the reactor. In closed systems, filters may be placed in the inlet or the exhaust side of the reactor or both, depending, among other things, on the operating temperatures. If two filters are used, the first, placed in the reactor inlet, helps to reduce the load on the second, which is subjected to highly active particles, and thus to reduce the frequency of replacing it.

Particular care in designing and manufacturing fuel and cladding is important in order to avoid ruptures and consequently contamination of gas coolants. This is important in gas-cooled reactors where the fuel elements operate at high temperatures.

7-4. COMPARISON OF GAS COOLANTS ON THE BASIS OF HEAT TRANSFER AND PUMPING POWER

Gas coolants have very high pumping-power-to-heat-transfer ratios, as compared with other fluids [1,2]. This ratio varies widely among the gases themselves. It is proposed here to compare the ratios of different gases as functions of operating parameters and physical characteristics. For simplicity, and since only comparative relationships are sought, it will be assumed that all gases flow and are heated in a circular tube of inside diameter D and length L.

From the Reynolds analogy between heat and momentum transfer [51]:

$$\frac{f}{8} = St = \frac{Nu}{Re\ Pr} \tag{7-1}$$

where f is the Darcy-Weisbach friction factor, and St, Nu, Re, and Pr are the Stanton, Nusselt, Reynolds, and Prandtl numbers, all dimensionless. This expression agrees well with experimental data in case where Pr is not far from unity, which is the case for most gases of interest.

Equation 7-1 may now be combined with the well-known Dittus-Boelter relationship for convective heat transfer:

$$Nu = 0.023\ Re^{0.8}\ Pr^{0.4} \tag{7-2a}$$

or

$$\frac{hD}{k} = 0.023 \left(\frac{DV\rho}{\mu}\right)^{0.8} \left(\frac{c_p\mu}{k}\right)^{0.4} \tag{7-2b}$$

to give

$$f = 8 \times 0.023\ Re^{-0.2}\ Pr^{-0.6} \tag{7-3}$$

and if the value of $Pr^{-0.6}$ is equated to unity,

$$f = 0.184 \, Re^{-0.2} \tag{7-4}$$

The last is an expression also derived from experimental tests on flow in smooth tubes over a wide range of Reynolds numbers.

The pumping work W, $lb_f ft/hr$, is related to the pressure drop Δp, lb_f/ft^2, by

$$W = \Delta p A_c V \tag{7-5}$$

where Δp is given by

$$\Delta p = f \frac{L}{D} \frac{\rho V^2}{2g_c} \tag{7-6}$$

The above equations are combined to give

$$W' = \frac{1}{Jg_c} \frac{LA_c}{D} (0.092 \, Re^{-0.2}) \rho V^3 \tag{7-7}$$

where h = heat-transfer coefficient, $Btu/hr \, ft^2 \, °F$

k = thermal conductivity of gas, $Btu/hr \, ft \, °F$

V = average gas velocity within length L, $ft/ħr$

ρ = gas average density, lb_m/ft^3

c_p = gas specific heat at constant pressure, $Btu/lb_m \, °F$

μ = gas absolute viscosity, $lb_m/hr \, ft$

W' = pumping work, Btu/hr

J = energy conversion factor = 778.16 $lb_f ft/Btu$

g_c = conversion factor = $4.17 \times 10^8 \, lb_m \, ft/lb_f \, hr^2$

The heat transferred between tube walls and gas is given by

$$\begin{aligned} q &= hA \, \Delta t_m \\ &= h(\pi DL) \, \Delta t_m \end{aligned} \tag{7-8}$$

where q = heat transferred, Btu/hr

A = circumferential area across which heat is transferred, ft^2

Δt_m = logarithmic mean temperature difference between coolant and tube walls, $°F$

h may be expressed by the Dittus-Boelter equation (7-2b). Combining Eqs. 7-8 and 7-2 and rearranging give

$$q = 0.023\pi kL \, \Delta t_m \, Re^{0.8} \, Pr^{0.4} \tag{7-9}$$

Dividing Eq. 7-7 by 7-9 gives

$$\frac{W'}{q} = \frac{(1/Jg_c)(LA_c/D)(0.092 \, Re^{-0.2})\rho V^3}{0.023\pi kL \, \Delta t_m \, Re^{0.8} \, Pr^{0.4}} \tag{7-10}$$

Substituting $\rho VD/\mu$ for the Reynolds number and $c_p\mu/k$ for the Prandtl number and rearranging give

$$\frac{W'}{q} = \frac{1}{Jg_c} \left(\frac{V^2}{\Delta t_m}\right) \left(\frac{Pr^{0.6}}{c_p}\right) \tag{7-11}$$

where the operating parameters and physical constants are separated in the two terms in parentheses. Equation 7-11 shows that, on the basis of the same coolant velocity and logarithmic mean temperature difference between wall and coolant, a good coolant should possess a low value of $Pr^{0.6}/c_p$.

A comparison between coolants including the heat transported by the coolant, is sometimes preferred. This is defined by the equation

$$q = \rho \, \frac{\pi D^2}{4} \, V c_p \Delta T \tag{7-12}$$

where ΔT = temperature rise of gas from tube entrance to tube exit, °F

Multiplying Eq. 7-11 by q^2 and dividing by its equivalent from Eq. 7-12 and rearranging give

$$\frac{W'}{q} = \frac{16}{\pi^2 Jg_c} \left(\frac{q^2}{\Delta t_m D^4 \Delta T^2}\right) \left(\frac{Pr^{0.6}}{c_p^3 \rho^2}\right) \tag{7-13}$$

By introducing the volume of coolant within the tube, $V_0 = L\pi D^2/4$, this relationship can be changed into the form

$$\frac{W'}{q} = \frac{1}{Jg_c} \left(\frac{q^2 L^2}{\Delta t_m \Delta T^2 V_0^2}\right) \left(\frac{Pr^{0.6}}{c_p^3 \rho^2}\right) \tag{7-14}$$

Equation 7-14 shows that, on the basis of the same heat transfer, the same coolant temperature rise, the same temperature difference between fuel and coolant, the same volume of coolant present, and the same coolant travel length, a good coolant should possess a low value of the physical group ($Pr^{0.6}/c_p^3 \rho^2$). The individual physical properties included in this group are strong functions of temperature and pressure. The dependence on temperature and pressure is different for different gases.

A different form of the physical group given above may be introduced by using the perfect-gas laws:

$$c_p = \frac{\bar{R}}{J} \, \frac{\gamma}{M(\gamma - 1)} \tag{7-15}$$

and

$$\rho = \frac{pM}{\bar{R}T} \tag{7-16}$$

where $\bar{R}$ = universal gas constant

M = molecular mass of gas

γ = ratio of specific heats of gas

p, T = average absolute pressure and temperature of gas

Combining with Eq. 7-14 and rearranging give

$$\frac{W'}{q} = \frac{J^2}{g_c \bar{R}} \left[\frac{1}{\Delta t_m} \left(\frac{qLT}{\Delta T V_0 p} \right)^2 \right] \left[\mathrm{Pr}^{0.6} M \left(\frac{\gamma - 1}{\gamma} \right)^3 \right] \qquad (7\text{-}17)$$

Thus, by introducing the pressure and temperature in the operating parameter group, the physical group becomes $\mathrm{Pr}^{0.6} M \left[(\gamma - 1)/\gamma \right]^3$. This group gives comparative results between different gases, identical to that of Eq. 7-14, if the properties are calculated at the same pressure and temperature for all gases.

Table 7-1 contains relative values for the physical group for different possible reactor gaseous coolants and for hydrogen, calculated at 80 and 600°F and atmospheric pressure (except for steam as noted). The value for hydrogen at 80°F is normalized to unity.

TABLE 7-1

Relative Values of W'/q for Different Gases Based on Their Physical
Properties at Atmospheric Pressure

Gas	80°F	600°F
Hydrogen	1.0	0.91
Helium	5.1	5.1
Steam	5.15*	2.65†
Carbon dioxide	11.1	4.84
Nitrogen	13.5	11.9
Air	13.9	11.9

* At 224°F, 1 atm.

† At 600°F, 40 atm.

Note that hydrogen would make an excellent reactor coolant except for its strong chemical activity and that it has a relatively high absorption cross section for neutrons. It is included in Table 7-1 only as a standard of comparison for other gases.

Table 7-1 contains some interesting information. While helium, steam, and CO_2 rank second, third, and fourth at low temperatures, the order is changed at high temperatures. This is because of the rather large increase in specific heats (and decrease in γ) with temperature for the triatomic gases relative to diatomic gases (Fig. 7-6). This is due to the large increase in molecular vibrations with temperature and energy that must be absorbed by the molecule to increase those vibrations.

Temperatures are a measure of translational energy only.)   Molecules
of monatomic gases practically have only translational energies and their
specific heats (and γ's) vary little with temperature.   It can be seen that
the relative values of the physical group change by about one-half or
more for H_2O and CO_2, by 9 to 15 percent for H_2, N_2, and air, and are
constant for He.

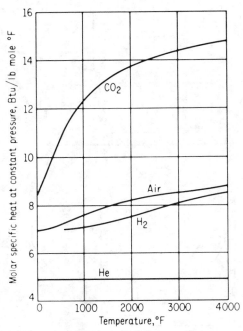

FIG. 7-6.   Variation of molar c_p with temperature for various gases.

Not all gases may operate at the same pressure and temperature.
Thus a complete W'/q evaluation of the different gases should take into
account both the operational and physical groups at different locations
in the cycle.   Also, the thermodynamic characteristics of the gases must
be taken into account.   This is dealt with in the next section.

7-5.   THERMODYNAMIC COMPARISON OF
GAS COOLANTS IN DIRECT-CYCLE APPLICATIONS

In direct-cycle applications, where the coolant gas is expanded in a
turbine (Sec. 7-2), the optimum operating conditions of the cycle depend
upon the physical and thermodynamic properties of the gas.
Figure 7-7 shows a block diagram as well as pV and Ts diagrams of an

ideal direct cycle.   The gas is compressed isentropically from point 1 to 2, heated in the reactor at constant pressure from 2 to 3, and then expanded isentropically through the turbine from point 3 to 4.   Cooling occurs from point 4 to point 1, either in a heat exchanger (closed cycle) or in the open atmosphere (open cycle).

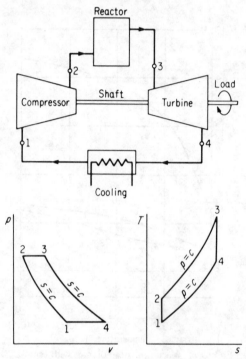

FIG. 7-7.   An ideal Brayton cycle.

The work done in the turbine (a steady-flow machine), with relatively negligible change in the kinetic energy of the gas, is W_T Btu/hr, equal to the change in its enthalpy.   Thus

$$W_T = H_3 - H_4 = \dot{m}(h_3 - h_4) \tag{7-18}$$

where H = total enthalpy of flowing gas, Btu/hr
h = specific enthalpy, Btu/lb$_m$
$\dot{m}$ = mass rate of flow of gas, lb$_m$/hr
For a gas, Eq. 7-18 may be rewritten in the form

$$W_T = \dot{m}c_p(T_3 - T_4)$$

where c_p is the specific heat at constant pressure of the gas (assumed constant) and T is the absolute temperature.   This equation can, with the help of the gas laws (Table 2-1), be written in terms of the pressure ratio across the turbine as

$$W_T = \dot{m}c_p T_3 \left(1 - \frac{T_4}{T_3}\right)$$

$$= \dot{m}c_p T_3 \left[1 - \frac{1}{(p_3/p_4)^{(\gamma-1)/\gamma}}\right] \tag{7-19}$$

Similarly, the work consumed by the compressor is

$$W_C = \dot{m}(h_2 - h_1)$$

$$= \dot{m}c_p T_2 \left[1 - \frac{1}{(p_2/p_1)^{(\gamma-1)/\gamma}}\right] \tag{7-20}$$

If the pressure ratios across the turbine and compressor are equal and if both are given the symbol r_p (in an actual cycle p_3/p_4 is slightly less than p_2/p_1 because of pressure losses in the reactor and cooling process), the net work of the cycle W_n, for the same specific heat, is

$$W_n = W_T - W_C = [\dot{m}c_p(T_3 - T_2)]\left[1 - \frac{1}{(r_p)^{(\gamma-1)/\gamma}}\right] \tag{7-21}$$

The first expression on the right-hand side of this equation is obviously the total heat added in the cycle, Q_A Btu/hr. The second term must then be the thermal efficiency η_{th}. It is a function of r_p and γ.

Although Eq. 7-21 pertains only to an ideal cycle, some of the trends predicted by it apply to actual cycles. In the following analysis it will be assumed that the maximum temperature in the cycle T_3 is fixed by metallurgical considerations of the reactor and turbine materials (the so-called *metallurgical limit* in gas-turbine work). Equation 7-21 is written in terms of the initial cycle temperature T_1, and T_3, as follows:

$$W_n = \{\dot{m}c_p[T_3 - T_1(r_p)^{(\gamma-1)/\gamma}]\}\left[1 - \frac{1}{(r_p)^{(\gamma-1)/\gamma}}\right]$$

$$= \dot{m}c_p T_1 \left\{[1 - (r_p)^{(\gamma-1)/\gamma}] + \frac{T_3}{T_1}\left[1 - \frac{1}{(r_p)^{(\gamma-1)/\gamma}}\right]\right\} \tag{7-22}$$

Examination of this equation shows the following:

1. Other things being equal, i.e., for the same T_1, T_3, r_p, and γ, the work per pound mass of gas, W_n/m, is a direct function of c_p.

2. Other things being equal, monatomic gases (He, A, etc.) with high values of γ produce slightly more net work per pound mass [slightly higher $(r_p)^{(\gamma-1)/\gamma}$] than diatomic or triatomic gases.

3. An increase in r_p from its lowest theoretical value of 1.0, where the corresponding net work is zero, decreases one part of Eq. 7-22 and

increases the other. The net work thus goes through a maximum, at an optimum value of r_p. This state of affairs can be shown graphically by the three ideal cycles of Fig. 7-8. These operate between the same

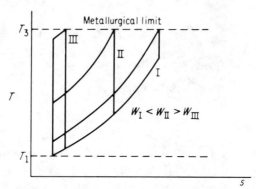

FIG. 7-8.   Effect of pressure ratio on net work
(T_1 and T_2 = const).

temperatures T_1 and T_3 and have the same exhaust pressure but different values of r_p. The net work in each case is represented by the enclosed area of the cycle.

The optimum pressure ratio can be evaluated for ideal cycles by differentiating the net work in Eq. 7-22 with respect to r_p and equating the derivative to zero. This gives a value of T_2 expressed by

$$T_2 = (T_1 T_3)^{1/2}$$

Thus $\qquad (r_p)_{opt} = \left(\dfrac{T_2}{T_1}\right)^{\gamma/(\gamma-1)} = \left(\dfrac{T_3}{T_1}\right)^{\gamma/2(\gamma-1)}$ $\qquad$ (7-23)

Note that the quantity $\gamma/2(\gamma-1)$ decreases as γ increases. Thus, for fixed initial and maximum cycle temperatures, the optimum pressure ratio for monatomic gases (He) is, in general, lower than for diatomic gases (air, N_2). These in turn have lower ratios than the triatomic gases (CO_2, steam). It follows that a monatomic gas, for example, may operate at lower maximum pressures or, if the pressure in the low-pressure sections of the cycle is increased (in a closed cycle), may operate with a larger average density. This is accompanied by a reduction in plant size and mass.

4. Because of the large effect of c_p, the same net work $W_n/\dot{m}$ can be obtained with lower pressure ratios by gases having high specific heats. Here is an obvious advantage of He over other gaseous coolants, to be added to its good heat-transfer and pumping characteristics.

Table 7-2 lists some properties of gaseous coolants at low pressures and temperatures.

TABLE 7-2
Some Properties of Gas Coolants at Low Pressures and Temperatures

Gas	Molecular mass	c_p, Btu/lb$_m$ °F	γ, dimensionless	k, Btu/hr ft° F	μ, lb$_m$/hr ft
H$_2$	2.016	3.421	1.405	0.125	0.024
He	4.003	1.250	1.659	0.090	0.050
Steam	18.016	0.440	1.335	0.013	0.030
CO$_2$	44.01	0.202	1.29	0.013	0.045
Air	28.97	0.240	1.4	0.015	0.046
N$_2$	28.02	0.2484	1.4	0.018	0.048
A	39.95	0.1250	1.668	.....	0.065

Example 7-1. Find the pressure ratio required to produce a net work of 328 Btu/lb$_m$ of helium coolant in an ideal Brayton cycle having initial and maximum temperatures of 500 and 1750°R, respectively. What is the pressure ratio corresponding to maximum net work? Use He properties at low pressures and temperatures.

Solution. For He, $\gamma = 1.659$ (Table 7-2). Thus $(\gamma - 1)/\gamma = 0.3972$. $c_p = 1.25$ Btu/lb$_m$ °F. Using Eq. 7-22,

$$\frac{W_n}{\dot{m}} = 328 = 1.25 \left\{ (500 + 1750) - \left[500(r_p)^{0.3972} + \frac{1750}{(r_p)^{0.3972}} \right] \right\}$$

This reduces to

$$[(r_p)^{0.3972}]^2 - 3.975 \, (r_p)^{0.3972} + 3.5 = 0$$

This equation yields two pressure ratios of 2.0 and 11.7.

That the solution contains two answers is in agreement with the concept of optimum pressure ratio discussed above. Using Eq. 7-23,

$$(r_p)_{opt} = \left(\frac{T_3}{T_1} \right)^{\gamma/2(\gamma - 1)} = \left(\frac{1750}{500} \right)^{1/2(0.3972)} = 4.85$$

Figure 7-9 shows results of calculations for $W_n/\dot{m}$ for ideal Brayton cycles using He, air, and CO$_2$ as coolants, where the initial and maximum temperatures are the same as those in Example 7-1, that is, 500 and 1750°R, respectively. Note that the specific power (Btu/lb$_m$) of helium is generally higher and, in the practical range of pressure ratios, occurs at much lower pressure ratios. To obtain specific powers on a pound-mole basis, the ordinates of Fig. 7-9 are multiplied by the molecular mass of the gas in each case.

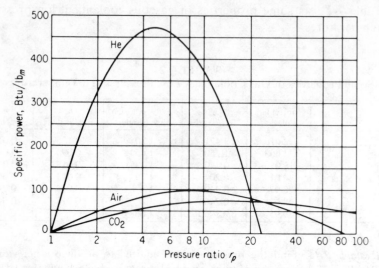

FIG. 7-9.   Specific power versus pressure ratio for three gases operating
on ideal Brayton cycle.

7-6.   THE ACTUAL CYCLE

The Brayton cycle with fluid friction is represented on the pV and Ts diagrams of Fig. 7-10 by 1-2'-3-4'.   Both the compression process with fluid friction 1-2' and the expansion process with fluid friction 3-4' show

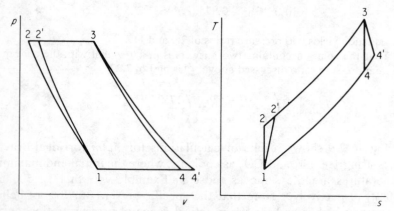

FIG. 7-10.   Brayton cycles with and without fluid friction.

an increase in entropy as compared with the corresponding ideal processes 1-2 and 3-4.   Pressure drops during heat addition (process 2-3) and heat rejection (process 4-1) are neglected. (In case of low pressure ratios the pressure drops may be significant and should be taken into account.)

The compression and expansion processes with fluid friction can be assigned adiabatic efficiencies as follows:

$$\eta_c = \text{adiabatic compression efficiency} = \frac{\text{ideal work}}{\text{actual work}}$$

$$= \frac{h_2 - h_1}{h_{2'} - h_1} \tag{7-24}$$

If we assume constant specific heats,

$$\eta_c = \frac{T_2 - T_1}{T_{2'} - T_1} \tag{7-25}$$

$$\eta_e = \text{adiabatic expansion efficiency} = \frac{\text{actual work}}{\text{ideal work}}$$

$$= \frac{h_3 - h_{4'}}{h_3 - h_4} \tag{7-26}$$

and for constant specific heats,

$$\eta_e = \frac{T_3 - T_{4'}}{T_3 - T_4} \tag{7-27}$$

where in each case the smaller work is always in the numerator.

The net work of the cycle is $W_n = $ (work of turbine) - (work of compressor). For constant specific heats,

$$W_n = \dot{m}c_p \left[(T_3 - T_{4'}) - (T_{2'} - T_1)\right] \tag{7-28a}$$

or

$$W_n = \dot{m}c_p \left[(T_3 - T_4)\eta_e - \frac{T_2 - T_1}{\eta_c}\right] \tag{7-28b}$$

This equation can be written in terms of the initial temperature T_1, a chosen metallurgical limit T_3, and the compression and expansion efficiencies (above) to give

$$W_n = \dot{m}c_p T_1 \left\{\left[\eta_e \frac{T_3}{T_1} - \frac{(r_p)^{(\gamma-1)/\gamma}}{\eta_c}\right]\left[1 - \frac{1}{(r_p)^{(\gamma-1)/\gamma}}\right]\right\} \tag{7-28c}$$

The second quantity in the braces can be recognized as the efficiency of the corresponding ideal cycle, i.e., one having the same pressure ratio and using the same coolant. As in the case of the ideal cycle, the specific power of the nonideal cycle, $W_n/\dot{m}$, attains a maximum value at some optimum pressure ratio and is a direct function of the specific heat of the gas used.

The heat added in the cycle, Q_A, is given by

$$Q_A = \dot{m}c_p(T_3 - T_{2'}) = \dot{m}c_p \left[(T_3 - T_1) - T_1 \frac{(r_p)^{(\gamma-1)/\gamma} - 1}{\eta_c} \right] \quad (7\text{-}29)$$

The efficiency of the nonideal cycle can then be obtained by dividing Eq. 7-28c by Eq. 7-29. While the efficiency of the ideal cycle is independent of the cycle temperatures (except as they may affect γ) and increases asymptotically with r_p, the efficiency of the nonideal cycle is very much a function of the temperatures. It also assumes a maximum value at an optimum pressure ratio for each set of temperatures T_1 and T_3. The two optimum pressure ratios, for specific power and for efficiency, are not the same, necessitating a compromise in design.

Figure 7-11 shows the specific power, Btu/lb$_m$, and efficiency of the

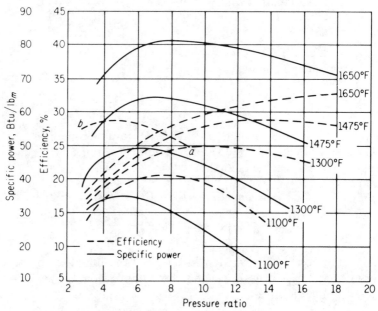

FIG. 7-11. Specific power and efficiency for nonideal Brayton cycle. Gas is air, $\eta_c = 0.84$, $\eta_e = 0.87$.

nonideal cycle calculated for constant initial temperature T_1, adiabatic compression and expansion efficiencies η_c and η_e, and various maximum temperatures T_3. The coolant is air. It can be seen that both the efficiency and specific power are strongly dependent on T_3; hence the necessity of operating with as high a reactor exit coolant temperature as possible. Also, it can be seen that the optimum pressure ratios increase with T_3 and that they are higher for efficiency than for power.

The Brayton cycle can be further modified by adding a *regenerator* in which the exhaust gases are used to preheat the compressed gas before entering the reactor or heat exchanger (Fig. 7-12). If the regenerator were 100 percent effective, the temperature of the gas entering the

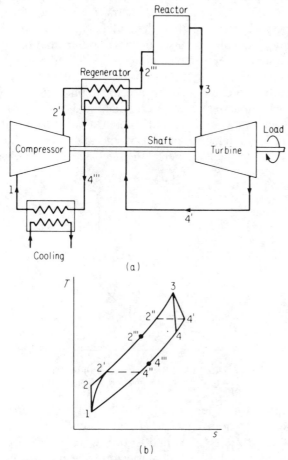

FIG. 7-12. Closed nonideal Brayton cycle with regeneration.

reactor would be raised from $T_{2'}$ to $T_{2''}$. The net work of the cycle would be maintained but the heat added would be reduced from $\dot{m}c_p$ $(T_3 - T_2)$ to $\dot{m}c_p(T_3 - T_{2''})$, with corresponding increase in cycle efficiency. Actually, the regenerator effectiveness is never 100 percent, and the compressed gases are heated instead to a lower temperature such as $T_{2'''}$. Regenerator effectiveness is defined as the ratio of the actual to theoretical amounts of heat transferred, that is, $\dot{m}(h_{2'''} - h_{2'})$ to $\dot{m}(h_{4'} - h_{4''})$, where point $4''$ is at the same temperature as point $2'$. For constant specific heats, the regenerator effectiveness is equal to $(T_{2'''} - T_{2'})/(T_{4'} - T_{2'})$. Figure 7-11 includes curve ab which represents the efficiency of the cycle with $T_3 = 1300°F$ and a regenerator of 75 percent effectiveness. It can be seen that considerable improvement occurs in efficiency at low pressure ratios. At pressure ratios higher than at a, the effect on efficiency is negative, because the exhaust gases become cooler than those after compression. The specific power is unaffected by the addition of a regenerator, except for the possibility of added pressure losses in the system.

Other modifications employ the use of a split compression (two or more compressors in series) with intercooling and split expansion (two or more turbines in series) with reheat. Such modifications are standard for large ground nonnuclear applications and are analyzed in more detail in most textbooks on heat power. Figure 7-13 shows a Brayton cycle

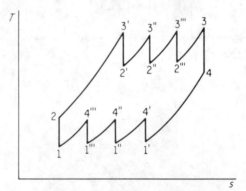

FIG. 7-13. Ideal Brayton cycle with several
stages of reheat and intercooling.

with several stages of reheat ($2'$-$3''$, $2''$-$3'''$, etc.) and intercooling ($4'$-$1''$, $4''$-$1'''$, etc.) and ideal expansions and compressions.

Regeneration may also be added to the above cycle. The efficiency of a plant operating on such a composite cycle may be calculated by considering all the work and heat addition terms, the various adiabatic efficiencies, and the regenerator effectiveness (if any). Note that, if the number of stages were large, and full regeneration is used, the Brayton cycle would approach an ideal (Ericsson) cycle where heat is added and rejected at constant temperature and which has the same efficiency as a Carnot cycle operating between the same temperature limits. The efficiency of the cycle thus approaches the ideal attainable between T_1 and T_3.

If all expanders have the same adiabatic efficiency η_e and the same pressure ratio r_{p_e} and if all compressors have the same adiabatic efficiency η_c and the same pressure ratio r_{p_c}, and if the regenerator effectiveness is η_R, the following general relationships apply:

$$\frac{W_n}{\dot{m}c_p} = T_3\eta_e(n_e + 1)\left[1 - \frac{1}{(r_{p_e})^{(\gamma-1)/\gamma}}\right] - T_1\frac{n_c + 1}{\eta_c}[(r_{p_c})^{(\gamma-1)/\gamma} - 1] \quad (7\text{-}30)$$

$$\frac{Q_A}{\dot{m}c_p} = T_3\left\{(n_e + 1) - (n_e + \eta_R)\left[1 - \eta_e\left(1 - \frac{1}{(r_{p_e})^{(\gamma-1)/\gamma}}\right)\right]\right\}$$

$$- T_1(1 - \eta_R)\left\{1 + \frac{1}{\eta_c}[(r_{p_c})^{(\gamma-1)/\gamma} - 1]\right\} \quad (7\text{-}31)$$

where n_e and n_c are the number of reheat and intercooling stages, respectively. The efficiency of the cycle may now be obtained by dividing Eq. 7-30 by Eq. 7-31. The greater the number of reheat and intercooling stages that there are, the higher the efficiency. However, this is attained at the cost of the capital investment and size of the plant. The design of the plant should be optimized, with consideration given to capital versus operating (fuel, etc.) expenses and to size. If size should be kept to a minimum, an improvement in efficiency may be achieved by increasing the pressure ratio of the cycle. This minimizes temperature variation during heat addition (see Fig. 7-8), but may also reduce the work (see Fig. 7-11). Remember that, in nuclear power plants, capital costs may exceed by far the operating and fuel costs and thus operation at or near maximum power rather than maximum efficiency may be desirable.

7-7. OTHER COMPARISONS OF GASEOUS COOLANTS

Cost and Availability

Air is, of course, free and abundant everywhere on earth. Other possible gas coolants vary in cost and availability, as shown in Table 7-3. An important part of the cost of any gas is that of purifying it to the degree desired for reactor work.

TABLE 7-3
Cost and Availability of Gaseous Coolants

Gas	Cost, $ /ft³ at 14.7 psia and 32°F	Availability
Hydrogen	0.10	Good
Helium	0.015	U.S., USSR, Italy
Carbon dioxide	0.003	Good
Nitrogen	0.008	Good
Argon	0.074	0.93 % by vol. in air
Neon	42.500	0.0018 % by vol. in air

Chemical Activity

To avoid chemical reaction with fuel elements and reactor structural members and to allow the use of conventional materials, it is best to use inert gases such as helium, argon, neon, etc. Of these, only helium is attractive, because argon suffers from induced radioactivity (Sec. 7-3) and neon is too costly.

Air, of course, contains oxygen and is corrosive. Its effects on bare fuels is principally in forming oxides. Above 350°C, uranium shows rapid reaction rates with air [52]. A penetration rate of air in bare uranium of 3.7×10^{-3} cm/hr has been reported. The values are less than a thousandth as much for CO_2 but are much higher for hydrogen. Results of experiments on cold-rolled thorium sheets [53] showed that at 500°C the mass increased by about 0.36 mg per square centimeter of exposed area per hour, with the rate increasing rapidly with temperature. Hydrogen is usually troublesome as a coolant since, even though the fuel may be clad, it is capable of diffusing through most cladding materials. When it gets to the uranium itself, it reacts with it, forming uranium hydride which causes swelling and subsequent clad rupture. Uranium hydride is an easily combustible material, making its handling difficult once it gets out of the ruptured element.

Carbon dioxide, used to a large extent as a coolant for thermal graphite-moderated reactors (the Calder Hall type, below), reacts at high temperature with carbon according to the chemical equation

$$CO_2 + C \rightleftharpoons 2CO \qquad (7\text{-}32)$$

The forward reaction causes the CO content of CO_2 to increase as the coolant is recirculated through the reactor. The reverse reaction, causing deposition of carbon, occurs at points of lower temperature. Carbon dioxide has also been found to react with stainless steel at very high temperatures.

Other gases such as methane (CH_4) and ammonia (NH_3) have enough promising characteristics to be considered as future reactor coolants. However, not all their characteristics, including their chemical stability under reactor radiations, have been firmly established.

Sealing and Containment

Gaseous coolants present a special problem is sealing and containment within pressure vessels, piping, etc. The mass-flow rate through fissures, fittings, and other possible openings is, other things being equal, inversely proportional to the molecular mass of the gas. Thus, hydrogen and helium are the most troublesome in this respect.

Effect on Mechanical Design

It was previously indicated that the type of gas influences the optimum pressure ratio (for power) of a direct cycle through its γ (Eq. 7-23). In this respect monatomic gases (helium) operate at much lower pressure ratios than other gases. Cycles using them may thus have lower pressures in the reactor, with consequent lighter and less expensive pressure vessels, or the gas may be pressurized (in a closed cycle) so that overall reduction in size and mass may be achieved.

In turbomachinery, the size and mass advantage of a low overall pressure ratio may largely be nullified if a large number of stages is required. The number of stages is an inverse function of the work (and pressure ratio) allowable per stage. This is a function of two things: the blade-tip velocity and the gas velocity. The former (a function of rpm and blade-tip diameter) is limited by centrifugal stresses to maximum values around 1,500 to 1,600 fps. The latter, the gas velocity, affects blade velocity (for maximum power per stage, blade velocity is a direct function of gas velocity). The gas velocity is limited by the blade-tip velocity as above or to values below sonic velocity, i.e., to Mach numbers less than unity, whichever is lower. The sonic velocity in a gas is given by

$$a = \left(\frac{\gamma}{M} \bar{R} g_c T \right)^{1/2} \tag{7-33}$$

where a = speed of sound in gas
$\quad M$ = molecular mass of gas
$\quad \bar{R}$ = universal gas constant
$\quad g_c$ = conversion factor
$\quad T$ = absolute temperature of gas

It can be seen that the speed of sound is directly proportional to the square root of γ. Thus, other things being equal, a monatomic gas has a slightly higher sonic velocity than a diatomic or triatomic gas. However, the effect of M is more pronounced because of the much wider variation in its value between different gases. The speed of sound is inversely proportional to the square root of M and thus is very high for such gases as hydrogen and helium. These gases are therefore limited by centrifugal stress considerations, and they operate with very low Mach numbers as compared with, say, air.

The work done by the gas in a single impulse stage is equal to the enthalpy drop (or rise) in that stage. The gas velocity V entering the turbine rotor stage is equal to that leaving the fixed blading (or nozzles) of that stage. The kinetic energy of that gas is therefore equal to the enthalpy drop of the stage (ignoring the gas velocity at the high-temperature end), or

$$\frac{V^2}{2g_cJ} = h_1 - h_2 = c_p(T_1 - T_2) \tag{7-34}$$

where the subscripts 1 and 2 denote inlet and exit for the stage, and V is the gas velocity impinging on the moving blades. Since, ideally

$$\frac{T_2}{T_1} = \left(\frac{p_2}{p_1}\right)^{\frac{\gamma-1}{\gamma}} = r_p^{\frac{\gamma-1}{\gamma}} \tag{7-35}$$

where r_p is the pressure ratio for the stage, then

$$r_p = \left(1 - \frac{V^2}{2g_cJc_pT_1}\right)^{\frac{\gamma}{1-\gamma}} \tag{7-36}$$

Thus where V is limited, the pressure ratio per stage is a function of c_p and $\gamma/(1-\gamma)$ for the gas, as well as the temperature. Note (Table 7-2) that helium has a higher value of c_p than all other gases (with the exception of hydrogen) but a lower value of $\gamma/(\gamma-1)$. The net effect is that helium can operate only at much lower pressure ratios *per stage* than other reactor coolants. Thus, while the overall pressure ratio is lower in the case of helium, this advantage may be nullified because the number of stages will have to be increased, increasing the cost and complexity of the turbomachinery. This difficulty may, however, be lessened by low-temperature operation (in a compressor with low pressure ratio) and by pushing tip speed to the allowable stress limits.

7-8. THE EFFECT OF FUEL-ELEMENT TYPE ON GAS-COOLED-REACTOR DESIGN

Probably the most outstanding feature in favor of gas cooling is the ability of gases to attain high temperatures without high pressurization and without appreciable change in physical characteristics. High gas temperatures mean increased cycle efficiencies and reduction in operating and capital costs. In gas-cooled reactors, the limitation on maximum operating temperatures is therefore only a metallurgical one, imposed by core materials, mainly the fuel or its cladding. Recall that in water-cooled reactors the limitation is that due to burnout [2].

In discussing the effects of fuel and cladding choice on reactor design, it will be helpful to reiterate the progress in the gas-cooled power-reactor field.

The first large gas-cooled power reactor, at Calder Hall, England, is an outgrowth of one built at Windscale and originally used for the production of plutonium. Because of the lack of large isotope-separation facilities at the time, natural uranium was used as the fuel. For similar reasons, conventional materials and techniques were employed in building it. Because the problem of conserving neutrons in a natural-uranium-fueled reactor is a difficult one, these materials had to have extremely low neutron absorption cross sections. Of the many materials suitable for cladding purposes, only magnesium satisfied the two requirements of conventionality and low neutron absorption and therefore was used as the cladding material. Beryllium, which also has low neutron absorption cross sections, was at the time an unknown quantity as to its machinability, properties, etc. However, it is now being considered for cladding purposes.

Magnesium is unfortunately not a high-temperature material. Its melting point is about 1200°F, and its maximum working temperature is about 850°F. Thus, in Calder Hall, maximum cladding temperatures were limited to about 750°F. When the maximum cladding temperature is so set, the maximum operating fuel temperature is determined by the fuel-element size and the specific power (kw/kg_m) at the point of interest. Natural uranium requires the use of a wide lattice pitch to improve the resonance escape probability. This means that relatively large-diameter fuel rods must be used. (Hinkley Point, one of the later Calder Hall-type plants, uses 1 1/8-in.-diameter fuel rods as compared with a hollow rod 0.75 OD and 0.32 ID for a partially enriched graphite-moderated fuel; see Table 7-4.) A large-diameter fuel element means a large temperature drop in the fuel. However, the maximum fuel temperature, at the center of the fuel rod, must be limited in the case of metallic uranium to values much lower than the α-β phase change and is determined partly by the desired fuel burnup. (In Hinkley Point the maximum fuel temperature is 775°F, and fuel burnup is 3,000 Mw-day/ton.) Thus a relatively low specific power, kilowatt per kilogram of fuel, and consequently a low core power density, kilowatt per liter, are necessitated in this type of reactor.

A relatively low limit to power density is also necessary to ensure reasonably high gas temperatures. Note that, for a fixed cladding temperature, large power density would require low gas temperatures (power density proportional to temperature difference between cladding and coolant). High gas temperatures can, however, be attained if the gas-coolant mass-flow rate is increased by pressurization (with consequent increase in cost and mass of the large pressure vessel required for such reactors) or by increased gas velocity (with consequent increase in pumping losses). Another point to be considered is the limitation due to the CO_2-C reaction (Sec. 7-7).

It can be seen that in the Calder Hall type of reactor the gas temperatures are limited, and the resulting steam conditions are only slightly better than those attainable in pressurized- or boiling-water reactors. The net plant thermal efficiencies are, however, less because of the larger pumping losses.

In later investigations, in the United States [54] and other countries, the use of partially enriched fuels was recommended for gas-cooled graphite-moderated reactors. With enrichment, ceramic fuels become feasible (neutron economy in natural-uranium-fueled and graphite-moderated reactors precludes the use of any but metallic uranium, since the added oxygen or carbon atoms in ceramic fuels lower the percentage of U^{235} in the fuel). Ceramic fuels (UO_2, UC, etc.) are capable of operating at much higher temperatures.

With partially enriched fuels, the problem of neutron economy ceases to become determining, and therefore higher-temperature materials with poorer nuclear characteristics (higher neutron absorption cross sections), such as the stainless steels and zirconium, may be used as cladding and structural materials. Also, the use of smaller fuel elements becomes possible with enrichment, thus further reducing the magnitude of temperature drop within the fuel element. With this and the increased cladding temperatures, higher power densities are possible. Further, with enrichment, the reactor vessel is smaller and more capable of withstanding higher pressures. This allows increased coolant mass-flow rate through pressurization, further contributing to increased power density.

It can be seen from the above that the choice of fuel type and cladding material has an important effect on reactor design and behavior. Where facilities for isotope separation and production of enriched fuels are limited, beryllium will have to be used for cladding if high temperatures are to be attained. In the United Kingdom, for example, beryllium is being investigated as to properties, fabrication, etc., for that purpose.

Table 7-4 is a comparison between some design characteristics of the Hinkley Point A power plant (one of the later in the Calder Hall series) and three optimum United States designs [54]. Other particulars of these and other gas-cooled power plants will be given in Chapter 8.

More recently, a reactor concept with heavy-water, instead of graphite, moderation has been introduced [48, 49]. In such a reactor, neutron economy is improved, and narrower lattices and smaller fuel elements can be used. An obvious problem here is the separation of the gas coolant and the liquid moderator. Another is the avoidance of moderator boiling by cooling it sufficiently to low temperatures or by pressurizing it. Because heavy-water-moderated reactors are relatively large, pressurization imposes a penalty on reactor-vessel size and cost. A light, relatively inexpensive reactor vessel can be used by sufficiently cooling the moderator to low enough temperatures, as indicated above, and by surrounding the fuel by pressure tubes through which the normally pressurized gas coolant passes. These pressure tubes may be made of a material of low neutron absorption cross section such as zirconium, if necessary. They must, of course, be sealed and may have double walls with a stagnant fluid in the narrow annular space to act as thermal insulation between coolant and moderator.

Another reactor concept in which exit gas temperatures of 1500°F and higher, and relatively high power densities, are possible involves the elimination of cladding altogether and the use of ceramic fuels (UO_2, UC, etc.) and ceramic structural materials (graphite, beryllium oxide, etc.). In such a reactor, the problem of leakage of the gas coolant, which would be contaminated with fission gases (because of lack of cladding),

TABLE 7-4

Some Characteristics of Gas-cooled Reactors as Affected by the Choice of Fuel

	British, Hinkley Point A	United States, natural uranium[†] (Kaiser-ACF)	United States, enriched (Kaiser-ACF)[†]	United States, enriched (ORNL)[‡]
Rated net output, Mw(e)	250	220	215	225
Coolant	CO_2	CO_2	CO_2	He
Fuel material	U metal	U metal	UO_2	UO_2
Enrichment atomic percent	0.71 (nat.)	0.71 (nat.)	2.5	2.0
Cladding material	Mg alloy	Mg alloy	304 ss	304 ss
Fuel-element geometry	Rod	Hollow rod	Hollow-sintered pellet	Hollow-sintered pellet
Dimensions, OD by ID, in	1.125 by 0	1.16 by 0.375	0.750 by 0.32	0.75 by 0.32
Cladding geometry	Helically finned	Helically finned	Finless tube	Capsule
Average burnup, Mw-day/ton	3,000	3,000	10,000	7,400
Specific power, Mw/ton (metric)	2.65	2.55	7.9	5.1
Core:				
Diameter, ft	49	50	20	20
Height, ft	25	29	26	30
Coolant pressure, psia	180	275	370	300
Coolant exit temperature, °F	700	800	1,000	1,000
Maximum steam conditions:				
psia	650	1,450	2,400	950
°F	685	750	950	950
Plant net thermal efficiency, percent	25.5	31.4	35.8	32.1
Plant cost, $/kw(e) net	.....	640	490	420
Power costs, mills/kwhr:				
Fixed charges	.....	12.8	9.8	8.8
Operation-maintenance	.....	1.2	1.1	1.1
Fuel	.....	2.2	2.5	2.6
Total	.....	16.2	13.4	12.1

[†] Kaiser Engineers, prime contractor; ACF Industries, Inc., subcontractor.
[‡] Oak Ridge National Laboratory.

must be solved. Thus either an intermediate coolant loop or continuous gas removal and purification, or both, must be provided. Unclad fuel elements (or moderators), especially those operating at high temperatures, require the use of inert-gas coolants, such as helium, to minimize chemical reactions (Sec. 7-7).

Because of the absence of strong neutron absorbers in such a reactor, breeding becomes feasible. Since most gas-cooled reactors are of the thermal type, only the $Th^{232} \longrightarrow U^{233}$ breeding series can be incorporated in them. (Recall that the $U^{238} \longrightarrow U^{239}$ breeding reaction can be used only with fast reactor whereas the $Th^{232} \longrightarrow U^{233}$ can be used with either fast or thermal reactors. (Also see Sec. 9-4.)

In one reference design, that of the *pebble-bed* reactor (Sec. 8-11), fuel in the form of spherical pellets 1.5 in. in diameter is composed of a UO_2-ThO_2-graphite mixture in the approximate mass ratio 1:11:108, respectively. Thus the fuel is in mixture with the moderator. The reactor consists of a pressure vessel containing a core and blanket (also of 1.5-in.-diameter pellets). Helium at 965 psia acts as coolant and passes through the voids between the spheres, cools both the core and blanket in two parallel streams, and has a combined reactor-exit temperature of 1250°F. The maximum fuel center and surface temperatures are 2440°F, respectively, and the power density is 27.7 kw/liter.

Gas-cooled fuel in *porous* form has been suggested [55]. Because of the intimate contact between coolant and large surface area of the porous fuel, good heat-transfer results. This is manifested in lower solid temperatures than surface cooling for the same power density; or in much higher power densities for the same temperatures. Design of the porous fuel to retain, or release of fission gases into the coolant can be made. The latter case results in higher fuel burnup but in the necessity to trap the fission gases from the coolant at some point in the plant.

Other designs for gas-cooled fuel in fibrous form [56] and in spiral-shaped plate [57] have been suggested.

PROBLEMS

7-1. A gas-cooled-reactor power plant uses CO_2 as coolant. The coolant spends half the time within the core and half in the rest of the primary system. The average neutron flux in the core is 10^{13}. The average coolant temperature and pressure are 500°F and 10 atmospheres. Calculate the activity of the coolant in curies/liter after (a) one week, and (b) one year of steady operation at the above flux.

7-2. List the ratios of pumping work to heat transported by the gases of Table 7-1 at low temperatures if the velocity in all cases is fixed by rod vibrational limitations and the logarithmic mean temperature difference between gas and surface is the same.

7-3. A direct closed cycle gas-cooled-reactor power plant is considered for aircraft nuclear propulsion of the turboprop variety. The plant is required to deliver 40,000 hp to the propellers. The coolant, helium, leaves the reactor at 1340°F and enters the compressor at 140°F. Assuming the cycle to be ideal and designed at optimum conditions for maximum power, find (a) the necessary coolant mass flow rate in lb_m/hr, (b) the cycle thermal efficiency, and (c) the reactor output in Mw(t).

7-4. A land-based indirect closed cycle, gas turbine reactor power plant generates 200 Mw(e). The working fluid is helium. It leaves the heat exchanger at 1240°F and enters the compressor at 140°F. The pressure ratio across the turbine is 2.6 and across the compressor 2.7. Both turbine and compressor have adiabatic efficiencies of 0.88. The turbine-generator efficiency is 0.85. Assuming no heat losses, find (a) the thermal efficiency of the cycle, (b) the helium mass-flow rate, lb_m/hr, (c) the necessary reactor output, Mw(t), and (d) the overall plant efficiency.

7-5. Repeat Prob. 7-4 but add a regenerator to the working cycle with a regenerator effectiveness of 0.80. Compare results.

7-6. A conceptual design for nuclear rocket propulsion (Resler and Rott, *ARS Journal,* November 1960) uses a closed regenerative gas-turbine cycle with pressurized helium as working fluid. The heat source is one region of a two-region high-temperature reactor. Instead of rejecting heat to space by radiation, heat is dumped into saturated liquid hydrogen flowing from a storage tank and receiving only latent heat of vaporization. The hydrogen vapor is then super-heated, first in the other reactor region, and then by dumping the net work of the cycle (electrically) into it. Gaseous hydrogen now enters the rocket chamber and exhausts through a nozzle to space. Hydrogen is used as propellent because of its high specific impulse (thrust per unit propellent mass flow rate, $lb_f sec/lb_m$). (a) Draw the flow diagram of the entire system. (b) Show that for an ideal system, the temperature of hydrogen entering the nozzle, T_n, is given by:

$$T_n = T_R + \frac{\lambda}{c_p} \left[\frac{T_R \left(1 - \dfrac{1}{r_p^{\gamma - 1/\gamma}} \right) - T_s (r_p^{\gamma - 1/\gamma} - 1)}{T_s (r_p^{\gamma - 1/\gamma} - 1) + \Delta T} \right]$$

where T_R and T_s are the temperatures of either gas leaving the reactor and the sink respectively, λ the latent heat of vaporization and c_p the specific heat at constant pressure of H_2, r_p the pressure ratio of both turbine and compressor, γ the specific heat ratio of He, and ΔT the temperature difference in the counterflow regenerator.

7-7. It is desired to compare the designs of helium and air turbines. Assume that both operate in an ideal Brayton cycle having initial and maximum temperatures of 500 and 1750°F respectively. The two cycles are operating at their respective optimum pressure ratios. Also assume that the maximum gas velocity in all stages of both turbines is 3,000 fps. Estimate the number of turbine stages required in each case.

chapter 8

Gas-Cooled Reactor Power Plants

8-1. INTRODUCTION

In the preceding chapter different cycles associated with gas-cooled reactors were presented and several characteristics of gas coolants and gas-cooled reactors were discussed. This chapter will be devoted to characteristics and descriptions of gas-cooled reactors and power plants using them. The majority of gas-cooled power plants built and being contemplated in the world today are of the indirect-closed-cycle gas-steam type using thermal reactors. The greater part of the material in this chapter will therefore deal with this type. Gas-cooled fast-breeder reactor power plants are presented in Chapter 10.

8-2. ANALYSIS OF THE GAS-STEAM SYSTEM: THE SIMPLE CYCLE

A simple gas-cooled indirect-closed-cycle power plant in which the

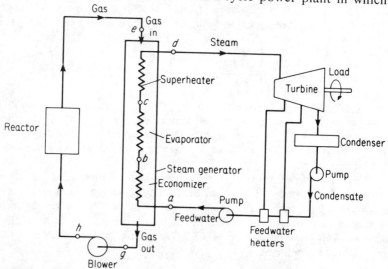

FIG. 8-1. Schematic of a simple-cycle gas-steam-reactor power plant.

working fluid is water-steam and the prime mover is a steam turbine or engine is shown diagrammatically in Fig. 8-1.　The processes occurring in the steam generator are represented by a temperature-enthalpy diagram (fig. 8-2).　This diagram contains two lines, one (*efg*) representing the

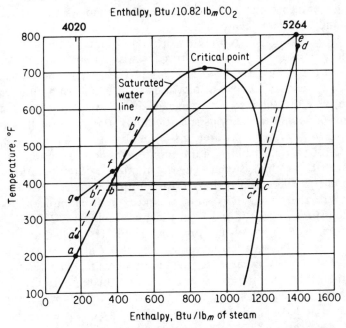

FIG. 8-2. Temperature-enthalpy diagram of a gas-steam heat exchanger in simple cycle.

gas and the other (*abcd*) representing water and steam.　The diagram is drawn for 1 lb_m of water (and steam), as can be verified from the steam tables, and for m lb_m of gas circulated for each pound mass of water.

In this cycle the primary circuit consists of gas entering the reactor at h and leaving it heated at e.　It gives up the heat received in the reactor to a steam generator, leaving it at g.　It is then pumped back to h by a blower, which overcomes the pressure losses in the primary circuit.

The secondary, or working-fluid, circuit consists of feedwater entering the steam generator at a in a counterflow fashion at a temperature t_a less than the saturation temperature corresponding to the generator pressure. It thus enters first an economizer that heats it to the saturation temperature t_b.　Saturated water now enters an evaporator where it receives latent heat of vaporization at substantially constant temperature and pressure and is transformed into saturated steam at c.　Because of the high temperatures possible with gas coolants, the resulting steam can be superheated. It thus enters a superheater, where its temperature is raised to t_d. Line *abcd* represents constant-pressure heating of the water-steam system. (Actually there is some drop in pressure throughout this line.) After leaving the steam generator, the steam completes the working fluid circuit by entering the turbine, condenser, condensate and feedwater pumps, and feedwater heaters.

Line *efg* of Fig. 8-2 represents the state points of the gas as it cools down the steam generator from *e* to *g*. This is approximately a straight line since, for a gas $\Delta H = mc_p \Delta T$, where the specific heat at constant pressure c_p is either a constant or varies slightly and smoothly with temperature over the temperature range in question. Taking into account the value of *m* and the various specific heats, the slope of the gas line is less than that of either the water $(a - b)$ or steam $(c - d)$ (Sec. 2-7).

The process taking place in the economizer determines the maximum steam pressure of the system. The economizer exit temperature t_b at the pinch point *fb* (Sec. 2-7) is the saturation temperature. The higher this temperature, the higher the system pressure. For example, if t_b is 400°F, the system pressure (as found from the steam tables) is 247 psia. However, if t_b is 375°F, the system pressure is only 184.3 psia, and the pinch point occurs earlier in the diagram. The effect of feedwater heating in this system becomes apparent: If the incoming water is heated to a higher degree than that given above, say to $t_{a'}$, and if the temperature difference at the pinch point remains unchanged, the pinch point occurs earlier. Less heat is added in the economizer, and the system pressure is reduced from that corresponding to t_b to that corresponding to $t_{b'}$. (See the dotted line $a'b'c'$ in Fig. 8-2.)

Thus while feedwater heating is beneficial from the standpoint of steam-cycle efficiency, it is associated in this case with lower system pressures. It should be recalled here that steam-cycle efficiencies increase with turbine inlet steam pressure as well as temperature. The important thing to remember here, then, is that thermodynamic advantages that are associated with higher feedwater temperatures are counterbalanced by the resulting lower steam pressures. Wootton, Taylor, and Worley [58] showed that, for each reactor inlet gas temperature, there occurs a feedwater temperature at which the steam-cycle efficiency is highest (Fig. 8-3).

We shall now discuss the effect of the temperature difference at the pinch point. While the overall enthalpy drops of the two lines of Fig. 8-2 are equal, the difference in height (i.e., in temperature) between the two could be varied without affecting this requisite of equal enthalpy drop. The average temperature difference affects the overall efficiency of the plant, the size of the steam generator, and the power consumed by the

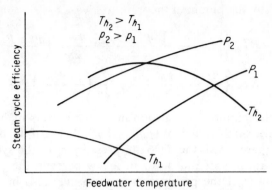

FIG. 8-3.   Efficiency of steam cycle in simple-cycle
gas-steam plant.   (Ref. 58.)

gas blower.   The larger this temperature difference, the greater the lost
work (work could be accomplished by an engine operating between the
high and low temperatures).   Also the larger the temperature difference,
the smaller, more compact, and lighter the heat exchanger, but the higher
the temperature of the gas at the exit of the steam generator, t_g.   A high
t_g means greater blower work.

The dependence of blower work W_b/m Btu per pound mass of gas on
gas temperature can be shown by the following analysis.   In the case of
irreversible adiabatic compression, W_b is given by

$$\frac{W_b}{\dot{m}} = \frac{\gamma}{\gamma - 1} RT_g \left[\left(\frac{p_h}{p_g} \right)^{(n-1)/n} - 1 \right] \qquad (8\text{-}1)$$

where p_g, p_h = absolute pressures across blower, lb_f/ft^2

n = polytropic exponent of gas ($n = \gamma$ in case of isentropic
compression), dimensionless

R = gas constant, Btu/lb_m °R

$\dot{m}$ = mass flow rate of gas, lb_m/hr

γ = ratio of specific heats of gas, dimensionless

T_g = absolute temperature of gas, °R

Equation 8-1 is obtained by steps similar to those leading to Eq. 7-20
except that n is used instead of γ and $R[\gamma/(1 - \gamma)]$ is used instead of its
equivalent, c_p.

The difference between the two pressures p_g and p_h is equal to the
sum of the gas-pressure drops in the reactor, heat exchanger, and piping.
Equation 8-1 can be written in the form

$$\frac{W_b}{\dot{m}} = \frac{\gamma}{\gamma - 1} RT_g \left[\left(1 + \frac{p_h - p_g}{p_g} \right)^{(n-1)/n} - 1 \right]$$

and expanded by the binomial theorem to give

$$
\frac{W_b}{\dot{m}} = \frac{\gamma}{\gamma - 1} RT_g \left\{ \left[1 + \frac{n-1}{n} \frac{p_h - p_g}{p_g} - \frac{n-1}{2n^2} \left(\frac{p_h - p_g}{p_g} \right)^2 + \cdots \right] - 1 \right\}
$$

$$
= \frac{\gamma}{\gamma - 1} RT_g \left\{ \frac{n-1}{n} \left[\frac{p_h - p_g}{p_g} - \frac{1}{2n} \left(\frac{p_h - p_g}{p_g} \right)^2 + \cdots \right] \right\} \qquad (8\text{-}2)
$$

Members of the binomial expansion involving powers higher than 2 can be neglected for small values of $(p_h - p_g)/p_g$, that is, if the pressure drops are small compared with the absolute pressure p_g of the system. If this pressure-drop ratio is of the order of a few percent, the term involving the second power of the pressure-drop ratio may also be dropped with little ensuing error, resulting in the simplified expression

$$
\frac{W_b}{\dot{m}} = \frac{\gamma}{\gamma - 1} \frac{n-1}{n} RT_g \frac{p_h - p_g}{p_g} \qquad (8\text{-}3)
$$

It can be seen that, for the same pressures and pressure-drop ratios, the work involved in the gas blower is a direct function of the absolute temperature T_g at the inlet of the blower, i.e., at the exit of the steam generator. The increased blower work associated with higher values of t_g is largely absorbed in the gas itself, except, of course, for the motor electrical losses and blower mechanical and heat losses. The work absorbed by the gas and represented by Eq. 8-2 shows up in an increase in gas enthalpy, or temperature across the blower (10 to 20°F), and should be taken into account in any heat-balance calculation. (Thus one may say that increased gas temperatures only increase blower "losses.") Because of the large fraction of gross plant output used by the gas blowers, a high overall plant efficiency is attained, therefore, with low values of temperature difference between the two lines of Fig. 8-2. As indicated before, the minimum gas temperatures are determined by the necessity of maintaining a sufficiently high gas temperature at the pinch point to allow adequate heat-transfer rates for a given-size steam generator.

The requirements of a small, compact, and thus less costly steam generator and of high power-plant efficiency are therefore conflicting. A compromise based on relative capital and operating costs is necessary. In practice, the approach temperature differences, both at the pinch point and at the gas inlet end *ed*, are chosen to be in the neighborhood of 20 to 30°F for low-temperature to 100 to 150°F for high-temperature gas-cooled reactor power plants.

It can also be easily seen, by studying Fig. 8-2, that, as the temperature of the gas leaving the steam generator is decreased, the steam pressure decreases for the same feedwater temperature, approach temperature differences, and gas temperature at the inlet to the steam generator.

It can also be shown that, if the feedwater temperature were to be fixed (say at the optimum values shown by Fig. 8-3), raising the temperature of the gas leaving the steam generator, t_g, results in increased steam pressures and increased steam-cycle thermal efficiency (Fig. 8-4). Also, raising t_g, although increasing blower losses, raises plant overall thermal

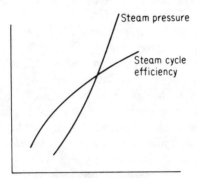

FIG. 8-4. Variation of steam-cycle efficiency and steam pressure with temperature of gas leaving steam generator in simple-cycle plant.

efficiency. This is defined as the net plant electrical output divided by the reactor thermal output. Since increased t_g means increased t_h, the temperature of the gas entering the reactor in this type of plant has therefore as much bearing on plant efficiency as does the temperature of the gas leaving the reactor [58].

Figure 8-5 shows the gross and net outputs and auxiliaries input of a

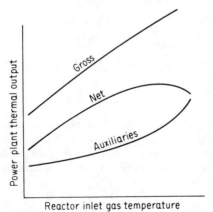

FIG. 8-5. Simple-cycle gas-steam-plant efficiency.

gas-cooled simple-cycle power plant as a function of the reactor inlet gas temperature for a fixed reactor thermal output. For fixed mass-flow rate of the gas, increased reactor inlet gas temperatures also mean increased reactor outlet gas temperatures, although the increase in the latter is much less than that in the former. Higher reactor outlet gas temperatures mean higher steam superheat temperatures and consequently higher overall plant efficiencies. The net curve in Fig. 8-5 deviates from the gross line more and more at higher reactor inlet gas temperatures because of the rising blower losses. Note that Eq. 8-3 predicts that blower work is directly proportional to both the gas absolute temperature and to the pressure drop. The latter, however, is not independent of temperature but rises with it because of the increased gas specific volume and the consequent increase in gas velocities. Thus blower work and consequently blower losses increase rapidly at high gas temperatures. Thus there is an optimum reactor inlet gas temperature at which a maximum net work is attained, as shown by Fig. 8-5.

Final choice of the operating conditions of the plant are determined not only by thermodynamic considerations (outlined above) but also by other considerations, such as increased capital cost (as, for example, when using very high gas temperatures) and of reactor nuclear design (for example, gas-flow-channel-configuration effects on reactor physics and size as well as gas temperature rise and pressure losses).

Example 8-1. A graphite-moderated CO_2-cooled reactor is utilized in a single-pressure cycle under the following operating conditions:

Temperature of gas leaving reactor, °F 800
Temperature of gas entering reactor, °F . . . 350
Gas pressure, psia 150
Feedwater temperature, °F 200

Find the steam pressure and temperature leaving the steam generator if the temperature differences are 30°F at the pinch point and 20°F at the superheat point.

Solution. The temperature of the steam leaving the generator is given by (refer to Fig. 8-2)

$$t_d = 800 - 20 = 780°F$$

At this point, assume the steam pressure to be 250 psia to obtain the enthalpy of the steam from the superheat steam tables. (Note that the enthalpy changes little with pressure in that range. Consequently a large error in assumed system pressure results in little error in the solution.) Thus at 250 psia and 780°F, the enthalpy of the steam is

$$h_d = 1412.4 \text{ Btu}/lb_m$$

The enthalpy of the feedwater can be obtained from the saturated steam tables at 200°F. (The error due to neglecting the enthalpy of the compressed liquid is small.) Thus

$$h_a = 167.8 \; \text{Btu}/\text{lb}_m$$

The change in enthalpy in the water-steam system is

$$(\Delta h)_{\text{H}_2\text{O}} = h_d - h_a = 1244.6 \; \text{Btu}/\text{lb}_m$$

The enthalpies of the incoming and outgoing CO_2 gas at 800 and 350°F, respectively, and 150 psia are (see Table B-10, Appendix B)

$$h_e = 486.5 \; \text{Btu}/\text{lb}_m$$
$$h_g = 371.5 \; \text{Btu}/\text{lb}_m$$

Thus

$$(\Delta h)_{\text{CO}_2} = h_e - h_g = 115.0 \; \text{Btu}/\text{lb}_m$$

For zero heat losses to the outside, $(\Delta H)_{\text{H}_2\text{O}} = (\Delta H)_{\text{CO}_2}$. Consequently the ratio of the mass-flow rates of CO_2 and H_2O should be equal to

$$\frac{1244.6}{115.0} = 10.82 \; \text{lb}_m \; CO_2/\text{lb}_m \; H_2O$$

A temperature-enthalpy curve is now plotted (see Fig. 8-2). The abscissa has two scales for the two fluids. The lower scale is in $\text{Btu}/\text{lb}_m \, H_2O$. The upper scale is conveniently in $\text{Btu}/10.82 \; \text{lb}_m$ and should be shifted to the left so that points e and g fall directly on top of points d and a, respectively.

Next, we plot the enthalpy curve abb'' of the liquid, beginning with a and increasing in temperature. Point b, the limit of the economizer process, is determined by the temperature approach at the pinch point which is 30°F (vertically upward). This determines the evaporation temperature and pressure. Thus

$$t_b = t_c = 399°\text{F}$$

and

$$p_b = p_c = 240 \; \text{psia}$$

Point c is determined by the enthalpy of saturated steam at 240 psia, which is $h_c = 1200.6 \; \text{Btu}/\text{lb}_m$. Straight line cd is then drawn from the superheat steam tables. At 240 psia and 780°F, $h_d = 1412.8 \; \text{Btu}/\text{lb}_m$, compared with 1412.4 Btu/lb_m for the previously assumed pressure of 250 psia. If the pressure thus found is markedly different from the one assumed, the new h_d may be sufficiently different to warrant recalculation.

8-3. ANALYSIS OF THE GAS-STEAM SYSTEM: THE DUAL-PRESSURE CYCLE

As has been shown, it is not possible in a low-temperature simple cycle to produce high-pressure steam in the steam generator and yet have

the gas leaving that generator cool enough for economical blower operation. The dual-pressure cycle, shown diagrammatically in Fig. 8-6, is
designed to offset this deficiency. The gas and steam processes of the
dual-pressure cycle are represented in Fig. 8-7.

The gas circuit is the same simple one used in the simple cycle wherein
the gas is heated in the reactor and goes through the steam generator,
the gas blower, and back to the reactor.

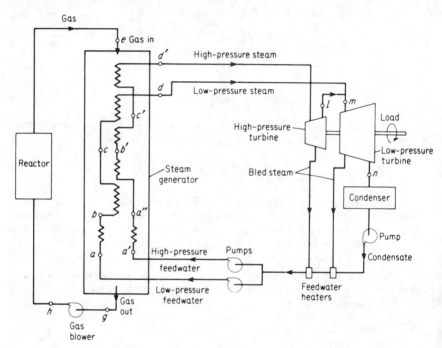

FIG. 8-6. Schematic of a dual-pressure gas-steam plant.

The steam cycle has two branches. Feedwater is heated in feedwater
heaters and then separates into the two branches where it is pumped by
two feedwater pumps to two different pressures. Except for a slight
difference in heating in the two pumps, the temperature of the feedwater
entering the steam generator is equal in both branches. The low-pressure water is further heated in the economizer from a to b, evaporated
at practically constant pressure and temperature from b to c, superheated
in the superheater from c to d, and then fed to a low-pressure turbine
stage. The high-pressure water is heated in a second economizer from
a' to a'' (subjected to the same downstream gas as the low-pressure water
economizer). This fails to heat this high-pressure water to its saturation
temperature. Thus it is further heated in a third economizer from a''

to *b'*, as shown.   This water is then evaporated from *b'* to *c'* and superheated from *c'* to *d'* by the incoming gas, after which it is fed to the turbine
inlet.

Figure 8-7 shows the paths of the different processes in the steam

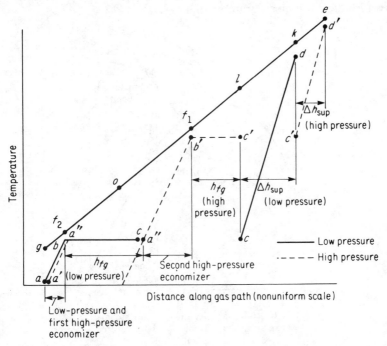

FIG. 8-7.   Temperature-enthalpy diagram of dual-pressure gas-steam
heat exchanger.

generator.   It is, however, drawn on a different basis from that of Fig. 8-2
of the simple cycle.   Figure 8-2 had enthalpy as abscissa, with the
enthalpy and generator distances (with respect to flow) roughly related.
The dual-pressure cycle, however, does not permit this kind of plot, and,
instead, the abscissa of Fig. 8-7 is distance along the steam generator from
bottom to top (and this scale is not necessarily uniform).   It is thus
imperative that both the low-pressure and the high-pressure steam processes be represented by broken lines.   We can see that there are two pinch
points, $f_1 b'$ and $f_2 b$, along the gas path, which may be chosen between 20
and 30°F, as suggested previously.   Also, in this case the following relationship must hold if the heat losses to the exterior are neglected:

$$(\Delta H)_{\text{gas}} = (\Delta H)_{\text{l.p.steam}} + (\Delta H)_{\text{h.p.steam}} \qquad (8\text{-}4)$$

The steam cycle is completed (Fig. 8-6) by having the high-pressure
steam expand in the turbine to the pressure of the low-pressure steam

wherein they mix and both enter the low-pressure turbine. Part of the steam in both the high- and low-pressure turbines is bled for feedwater heating. The mixing at point l causes the steam entering the low-pressure turbine to acquire a temperature somewhere between t_l and t_d, as shown by the Mollier diagram of Fig. 8-8.

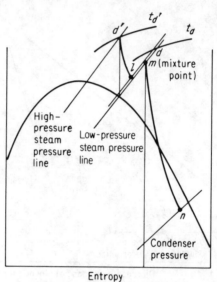

FIG. 8-8. Mollier diagram for steam expansion in turbine of dual-pressure plant.

The main advantage of the dual-pressure cycle is now evident. Thermodynamically beneficial high-pressure steam is generated. At the same time the temperature of the gas leaving the steam generator (which may have been left at too high a value, approximately equal to t_0, to allow economical blower operation) is further lowered to t_g by the low-pressure water-steam branch.

As in the simple cycle, the plant performance is affected by the gas temperatures, temperature differences (at pinch and superheat points), the feedwater temperature, and the steam pressure. It is further affected by the ratio between high-pressure and low-pressure steam flow rates. The relationships among all but the last are similar to those for the simple cycle. For example, relationships similar to those shown in Figs. 8-3 to 8-5 apply for the dual-pressure cycle, with the pressure lines those of the high-pressure branch but with the efficiencies higher than those of the simple cycle. The ratio of high-pressure to low-pressure flow rates decreases as the pressure of the low-pressure steam increases for constant steam-generator inlet gas temperature t_e.   The same ratio increases with t_e, other things being equal.

It should be pointed out that the advantage of the dual-pressure cycle over the single-pressure cycle diminishes as the temperature of the coolant entering the reactor (or leaving the steam generator) is increased. This is obviously true since an increased t_g (Fig. 8-2) in the simple cycle allows an increased t_b and consequently higher pressures.   It is to be expected, therefore, that reactors designed for high-temperature operation such as the AGR and the HTGR types (Secs. 8-6 and 8-7) operate efficiently on the single-pressure cycle.

8-4. THE U.K. GAS-COOLED REACTOR PROGRAM

One of the largest single programs for civilian nuclear power is the British development and construction of a series of graphite-moderated, and CO_2-cooled reactor power plants, Table 8-1.   This effort started in October 1956, when the first power plant, using natural-metallic-uranium, magnesium-clad fuel was put in operation at Calder Hall, England [59]. This power plant type originally known as "Calder Hall" but subsequently as the "Magnox" type, because of magnesium cladding, was devised because of economic and military (production of Pu^{239}) necessity. It had the advantages of using relatively familiar and consequently economical materials and fuels which do not require enrichment.   The coolant is cheap and plentiful.   The steam temperatures and pressures are, however, rather low, being a few hundred degrees Fahrenheit and a few hundred psi, well below current, highly efficient fossil-fueled power plants.   Because of their low temperatures, the Magnox stations are of the dual-pressure type.

Because of the use of natural uranium (low in fissionable U^{235}) and graphite, the Magnox fuel can be used only in metallic form so that no additional neutron absorbers are present (natural uranium in UO_2 can be used with heavy water as moderator).   Other materials of construction, such as the magnesium alloy used for cladding, were chosen with this requirement in mind.   This limitation in the choice of materials has imposed severe restrictions on the performance of the plant.   For example, because the maximum allowable operating temperatures of the magnesium-alloy cladding is around 850°F, the CO_2 reactor exit temperature was limited to 700°F.   The specific power attained was initially low, about 1.5 Mw/ton.   Improvements in heat transfer by better finning and heat exchangers, however, have raised this figure in later designs to about 2.3 Mw/ton.   The maximum center temperature of the metallic fuel has also limited fuel burnup to about 3,000 Mw-day/ton.

A second nuclear power program, commonly called the AGR (advanced gas-cooled reactor) was started in the United Kingdom in the late 1960's. The objectives of this program, which is based on the experimental prototype at Windscale, Table 8-1, are to construct nuclear power stations that supply steam at conditions comparable with those in modern fossil-fueled power stations, and with a degree of integrity that would permit siting nearer centers of population. AGR power plants use enriched ceramic fuels clad in stainless steel, but otherwise still use graphite moderators and CO_2 gas coolant. They operate at clad surface temperatures of about 1520°F (~ 825°C) with gas outlet temperatures of up to 1230°F (~ 665°C) resulting in steam temperatures of up to 1050°F (~ 565°C). Because of the high temperatures, the AGR stations are of the single-pressure type. They are also characterized by prestressed-concrete pressure vessels, double containment of all access penetrations, and provision for refueling the reactors on load for high availability.

TABLE 8-1
The British Gas-Cooled Reactor Program

Program	Station	Date on Power	Number of Reactors	Station Output, Mw(e)	Thermal Efficiency, percent	Capital Cost, £/Kw(e)	Generating Cost, d/kwhr
Experimental	Calder Hall	1956-7	4	180			
	Chapelcross	1959	4	180			
	Dounreay F. R. .	1960-4	1	13			
	Windscale AGR .	1962-3	1	33			
	SGHWR	1967	1	100			
	Prototype Fast .. Reactor	1971	1	240			
1st Program, "Magnox"	Berkeley	1962	2	275	24.4	160	1.27
	Bradwell	1962	2	300	28.2	159	1.11
	Hunterston	1964	2	360	28.2	—	—
	Hinkley Point A .	1965	2	500	26.4	133	1.01
	Trawsfynydd	1965	2	500	26.4	123	0.94
	Dungeness A ...	1965	2	550	32.9	110	0.72
	Sizewell	1966	2	580	30.5	107	0.70
	Oldbury	1968	2	600	33.6	111	0.68
	Wylfa	1969	2	1180	31.5	103	0.64
2nd Program "AGR"	Dungeness B ...	1970	2	1200	41.5	92	0.52
	Hinkley Point B .	1973	2	1250	—	—	0.48
	Hunterson B	1973	2	1250	41.7	—	—
	Seaton Carew ...	–	2	1250	—	—	0.43
	Heysham	–	4	2500	—	—	—

The following two sections present one of the Magnox and one of the AGR types.

8-5. THE HINKLEY POINT A STATION

Hinkley Point A is one of the more advanced Magnox type power stations [60].   Figure 8-9 shows a cross section and plan view of one of two reactors comprising the station.   Each reactor is rated at 966 Mw(t) and each supplies 10,000 lb_m/sec of 700°F CO_2 to six dual-pressure steam generators.   The steam conditions are 650 psi and 685°F in the high-pressure branch and 180 psi and 670°F in the low-pressure branch.   The

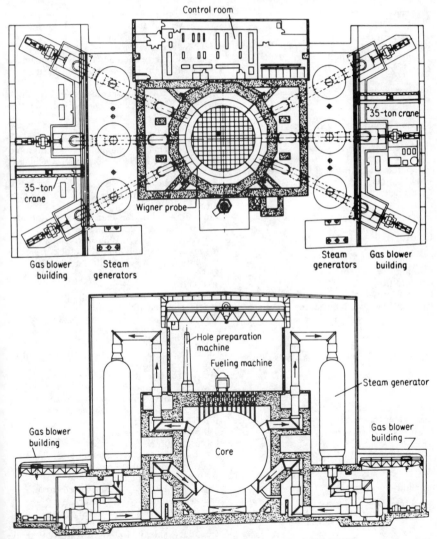

FIG. 8-9. One of two Hinkley Point reactors. (Ref. 60.)

elevation of the steam generators with respect to the reactor is so that maximum natural circulation of the coolant occurs in case of blower failure. The station guaranteed net output is 500 Mw(e). These and other data of Hinkley Point A are given in Table 8-2.

It is instructive to note the progress in clad fin design in the Magnox

TABLE 8-2
Some Design Data of Hinkley Point A Station

Plant capacity	500 Mw(e) guaranteed net; 551 Mw(e) + 99 Mw(e) auxiliaries installed
Reactor output	980 Mw(t) per reactor
Reactor description	Thermal heterogeneous; core 49 ft in diameter and 25 ft high; fuel elements on $7\frac{3}{4}$-in.-pitch square lattice
Pressure vessel	Spherical, 67 ft in diameter and 3 in. thick; mass, 1,700 long tons supported on 30-ft-diameter skirt
Reactor foundation load	5,500 tons
Fuel	Natural uranium, 370 long tons per reactor
Fuel elements	Eight elements per channel, stacked on top of one another, each $1\frac{1}{8}$ in. diameter by 36 in. long
Estimated maximum fuel temperature	775°F
Fuel channels	4,500 per reactor, each 3.85 in. diameter (mean)
Cladding	Machined or extruded extended surface magnesium alloy
Moderator	Graphite, 3,500 tons per reactor, supported on ball bearings
Control rods	108 control rods + 12 shutoff rods per reactor, in channels same diameter as fuel; driven by variable-frequency induction motors
Coolant	CO_2
Coolant flow	10,000 lb_m/sec per reactor
Coolant pressure	200 psi design; 180 psi operating
Coolant reactor temperatures	355°F inlet and 700°F outlet
Gas blowers	Six per reactor, single-stage, axial-flow type with overhung rotors with only one gland seal each, driven by variable-frequency motors supplied from variable-speed auxiliary turbo-alternators (below)
Steam generators	Six per reactor, dual-pressure type, each vessel 21.5 ft diameter and 90 ft high; $2\frac{3}{8}$ in. thickness; tube finning in form of studs
Steam generated	460,000 lb_m/hr per steam generator
Feedwater temperature	160°F
High-pressure steam	650 psi and 685°F
Low-pressure steam	180 psi and 670°F
Turboalternators	Six for station, hydrogen-cooled, impulse-reaction type, 3,000 rpm, 13.8 kv, 93.5 Mw continuous rating each
Auxiliary alternators	Three, 33 Mw each, variable-speed (for blower) operation, same type as above
Shielding	Dense reinforced concrete, air-cooled; main shield: 75 ft diameter, 75 ft high, and 7 ft thick; secondary shield: 100 by 140 ft, 4 ft thick, enclosing complete reactor

program and its effect on plant performance. The early-type Calder Hall plants used magnesium fuel cladding with transverse (circumferential) finning. The fuel elements were then placed in circular channels with axial coolant flow (Fig. 8-10). This design was the result of experiments

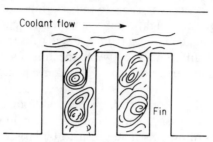

FIG. 8-10. Transverse finning with axial flow used in early Calder Hall plant showing coupled vortexes between fins.

in which electrically heated finned rods of various configurations were tried. The arrangement of transverse fins with axial flow was found, by smoke tests, to lead to the formation of two coupled vortexes of gas between adjacent fins [59], as shown. It was claimed that there was sufficient transfer of mass and heat between the gas in the space between the fins and in the main stream to render such an arrangement efficient when compared with many other standard arrangements. However, the eddy formations were found to be quite sensitive to the Reynolds number, and the heat-transfer coefficient dropped off rapidly when this number was increased beyond that employed in Calder Hall. Thus for higher-performance reactors, a different type of finning was necessary.

A different and more advanced design of fuel-element cladding was utilized in the Hinkley Point A power station. This is the so-called "polyzonal" fuel element [61] shown in Fig. 8-11a. The fins, made of magnesium alloy, are in the form of multistart helixes of long lead. Three equally spaced radial splitters are inserted in slots cut into the helical fins along the axis of the elements, as shown. The radial length of the splitters is such that they extend almost to the wall of the gas-coolant flow channel. This design, as contrasted with the early Calder Hall design, has allowed a considerable increase in power density without significantly raising the maximum temperature of the fuel and without the high frictional losses associated with the enforced double vortexes.

The gas flow pattern developed within the polyzonal can is as follows: Part of the gas entering a channel at the upstream (lower) end becomes the main stream, filling the space between the radial splitters and around the fin tips. The other part fills the flutes between the spiral fins and forms intense vortexes. This latter portion follows a helical path in the flutes until it collides with a splitter and is then deflected radially out of the flute and into the main gas stream. Thus the hot gas in the flutes continually mixes with the cooler gas of the main stream. The process is repeated every time the gas reaches a new fuel element (there are eight fuel elements in each channel). The continuous·replacement of the gas in contact with the hot fins results in higher temperature differences between the fins and the gas in contact with them than in the earlier design. This consequently results in higher heat-transfer rates.

It may be useful to reiterate here that, although magnesium has a lower conductivity than, say, aluminum, its much lower neutron absorption cross sections allow fins made of it to be larger, resulting in larger heat-transfer surfaces. For high-performance reactors, the fin length is short, making the use of magnesium less attractive.

Figure 8-11b shows a studded tube of the type used in the steam generators of later Magnox reactor designs. The fins are approximately 1 in.

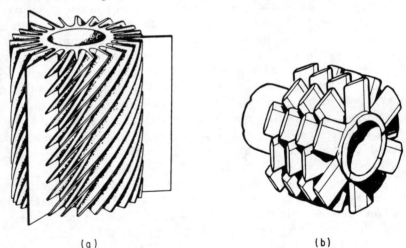

(a)   (b)

FIG. 8-11.   Finning used in Hinkley Point A power plant.   (a) "Polyzonal" fuel-element cladding; (b) steam-generator tubing.

wide, tapering to sharp edges. This design, together with an improvement in the use of space inside the steam generators, has allowed a 250 percent increase in steam production over the earlier designs (shorter fins of different shape) with only a 20 percent increase in height and a 25 percent increase in diameter of the steam generators— a good example of what can be done with heat exchangers when particular attention is paid to detail.

8-6. THE HINKLEY POINT B STATION

Hinkley Point B [62] belongs to the second nuclear power program in Great Britain, Table 8-1, and is therefore of the advanced gas-cooled reactor (AGR) type. The station contains two reactors and two turbo-alternators producing a total of 1250 Mw(e). The Station was designed and constructed by The Nuclear Power Group, Ltd. (TNPG).

Figure 8-12 shows an isometric cutaway of the two-reactor station.

FIG. 8-12. The Hinkley Point B power plant. (By permission from Nuclear Engineering International.)

Table 8-3 lists some of the design data of the station. Figure 8-13 shows a cross section of one of the reactors. The reactor and turbine house are combined in a single complex. The twin reactors and twin turbines are served by a central control and instrumentation block and the two reactors are refueled by the same refueling machine operating in a common charge hall.

The core is a 16-sided stack of graphite blocks, radially restrained by restraining rods and a steel tank. It is divided into twelve horizontal

TABLE 8-3
Some Design Data of Hinkley Point B Station

Number of reactors	2
Plant capacity	1,250 Mw(e) net, 1,320 Mw(e) gross
Reactor output	1,500 Mw(t) per reactor
Reactor description	Thermal heterogeneous, core 30 ft diam, 27.2 ft high; fuel elements on 18.1 in. pitch square lattice
Pressure vessel	Cylindrical, prestressed concrete, 62 ft diam, 63.5 ft high
Fuel	Enriched UO_2 pellets, stainless steel clad, enrichment 1.4% and 2.1% first charge, 1.7% and 2.6% feed; 122.5 ton U
Fuel element	Eight elements per channel, each 36, 0.6-in. diam pin clusters in graphite sleeve, 40.8 in. long, 7.5 in. inner diam
Fuel channels	308 per reactor, each 10.375 in. diam
Moderator	Graphite
Control rods	81 channels, 5 in. diam
Coolant	CO_2
Coolant flow	7,743 lb_m/sec
Coolant pressure	615 psia max
Coolant temperatures	310°C core mean inlet 665°C core mean outlet
Circulators	Eight per reactor, 2,970 rpm encapsulated centrifugal type, constant speed electric motor drive, variable guide vane flow control; power consumption 28 Mw(e) per reactor
Boilers	Twelve units per reactor, single-pressure type; bulk gas temperatures 654°C inlet, 276°C outlet; superheater outlet steam 2,418 psia and 541°C; reheater steam 977 lb_m/sec flow, 610 psia inlet, 590 psia outlet, 343°C inlet, 541°C outlet
Thermal efficiency	41.7%

layers, the top and bottom of which constitute the top and bottom graphite reflectors. The graphite moderator is the other ten. The large polygonal graphite blocks, Fig. 8-14, have large vertical bores which form the vertical fuel channels. The graphite blocks are interconnected by graphite cross keys to maintain stability and proper pitch. Square interstitial graphite bricks are placed between the polygonal blocks. Most of these have small holes for cooling. Some (one in four) have large holes to accommodate the control rods. Others have varying hole sizes to ac-

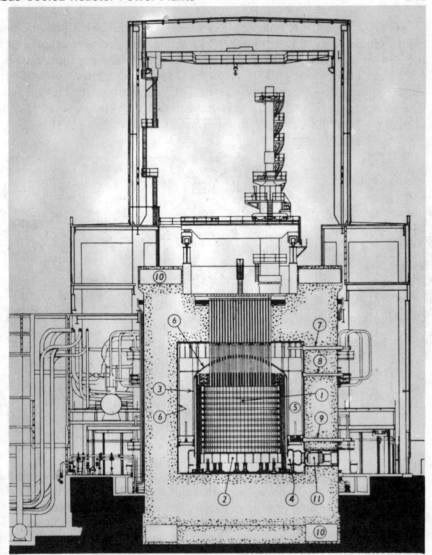

FIG. 8-13. Elevation of the Hinkley Point B reactor building. 1. Reactor core. 2. Supporting grid. 3. Gas baffle. 4. Circulator outlet gas duct. 5. Boiler 6. Thermal insulation. 7. Reheat steam penetrations. 8. Main stream penetrations. 9. Boiler feed penetrations. 10. Cable stressing gallery. 11. Gas circulators. (Courtesy TNPG, Limited.)

commodate neutron sources, graphite samples and flux monitorizing instruments. The whole structure is surrounded by an annulus, two polygonal blocks wide that contain only small holes for cooling and which act as the radial reflector. Three layers of rectangular graphite bricks above the graphite core form a neutron shield.

There are 308 fuel channels per core.   Each fuel channel contains eight fuel elements, linked together by a tie bar to a fuel unit extending to the top of a refueling standpipe and terminating in a pressure closure. Each fuel element is 40.8 in. (1039 mm) long and consists of a 36-pin cluster of stainless-steel-clad, 0.57 in. (14.5 mm) diameter UO_2 pellets. The pins are 0.6 in. (15.25 mm) in diameter.   Each fuel element is contained in a graphite sleeve, 190 mm inside diameter.

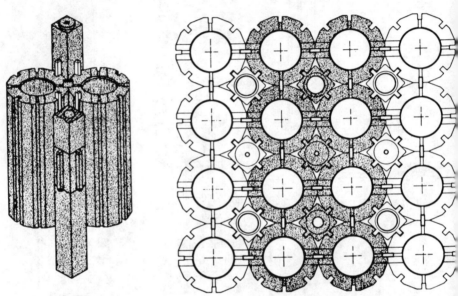

FIG. 8-14.   Cross section of the Hinkley Point B reactor core.   (Ref. 62.)

The coolant gas enters the fuel channels at the bottom at 590°F and 42 kg_f/cm² abs.   It flows upward through the fuel channels and guide tubes to the space between the gas baffle and the top pressure vessel liner, Fig. 8-15.   From there, at 1210°F and 39.9 kg_f/cm² abs, it flows downward through the boilers leaving at 276°F and 39.7 kg_f/cm² abs.   It is then pumped by the circulators and channeled up between the boiler shield wall and gas baffle to the domed region above the core.   A small quantity flows down between the core and boiler shield wall.   The remainder of the gas enters the core at the top and flows downward between the core bricks, through the annular gaps in the graphite moderator and through parts in the lower graphite to the diagrid area. Another small quantity of gas leaving the circulators passes underneath the boiler shield wall to provide a path for natural circulation in case of circulator failure.   This gas combines with that from the boiler shield wall and passes through the diagrid and support plates at the base of the fuel. The total gas then enters the fuel channels and the gas flow circuit is completed.

There are twelve boiler units in the annulus between the reactor gas baffle and the pressure vessel. The annulus is partitioned into four

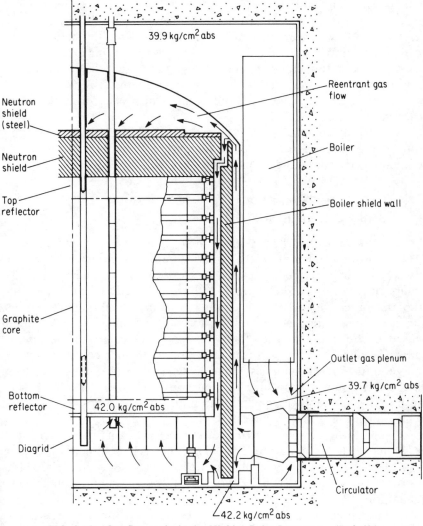

FIG. 8-15. Gas-flow paths in the Hinkley Point B reactor. (Ref. 62.)

quadrants each containing three boiler units and two gas circulators. The vertical boilers contain (from bottom to top) two economizers, one evaporator, two superheaters and one reheater, each.

The eight circulators per reactor are driven by constant-speed electric motor drives.   Coolant flow is controlled by variable inlet guide vanes. The circulator-motor combinations are located in horizontal penetrations at the bottom of the pressure vessel wall.

The pressure vessel is a concrete vertical cylinder with helical multi-layer prestressing cables in the walls.   The vessel inner surface is insulated and cooled to maintain concrete temperatures below 158°F (70°C).

Two 660 Mw(e) turbogenerators are provided for the station.   The turbines are horizontal, single shaft, close-coupled, single reheat, disk and diaphragm type with one single-flow high-pressure unit, one double-flow intermediate-pressure unit, and three double-flow low-pressure expansions in an outer low-pressure casing which also forms the condenser shell.

A relatively compact 12 × 12-m control room, manned by three operators, is provided for the station.   Wide use is made of data-processing equipment for control and monitoring.

8-7.   THE HIGH-TEMPERATURE
GAS-COOLED REACTOR (HTGR)

In contrast to British development, ordinary-water reactors (PWR, BWR) constitute the bulk of the American effort in commercial nuclear power, and only a modest effort in gas cooling is under way.   The American concept is usually referred to as the HTGR.   In that concept, graphite is used for fuel particle coating (primary fission product barrier), fuel structural material, moderator and coolant channel walls.   Helium is used as coolant.   The use of an all ceramic fuel element results in low parasitic neutron capture in the core and therefore high conversion ratios and good fuel cycle economics.   The coated fuel particle design results in high specific powers and high fuel burnups.   Furthermore, because of the use of graphite, high coolant exit temperatures are possible resulting in high plant thermal efficiencies.

Figure 8-16 shows a coated fuel particle and a graphite fuel block design in which it is contained [63].   The block is hexagonal, 14 in. across sides and 2.5 ft high.   The coated particles are contained in over 200 fuel holes interspersed with a lesser number of coolant holes through which helium flows.

Conversion ratios in such designs depend of course on reactor size and are expected to reach 0.7 to 0.8 for 1,000-Mw(e) reactors.   The 40-Mw(e) Peach Bottom and the 330-Mw(e) Fort Saint Vrain HTGR reactors have conversion ratios of 0.44 and 0.61 respectively.   Replacement of some of the graphite with BeO is expected to result in conversion ratios above 0.9.   In one concept breeding (with Th^{232} as the fertile material in this thermal reactor type) is expected with conversion (or breeding) ratios 1.0 to 1.1, by allowing fission gases to escape into the coolant. The coated fuel particles, are miniature spherical fuel elements containing UC_2 fuel and ThC_2 fertile material, with diameters of 200 ± 50 microns for the fuel and 400 ± 100 microns for the fertile material. They are clad with a coating of pyrolytic carbon 100 microns thick for the fuel and 125 microns thick for the fertile material. This cladding here, as elsewhere, isolates the fuel and prevents chemical reactions and minimizes the release of fission products. The two sizes for fuel and fertile materials make it easy to separate the two for fuel recycle purposes.

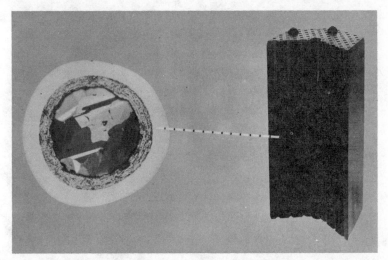

FIG. 8-16. BISO-coated particles in hex-block fuel element. (Courtesy Gulf General Atomic Corp.)

Because of their small size and spherical shape, coatings have been developed which are able to withstand mounting pressures due to fission gas release and expanding fuel and therefore maintain integrity for long periods at high temperatures—i.e., result in high fuel burnups. Multilayer (up to 3) coatings of different structures and properties have been developed which have good irradiation resistance. In one case a two-layer coating (Fig. 8-16) called Biso (for Buffer-Isotropic) consists of an inner, low-density buffer layer which serves to protect the outer layer from fission-product recoils, relieves stresses by accommodating dimensional changes of fuel and outer layer and reduces pressure buildup by absorbing some fission gas. The outer layer is a high-strength isotropic layer which acts as the pressure vessel for the fuel particle.

8-8. THE PEACH BOTTOM HTGR PLANT

The Peach Bottom HTGR is a 40 Mw(e) power plant located on the Susquehanna River midway between Philadelphia and Baltimore [64]. The reactor system was developed by the General Atomic Division of General Dynamics Corporation (now Gulf General Atomic).

The reactor uses U^{235} as fissionable fuel, Th^{232} as fertile material, graphite as moderator, cladding, structure and reflector, and helium as

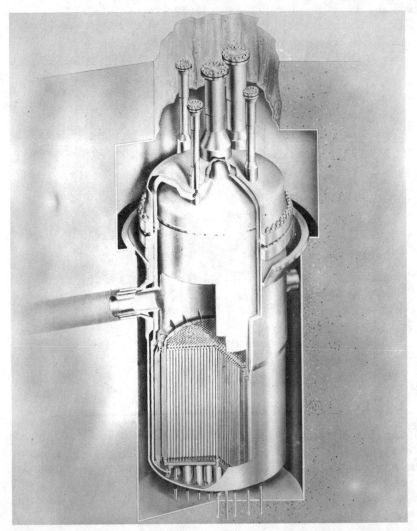

FIG. 8-17.   The Peach Bottom reactor.   (Courtesy Gulf General Atomic Corp.)

coolant. The active core, 9.2 ft in diameter and 7.5 ft high, contains 804 fuel elements, 36 control rods, and 19 shutdown rods. Figure 8-17 shows a cutaway of the Peach Bottom reactor.

The fuel elements are solid semihomogeneous in that graphite serves the multiple functions of fuel matrix, cladding, moderator, and structure. The fuel element, Fig. 8-18, is 3.5 in. in diameter and 12 ft high.

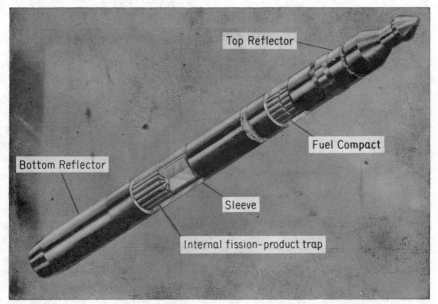

FIG. 8-18. The Peach Bottom reactor fuel element. (Courtesy Gulf General Atomic Corp.)

It consists of a middle fuel-bearing section between upper and lower reflector sections and an internal fission-product trap, encased in a graphite sleeve. The fuel is in the form of annular grooved fuel compacts stacked on a cylindrical graphite spine. The overall length of the compacts is 7.5 ft. The compacts are 2.75 in. in diameter and 1.5 in. long and contain a mixture of highly enriched uranium carbide and thorium carbide particles 150 to 400 microns in diameter, coated with 50-60 micron-thick dense graphite. The particles are dispersed in a graphite matrix. The coating here serves as a protector against fuel oxidation during fabrication as well as barrier to increase fission product retention during reactor operation.

A feature of this reactor is the continuous purge of those fission products that are not retained by the pyrolytic coating of the fuel particles. The nonvolatile fission products will be retained in the fuel. The volatile ones are retained long enough until many of their short-lived isotopes have decayed.   Of those products made volatile by the high fuel temperatures that escape, the lower volatility ones will condense in the internal trap which is located in the bottom (and therefore colder) end of the fuel element.   The fission products finally escaping the internal trap with the purge gas to the external traps are principally krypton and xenon as well as delayed iodine and bromine.

In the external trap system, the purge gas is cooled to room temperature.   The halogens are removed by a charcoal bed.   The krypton and xenon are delayed in charcoal beds.   Krypton and other remaining contaminants are removed in liquid-nitrogen-cooled traps.   The purge helium is then returned to the primary coolant loops.   Components of this system are individually shielded and cooled as necessary, and maybe removed separately.   The design, however, is for the system to operate throughout the life of the reactor without replacement.

The helium coolant primarily flows outside, and parallel to the fuel element.   A small fraction of coolant, called the *purge gas,* enters the element through the porous top reflector, flows downward between the fuel compacts and the graphite sleeve, thus sweeping fission products. It then flows through the internal fission-product trap where some of the fission products are either deposited or absorbed.   This trap is located in the lower reflector and consists of a slotted graphite cylinder containing silver-coated charcoal reagent in granular form.   The purge gas then goes through a standoff pin at the bottom of the element, and finally to a double-walled purge line leading to the external fission-product traps.

The primary coolant at 20 atmospheres enters the reactor at 634°F at the rate of 439,600 lb_m/hr in two loops through two annular ducts formed between inlet and exit pipes, Fig. 8-17.   It flows half upward and half downward, cooling the reactor vessel walls.   The upward flow further goes down through the reflector blocks.   The downward flow passes between the vessel and a steel thermal shield situated between the reflector blocks and the vessel wall.   The divided flow now recombines in the lower plenum, enters the core at 654°F, the upper plenum at 1380°F and leaves the reactor at 1354°F.   Purge gas flows through the elements as indicated above at the rate of 1,000 lb_m/hr.

The control rods are made of boron-carbide-bearing graphite and operate within graphite guide tubes 12 ft long.

Each of the two primary loops consists of a steam generator, a helium compressor, and related piping and auxiliaries.

The helium compressors are of the horizontal, single-stage centrifugal type.   An overhung impeller design results in only one bearing.   Shaft sealing is provided by helium-buffered, oil-flooded floating bushing. Each compressor is driven by a 2,500-hp, 3,500-rpm electric motor through a variable-speed coupling to regulate the main coolant flow.

TABLE 8-4
Main Design Parameters of Peach Bottom HTGR

Reactor heat output	115.3 Mw(t)
Plant net power	40 Mw(e)
Core active dimensions	9.16 ft diam, 7.5 ft high
Fuel elements	804 (U, Th)C_2 fueled, graphite clad, (U + Th) 1 atom %, 20 mass % in compacts
Core fuel initial loading	200 kg_m U^{235}, 13 kg_m U^{238}, 1897 kg_m Th
Heat flux, Btu/hr ft²	8,000 av., 102,000 peak
Power density, kw/liter	8.24
Specific power	580 kw(t)/kg_m U^{235}
Average conversion ratio	0.50
Excess reactivity	19% cold clean, 8% hot clean, 5% hot poisoned
Control rods	36 B_4C rods
Coolant	He
Coolant conditions	634°F reactor inlet, 1354°F outlet, 335 psig vessel pressure, 10 psi rise in circulators, 439,600 lb_m/hr total flow rate, 1,000 lb_m/hr purge rate
Circulators	2, 33,800 cfm horizontal single-stage centrifugal, 2,500-hp motor each
Steam-cycle conditions	1000°F, 1,450 psig at turbine throttle, 425°F economizer inlet
Steam-cycle equipment	2 forced recirculation steam generators, 1 tandem-compound double flow 46 Mw(e) gross turbine, 1 single-pass divided box condenser, 3 horizontal U-tube feedwater heaters
Plant efficiency	39.9% gross, 34.7% net

Auxiliary drive motors using an emergency power supply are used in the event of power failure.

The steam plant is of the simple cycle type containing interconnected steam generators, turbine-generator, condenser, two feedwater pumps, and three feedwater heaters. The steam generators are 8 ft diameter and 30 ft high each and are of the shell-and-tube once-through type containing economizer and evaporator sections of carbon-steel, and superheater of type 304H stainless steel. The cooled helium is recirculated back to the reactor via the annulus between the carbon-steel shell and a stainless-steel shroud to keep the shell cool. Feedwater enters the economizer section, then mixes with recirculated flow. Recirculation pumps are designed for 4:1 recirculation ratio. The mixture flows through the downcomer and is then pumped to the evaporator section. The steam-water mixture then enters the steam drum through a riser. The steam separates, passes through a steam-purification system in the drum, goes to the superheater section, and finally leaves the steam generator to the turbine.

The turbine-generator is an outdoor, 3,600 rpm, tandem-compound double-flow unit using 1,450 psig, 1000°F throttle steam and generating 40 Mw(e) net. Reactor decay heat is removed during shutdown by a subboiling heat exchanger via the evaporator section of the steam generators. In the event of steam generator recirculation pump failure, sufficient natural circulation is obtained by placing the steam drum liquid level 35 ft above the top of the main generator tubesheet. Emergency cooling for the reactor vessel is provided by a series of coils welded to a steel plate shroud surrounding the pressure vessel. Cooling water is supplied to the coils when the system is activated. The system is capable of keeping the vessel temperature below 900°F, corresponding to a maximum core temperature of 3600°F.

Load following control in Peach Bottom is obtained automatically at a design rate of 3 percent per minute and involves a combination of reactor regulating control rod motion and helium flow variation by the variable-speed circulators to maintain constant steam pressure and temperature. A demand for increased power opens the turbine admission valves. This causes a momentary drop in steam pressure, which in turn causes the speed of the helium circulators to increase. The resultant increased helium flow causes a drop in reactor exit helium temperature which activates the regulating rods to increase reactor power.

8-9. THE FORT SAINT VRAIN HTGR

The Fort Saint Vrain Nuclear plant is a commercial-size HTGR-type power station built near Platteville, Colorado, about 35 miles north of Denver [65]. Built by Gulf General Atomic Corp. under the U.S. advanced reactor program, the station is a 330 Mw(e) net version of the 40 Mw(e) prototype HTGR at Peach Bottom.

Like Peach Bottom, Fort Saint Vrain uses graphite for moderator, cladding, core structure and reflector, and helium as coolant. Fort Saint Vrain, however, deviates from Peach Bottom principally in the use of hexagon-shaped graphite fuel elements, a prestressed concrete pressure vessel housing reactor, steam generators, and circulators, a reheat steam cycle, steam-driven axial flow (instead of electrically driven centrifugal flow) helium circulators, and an indoor turbine-generator. Table 8-5 lists some design parameters of Fort Saint Vrain.

TABLE 8-5

Main Design Parameters of The Fort Saint Vrain Power Plant

Plant heat output	337.5 Mw(t)
Plant electrical output	341 Mw(e) gross, 330 Mw(e) net
Core	Graphite, 19.5 ft diam, 15.5 ft high
Fuel elements	1,500, hexagonal graphite 14 in. across flats, 31 in. high; vertical holes contain fuel and fertile material; vertical holes for coolant flow
Fuel and fertile material	BISO coated particles. Fissile particles contain 93% enriched UC_2 and ThC_2; Fertile particles contain ThC_2 only
Coolant	Helium
Coolant flow rate	3.4×10^6 lb_m/hr
Coolant conditions	700 psia and 760°F core inlet, 1430°F core outlet, 14 psi pressure drop
Coolant circulators	4, steam-driven vertical single stage axial flow compressors
Steam-flow rate	2.3×10^6 lb_m/hr
Steam conditions	2,400 psig and 1000°F throttle inlet, 600 psia and 1000°F reheat
Plant net efficiency	39.4%

The fuel element, Fig. 8-16, is a hexagonal graphite structure, 14 in. across the flats and 31 in. high. Six elements are stacked on top of each other to form the active height of the core. There are 1,500 total elements in the core, divided into 37 axial and radial regions with different fuel loading for uniform temperature distribution. The active fuel, in the form of coated Biso particles, is contained in an array of small-diameter holes within each fuel element. The element also contains a lesser number of coolant holes parallel to the fuel holes. Three graphite dowels serve to align the individual fuel elements and to insure that the coolant holes in any stack are also aligned. A graphite reflector surrounds the active core which is 19.5 ft in diameter and 15.5 ft high. The fuel material contains 93 percent enriched U^{235} and thorium in the fissile particles and thorium only in the fertile particles. The Biso coating is improved and highly impervious. No fission gas purge is incorporated in this plant. One-sixth of the fuel regions are expected to be refueled each year. The average fuel burnup is expected to be 100,000 Mwd/ton of U and Th. Figure 8-19 shows the reactor arrangement. The core, composed of the hexagonal fuel elements described above, is situated in the upper half of a prestressed-concrete pressure vessel.

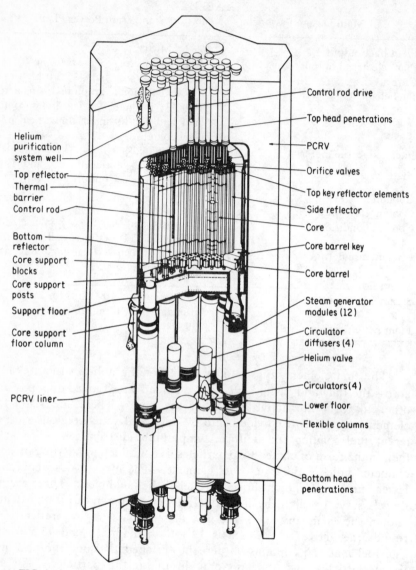

Helium purification system well

Top reflector
Thermal barrier
Control rod

Bottom reflector
Core support blocks
Core support posts
Support floor

Core support floor column

PCRV liner

Control rod drive

Top head penetrations

PCRV

Orifice valves

Top key reflector elements

Side reflector

Core

Core barrel key

Core barrel

Steam generator modules (12)

Circulator diffusers (4)

Helium valve

Circulators (4)

Lower floor

Flexible columns

Bottom head penetrations

FIG. 8-19. The Fort Saint Vrain reactor arrangement. (Courtesy Gulf General Atomic Corp.)

Prestressed concrete reactor pressure vessels, PCRV, have been used in 14 gas-cooled reactors in France and Great Britain, both for Magnox and AGR systems. All of these but three housed the entire primary coolant system. In the United States, PCRV is also considered for gas-cooled fast-breeder reactors, Sec. 10-6. Horizontal and vertical cylindrical as well as spherical shapes are used. There are many technical advantages of prestressed-concrete reactor pressure vessels over steel ones. They can be built for large sizes and reasonably high pressures (up to about 1,250 psi), while steel vessels of comparable sizes would be difficult to fabricate with the required thicknesses and would have to be field-welded. Since there are no size limitations for PCRV, the entire primary circuit can be housed within it, thus eliminating the possibility of large duct failures. Economic advantages can also be expected because of easier construction and scheduling and the use of largely local materials and manpower.

The Fort Saint Vrain vessel contains and shields the entire primary coolant system. It is made of high-strength concrete reinforced with bonded reinforcement steel and prestressed with steel tendons. The tendons, located in conduits embedded in concrete, are used to prestress the vessel prior to pressurization. The prestressing forces are oriented so that they oppose those due to internal pressure. The vessel exterior approximates a hexagonal prism with flat ends. The interior is a cylindrical cavity 31 ft in diameter and 75 ft high. The top head has refueling penetrations which also house the control-rod drives, an access penetration for removal of reactor components and penetrations, and wells which house a helium purification system and neutron chambers. The bottom head has penetrations to accommodate steam generator modules and helium circulators as well as a central hole for access to the main cavity. The vessel is lined with a carbon steel liner which provides a helium-tight membrane seal. The liner is protected with a thermal shield on the reactor side. In addition, a system of coolant tubes is welded to the concrete side of the liner. Water coolant in these tubes limits liner and concrete temperatures.

Four vertical helium recirculators consisting of a single-stage axial-flow compressor are driven by a single-stage steam turbine each. The circulators housings penetrate the bottom head of the vessel, Fig. 8-19. They are surrounded by 12 steam generator modules whose housings also penetrate the bottom head as well as the core support floor. The steam generators are once-through units with integral reheaters. Water flows upward in the generators through economizer-evaporator tube bundles which are in the form of helically coiled tubes, then steam flows downward through the superheater tube bundles. Steam flows upward through the reheater tube bundles.

Hot helium, traveling downward through the core, enters the top of each steam generator module at 1430°F, flows downward through the reheater, superheater and finally through the economizer-evaporator tube bundles. Cooled helium then leaves the bottom of the generators, turns and flows horizontally through a plenum to the circulator inlets. The

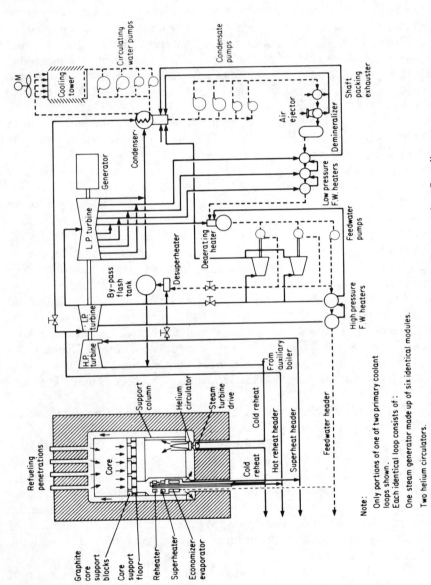

Note:

Only portions of one of two primary coolant
loops shown.
Each identical loop consists of:
One steam generator made up of six identical modules.
Two helium circulators.

FIG. 8-20. The Fort Saint Vrain power-plant flow diagram.

helium is then returned to the top where it enters the core at 760°F and 700 psia.

Superheated steam leaves the steam generator modules at 2,400 psig and 1000°F to the high-pressure turbine, Fig. 8-20. Exhaust from the high-pressure turbine at 670°F is first used to drive the four circulator turbines. It then enters the reheaters in the steam generator modules and returns to the intermediate pressure turbine at 600 psia and 1000°F. It finally leaves the double-flow low-pressure turbine at 2.5 in. Hg abs to the main condenser. The condenser cooling water is cooled in an induced-draft cooling tower. There are four condensate pumps and four condenser cooling-water circulating pumps. There are 6 stages of feedwater heating, bringing up the feedwater temperature to 400°F. A full-flow demineralizer is situated just before the low-pressure feedwater heaters. Drains from the two-high-pressure heaters cascade to a deaerating heater (where the oxygen content is reduced to 0.005 cm³/liter or less.) Two of three feedwater pumps are steam driven and one is electrically driven.

The net thermal efficiency of this high-temperature, reheat-cycle plant is 39.4 percent, comparable to modern fossil-fuel plants.

Seventy-four control rods, operated in pairs (to reduce penetrations) by 37 control-rod drives through penetrations in the top head of the prestressed-concrete pressure vessel, enter the cooler top part of the core through guide tubes. Control-rod material is made of graphite containing 30 mass percent natural boron in boron carbide form.

8-10. PEBBLE-BED REACTORS

In *pebble-bed* reactors, the fuel and some or all of the moderator material are mixed together and made into pebbles of some simple geometrical shape, usually spherical. The pebbles are then randomly packed into a suitable vessel or bed to form the reactor core (Fig. 8-21). The core is cooled by gas flowing through the void space between the pebbles. In one such reactor (Sec. 8-11), the fuel elements are in the form of a sphere 2.4 in. in diameter.

The fuel elements may be unclad, making it possible for the reactor to operate at much higher temperatures than one using clad fuel elements and making possible the use of breeding in such a system. Because of the simple shape of the fuel elements of a pebble-bed reactor and the possible absence of cladding and because it is not necessary to maintain very close tolerances in fuel-element dimensions, the cost of fuel-element manufacture is materially reduced. Also, the fuel loading and unloading procedures are simplified, making it possible to load and unload more frequently than in other reactor types without reactor shutdown.

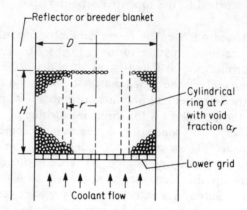

FIG. 8-21.   Elements of a pebble-bed
reactor.

Unclad fuel elements in which the fuel and graphite moderator are
mixed, i.e., a homogeneous fuel element (Fig. 8-22a) may be manufactured either by impregnating the graphite with the fuel in a nitrate solution
followed by denitration or by blending particles of the fuel and graphite
powder with a binder, followed by curing the binder.   Another kind of
pebble-bed fuel element is the heterogeneous type in which a fuel
lump is inserted within a graphite sheath, which then acts as cladding,
Fig. 8-22b [66].

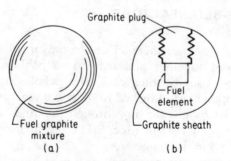

FIG. 8-22.   Spherical fuel pebbles.   (a)
Homogeneous; (b) heterogeneous.

Spherical fuel elements are structurally strong and have been irradiated to 20,000 Mw-day/ton of contained uranium with little change in
structural properties [67].   Fission-gas release from such unclad elements
is a problem, however, and necessitates the use of systems to remove them
from the coolant.   Fission-gas release may be somewhat minimized by
the use of ceramic coatings on the fuel elements.

Of importance in this type of reactor is the effect of the fuel-element packing on *void fraction* or *porosity* α, defined as

$$\alpha = \frac{\text{free volume}}{\text{total volume}} \tag{8-5}$$

and consequently on core physics and on heat-transfer and flow characteristics of the coolant. The most loosely packed beds have their spheres packed in a cubic arrangement with a constant void fraction of 0.476 throughout the bed. Random packing, however, results in void fractions that vary with the ratio of bed diameter D to pebble diameter d, thus reflecting the effect of the wall. Carman [68] experimentally obtained the variation of average void fraction in a bed, $\bar{\alpha}$, with D/d for randomly packed spheres that were shaken down to the closest possible packing, shown in Fig. 8-23. It can be seen that, for D/d of 15 or greater,

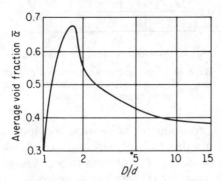

FIG. 8-23. Average void fraction in a pebble bed versus D/d. (Ref. 68.)

$\bar{\alpha}$ becomes a constant 0.39 for the entire bed. It has also been experimentally shown [69] that, for any one D/d, local void fractions at various radial distances, α_r, deviate from this average value, being 1.0 at the wall and fluctuating (with decreasing amplitude) around α away from the wall (Fig. 8-24). α_r here is defined as the average void fraction of a high axial (concentric) thin cylindrical ring at r.

As indicated above, the void fractions have their effects on fluid flow and heat transfer and on core nuclear design. A compromise between these two requirements is necessary. In one case (described below), the core average void fraction was reduced from 0.39 to 0.29 by the addition of pure moderator (graphite) columns in the core. The columns served the double purpose of extra moderator and housing for the control rods.

A feature of pebble-bed reactors is the radial interconnection between

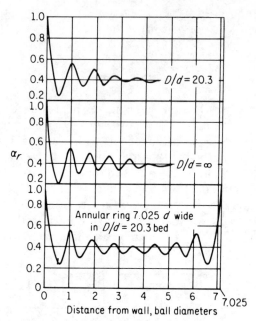

FIG. 8-24. Void fractions at various distances from wall in two-cylindrical pebble bed and in an annulus. (Ref. 69.)

gas passages, with consequent radial-flow components, due to radial temperature gradients (because of variations of radial neutron flux) and pressure gradients (because of variations in α_r). A complete analysis of fluid flow and heat transfer is therefore complex and requires the use of digital computers. A simplified analysis may, however, be made by dividing the core of the pebble-bed reactor into a number of regions where the void fractions may be assumed constant. For example, a cylindrical (or nearly cylindrical) core may be divided into a number of concentric rings. Neglecting cross flow, the pressure drop in turbulent flow in one ring (Fig. 8-21) may be expressed by the Darcy formula, written in the form

$$\Delta p = f \frac{H_e}{D_e} \frac{G_e^2}{\rho 2 g_c} \tag{8-6}$$

where Δp = pressure drop, lb_f/ft^2
$\qquad f$ = friction factor, dimensionless
$\qquad H_e$ = effective path length of gas through bed, ft
$\qquad D_e$ = equivalent diameter of flow, ft
$\qquad G_e$ = effective mass-flow rate of coolant, based on actual velocity

of coolant in void space between pebbles, lb_m/hr ft² of free area

ρ = density of coolant in ring, lb_m/ft³

g_c = conversion factor = 4.17×10^8 lb_m-ft/lb_f hr²

The path followed by the gas in a randomly packed granular bed has been shown by Carman [68] to have an average inclination of 45°. Thus

$$H_e = \sqrt{2}\,H \tag{8-7}$$

where H is the actual height of the bed in feet.

The equivalent diameter of the bed is given by

$$D_e = 4 \frac{\text{free volume within bed}}{\text{total surface area of pebbles}} \tag{8-8}$$

For spherical pebbles, this becomes

$$D_e = 4 \frac{\alpha_r V_r}{V_r(1 - \alpha_r)(\pi d^2 / \frac{1}{6}\pi d^3)} = \frac{\alpha_r}{1.5(1 - \alpha_r)}\,d \tag{8-9}$$

where α_r = average void fraction of ring at r in axial direction, dimensionless

V_r = total volume of ring, ft³

d = diameter of pebble, ft

The effective mass-flow rate is related to an apparent mass-flow rate and the void fraction. Thus for any one ring

$$G_e = \frac{G}{\alpha_r} \tag{8-10}$$

where G is the apparent mass-flow rate, in lb_m/hr ft² of ring cross-sectional area.

Equations 8-7 to 8-10 may now be combined with Eq. 8-6 to give

$$\Delta p = 0.254 \times 10^{-8} f \frac{H}{d} \frac{G^2}{\rho} \frac{1 - \alpha_r}{\alpha_r^3} \tag{8-11}$$

Robinson [67] has indicated that a conservature correlation for the friction factor in a pebble bed (25 percent higher than the mean of the best available data) may be given by

$$f = \frac{67.5}{Re^{0.27}} \tag{8-12}$$

This can now be modified to the form

$$f = \frac{67.5}{(D_e G_e/\mu)^{0.27}} = \frac{67.5}{(dG/\mu)^{0.27}\,[1/1.5(1 - \alpha_r)]^{0.27}} \tag{8-13}$$

Combining with Eq. 8-11 gives

$$\Delta p = 19.13 \times 10^{-8} \frac{H}{d^{1.27}} \frac{\mu^{0.27}}{\rho} G^{1.73} \frac{(1 - \alpha_r)^{1.27}}{\alpha_r^3} \tag{8-14}$$

This equation does not take into account gas dead spots which may tend to increase the pressure drop as well as cross flow. For a known operating pressure, the values of $\mu^{0.27}$ and ρ can be written in terms of an average axial temperature in the ring. These average axial temperatures vary radially from ring to ring, because of the variation in neutron and therefore heat fluxes. Since Δp will be the same for all core rings (or regions), a number of simultaneous equations can be written for all rings: Knowing the values of H, d, and α_r, one can then compute G for each ring. Since the radial variation in temperature is a function of the nuclear characteristics of the core, fluid-flow and nuclear characteristics are closely coupled.

A more exact treatment obviously involves taking into account axial as well as radial variations in heat flux and the radial interconnections (cross flow, etc.) between the different core regions.

Robinson [67] has suggested the use of the following conservative heat-transfer correlation for randomly packed beds. This correlation is some 25 percent lower than the mean of the best available data:

$$h = 0.43 G_e c_p \, \text{Re}^{-0.30} \text{Pr}^{0.66} \tag{8-15}$$

where h = coefficient of heat transfer, Btu/hr ft^2 °F

c_p = specific heat of gas, Btu/lb$_m$ °F

Experimental data [70] on a full-scale, 4-ft-diameter bed packed with $\frac{3}{8}$-in.-diameter alumina spheres, heated by combustion gases, and cooled by air with a maximum outlet temperature of 2500°F resulted in the following heat-transfer correlation:

$$\text{St Pr}^{\frac{2}{3}} = 0.40 \, \text{Re}^{-0.437} \tag{8-16}$$

where St is the Stanton number ($= \text{Nu}/\text{Re Pr}$). The specific heat and viscosity of the coolant (air in the experiments) are to be determined at the mean film temperature. The characteristic length to be used in the Reynolds number here is the average diameter of a pebble.

8-11. THE ARBEITSGEMEINSCHAFT VERSUCHSREAKTOR (AVR).

The AVR is a small 15 Mw(e) gross, 13.2 Mw(e) net, 49 Mw(t), helium-cooled pebble bed reactor, built at Jülich, West Germany by Arbeitsgemeinschaft Brown/Boveri/Krupp-Reaktorbau [71]. AVR is the only pebble-bed reactor in operation.

The fuel elements are graphite spherical pellets of the type shown in Fig. 8-22b. They measure 2.36 in. in diameter, each containing an insert of 20 percent enriched uranium carbide and graphite in the center for a total of 3.5 g_m of uranium in each. The pellets therefore contain both fuel and moderator. They are introduced to the 12-ft-diameter, 14-ft-high core at the top and removed from the bottom while the reactor is in operation. In order to assure an even flow pattern of the pellets through the core, experiments using colored glass beads on a scale model of the core were conducted. The core is graphite reflected.

Because the fuel elements are not completely impermeable, some radioactive fission gases will diffuse into the coolant. The reactor vessel, Fig. 8-25, is therefore designed to contain the core, the entire primary gas loop and the steam generators. The vessel is a double-walled steel tank. The steam generators (2) are situated above the core (7), while the blowers (11) are below it. Helium flow is such that the inner wall (13) of the double-walled pressure vessel is kept cool, at temperatures not higher than 570°F.

1.07 × 10⁶ lb_m/hr of helium flow through the core at 40 fps and 162 psia. It is heated in the core from 392 to 1560°F. 121,000 lb_m/hr of steam is produced at 1045 psia and 940F. Table 8-6 lists some of the operational parameters of AVR.

TABLE 8-6
Some Design Parameters of AVR

Heat output	49 Mw(t)
Electrical output	15 Mw(e) gross, 13.2 Mw(e) net
Fuel elements	Spherical, 2.4 in. diameter
Fuel composition	UC_2, 20% enriched, graphite clad
Fuel loading	17 kg_m U^{235}
Core dimensions	12 ft diam, 14 ft high
Average heat flux	180,000 Btu/hr ft²
Average specific power	3000 kw(t)/kg_m U^{235}
Coolant	He
Coolant temperatures and pressure .	392°F inlet, 1560°F outlet, 162 psia
Coolant flow	1.07 × 10⁶ lb_m/hr, 40 fps
Moderator	Graphite
Pressure vessel	19 ft diam, 82 ft high, 1.6 in. thick
Steam	121,000 lb_m/hr, 940°F, 1,045 psia

Because of the in-vessel design of AVR, one of the major safety considerations is in the case of a steam boiler-tube failure resulting in water entering the reactor (water is at a higher pressure than helium) with consequent water-graphite chemical reaction, enhanced by

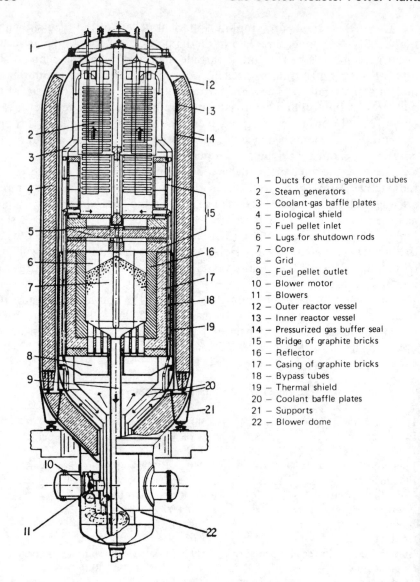

1 — Ducts for steam-generator tubes
2 — Steam generators
3 — Coolant-gas baffle plates
4 — Biological shield
5 — Fuel pellet inlet
6 — Lugs for shutdown rods
7 — Core
8 — Grid
9 — Fuel pellet outlet
10 — Blower motor
11 — Blowers
12 — Outer reactor vessel
13 — Inner reactor vessel
14 — Pressurized gas buffer seal
15 — Bridge of graphite bricks
16 — Reflector
17 — Casing of graphite bricks
18 — Bypass tubes
19 — Thermal shield
20 — Coolant baffle plates
21 — Supports
22 — Blower dome

FIG. 8-25.   The AVR pebble-bed reactor.

high temperatures.   Another consequence: If a sufficient amount o
water enters the reactor, the thermal capacity of the core contents an
reflector would vaporize all the water resulting in a pressure rise beyon
the vessel design pressure.   To guard against this event, the steam gene
rator is divided into four independent units, the loss of water from anyon
would not raise the reactor pressure beyond the design pressure.   A rea

tor safety valve and special valves to close off the leaky unit are also provided. A spray-cooled vessel is also provided to receive the steam-gas mixture should the pressure rise above the safety-valve setting.

Another safety consideration is the possible contamination of the steam loop and turbine by radioactive fission gases in case of boiler-tube leakage and loss of pressure. This is prevented by a series of valves that are set to close if the boiler pressure falls below the gas pressure.

Load following control of AVR relies upon the negative temperature coefficient of reactivity, which is large because of the near-homogeneous structure of the fuel and moderator. Control is accomplished by varying the coolant flow. Four control rods are provided to compensate for slow reactivity changes and for shutdown.

8-12. THE PEBBLE-BED-REACTOR STEAM POWER PLANT

Another pebble-bed-reactor power plant is one designed, but never built, by Sanderson & Porter for the U.S. Atomic Energy Commission [67, 72]. It is a single-pressure cycle utilizing a helium-cooled thermal-breeder 300-Mw(t) reactor. The reactor (Fig. 8-26) consists of an active core of randomly packed spheres which are mixtures of graphite and fuel. The core is surrounded by a breeder blanket, composed of spheres of graphite and fertile fuel. The core and breeder are separated by a graphite cylindrical wall. The reactor pressure vessel is cooled on the inside by the incoming coolant and thus is thermally insulated from the rest of the reactor. Because of the materials used, the reactor may operate at temperatures sufficiently high to generate steam suitable for modern turbines.

The reactor core is 9.0 ft in diameter and 8.1 ft high. The graphite cylinder surrounding it is 4 in. thick. It contains 228,000 1.5-in.-diameter spherical fuel-graphite elements containing 10 percent by mass of ThO_2 and UO_2 and graphite. The breeder blanket is 16 in. thick. It is composed of 296,300 elements of the same size as the core but containing 50 percent by mass of ThO_2.

Some characteristics of this pebble-bed-reactor power plant are given in Table 8-7.

Helium enters the reactor vessel via the annular spaces of three coaxial pipes (two shown in Fig. 8-26) connecting to the reactor-vessel lower hemisphere. It flows upward in the annular space between the reactor vessel and a single thermal shield 2 in. thick. It then flows downward through both core and blanket (in parallel), past a bottom core grate, into a plenum and leaves the reactor vessel via the three central pipes of the three coaxial pipes in the lower hemisphere.

Control is accomplished by six control rods operating axially in graphite posts that penetrate the core. Fuel loading and unloading (at about three elements per day) is done during reactor operation by the use of double-valved loading and unloading tubes. These tubes, placed

TABLE 8-7
Some Characteristics of the Pebble-Bed-Reactor Power Plant

Reactor power:	
Core	290.4 Mw(t)
Blanket	45.3 Mw(t)
Total	335.7 Mw(t)
Plant net power	125.0 Mw(e)
Coolant pressure	965 psia
Reactor inlet temperature	550°F
Reactor outlet temperature	1233°F
Coolant flow	1.343×10^6 lb_m/hr
Total blower power (three blowers) .	8.536 Mw
Steam pressure	1,450 psia
Steam temperature	1000°F
Steam flow (three steam generators) .	1.056×10^6 lb_m/hr
Core diameter	9.0 ft
Core Height	8.1 ft
Core loading:	
U^{233}	68.2 kg_m
Th^{232}	750.2 kg_m
Graphite	11,085 kg_m
Average neutron flux (thermal) . . .	Starts up at 1.3×10^{14}; after 100 days 2.2×10^{14}
Maximum fuel-surface temperature .	1800°F
Core power density	0.915 Mw/ft³ of core
Blanket equivalent height	11.1 ft
Blanket thickness	1.66 ft
Blanket loading:	
Th^{232}	13,400 kg_m
Graphite	13,400 kg_m
Breeding ratio:	
Core	Initial 0.477
Blanket	Initial 0.363
Reactor vessel shape and size	Cylindrical with hemispherical ends, 13 ft inside diameter and 21 ft overall height

vertically on the top and bottom of the core, respectively, have a diameter such that only one element passes at a time. The discharged elements are held several days in a storage tube to allow for sufficient decay of radioactivity before removal from the plant.

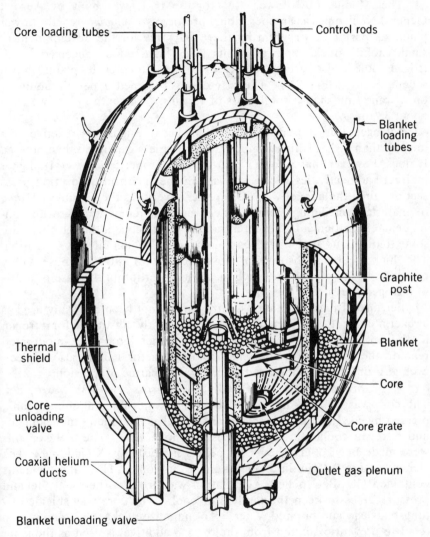

Core loading tubes

Control rods

Blanket loading tubes

Graphite post

Thermal shield

Blanket

Core

Core unloading valve

Core grate

Coaxial helium ducts

Outlet gas plenum

Blanket unloading valve

FIG. 8-26. Pebble-bed reactor. (Courtesy Sanderson & Porter, Inc.)

The reactor, heat exchangers, and blowers are all mounted in the same elevation to minimize thermal-expansion problems. In the heat exchangers, as in the reactor, the cooler gas is made to sweep the walls, thus maintaining them near the lowest temperatures of the system.

8-13. A GAS-COOLED MOBILE LOW-POWER REACTOR

The Mobile Low-Power Reactor-1, ML-1, was built by Aerojet General Nucleonics as an operational prototype of a *mobile* reactor power plant, as part of a program by the U.S. Army for the development of small mobile nuclear power plants [73]. ML-1 was conceived as nitrogen-cooled, gas-turbine powered, direct closed-cycle plant, shown schematically in Fig. 7-3. The reactor was designed to permit mounting on a trailer and had a total mass of 15 tons. The entire plant weighed about 38 tons.

Nitrogen was chosen as the coolant because it is less reactive chemically than CO_2 at high temperatures, and less expensive than helium. It does, however, have poorer heat-transfer properties (Sec. 7-4), higher indirect radioactivities (Sec. 7-3), and higher neutron absorption cross sections than the above two. Of greater severity is the possible *nitriding* of materials, including stainless steel, at high temperatures, with consequent embrittlement and intergranular corrosion, although there are some indications that additives can be used to inhibit this corrosion. The high neutron absorption cross sections of nitrogen pose safety problems because of the positive reactivity insertion in the case of a loss of nitrogen pressure in the core.

Because of these limitations, nitrogen has not been generally used as a reactor coolant. Helium has been subsequently substituted for nitrogen in the program. Under consideration is a beryllium-oxide-moderated reactor, the Experimental Beryllium Oxide Reactor, EBOR, also a direct-cyel, gas-turbine plant intended primarily for marine propulsion.

ML-1 produced 3.3 Mw(t), 0.4 Mw(e) gross and 0.33 Mw(e) net. The core was a 22-in.-diameter, 28-in.-high water-filled aluminum tank, pierced by 73, 1.875 in. diameter pressure tubes containing fuel elements and nitrogen coolant. The water is the moderator. The fuel elements were made of 93%-enriched UO_2 pellets in Hastelloy-X cladding. The elements were 0.25 in. in diameter and 28 in. long. The outer elements contained UO_2-BeO pellets. Insulation was provided between fuel and pressure tubes to keep the moderator cool. The core was shielded by tungsten, lead and borated water. To minimize package mass, tungsten (clad to prevent oxidation from the oxygen additive) is used as the inner shield. Tungsten is more dense than lead, has good resonance capture of epithermal neutrons with resultant weak gamma photons, and results in a shield lighter than lead. An outer layer of lead is used, however, to attenuate gamma radiation. The reactor was controlled by 5 semaphore blades containing Ag-In-Cd.

The nitrogen coolant entered the reactor at 800°F and 308 psia and left at 1200°F.   It entered the gas turbine at 1200°F and 285 psia.   Table 8-7 contains some design data of ML-1.

TABLE 8-7
Some Design Data of ML-1

Heat output	3.3 Mw(t)
Electrical output	0.4 Mw(e) gross, 0.33 Mw(e) net
Fuel elements	61, 19-rod clusters
Fuel composition and loading	UO_2-BeO, 93% enriched, clad in 30-mil Hastelloy-X; 49 Kg_m U^{235}
Core dimensions	1.8 ft diameter, 1.8 ft high
Average heat flux and specific power .	80,500 Btu/hr/ft², 67.5 kw/kg_m U^{235}
Coolant and flow	N_2, 89,000 lb_m/hr, 156 fps
Coolant temperatures and pressure .	800°F inlet, 1200°F outlet, 308 psia
Moderator	H_2O
Working fluid	N_2 coolant in direct-cycle gas turbine

8-14.   OTHER GAS-COOLED REACTOR PLANTS

(a)   An Indirect-Closed Cycle

The plant shown in Fig. 8-27 was proposed by Daniels [74] and is believed to be one of the very first proposed gas-cooled cycles.   The indirect-closed-cycle plant was conceived as a small unit (5 Mw) of low initial cost and easy transportability.   It incorporates many of the features proposed for later designs.   The fuel elements, ceramic and clad in graphite, were suggested for operation at maximum temperatures around 3600°F.   The reactor is graphite-moderated, reflected, and helium-cooled.   No metals are used within the core.   An indirect cycle was chosen because of the uncertainty as to the extent of the diffusion of the radioactive fission gases into the helium primary coolant.   The secondary coolant suggested for this plant was air.

The fuel is 10 percent enriched uranium carbide (UC_2).   The reactor flux may be flattened either by varying the enrichment around this value (as an average) so that the outer rings contain richer fuel or by varying the UC_2-graphite ratio so that the same rings contain more fuel.   The graphite reflector contains some thorium carbide rods for breeding purposes (to U^{233}).   Control and safety rods are of molybdenum-bearing boron or boron carbide (B_4C).   Molybdenum was chosen because it is cheap and can stand high temperatures (melting point 4750°F).   It cannot, however, be used in the presence of oxygen or nitrogen.

The cycle flow diagram (Fig. 8-27) shows that helium gas, pressurized to 225 psi, is heated in the reactor to 1350°F.   It then passes through a

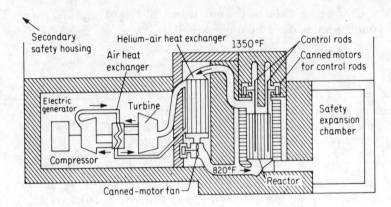

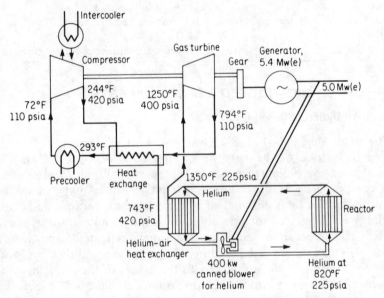

FIG. 8-27. The Daniels gas-turbine power plant. (Ref. 74.)

He-air heat exchanger from which it is pumped back to the reactor at
820°F. This heat exchanger is in the form of a stainless-steel shell 5 ft
in diameter and 25 ft long. Helium passes through 3,000 parallel $\frac{1}{2}$-in.
tubes with a total surface area of 10,000 ft², while air counterflows outside
the tubes at 420 psi and is heated from 743 to 1350°F.

The secondary coolant, air, expands in the gas turbine to 110 psi and
794°F. It then gives up further heat to the same air leaving the compressor, in an air-to-air heat exchanger. This self-interchange serves to
cool the air before entering the compressor, reducing the compressor
losses but at the same time reserving the heat removed for reuse in the

cycle. Heat rejection (necessary to complete a thermodynamic cycle) takes place in the precooler, which cools the air further to 72°F before it enters the compressor. The compressor is equipped with an intercooler for further reduction of compressor work load. Compressed air leaves the compressor at 420 psi and 244°F, picks up the heat rejected in the air-to-air heat exchanger, and reenters the He-air heat exchanger at 743°F, completing the cycle.

A helium purification system (not shown) removes dust, moisture, air, and fission products from the coolant. The reactor building is equipped with an expansion chamber that has a volume roughly fifteen times that of the pressurized helium in the reactor circuit. This chamber is designed to contain the entire helium gas (in which some of the radioactive fission products may be present) at low pressure in case of a system leakage.

Other gas-cooled power plants under study and employing gas turbines are the marine gas-cooled reactors (MGCR) designed by General Dynamics Corporation for the AEC[75], the merchantship reactor designed by American-Standard Corporation for the AEC[76], the Gascooled Reactor Experiment (GCRE) of the Army program designed by the Aerojet-General Nucleonics Corporation [73], and others.

The MGCR is a helium-cooled, graphite-moderated heterogeneous reactor utilized in a direct-closed-cycle power plant. It will supply 20,000 hp to the propeller shaft and 750 hp for ship auxiliaries. Helium at 790°F and 970 psi will enter the reactor through the annulus of coaxial pipes, pass between the pressure vessel and moderator, through the fuel, and leave the reactor at 1300°F through the inner coaxial pipe. In marine application, much attention is devoted to plant mass and consequently to shielding. The American Standard plant is similar to MGCR except in the fuel-element design.

The GCRE is a nitrogen-cooled direct-closed-cycle power package designed for easy portability, operation under extreme environmental conditions, and independence of special supporting facilities. GCRE was a prototype for ML-1 (Sec. 8-13).

(b) The Heavy-Water-Moderated, Gas-Cooled Reactor

Such reactors [49, 77] employ pressure tubes for separation of the gas and water. Their main advantage is in the ability to attain high neutron economy.

(c) The Nuclear Gas Engine

In this concept, proposed by Fraas [78], a reciprocating two-cycle engine with valves in the head and ports in the liner is used. On the compression stroke, the piston forces gas at 100 atm into the reactor via a valve in the head. At top center, this valve closes and another opens, admitting hot gas from the reactor. After expansion, ports are uncovered near bottom center and cool gas is forced into the cylinder scavenging the expanded gases to a cooler and blower, and later back to the cylinder. Heat is added and rejected (ideally) at constant pressure. Reciprocating engines of this type have higher thermal efficiencies than gas turbines at lower average temperatures. The conceptual design calls for nitrogen at 1200°F reactor exit temperature. Problems here include breathing losses, leaks, lubrication, and oil carryover.

PROBLEMS

8-1. An indirect closed cycle gas-cooled reactor power plant using helium as coolant and steam as working fluid generates 900 Mw(e). Steam is produced at the rate of 6×10^6 lb_m/hr at 1,600 psia and 1000°F from feedwater at 250°F. Helium at an average pressure of 350 psia enters and leaves the steam generator at 1350 and 650°F respectively. The gas pressure drop throughout the primary circuit is related to gas flow rate $\dot{m}$ (lb_m/hr) by 0.132×10^{-12} $(\dot{m})^2$ (psi). The helium blowers have mechanical efficiency of 73.2 percent, motor efficiency of 90 percent and polytropic coefficient of expansion of 1.50. Ignoring all auxiliaries but the blowers, find (a) the steam plant thermal efficiency, (b) the thermal output of the reactor, Mw(t), and (c) the thermal efficiency of the entire plant.

8-2. A simple gas-steam cycle reactor power plant uses helium as a reactor coolant. 7×10^6 lb_m/hr of helium enter the steam generator at 1420°F and leave at 750°F. 1200°F steam is generated at 2,500 psia from feedwater at 440°F. Assuming no heat losses, find the temperature difference at the pinch point in the steam generator.

8-3. The steam turbine in Prob. 8-2 has an adiabatic efficiency of 90 percent and a mechanical efficiency of 95 percent. The efficiency of the electric generator is 88 percent. The condenser operates at a pressure of 1 psia. Find the gross electrical output of the plant and the corresponding thermal efficiency of the steam cycle. Assume that steam bleeding for feedwater heating reduces turbine work by 10 percent.

8-4. A dual-pressure gas-steam cycle uses 12,000 lb_m/sec of CO_2 as reactor coolant. The gas enters the steam generator at 700°F. One third of the steam is generated at 700 psia and 680°F, and the remainder at 180 psia and 660°F. The feedwater temperature is 160°F. The economizer pinch point temperature difference is kept at 27°F. Find (a) the CO_2 temperature at the generator exit, (b) the amount of steam generated in lb_m/hr, and (c) the Mw(t) input to the steam cycle.

8-5. Repeat the above problem but consider steam to be generated only at the higher pressure. Would you expect the same differences between the two problems had the reactor been of the AGR or HTGR types? Discuss.

8-6. An indirect closed-cycle, gas-to-water reactor power plant generates

2×10^6 lb_m/hr of steam at 1,000 psia and 800°F from feedwater at 200°F. Helium coolant at 200 psia leaves the reactor at 840°F and the boiler at 540°F, is pumped with a polytropic exponent $n = 1.50$, and undergoes a 60-psi pressure drop throughout the primary loop. Assuming no heat losses, find (a) the mass flow rate of coolant, (b) the reactor power output in Mw(t), and (c) the coolant pumping requirements in Mw, exclusive of blower and motor losses.

8-7. A helium-cooled pebble-bed-reactor core is in the form of a cylinder 12-ft diam and 10-ft high. It contains 2-in.-diam fuel elements. Steam is generated at 1,000 psia and 950°F from feedwater at 400°F. 1.81×10^6 lb_m/hr of coolant at 60 atmospheres leave the reactor at 1150°F and the steam generator at 500°F. Assuming no heat losses, find (a) the reactor power in Mw(t), (b) the coolant blower requirement in Mw, and (c) the percentage of plant electrical output used by the blowers if the plant efficiency is 44 percent, the motor-blower efficiency is 80 percent, and the pressure losses in the rest of the coolant loop are 20 percent of those in the core.

chapter **9**

The Fast-Breeder Reactor

9-1. INTRODUCTION

The growth of nuclear power in the world's electric utility industry in the late 1960's has primarily relied upon water-cooled and gas-cooled thermal neutron reactors. These are burners which in effect consume fissionable fuel with low conversion ratios (Sec. 9-3). The expected continued growth in the use of nuclear power and consequent consumption of available nuclear fuels have prompted the reevaluation of long-term goals. Ways of conserving uranium reserves and of keeping fuel-cycle costs down as fuel costs go up are therefore sought. The fast breeder reactor is the logical step in that direction.

Present-day burner reactors produce some Pu^{239} from the U^{238} in their fuel. Their fuel-cycle costs are lowered by the so-called "plutonium credit," or government buyback. This policy will, however, be discontinued in the early 1970's and a market for a growing supply of converted plutonium must be found. Plutonium-fueled reactors could provide this market. Plutonium recycle systems, i.e., the use of thermal reactor-produced plutonium in plutonium-fueled thermal reactors has the advantages of not developing new reactor systems but the overriding economic disadvantage of boosting plutonium prices and fuel-cycle costs or both. Plutonium used in plutonium breeders therefore is the logical answer. These are fast reactors whose principal attraction is their ability to potentially attain high breeding ratios which would result in the reduction of fuel costs and postpone the need for expensive mining of low-grade fuel ores. Despite this potential, much development work remains to be done to make individual fast reactors economic.

Ways of integrating high conversion reactors and breeder reactors, as they are developed to an economical state, into the thermal reactor economy are therefore to be sought. One of the schemes proposed [79] is that of a reactor "mix" in which an integrated system of burners, converters, advanced converters and low- and high-gain breeders might coexist or succeed each other. The advanced converters would probably be represented by heavy-water moderated and high-temperature thermal gas-cooled reactors in which structural materials have low neutron cross sections. High-temperature thermal gas-cooled reactors, for example, can attain conversion ratios between 0.6 and 1.0. This mix is believed to guarantee continuing low fuel-cycle costs and provide optimum economic generating plants for the electric utility industry.

Fast reactors are those whose neutrons are not slowed down to thermal energies by a moderator. Coolant and other reactor materials, however, moderate the neutrons to a certain extent in a fast reactor so that the neutron spectrum extends from fission energies that may be as high as 17 Mev but average about 2 Mev down to about 0.05 or 0.1 Mev. A *hard-spectrum* fast reactor is one in which the neutron density distribution extends over a narrow range nearer the high end of the energy spectrum. A *soft-spectrum* fast reactor, on the other hand, is one in which more moderation occurs, such as by liquid sodium and oxide fuels and therefore the distribution extends to lower energies. A small amount of moderator such as BeO or zirconium hydride may be added to a fast reactor to assure a large negative Doppler coefficient of reactivity (Sec. 9-9). A breeder reactor is one in which more fissionable fuel is produced than is consumed, or one that has a conversion, or breeding, ratio greater than I.

This chapter discusses the characteristics of fast reactors in general. The next chapter will present some representative fast-reactor power plants that have been built or are in the design stage.

9-2. NUCLEAR REACTIONS IN FAST-BREEDER REACTORS

A typical neutron-flux spectrum in a liquid-metal-cooled fast reactor compared to that in a water-cooled thermal reactor is shown in Fig. 9-1. The objective in the latter is to maintain a chain reaction with thermal neutrons having energies below 1 ev. In a fast-breeder reactor, the objective is to maintain a chain reaction with fast neutrons having an average energy of about 1 Mev, by fission in U^{235} and Pu^{239}. It also must provide additional fast neutrons sufficient to convert U^{238} to Pu^{239}.

Typical fission reactions in U^{235}-fueled fast reactors are the same as those in thermal ones, as for example

$$_{92}U^{235} + {}_0n^1 \longrightarrow {}_{56}Ba^{137} + {}_{36}Kr^{97} + 2{}_0n^1 \qquad (9\text{-}1)$$

The fission product yield in fast fission is flatter than that in thermal fission, Fig. 9-2, i.e., there are more products with intermediate mass numbers, A, between 105 and 130 in fast fission.

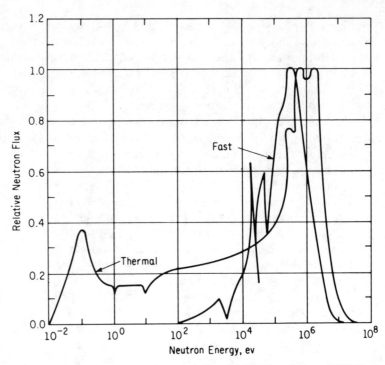

FIG. 9-1. Typical neutron spectra in fact (LMFBR) and thermal (PWR)
reactors.

U^{233} and Pu^{239} typical fission reactions are

$$_{92}U^{233} + _{0}n^{1} \longrightarrow {}_{56}Ba^{136} + _{36}Kr^{96} + 2_{0}n^{1} \tag{9-2}$$

and

$$_{94}Pu^{239} + _{0}n^{1} \rightarrow {}_{56}Ba^{137} + _{38}Sr^{100} + 3_{0}n^{1} \tag{9-3}$$

The average number of neutrons emitted in the above reactions, v, depend upon the fissile isotope used and to a small degree, neutron energy. v is highest for fast fission in plutonium, being about 3 (Table 9-1).

A typical nonfission capture reaction by a fissionable nucleus is

$$_{94}Pu^{239} + _{0}n^{1} \longrightarrow {}_{94}Pu^{240} \tag{9-4}$$

Pu^{240}, as U^{238}, is fertile and also fissionable by fast neutrons. However, it has a high ratio of nonfission capture-to-fission, α, so that its production from Pu^{239} represents a net loss to the overall neutron balance.

Other parasitic reactions in fast reactors are those due to coolant and structural materials. In sodium-cooled, stainless-steel cores, two of the reactions are

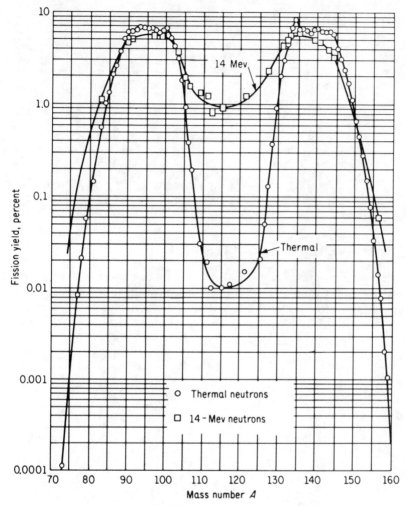

FIG. 9-2. Fission product yield for U^{235} by thermal and 14-Mev neutrons. (Ref. 80.)

$$_{11}Na^{23} + {}_0n^1 \longrightarrow {}_{11}Na^{24}$$
$$_{11}Na^{24} \xrightarrow{\text{14.8 hr}} {}_{12}Mg^{24} + {}_{-1}e^0$$
$$\left.\right\} \quad (9\text{-}5)$$

and

$$_{26}Fe^{56} + {}_0n^1 \longrightarrow {}_{26}Fe^{57} \qquad (9\text{-}6)$$

where the end products Mg^{24} and Fe^{57} are stable. The intermediate product Na^{24} is highly radioactive (Sec. 9-11).

TABLE 9-1
Fuel Production Constants

Constants	Fissionable Fuels								Fertile Materials	
	Thermal				Fast				Fast	
	U^{233}	U^{235}	Pu^{239}	Pu^{241}	U^{233}	U^{235}	Pu^{239}	Pu^{241}	Th^{232}	U^{238}
σ_f^*	527	577	790	1000	2.2	1.4	1.78	2.54	0.025	0.112
β	0.0026	0.0065	0.0020		0.0027	0.0065	0.0020	0.0053	0.0204	0.0147
ν	2.51	2.40	2.90	2.98	2.59	2.50	3.0	3.04	2.04	2.6
η	2.28	2.06	2.10	2.13	2.42	2.20	2.6	2.73	2.0	2.27
α	0.10	0.17	0.5	0.40	0.068	0.15	0.15	0.114		1.44
f_b	0.561	0.515	0.524	0.531	0.587	0.545	0.615	0.634	0.500	0.615
C_{max}	1.28	1.06	1.10	1.13	1.42	1.20	1.6	1.73	1.0	1.17

*barn

9-3. CONVERSION AND BREEDING

Breeding reactions begin with the nonfission capture of a neutron by a fertile nucleus such as U^{238} and Th^{232} and result in a series of reactions culminating in fissionable Pu^{239} and U^{233} respectively. The reactions are

$$\left. \begin{array}{l} {}_{92}U^{238} + {}_0n^1 \longrightarrow {}_{92}U^{239} + \gamma \\[4pt] {}_{92}U^{239} \xrightarrow{\text{24 min}} {}_{93}Np^{239} + {}_{-1}e^0 \\[4pt] {}_{93}Np^{239} \xrightarrow{\text{2.3 days}} {}_{94}Pu^{239} + {}_{-1}e^0 \end{array} \right\} \tag{9-7}$$

and

$$\left. \begin{array}{l} {}_{90}Th^{232} + {}_0n^1 \longrightarrow {}_{90}Th^{233} + \gamma \\[4pt] {}_{90}Th^{233} \xrightarrow{\text{22 min}} {}_{91}Pa^{233} + {}_{-1}e^0 \\[4pt] {}_{91}Pa^{233} \xrightarrow{\text{27 days}} {}_{92}U^{233} + {}_{-1}e^0 \end{array} \right\} \tag{9-8}$$

The rate of change of fissionable nuclei is given by

$$\frac{dN_{ff}}{d\theta} = (\text{rate of production}) - (\text{rate of depletion}) \tag{9-9}$$

where N_{ff} is the number density of fissionable fuel nuclei in the reactor at time θ.   The above rate is positive if the rate of production exceeds the rate of depletion.   This cannot happen, of course, in reactors containing highly enriched fuels, but only when large amounts of fertile material is present.

We shall first discuss the rate of production.   For this, the most important parameters are ν, η, and α. ν, it is to be recalled [2] is the average number of neutrons produced in a single fission reaction. η, the fission factor, is the average number of neutrons produced per neutron absorbed, and not necessarily causing fission, in a fuel nucleus. α is the ratio of the rate of neutron nonfission capture to the rate of fission, given by

$$\alpha = \sigma_c / \sigma_f \qquad (9\text{-}10)$$

where σ_c and σ_f are microscopic cross sections for nonfission capture and fission respectively.

Table 9-1 gives, among other parameters, typical values of v, η and α for thermal and fast reactors, averaged over ranges of energies typical of these reactors. It should be noted here that, in design, much finer averaging is used. Nonetheless the values given are useful for discussion purposes.

v varies with the fuel nucleus but little with neutron energy. η is a stronger function of neutron energy, Fig. 9-3, and is generally higher in a fast neutron spectrum than in a thermal neutron spectrum. α generally decreases with neutron energy (Table 9-1) so that there is a smaller loss of neutrons by parasitic capture in the fuel in a fast reactor than in a thermal reactor.

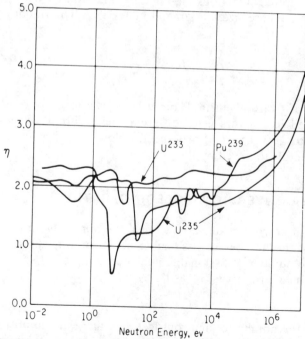

FIG. 9-3. η as a function of neutron energy for three fissionable nuclei.

η is the most important factor in breeding, for it is the number of neutrons necessary to sustain fission, parasitic absorptions, leakage, as well as breeding. It is related to v and α by

$$\frac{v}{\eta} = \frac{\text{rate of neutron absorption in the fuel}}{\text{rate of fission reactions in the fuel}}$$

$$= 1 + \frac{\text{rate of nonfission capture}}{\text{rate of fission}} = 1 + \alpha \qquad (9\text{-}11a)$$

or
$$\eta = v \left(\frac{1}{1 + \alpha} \right) \qquad (9\text{-}11b)$$

The rate of production is dependent upon the number of excess neutrons available, in a given time, for conversion or breeding. These are equal to the number produced, in the same time, by fission minus those used up in fission and lost by leakage and nonfission capture in the fuel and all other core materials. The fraction of neutrons needed to sustain a steady state is $1/\eta$. In a Pu^{239}-fueled fast reactor, for example, this fraction is $1/2.6$ or 38.5 percent. The balance, called f_b, is the fraction available for breeding (or conversion) and losses by leakage and nonfission capture reactions in all core materials. These losses are functions of the core size, design and materials. The fraction f_b is given by

$$f_b = 1 - \frac{1}{\eta} \qquad (9\text{-}12)$$

Values of f_b are given in Table 9-1. f_b is a measure of what can be accomplished in reactors using different fuels. For Pu^{239}, for example, it is equal to $1-0.385$ or 61.5 percent. Note that f_b is highest for Pu^{241} in a fast reactor, followed by Pu^{239}, etc.

9-4. THE CONVERSION (BREEDING) RATIO

We shall now introduce a new term, the *conversion ratio, C*, equal to the ratio of the number of fissionable nuclei produced from fertile material to the number of fissionable nuclei consumed in fission and nonfission reactions. It is given by

$$C = \eta - 1 - L \qquad (9\text{-}13)$$

where L is the number of neutrons lost by leakage and nonfission capture per neutron absorbed in a fissionable nucleus. (It is to be noted that in multigroup calculations, C is more complex than above. The present iscussion is, however, adequate for our purposes.) The maximum ossible value of C, C_{max}, is obtained if L were reduced to zero or

$$C_{max} = \eta - 1 \qquad (9\text{-}14)$$

Depending upon η and L, C can be much less than unity, and the eactor is called a *burner*. A reactor with a low value of C is generally alled a *converter*. One with high C but less than 1.0 is called an dvanced converter. For C less than 1.0 it can be shown [81] that there s a maximum theoretical limit to the amount of fertile nuclei that can e converted to fissionable nuclei. This maximum depends upon both C and the initial number of fissionable nuclei.

$C = 1$ means that the reactor is producing a number of fissionable uclei equal to what it consumes. $C > 1$ means that there is no limit to onversion. In both these cases it is theoretically possible to consume ll fissionable and fertile nuclei present. In practice, however, as with urner reactors, the fuel elements must be reprocessed and replaced peri- dically because fission products absorb neutrons and reduce reactivity or ecause of metallurgical considerations.

All but reactors using fully enriched fuel (no fertile isotopes) are roadly classified as (1) *converters* and (2) *breeders*. A converter is efined in two ways: it is either (a) a reactor that produces less fissionable uel than it consumes, or (b) a reactor that consumes one type of fuel and roduces another, such as one consuming U^{235} and producing Pu^{239} rom U^{238}, regardless of the conversion ratio. A breeder is also defined n two ways: (a) It is a reactor that produces more fissionable atoms han it consumes (and which consumes and produces fuels of any kind), e., one with $C > 1$, or (b) it is a reactor that produces the same kind of uel that it consumes, such as one consuming Pu^{239} and producing Pu^{239} rom U^{238} (or consuming U^{233} and producing the same from Th^{232}).)efinitions (a) are more commonly accepted.

The term *breeding ratio* has the same meaning as conversion ratio. *reeding* (or *conversion*) *gain G* is the gain in fissionable nuclei per fis- onable nucleus consumed. Thus

$$G = C - 1 = \eta - 2 - L \qquad (9\text{-}15)$$

nd maximum gain, G_{max}, is

$$G_{max} = C_{max} - 1 = \eta - 2 \qquad (9\text{-}16)$$

Within the above-mentioned two broad categories one can find in he literature references to advanced converters, high-gain converters, ow-gain breeders, high-gain breeders, etc.

The last row in Table 9-1 gives values of C_{max}. Allowing for the fact hat L is finite and therefore C is less than C_{max}, it can be seen that best breeding can occur in a fast reactor fueled with fissionable plutonium (Pu^{239} and Pu^{241}) followed by fast reactors fueled by U^{233} and U^{235} in that order. In the case of thermal reactors, substantial breeding can be expected only from U^{233} as a fuel.

Pu-U mixed fuels, in oxide form, are therefore used widely as fuel material in fast-breeder reactors. In the SEFOR fast reactor (Sec. 10-4 the fuel composition is 18.7 percent fissile plutonium, Pu^{239} and Pu^{241}, 1.4 percent nonfissile plutonium, Pu^{240} and Pu^{242}, and the rest is depleted uranium, mostly U^{238} plus minute amounts of U^{235} and U^{236}, all in oxide form. Another reason for the use of plutonium-uranium mixed fuels will be given in Sec. 9-7.

An accurate evaluation of L depends upon reactor core materials and design, and will not be treated here. It is, however, instructive to discuss some of the effects of reactor materials on breeding.

It has been shown (Table 9-1) that α, the ratio of nonfission capture to fission reaction rates, decreases with neutron energy for all fuels. Thus there is less parasitic capture by the fuel, in a fast reactor than in a thermal reactor. Coolant and structural materials, such as sodium iron and zirconium, however, behave differently. While the fission cross sections of fuels drop materially with neutron energy (Table 9-1) the capture cross sections of Na, Fe and Zr do not drop as fast, and the ratio of their capture cross sections to fuel fission cross sections actually go up with neutron energy (Table 9-2). There is therefore relatively more parasitic capture by coolant and structural materials in a fast reactor than in a thermal reactor. All core materials also scatter neutrons to some extent, so that they tend to soften the spectrum and lower the value of η and therefore the excess neutrons available for breeding, somewhat.

TABLE 9-2
Ratios of σ_c of Some Core Materials to σ_f of Fuels

| Fuel | Core Materials | | | | | |
| | Thermal Reactor | | | Fast Reactor | | |
	Na	Fe	Zr	Na	Fe	Zr
U^{233}	0.00096	0.0048	0.00034	0.00118	0.0040	
U^{235}	0.00087	0.0044	0.00031	0.0018	0.0061	
Pu^{239}	0.00068	0.0034	0.00024	0.00146	0.0049	0.0033

Because of these effects and other considerations (Sec. 9-6) a higher concentration of fissionable fuel must be loaded in a fast reactor than in a thermal reactor core. The usual ratio is about four or five to one This has both advantages and disadvantages. The poisonous effects of fission products during the life of the core are lessened relative to the fuel nuclei, and therefore high fuel burnups (up to and exceeding 100,000 Mwd/ton) are possible. On the other hand, this high concentration gives rise to safety and other problems (Sec. 9-6).

9-5. DOUBLING TIME

Of great economic importance to a breeder reactor is its *doubling time*.* This is the time required for a breeder reactor to produce as many new fissionable nuclei as the total number of fissionable nuclei that are both normally contained in the core and tied up in the reactor fuel cycle (i.e., in fabrication, reprocessing, etc.). In general, doubling times range between 10 and 20 years, the shorter the better.

The number density of new fissionable nuclei produced in a breeding reactor during time θ, ΔN_b nuclei/cm^3, may be given by

$$\Delta N_b = \Delta N_{ff} G = \Delta N_{ff}(\eta - 2 - L) \tag{9-17}$$

where ΔN_{ff} = **number of original fissionable fuel nuclei consumed (by neutron absorption) during time θ per cm^3**

but

$$\Delta N_{ff} = F_c(N_0)_{ff}(\sigma_a)_{ff}\overline{\varphi}\theta \tag{9-18}$$

where $(N_0)_{ff}$ = number of fissionable fuel nuclei present in the core and tied up in the fuel cycle at arbitrary time 0, nuclei/cm^3
  F_c = fraction of $(N_0)_{ff}$ that is in the core, dimensionless
  $(\sigma_a)_{ff}$ = microscopic absorption cross section of fissionable fuel, cm^2
  $\overline{\varphi}$ = average reactor neutron flux, neutrons/sec cm^2

Thus

$$\Delta N_b = F_c(N_0)_{ff}(\sigma_a)_{ff}\overline{\varphi}\theta(\eta - 2 - L) \tag{9-19}$$

θ_d, called the *simple doubling time,* by definition, occurs when $\Delta N_b = (N_0)_{ff}$. Thus

$$\theta_d = \frac{1}{F_c(\sigma_a)_{ff}\overline{\varphi}(\eta - 2 - L)} \tag{9-20}$$

Note that θ_d is shortened by operating at high power levels (high $\overline{\varphi}$) and that it is also inversely proportional to the breeding gain, $(\eta - 2 - L)$

---

* The term "doubling time" here should not be confused with that for an unsteady-state reactor which equals approximately 73.58 percent of the reactor period.

(Fig. 9-4). In practice θ_d slightly increases with reactor life because L increases with fission product buildup, because of the finite time required after start up to build up the fissile inventory in a breeder-reactor blanket, the time taken for fuel and blanket element reprocessing, the economics of reprocessing, etc.

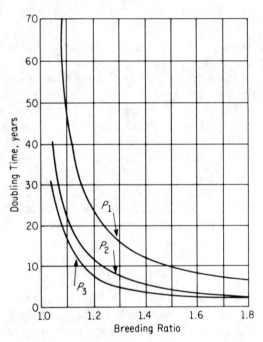

FIG. 9-4.   Fast-breeder-reactor doubling times.
Specific powers $P_3 > P_2 > P_1$.

It is of course economically desirable that doubling times be short and, ultimately, short enough so that new Pu²³⁹ inventories are continually provided for new Pu²³⁹-U²³⁸ breeders. In this case it is thought of as being reinvested and plutonium dividends are compounded. This gives rise to a *compound doubling time*, θ_{dc}, related to the simple doubling θ_d by

$$\theta_{dc} = \frac{0.6931}{\ln\left(1 + \dfrac{1}{\theta_d}\right)} \tag{9-21}$$

Simple and compound doubling times may be based on core inventory only (excluding fuel cycle inventory) and $F_c = 1.0$, in which case their values would be shortened.

9-6. SOME SAFETY ASPECTS OF FAST REACTORS

Numerous nuclear power reactors of various types and sizes have already been built and operated without endangering public safety. The continuing need for more power gives rise to new reactor designs, each bringing its own characteristic safety problems with it. Boiling- and pressurized-water reactors, for example, operate with a high-pressure coolant and are potentially subject to primary system rupture, coolant depressurization, and loss of cooling. When heat flux and cooling are no longer properly matched, departure from nucleate boiling (DNB) will cause fuel element over-heating and failure [2].

Liquid-metal-cooled fast breeder reactors (LMFBR), on the other hand, operate with low pressure coolants (sodium) which will not completely flash on depressurization. Potential sodium-air and sodium-water reactions (Sec. 9-11) pose problems but the primary system is usually surrounded with an inert atmosphere such as argon. The LMFBR is subject to positive void coefficients and fast reactors in general have short prompt neutrons lifetimes and operate with enriched fuels which may cause large reactivity insertion rates in the event of core collapse.

Because parasitic capture by core materials is higher in fast reactors (Sec. 9-4) substantial fuel enriching of 15 percent or higher (compared to 1–3 percent in water-cooled thermal reactors) is essential in fast reactors. Also because fast-fission cross sections are a few hundred times lower than those for thermal neutrons, much more fissionable fuel is required in a fast reactor core than in a thermal core of the same volume and power output. Because of the absence of moderator, however, fast reactor cores are much more compact than thermal cores of the same power. They, therefore, have much higher power densities (up to 600 kw/liter) and specific powers. The cores have more restricted coolant flow passages and more severe heat-removal problems.

While coolants such as liquid metals, gases, and steam are used because of their low moderating ability, the large specific powers require that large heat-transfer surfaces and large heat-transfer coefficients be used to reduce fuel centerline and cladding temperatures and avoid melting. Fuels are therefore divided into small-diameter rods, usually referred to in fast-reactor terminology as fuel "pins". In addition, the use of gases or steam may require the use of roughened cladding surfaces or turbulence promoters to increase the heat-transfer coefficients between coolants and cladding surfaces. Because of core compactness, core orificing becomes more difficult but also more essential in fast reactor cores to avoid the effects of power peaking.

Because of their high enrichments and because fission products have low cross sections for fast neutrons, fast-reactor fuels are less sensitive to fission-product poisoning. Thus, unlike thermal reactor fuels, fast-reactor burnup is not fission product (reactivity) limited and fast reactors operate on less excess reactivity for a given burnup requirement. Fuel burnups in fast reactors are limited more by radiation damage to the fuel. This, and good internal breeding, have resulted in high fuel burnups, of the order of 100,000 Mwd/ton and higher in fast-reactor fuels. Also reprocessing of these fuels is not as elaborate since not all fission products need be removed. With high burnups, good fission-product gas retention, or release become a major consideration in fast-reactor fuel system design. Unvented and vented fuel elements can be found in various designs. In reactor design, the possibility of dense hydrogenous material leakage into the core must be avoided, since it can cause prompt critical conditions due to the positive reactivity usually associated with the extreme softening of the neutron spectrum.

Potential accidents in a LMFBR can be of many types. To be particularly avoided are those that cause large reactivity insertion rates. Reactivity insertions caused by such malfunctions as control rod ejection are not likely to be high. Of more concern is that local failures may, due to lack of design precautions, *propagate.* If propagation becomes autocatalytic, large-scale fuel failure and sodium boiling and voiding could occur, causing the large reactivity insertion rates.

Local failure of a single pin (one of hundreds of thousands in a typical power core) rarely results in propagation to more than just a few other pins, and does not normally even require reactor shutdown. Local failure in the form of coolant *flow blockage* to a single fuel subassembly (flow blockage to more than one subassembly occurred in the Enrico Fermi reactor) could result in coolant boiling and voiding and fuel melting. Propagation here is a possibility that can be caused by pressure pulses from vapor explosion (violent boiling), reactivity effects due to fuel movement, fuel melting through the subassembly wall (though this is slow and therefore detectable), coolant reactivity effects, and others. Subassemblies without walls have the advantage of allowing the coolant to enter a blocked subassembly but the disadvantage of eliminating a strong barrier to propagation.

Other failures, termed whole-core failures, could be caused by control system malfunction, coolant pump failure, and others. They are usually detectable and their effects can be minimized or prevented by reactor shutdown.

Questions on which more information and data are needed concern the phenomenon of vapor explosion, the behavior of fuel pins in failure, the mechanism of propagation, the mechanism of motion of molten fuel, more exact calculations of the energy released in prompt critical excursions, and others.

One of the main safety considerations in fast reactors is the possibility of core *meltdown*. Meltdown can result in the formation of a supercritical mass if sufficient enriched fuel assembles in a solid mass surrounded by a good reflector. In the Enrico Fermi reactor, for example, such meltdown, if it occurs, can result in a mass which would be 6 times critical. Good reactor design ensures that melted fuel would be dispersed at the bottom in a less than critical configuration. In thermal reactors this problem is practically nonexistent because of low enrichments and fuel dilution by moderator and other core materials.

Core meltdown can be caused by (a) *power excursion,* or (b) *loss of cooling*. (a), which can result from (b), is avoided by proper controls. (b) is a more serious problem and can cause meltdown, even if the reactor is immediately shut down, because of decay heat. A rapid reinstatement of cooling is therefore essential in all situations. In sodium-cooled reactors, sodium boiling following loss of flow, and/or a sufficient sodium pool above the core, to increase time required for drain in case of pipe rupture, can reduce the seriousness of the accident. The problem is more difficult in gas- and steam-cooled fast reactors.

Most safety considerations are associated with reactor kinetics, partial loss of liquid (sodium) coolant by voiding (boiling), and the Doppler coefficient of reactivity. These are discussed respectively in the next sections.

9-7. FAST-REACTOR KINETICS

Effective reactor control is an important safety consideration. This is closely related to the delayed neutron fraction β, given for the pure fuels in Table 9-1. The effective values in typical reactors are given in Table 9-3. For the same primary fuel (such as U^{235} in enriched uranium, or Pu^{239} in a Pu^{239}-U^{238} mixture), β is larger in fast reactors than in thermal reactors, and in turn, larger in thermal heterogeneous than in thermal homogeneous reactors. This is because of the added fast fission of the fertile isotopes U^{238} and Th^{232} which have large values of β (Table 9-1). Fast U^{238} fission, for example, accounts for 15–30 percent of the total energy generated in fast reactors, compared to only 4–8 percent in thermal heterogeneous reactors. Recall that both U^{238} and Th^{232} are fissionable only by fast neutrons with a threshold about 1.4 Mev. In a thermal homogeneous reactor this percentage is even lower because of the rapid thermalization of newly born neutrons by the ever-present moderator. Thus a typical fast reactor fueled by a Pu^{239}-U^{238} mixture has its β effectively raised from 0.002 for Pu^{239} alone (Table 9-1) to 0.0034 (Table 9-3).

TABLE 9-3
Effective Values of β and θ_0 for Typical Reactors

Primary Fuel	Reactor					
	Thermal homogeneous*		Thermal heterogeneous		Fast	
	β	θ_0	β	θ_0	β	θ_0
U^{233}	0.0006	0.010	0.0029	0.051	0.0038	0.055
U^{235}	0.0013	0.020	0.0067	0.083	0.0076	0.085
Pu^{239}	0.0020	0.007	0.0025	0.035	0.0034	0.039

* From Ref. 82. θ_0 in sec.

While delayed neutrons are few, they have such long lifetimes (0.44–55 sec in all reactors) compared to those of prompt neutrons (10^{-3}–10^{-5} sec in thermal reactors and 10^{-7} sec in fast reactors) that the average neutron lifetimes, θ_0, are long enough to permit control rods to be moved and control power level. While the lifetimes of delayed neutrons in fast reactors are shorter than those in thermal reactors, their larger abundance makes fast reactor average neutron lifetimes comparable to those of thermal reactors (Table 9-3) and makes fast reactors as manageable. (Note that thermal homogeneous reactors have very short average neutron lifetimes, Table 9-3. They are, however, inherently safe because of their large negative temperature coefficient of reactivity.)

Comparing different fuels, the small β for U^{233}- and Pu^{239}-fueled reactors make their average neutron lifetimes shorter than U^{235}-fueled ones. This is a disadvantage for the former which makes their control problems more difficult, though not insurmountable. If $\bar{\varphi}_0$ and $\bar{\varphi}$ are average neutron fluxes at arbitrary times 0 and θ in a reactor subjected to a reactivity insert k_{ex}, then |1|

$$\frac{\bar{\varphi}}{\bar{\varphi}_0} = e^{k_{ex}\theta/\theta_0} \tag{9-22a}$$

The reactor period is $\theta_p = \theta_0/k_{ex}$ so that

$$\frac{\bar{\varphi}}{\bar{\varphi}_0} = e^{\theta/\theta_p} \tag{9-22b}$$

If $k_{ex} = 0.1$ percent, the flux increases by factors of about 2.17 per minute in a U^{235}-fueled reactor and 4.64 per minute in a Pu^{239}-fueled reactor. The corresponding reactor periods are 85 and 39 sec respectively.

A reactivity insert numerically equal to β results in the reactor being critical on prompt neutrons alone, the condition referred to as prompt critical.* The flux and power rise at a rate essentially determined by the prompt neutrons alone. The lifetimes of these neutrons are extremely short (above) and the flux increases so rapidly that the reactor becomes difficult or impossible to control. All reactors must therefore be operated under delayed critical conditions. Fast reactors have prompt neutron lifetimes that are 100 to 1000 times shorter than those for thermal reactors, making it particularly important to avoid prompt critical conditions in fast reactors. Power excursions associated with prompt criticality result in heat and kinetic energy of damaged reactor core materials and immediate surroundings. The reactivity transient is terminated by the Doppler effect (Sec. 9-9) or by material motion. Under certain conditions, the reactor may go through more than one prompt criticality before the excursion is finally ended.

Because β of pure Pu^{239} is much smaller than that of pure U^{235}, a fast reactor fueled with pure Pu^{239} does not have as large a leeway of operation before it reaches prompt critical conditions. This leeway is called a *control band*. It is the band between conditions wherein the reactor is self sustaining on prompt and critical neutrons (delayed critical) and on prompt neutrons alone (prompt critical). This control band should be as wide as possible from a safety standpoint since all reactors are put on a positive period (reactivity is inserted) during startup and when power is increased. The narrow control band of Pu^{239} is increased by mixing with U^{238}, a condition required for breeding in any case. Note that β of pure U^{238} is more than 7 times that of Pu^{239}, (Table 9-1). Since, in a fast-breeder reactor, U^{238} engages in fast fission and contributes a large portion of the energy generated (about 20 percent in a typical reactor), it will result in a fraction of delayed neutrons for the fuel mixture of about 0.00454, more than double that of pure plutonium alone.

9-8. THE SODIUM VOID COEFFICIENT

The void coefficient $d\rho/d\alpha$ where ρ is the excess reactivity and α is the void fractions has been defined and evaluated for water-cooled reactors (Sec. 3-7). In water-cooled reactors void formation displaces part of the coolant, and since this coolant doubles as moderator, this usually results in a negative void coefficient, which is desirable from a

---

* The unit of reactivity of one *dollar* ($) is the one that would render a reactor prompt critical. The number of dollars of reactivity are therefore equal to k_{ex}/β. One *cent* (¢) of reactivity equals one hundredth of a dollar of reactivity.

safety point of view, since increased voids often result from increased reactivity.

In sodium-cooled fast reactors the coolant degrades the neutron spectrum at the high-energy end because of inelastic scattering by sodium nuclei, and enhances the low energy end because of elastic scattering (Fig. 9-1). An increase in liquid-sodium temperature lowers its density and contributes to the overall reactor temperature coefficient of reactivity. The limiting case is complete loss of sodium, i.e., void formation either by boiling or by loss of coolant from the core.

The main effect of sodium removal, by void generation, or decrease in density, is that it hardens the neutron flux because of decreased moderation by the sodium coolant and, to a lesser extent, decreased absorption in the sodium resonance peak region around 3 kev (see Fig. 9-8). Flux hardening results in an increase in the fission factor, η. In the center of large sodium-cooled fast reactors, where neutron leakage is not felt, sodium void coefficients are therefore invariably positive. The increase in η with energy is more pronounced at lower fuel enrichments so that higher enriched reactors have lower void coefficients.

The effect of increased sodium voids, however, is also to increase neutron leakage. This contributes a negative void coefficient. In small-sized sodium-cooled fast reactors the effect of leakage predominates and the sodium void coefficient is negative. The effect of leakage, however, decreases with core size. In large-sized reactors the effect of flux hardening is predominant resulting in a net positive void coefficient. The spatial distribution is such that a positive void coefficient exists near the center, and a negative coefficient, due to increased leakage, exists near the boundaries. The overall effect is therefore dependent on core configuration. This is shown by the components and net local void coefficients in a 800-liter core (Fig. 9-5) and by the overall void coefficient versus size in carbide and oxide-fueled reactors (Fig. 9-6).

With growing emphasis on large reactors, the relative effect of leakage becomes less and less and reactor safety becomes more and more of a problem. The problem is tackled both analytically and empirically. The analytical work is complicated by uncertainties in multigroup neutron cross-section sets. Data is obtained from costly experimental measurements made in nonprototype low-power critical assemblies.

Two design philosophies seem to emerge from the consideration of positive sodium-void coefficient for large cores. One is to design a core with an enhanced leakage component, thus resulting in an overall negative sodium-void coefficient. This approach is safety motivated and is the one that has gained more favor in the United States. It suffers, however, from penalties on fuel inventory, blanket and internal breeding ratios and fuel-cycle economics. A high-leakage core also increases reactivity

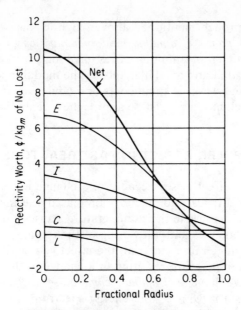

FIG. 9-5. Components and net reactivity
worths of the local sodium coefficient in
a 800-liter plutonium-fuel fast reactor.
E: elastic scattering; I: inelastic scattering;
C: capture; L: leakage. (Ref. 83.)

losses with fuel burnup. The other philosophy, favored in Europe and
the Soviet Union is not to pay these penalties, design a core with an
expected positive sodium coefficient but one with sufficient insured
mechanical integrity to guard against void formation or loss of coolant
and to counteract possible void formation with early instrumentation
detection and feedback control procedures. There are indications (1970)
of possible shifts in position toward a middle ground.

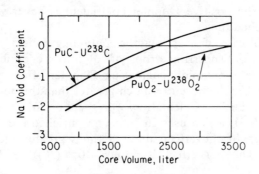

FIG. 9-6. Overall sodium coefficient as a
function of core volume for carbide and
oxide-fueled reactors. (Ref. 83.)

The first approach mentioned above –high leakage– is characterized by the so-called *pancake, annular,* and *modular* designs. The pancake (a very flat cylinder) and annular cores have high surface volume ratios surrounded by radial and axial blankets. The modular core is essentially an assembly of high-leakage small cores or modules separated by blankets that surround each module. These approaches are discussed in Sec. 10-5.

9-9. THE DOPPLER EFFECT IN FAST REACTORS

The Doppler effect arises because of a change in the relative energy distribution of neutrons and nuclei. Increased thermal motion of the nuclei, due to an increase in their temperature, causes a broadening of the cross-section resonances and a lowering of their peaks. Because of self-shielding of the nuclei in a material, the effective reaction rate increases with increasing temperature resulting in a change in the reactivity of the system. Thus we have a *Doppler temperature coefficient of reactivity.*

In a fast reactor with a soft energy spectrum, the neutron energy spectrum spans the resonance regions of both fissionable and fertile fuels. If the temperature increases, the Doppler effect results in increased production of neutrons from the fissionable nuclei U^{233}, U^{235}, or Pu^{239}, as well as increased neutron capture by fuel nuclei, nonfissionable in the resonance region, such as U^{238}. Thus there may be a net increase or decrease in reactivity, or a positive or negative Doppler temperature coefficient, depending upon which of the above two effects predominates.

A net positive Doppler temperature coefficient requires a large concentration of fissionable nuclei or fuels of high enrichment. However, such a high concentration results in a harder neutron spectrum where more of the neutrons are absorbed before they have had a chance to slow down to resonance energies where the Doppler effect is pronounced. Consequently, the positive Doppler coefficient is never very large even for highly enriched fuels.

In the usual case of a fast reactor containing fuels of normal enrichment, the neutron spectrum drops off at lower energies but not fast enough to offset the rapidly increasing Doppler effect. The negative contribution of U^{238} is therefore much greater than the positive contribution of U^{235} or Pu^{239}. Calculations for the Enrico Fermi fast-breeder reactor (Sec. 10-3), which uses 26.7 percent enriched fuel pins, show the negative portion of the effect about 8 times greater than the positive portion [84]. The net effect is about -10^{-6} (°F)$^{-1}$ at 1020°F—not a large value.

A large concentration of U^{238} or fuels of low enrichment, coupled with a soft spectrum would produce a large negative Doppler coefficient.

Oxide fuels soften the spectrum because of the scattering effect of oxygen. Another way of producing a large negative Doppler coefficient of reactivity is to include a small amount of moderator such as zirconium hydride or beryllium oxide in the core. It is estimated [84] that in large fast oxide-fueled breeder reactors the net effect may be as high as -5×10^{-6} (°F)$^{-1}$. Large negative Doppler coefficients of reactivity are of importance from a fast reactor safety standpoint, and might actually preclude the possibility of explosive energy release in such reactors.

In sodium-cooled fast reactors, a *sodium Doppler coefficient*, though small and not prompt because of heat-transfer lag, arises because of sodium resonances. The main sodium resonance peaks at about 3 kev (Fig. 9-8).

9-10. FLUX DISTRIBUTIONS IN FAST REACTORS

In reactor physics [2] neutron flux distributions and critical size relationships were developed for a *thermal* reactor core. The processes of slowing of thermalized neutrons, the neutron life cycle from birth to fission including fast fission, fast leakage, resonance absorption, thermal leakage, nonfission absorption, and fission were all discussed in the progress of that development. Two important equations were developed, the four-factor formula

$$k_\infty = \epsilon p f \eta \tag{9-23}$$

and the reactor equation

$$\nabla^2 \varphi + B^2 \varphi = 0 \tag{9-24}$$

where k_∞ = infinite multiplication factor
ϵ = fast-fission factor,
p = resonance escape probability
f = thermal utilization factor
η = thermal fission factor
φ = neutron flux
B^2 = buckling

The reactor equation was solved with the help of the four-factor formula assuming a two-group model, in which the neutrons were assumed to belong to two energy groups, fast and thermal. Other approximations such as the one-group, the modified one-group, the Fermi-age, and the more accurate multigroup models were briefly discussed. The solution resulted in evaluating critical sizes and neutron flux distributions in different core geometries. The latter were subsequently used to evaluate the heat-generation and removal processes and temperature distributions in reactor cores [2]. In order to ascertain that these he generation relationships apply to *fast* reactors, it is necessary to sho that the flux distributions are similar. Again the student must cautioned that only an approximate and introductory treatment of t subject of fast reactor physics can be given here.

The most exact treatment involves, as with thermal reactors, tedious multigroup approach that invariably requires the use of hig speed digital computers. In this, several neutron-energy groups are us in which the fluxes are represented by $\varphi_1(r), \varphi_2(r), \cdots, \varphi_j(r), \cdots, \varphi_n(r)$, t flux within each group at some position r. Neutrons enter the j_{th} gro by elastic and inelastic scattering from a higher-energy group and birth of fission neutrons having energies within this group. The neutro leave the j_{th} group by elastic and inelastic scattering to a lower-ener group, by nonfission capture, and by fission within this group.

The number of groups chosen depends upon the capacity of t computer available as well as other difficulties. Lattice parameters ai various neutron cross sections are averaged for each group. In therm reactors the problem is somewhat simplified by the fact that the neutr spectrum is not highly dependent on core composition, the therm spectrum is roughly Maxwellian and the thermal cross sections exhi the well-known $1/V$ energy dependence [2]. In fast reactors, on the oth hand, the neutron spectrum is highly dependent on core compositic no simple relations exist between cross sections and energy over a portion of the spectrum, and only a partial listing of the necessary latti parameters and neutron cross sections of different materials as a functi of fast neutron energy is known [85]. In particular, there is a shorta of inelastic scattering cross-section data. Another difficulty is in av aging cross sections for each group. It is desirable, for example, choose the groups so that cross sections are relatively constant with each and that severe changes of these cross sections occur at the bou daries between these groups.

Very rough solutions of the reactor equation can be made, howev using a one-group approach, which, though not very accurate, is no etheless instructive. One energy group of fast neutrons is assumed exist at all times, and lattice and fuel parameters and cross sections a given single values by empirically averaging them over the fast neutr spectrum.

In a fast, hard-spectrum reactor, the neutron life cycle is mu simpler than that in a thermal reactor. ϵ and η combine in the o parameter η, simply called the *fission factor* and already used in tl chapter. There is no resonance absorption for neutron energies above few kev so that p can be assumed to be unity. f becomes simply t utilization factor. The four-factor formula thus reduces to

$$k_\infty = f\eta \qquad (9\text{-}25)$$

Newly born neutrons in a fast reactor are slowed down and diffuse as in a thermal reactor but are absorbed while still fast. They will possess a *fast-diffusion length,* L_f, which is given by

$$L_f^2 = \frac{1}{3(\Sigma_{tr})_f(\Sigma_a)_f} \qquad (9\text{-}26)$$

where $(\Sigma_{tr})_f$ and $(\Sigma_a)_f$ are fast macroscopic transport and absorption cross sections. The concept of the diffusion of neutrons in media, still applies and a diffusion equation may be written as

$$\nabla^2\varphi - \frac{1}{L_f^2}\,\varphi = -\frac{S_f}{D_f} \qquad (9\text{-}27)$$

where D_f is a *fast-diffusion coefficient,* S_f, the source term, is simply equal to

$$S_f = k_\infty (\Sigma_a)_f\varphi \qquad (9\text{-}28)$$

which is different from that of the fast group in the thermal two-group method in that Σ_a and φ are for fast neutrons here. D_f is given by

$$D_f = \frac{1}{3(\Sigma_{tr})_f} \qquad (9\text{-}29)$$

so that

$$L_f^2 = \frac{D_f}{(\Sigma_a)_f} \qquad (9\text{-}30)$$

The diffusion equation now becomes

$$\nabla^2\varphi - \frac{(\Sigma_a)_f}{D_f}\,\varphi + \frac{k_\infty(\Sigma_a)_f}{D_f}\,\varphi = 0$$

or

$$\nabla^2\varphi + \frac{(\Sigma_a)_f(k_\infty - 1)}{D_f}\,\varphi = 0 \qquad (9\text{-}31)$$

As with thermal reactors, Eq. 9-31 is written in the form

$$\nabla^2\varphi + B^2\varphi = 0 \qquad (9\text{-}32)$$

where B^2 is given by

$$B^2 = \frac{(\Sigma_a)_f(k_\infty - 1)}{D_f} \qquad (9\text{-}33)$$

or

$$B^2 = \frac{k_\infty - 1}{L_f^2} \qquad (9\text{-}34)$$

The solution of Eq. 9-32, the resultant neutron flux distributions, and expressions for the critical size as functions of the buckling are all identical to those for thermal reactors [1]. The values of B^2 however, will be different because of the differences in k_∞ and L_f^2. The homogeneous flux *distributions* in fast reactors are therefore identical to those in thermal reactors.

Figure 9-7 shows multigroup calculations of Pu[239] critical masses of Pu[239]-U[238]-fueled cores, versus the volumetric percentage of Pu[239] in the total fuel, for three core compositions, *A*, *B*, and *C*, all reflected by a 20-in.-thick depleted uranium blanket. Curve *D* has the same composition as *C*, but is for U[235] instead of Pu[239]. As can be seen, the critical mass **decreases** with fissile material concentration and with total fuel

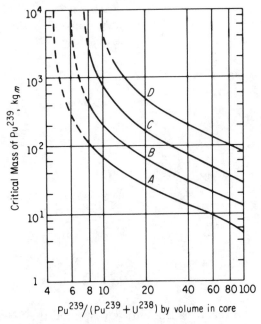

FIG. 9-7. Critical mass of Pu[239] in Pu-U-fueled fast reactors. A: fuel only in core; B: fuel 50%, Na 33.3%, stainless steel 16.67% by volume; C: fuel 25%, Na 50%, stainless steel, 25%; D: same as C but U[235] instead of Pu[239]. Blanket: depleted uranium 40%, Na 40%, stainless steel 20%. (Ref. 86.)

percentage, and that U^{235} critical masses are higher than Pu^{239}. Also, it should be recalled that critical masses in fast reactors are much greater than those in thermal reactors.

From a physics point of view then it would be desirable to design a core with a large Pu^{239} percentage in the total fuel and a large percentage of fuel in the core to reduce critical mass and fuel inventory. Against this, however, are heat-transfer considerations–namely the necessity to provide adequate cooling and sufficient fuel to reduce heat fluxes and maximum fuel temperatures to tolerable levels for a given core thermal output.

9-11. FAST-REACTOR COOLANTS, LIQUID METALS

Fast-reactor coolants must of course have no or low moderating capabilities. Water (ordinary and heavy) and organic coolants are therefore excluded. Many liquid metals and gases or vapors are considered.

Liquid (or molten) metals generally have low moderating ratios. Most liquid metals also have low absorption cross sections for neutrons in the fast- and intermediate-energy ranges, Fig. 9-8. Some, like Na, also have low neutron cross sections in the thermal-energy range and therefore are also suitable for use in thermal reactors.

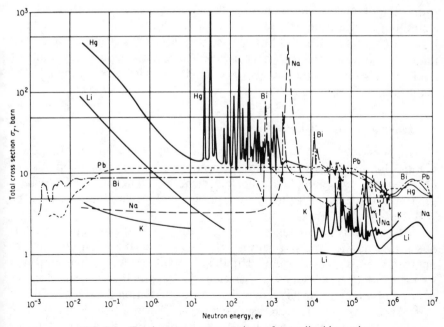

FIG. 9-8. Total neutron cross sections of some liquid metals.

In general, liquid metals are not readily available in sufficient quantities in all parts of the world and are costly. They are particularly resistant to radiation damage so that no appreciable replenishment costs are incurred (as in the case of organic coolants), although some degree of purification to remove corrosive oxides may be necessary. Liquid metals have relatively low melting points and high boiling points (or low vapor pressures) and therefore remain in liquid form over a wide range of temperatures. High reactor exit temperatures are therefore possible with low system pressures. This results in both high power cycle efficiencies (lower heat rejection and, therefore, thermal pollution) and low pressure in the reactor vessel than in water-cooled reactors.

Liquid metals have excellent heat-transfer and fluid-flow characteristics. Their high thermal conductivities result in reduced hot-spot factors and in lowering the temperature gradients in the core and consequently in a reduction of the probability of structural warpage, though a problem of thermal shock exists.

Of the many available liquid metals, the alkalies, which include sodium, potassium, sodium-potassium mixtures (NaK), rubidium, cesium, and lithium, have received the greatest attention. Sodium, which is commercially available in solid-brick form, is the least expensive. It has a good temperature range: 208°F melting point and 1621°F normal boiling point. The somewhat high melting point causes it to solidify at room temperatures, necessitating the use of external heating of the sodium system during extended shutdowns. Sodium has excellent heat-transfer characteristics and can be used with thermal as well as epithermal reactors. It has well-known corrosive characteristics and, by contrast with many other liquid metals, is compatible with a large number of materials.

Induced Radioactivity in Liquid Metals

In general, liquid metals become intensely radioactive. In power plants, this necessitates the use of an intermediate coolant loop, also of liquid metal for good heat-transfer purposes. The intermediate loop separates the reactor primary loop, where induced radioactivity exists and the working fluid (steam).

All naturally occurring sodium is made up of the isotope Na^{23}. It has a thermal-neutron absorption cross section of 0.53 barn and a fast-neutron cross section of about 1 millibarn at 0.25 Mev. When subjected to neutrons it undergoes the reaction $Na^{23}(n, \gamma)Na^{24}$. Na^{24} is a radioisotope of about 15 hr half-life which emits γ radiation mainly of 2.76 and 1.38 Mew energy. It β-decays into stable $_{12}Mg^{24}$, which has a low neutron cross section. The level of activity of sodium, of course, depends upon the relative magnitude of the time spent within and outside the core. Naturally occurring potassium consists of the two isotopes K^{39} ($\sim$93.10 percent) and K^{41} ($\sim$6.88 percent). The remainder (0.0118 percent) is radioactive K^{40} which is a β and γ emitter with a long half-life (1.32×10^9 yr) and thus has a very low level of activity. Of the three isotopes K^{41} is converted to a radioactive isotope upon neutron absorption. It converts to K^{42}, a β and γ emitter of 12.40-hr half-life, in turn converting to stable Ca^{42}. While K^{40} converts to K^{41} which is susceptible as above, its abundance in nature is so low that its contribution to activity is unimportant.

The isotope Li^7 (92.58 percent of all lithium) converts to Li^8 of 0.85-sec half-life.

Hg^{196}, Hg^{198}, and Hg^{202}, with abundances of 0.146, 10.02 and 29.8 percent, convert to radioisotopes which have half-lives of varying length, from 44 min to 47.9 days, and γ energies varying between 0.13 and 0.37 Mev. Other reactions in heavy liquid metals occur in lead, $Pb^{208}(n,\gamma)Pb^{209}$ (3.3-hr half-life) and in bismuth, $Bi^{209}(n,\gamma)Bi^{210}$. Bi^{210} is a β emitter with a 5-day half-life, converting to Po^{210}, an α emitter of 140-day half-life converting to stable Pb^{206}. These reactions are serious enough to be taken into account if these liquid metals were used as reactor coolants.

Liquid-Metal Compatibility with Materials

Most materials suffer from attack or corrosion when subjected to liquid metals, [87, 88, 89]. Some of the materials that are particularly soluble in hot sodium, potassium, and NaK are cadmium, antimony, bismuth, copper, lead, silicon, tin, and magnesium. On the other hand, nickel, Inconel, Nichrome, the Hastelloys, and nickel- and chromium-bearing steels (such as the Series 300 stainless steels and the columbium-bearing Type 347 stainless steel) are well suited for high-temperature use with the above liquid metals.

The chief reason for corrosion is the ability of sodium (and sodium in NaK) to dissolve oxygen. This solubility increases rapidly with temperature, approaching a saturation of about 0.1 percent by mass at 900 to 1000°F. Oxygen reacts with sodium to form Na_2O which is highly corrosive. It is also relatively insoluble in sodium or NaK, especially at low temperatures, causing deposition in cooler passages. A little deposition acts as a nucleus around which crystal growth of a mixture of approximately 20 percent Na_2O and 80 percent Na takes place, causing plugging in narrow cool passages. Because of this, a minimum of narrow passages should be designed into a sodium system, and when necessary they should be streamlined, restricted to the hotter regions of the system, or placed vertically to minimize precipitation.

The most troublesome effects of corrosion are the so-called *self welding* and *thermal-gradient transfer* [87]. Self-welding results in the malfunction of such components as pumps and valves. It is believed to be caused by the sodium reducing the oxides of surfaces in contact, when it diffuses between these surfaces. Self-welding increases with sodium temperature and with contact pressure between the surfaces. In thermal-gradient transfer, materials are dissolved by the liquid metal in a high-temperature region, where the solubility is high, and precipitated in a cooler region, where the solubility is low. While the rates of such mass transport are low, extended operation and coolant circulation may result in noticeable corrosion of the hotter regions and plugging in the cooler regions of the system.

It is obvious that absorption of oxygen by the coolant should be avoided. An inert-gas blanket should be provided over all free-sodium surfaces. The inert gas may be helium, which has the advantage of no induced radioactivity, or argon, which has the advantage of being heavier than air, thus facilitating the blanketing problem. Nitrogen is less expensive but is also soluble in sodium. This solubility, though low (less than 1 ppm), results in the mass transport of N_2 through the system (thought to be aided by the presence of calcium or carbon). This in turn results, at high temperatures, in nitriding and damage to thin-walled components such as cladding, valve bellows, etc.

Oxygen may enter the system via impurities in the gaseous blanket and during refueling or repair periods. Thus it is usually necessary to sample and analyze the coolant. Purification is done by bypassing a portion of the hot coolant (when oxide solubility is highest) and depositing the oxide in a cold trap. An oxide concentration below 30 ppm appears to give satisfactory results, although lower concentrations are easily attainable. Also purification of the gaseous blanket to remove air and water vapor may be necessary.

Sodium reacts vigorously with most noninert gases and liquids. In the solid state, it is soft and silvery white but tarnishes readily when exposed to air, because of the formation of an oxide film on its surface. In the liquid state, also silvery white, when exposed to air it burns with a thick smoke of Na_2O. If such exposure is due to a fissure or hole in the container, the heat of combustion causes the sodium to reach its boiling point and to form a highly corrosive spongy mixture with its oxides. This mixture enlarges the original fissure, aggravating the situation.

Sodium also reacts vigorously with water:

$$Na + H_2O \longrightarrow NaOH + \tfrac{1}{2}H_2 + 3600 \text{ Btu}/lb_m Na$$
$$Na + \tfrac{1}{2}H_2O \longrightarrow \tfrac{1}{2}Na_2O + \tfrac{1}{2}H_2 + 1610 \text{ Btu}/lb_m Na$$
$$Na + \tfrac{1}{2}H_2 \longrightarrow NaH$$

$$(9\text{-}35)$$

These reactions become explosive if they take place in a confined volume. If air is also present, the hydrogen liberated will burn, resulting in an additional 2300 Btu/lb_mNa.   In sodium-water heat exchangers, care must be taken to choose materials of construction that will not be attacked by either fluid.   One design, that of the Sodium Reactor Experiment, uses a double-walled tube construction with an intermediate high-thermal-conductivity fluid (mercury) in the annulus.

If H_2 and H_2O enter a sodium system in nonexplosive quantities (via impurities or leakage) they form NaH and NaOH.   This appreciably increases the moderating ratio of the coolant, an effect that may not be desirable, in fast reactors.

Alkalies, other than sodium, have chemical reactions similar to those given above.   Potassium, however, is more reactive than sodium.   NaK, with many of its mixtures liquid at room temperature, is more reactive than either of its components.

Fire-extinguishing materials for alkali metals include soda ash ($NaCO_3$) and rock salt ($CaCO_3$).   Alarm systems, such as that shown in Fig. 9-9, should be installed to warn against leakage.

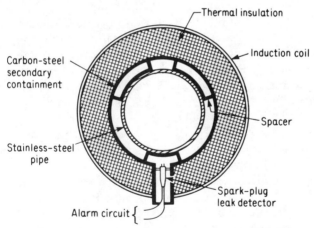

FIG. 9-9.   A liquid-metal-pipe alarm system.   (Ref. 61.)

9-12.   FAST-REACTOR COOLANTS, GASES AND STEAM

Besides liquid metals, gases (He and CO_2) and steam have been proposed for cooling fast reactors.   Though they contain moderating nuclei (C and H), CO_2 and steam have low moderating power because of their low densities and are therefore suitable for use in fast reactors.

There are several advantages in using a gas or steam as coolant for fast reactors [90]. The main advantages are:

1. The absence of reactivity excursions due to the formation of voids or gas entrainment in sodium-cooled reactors.
2. There is no necessity to pay the penalty of special core design to avoid positive sodium void coefficients.
3. Gas-cooling results in a harder neutron spectrum than sodium, resulting in higher breeding ratios than with sodium, although steam does not share in this advantage.
4. High specific power.
5. Compatibility with water (unlike sodium).
6. Gases could be used in closed cycle and steam in direct cycle power plants, reducing capital costs.
7. There are no thermal shock problems often feared with sodium.

Other advantages such as inertness, low induced activities and the ability to design a system for visual rather than remote repair are similar to those for gas-cooled thermal reactors.

On the debit side, gases and steam suffer from low heat-transfer coefficients resulting in higher clad temperatures, and high pumping losses (Sec. 7-4), a disadvantage in the high-power-density fast reactors. Of more importance are the difficulties of providing adequate cooling in the event of an emergency and for decay heat in the event of loss of power. A large amount of core sensible heat can be rejected to a liquid coolant by boiling it in the event of the loss of external cooling (latent heats of vaporization are large). Also natural circulation effects in gases or steam are not as adequate as those obtained with liquid coolants, although complete loss of coolant situations do not occur with gases. A cooling system should therefore be designed so that cooling must be quickly reinstated after its loss, a problem made difficult by the usually compact, restricted flow-passage fast cores.

Helium, carbon dioxide and steam have been considered as fast reactor coolants. Other gases are not suitable because of chemical activity or metallurgical considerations (CO, H_2, N_2, air), low heat transfer (Ne and A), cost (Ne) activation (A) or neutron cross sections (N_2).

Of the three gases considered, (He, CO_2, and steam), helium has the lowest pressure-drop-to-heat-transfer ratio, thus reducing flow-induced vibrations in the normally high-speed coolant flows necessary to obtain adequate heat transfer in fast reactors. Steam is less costly and offers the possibility of direct cycle applications but suffers from lower breeding ratios than gases because of the softer spectrum associated with it. A breeding ratio between 1.1 and 1.2 is usually mentioned for steam, compared to 1.55 for oxide and 1.65 for carbide fuels for helium. The figures mentioned for oxide-fueled, sodium-cooled reactors are about

1.2 to 1.4 where the lower and higher figures represent American and European designs respectively and reflect differing philosophies on reactor design (neutron leakage) to insure a safe sodium-void coefficient.

Steam has lost much favor because of the inability to find a suitable fuel-cladding material in the highly corrosive atmosphere of steam at high temperatures.

9-13. MATERIAL SWELLING IN FAST REACTORS

Materials that are subjected to fast neutron bombardment undergo swelling or increase in dimension. The problem was first discovered in 1967 by UKAEA workers who observed it on stainless steel claddings of mixed oxide fuel elements used in the Dounreay Fast Reactor (DFR). It was subsequently confirmed by US workers on stainless steel irradiated in EBR-II. Since austenitic stainless steels are widely used in sodium-cooled fast reactors, their swelling poses a serious problem for the metallurgist and designer of such reactors.

The degree of swelling, defined as the change in volume per unit volume $\Delta V/V$, is a function of both temperature and fast neutron fluence (integrated flux with time, $\phi\theta$), but particularly the latter. It may increase with increasing temperature, as in the case of cold-worked 316 stainless steel; or it may increase, reach a peak and then decrease with temperature as in the cases of solution-treated 304 and 316 stainless steels. It increases rapidly with an increase in fluence, for most materials, according to

$$\frac{\Delta V}{V} \propto (\varphi\theta)^n \tag{9-36}$$

where φ is flux and θ is time. n varies slightly according to material, but averages about 1.7 for stainless steels.

One of the problems facing the designer is that present swelling data (1970) is at fluences of the order of 8×10^{22} neutrons/cm^2. These are not representative of those levels that must be reached by fast-breeder reactors if they are to be commercially competitive with ordinary-water reactors. Those levels are of the order of 2×10^{23} or higher. Figure 9-10 shows the dependence of swelling on temperature and fluence of 304 and 316 stainless steels, based on a calculational model [91].

Swelling of fuel elements in a typical commercial liquid-metal-cooled fast-breeder reactor might result in *lengthening, dilation* (cross-sectional diameter increase), and in *bowing*. These result in complicating fuel handling, fuel management, and design, in reducing breeding gain (by as much as 20 percent), and other effects. Swelling of core-structural material such as that of guide tubes and guide plates result in misalignment, frequent replacements, and other undesirable effects.

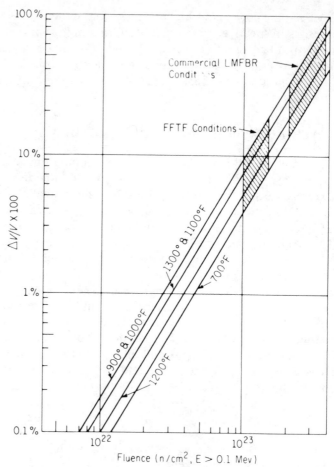

FIG. 9-10. Predicted swelling of 304 and 316 stainless steels
as a function of fluence and temperature. (Ref. 91.)

The problem of swelling must be minimized by the metallurgist who
must find stainless steels and other materials that are resistant to swelling,
and by the core designer who must find design solutions to cope with the
swelling problem. Such solutions may fall into the following categories:

(a) *Mechanical.* This might involve the use of articulated, flexible
or undercut ducts, local clamping, replaceable structures, smaller or fewer
fuel elements, etc.

(b) *Neutronic.* This involves a reduction in neutron energy by introducing core moderators (also results in a large negative Doppler coefficient, but at the expense of breeding gain), higher fuel enrichment
(again resulting in smaller or fewer fuel elements), etc.

It now appears that core lifetime (really fluence) of a fast reactor might be limited by the extent of swelling of fuel and nonfuel materials that can be tolerated, rather than by fuel-burnup considerations.

PROBLEMS

9-1. A fast reactor is cooled with liquid sodium. The average neutron flux in the core is 10^{14}. The average sodium temperature in the primary system is 1200°F. The sodium spends half the time within the core and the remainder in the rest of the primary system. The neutron energy distribution in the core is such that the neutron absorption microscopic cross section for sodium is 2 barn. Calculate the activity of the coolant in curies/liter after several months of steady operation.

9-2. Estimate the compound doubling time in years of a very large gas-cooled fast-breeder reactor that is fueled only with Pu^{239}. The average neutron flux is 10^{15} and the core contains 80 percent of the total fuel inventory of the plant. Make and state any reasonable assumptions you may need.

9-3. A fast reactor with a soft neutron spectrum uses a Pu^{239}-U^{238} fuel mix. Average neutron flux is 2×10^{15}. The number of neutrons lost in nonfission and nonbreeding reactions only are 0.03 per neutron born in fission. Because of the soft spectrum, fission in U^{238} may be ignored. Using data from Table 9-1 for simplicity, determine (a) reactor simple doubling time and (b) nonfission and nonbreeding reaction losses that would render the reactor a nonbreeder. Take $P = 0.95$.

9-4. Show that in a uranium-fueled thermal reactor where the fast neutron non-leakage probability is given (Fermi-age) by

$$P_f = e^{-B^2 L_s^2}$$

where B^2 is the buckling and L_s is the fast-neutron slowing-down length, and where all neutrons absorbed in U^{238} result in the production of Pu^{239}, the conversion factor is given by

$$C = \left[\frac{(\Sigma_a)_{238}}{(\Sigma_a)_{235}} + 1 \right] \eta \epsilon (1 - p) e^{-B^2 L_s^2} + \frac{(\Sigma_a)_{238}}{(\Sigma_a)_{235}}$$

9-5. Control of fast reactors is often accomplished by moving portions of the fuel in the core or portions of the reflector, instead of a control rod made of a neutron absorber as in a thermal reactor. Explain. What is the effect on multiplication and nonleakage probability in the two cases?

9-6. In a reactor in which fissionable nuclei are consumed and others are produced, the change in reactivity due to fissionable material change may be given by

$$\Delta \rho_f = \left[\frac{(\sigma_a)_p \eta_p}{(\sigma_a)_0 \eta_0} C - 1 \right] R$$

where the subscripts p and 0 refer to produced and original fissionable nuclei respectively, C is the conversion ratio and R is the fraction of all fissionable nuclei consumed in fission and nonfission absorption.   Derive the above expression.

9-7.   A reactor is fueled with 60 tons of 1.5 percent enriched uranium metal (density 19 g_m/cm³).   It has a conversion ratio of 0.55.   Determine the change in reactivity due to fissionable material change only after one year of steady operation at 300 Mw(t).   Assume that fission occurs only in U^{235} and that each fission reaction generates 180 Mev of useful energy in the reactor.   Use the expression in Prob. 9-6.

chapter 10

Fast-Breeder
Reactors and Power Plants

10-1. INTRODUCTION

Fast reactors have been built in the United States, the United Kingdom, the Soviet Union, and France, and German and Italian efforts are underway (Table 10-1). The reactors were cooled by sodium, NaK, or mercury and fueled with uranium and plutonium in the form of metal alloys, oxides or carbides. It is expected that a new generation of demonstration fast-breeder reactor plants will appear in the

TABLE 10–1
The World's Fast Reactors, by Early 1970's

Country	Reactor	Power, Mw(t)	Coolant	Fuel	Status*
U.S.A.	Clementine	0.025	Hg	Pu metal	O & D
	LAMPRE-1	1	Na	Molten Pu	O & D
	EBR-1	1.4	NaK	U metal	O & D
	EBR-11	62.5	Na	U metal	O
	Enrico Fermi	200	Na	U metal	O
	SEFOR	20	Na	Pu-U oxide	O
	FFTF	400	Na	Pu-U oxide	D
U.S.S.R.	BR-1	0		Pu-U metal	O & D
	BR-2	0.100	Hg	Pu metal	O
	Br-5	5	Na	U C	O
	BOR-60	60	Na	Pu-U oxide	O
	BN-350	1,000	Na	Pu-U oxide	C
	BN-600	1,440	Na	UO_2	C
U.K.	Dounreay	60	NaK	U metal	O
	PFR	600	Na	Pu-U oxide	C
France	Rapsodie	20	Na	Pu-U oxide	O
	Phénix	600	Na	Pu-U oxide	D
Germany	SNR	750	Na	Pu-U oxide	D
Italy	PEC	140	Na	UO_2	D

* O, operating; O & D, decommissioned; C, under construction; D, in design.

mid 1970's and large economical commercial plants in the early or mid 1980's.

The first fast reactor was built in the United States in 1946. Called Clementine, it was plutonium fueled, produced 25 kw(t), was mercury-cooled (one of only two to date) and has long been shut down. The main initial effort was the Experimental Breeder Reactors EBR-I and EBR-II [92]. EBR-I was a 1200-kw(t), 200-kw(e) fast-breeder reactor based on the concepts of Fermi and Zinn in 1945. Construction began in November 1947; it achieved criticality in August 1951 and produced electric power in December 1951. It was NaK-cooled and used stainless-steel-clad fully enriched 2 percent zirconium-uranium alloy for fuel and stainless steel clad natural uranium external blanket. It was discontinued after successfully demonstrating the feasibility of fast reactors and breeding. EBR-II is a 62.5-Mw(t), 20-Mw(e) fast breeder. It is sodium-cooled and is fueled with stainless-steel-clad 49.5 percent enriched-uranium-alloy pins. The blanket contains stainless-steel-clad depleted uranium pins. The EBR-II plant has an integrated fuel manufacturing and reprocessing facility and will operate on a fuel-recycle basis, ultimately with plutonium. Other U.S. reactors built or contemplated are discussed in more detail later in this chapter.

A large British fast-reactor effort is centered around Dounreay [93] and PFR-Dounreay. The first is a NaK-cooled fast breeder with tubular fuel clad in vanadium on the inside and partly finned niobium on the outside. Natural uranium is used in external breeder blankets completely surrounding the core. Control is achieved by axially moving 12 fuel subassemblies (of 10 fuel elements each), situated around the periphery of the core. These are divided into two safety rods, six control rods, and four shutoff rods.

In the USSR three experimental fast reactors have been built since 1955. The Br-1 is a PuU-fueled and U-reflected reactor designed to carry out cross-section measurements at various neutron energies. The BR-2 is a Pu-fueled and mercury-cooled reactor used to determine various nuclear constants, breeding ratios, and operating experience with liquid metals. The BR-5 is a UC-fueled, sodium-cooled reactor with a uranium-nickel blanket. The program includes three power reactors. The BOR-60 is operating. BN-350, under construction, is a dual-purpose reactor intended to produce 150 Mw(e) and process steam for a 120,000 m^3/day water desalination plant. BN-600, under construction, is a 600 Mw(e) power reactor.

In France the effort is centered around Rapsodie, a 20 Mw(t) sodium-cooled and plutonium-uranium-oxide fueled reactor. A 600 Mw(t), 250 Mw(e) reactor, the Phénix, is planned for the early 1970's.

10-2. REACTOR-PLANT ARRANGEMENTS

Because sodium and other liquid metals suffer from high induced radioactivities and are generally chemically active (Sec. 9-11), intermediate coolant loops are used between the primary radioactive coolant and the steam cycle, Fig. 10-1. The intermediate coolant is usually also a liquid metal, often Na or NaK. The intermediate loop guards against

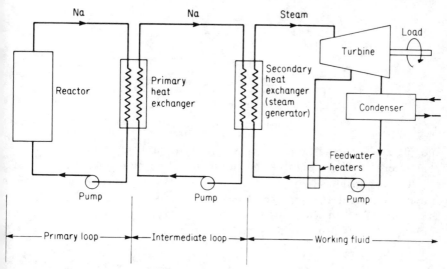

FIG. 10-1.   Schematic arrangement of a liquid-metal-cooled reactor power plant.

reactions between the radioactive primary coolant and water. Such reactions result, among other things, in the radiolytic decomposition of steam-generator water by the strong γ radiations emitted by Na[24]. The intermediate loop also ensures against water or hydrogen entering the reactor.

There are two primary-loop designs that are being considered. These are: (a) the *loop,* or *pipe* type, and (b) the *pool, tank* or *pot* type.

The *loop* type, represented schematically in Fig. 10-2, is the more conventional of the two, being the design used in all operating sodium-cooled plants to date, with the possible exception of EBR-II. In it, the reactor vessel, heat exchangers, liquid metal pumps, and other components of the primary system and their interconnecting piping are separated within a large building containing an inert atmosphere to preclude sodium fires in case of a sodium leak. The major advantages of this design are the accumulated experience with it and the separation or decoupling of the components of the primary system. It has the disadvantage of large and multiple shielding of pipeways, equipment cells,

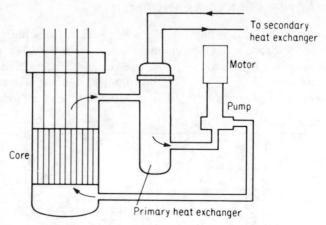

FIG. 10-2. Schematic of a loop-type arrangement.

and of the large and complex structure resulting from the spread of the components. The design of the interconnecting piping is complex and requires expansion loops to accommodate thermal expansion. Stress concentrations at the pipe-reactor vessel joints pose critical problems. Breaks or leaks in the piping system may seriously affect reactor core cooling. Leaks also would necessitate extended shutdowns for repairs.

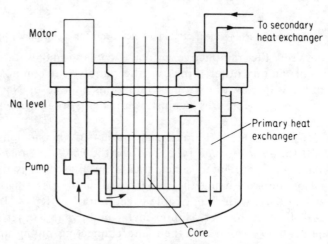

FIG. 10-3. Schematic of a pool-type arrangement.

In the *pool* system, represented schematically in Fig. 10-3, the entire primary system, including reactor, primary heat exchangers and pumps, is submerged in a large tank filled with molten sodium. That tank is part of the primary coolant loops. The heat exchangers discharge coolant directly into the tank, and the pumps receive coolant directly from it. The main advantages of the pool design are the relative insensitivity to sodium leaks in the primary system, and a more compact primary-system arrangement. The disadvantages are due to the close coupling of the various components which leads to accentuated mechanical and thermal interactions, and the rather complex structure of the pool closure which must serve the multiple functions of shield, inert gas closure, and support of equipment above, and must contain all the necessary penetrations to the components.

In general, it is now believed that the pool system has the edge in safety and economy while the loop system has the edge in that it is a straightforward mechanical design.   Both types are being used in current designs.   The pool is used in 1,000-Mw(e) follow-on designs by G.E., B&W., (Sec. 10-5) and in French and British designs.   The loop is used in follow-on designs by Westinghouse, Combustion Engineering, Atomics International, and in Russian and German designs.

10-3.   THE ENRICO FERMI NUCLEAR POWER PLANT

The first fast-breeder commercial power plant built in the United States is the Enrico Fermi nuclear power plant [94].   It is a 300 Mw(t), 94 Mw(e) net unit built for a group of Midwestern United States power companies that formed the Atomic Power Development Associates, Inc. (APDA).   It is located on Lake Erie at Lagoona Beach, about 30 miles southwest of Detroit, Mich.   Some design data of the plant are given in Table 10-2.   A perspective view of the reactor is shown in Fig. 10-4.

The core has an estimated average neutron energy of 0.25 Mev and contains fuel pins made of zirconium-clad, 10 percent molybdenum-uranium alloy of 26.7 percent enrichment.   It is cylindrical, 30.5 in. in diameter and 31.2 in. high and is surrounded on all sides by a breeder blanket of stainless-steel-clad depleted uranium.   The blanket is divided into a radial section and two axial sections.   The radial section is an annulus whose outside diameter is 78.5 in. and height is 67 in.   The axial sections occupy short cylindrical volumes on top and below the core, each 18 in. high.

The core fuel subassembly shown in Fig. 10-5 has $32\frac{57}{64}$ in. of actual active fuel in the core portion and 17 in. of blanket material on the top and bottom.   Figure 10-5 shows this subassembly divided into three parts for convenience (*Note:* match lines A and B).   The radial-blanket subassembly, shown in Fig. 10-6, contains $71\frac{1}{2}$ in. continuum of blanket material.   The fuel pins have a diameter small enough to afford a large heat-transfer surface of 1,378 sq ft within the small core volume of

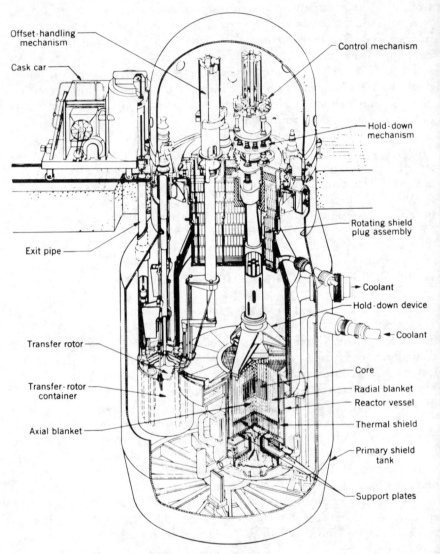

FIG. 10-4.   Perspective view of Enrico Fermi Reactor.   (Courtesy Atomic Power
Development Associates.)

11.65 cu ft, to take care of the large power density and to keep the fuel
maximum temperatures within a predetermined safe limit of 1235°F.

Figure 10-7 shows a flow diagram of the Enrico Fermi power plant.
It consists of primary, intermediate, and working-fluid loops.   Heat is

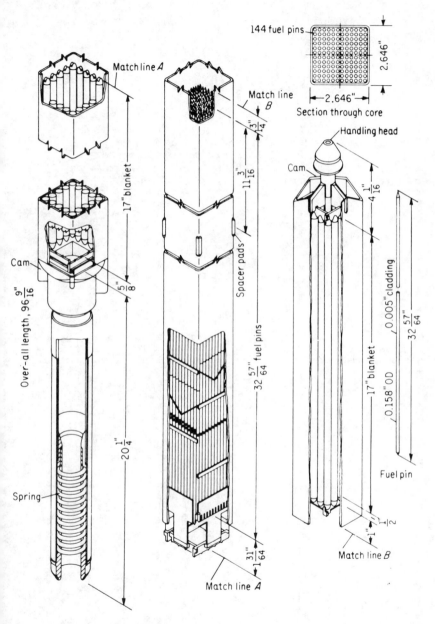

FIG. 10-5. Enrico Fermi core subassembly. (Courtesy Atomic Power Development Associates.)

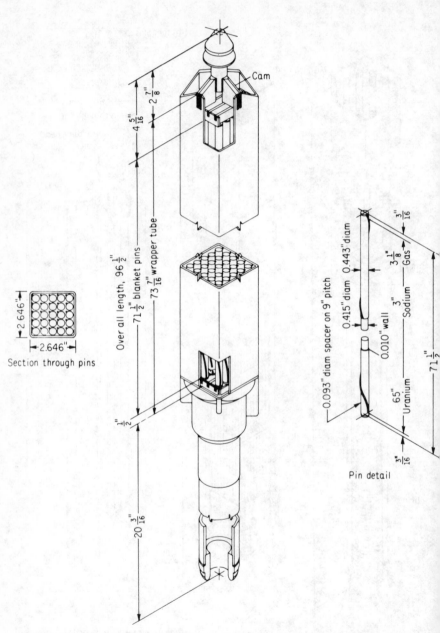

FIG. 10-6.   Enrico Fermi radial-blanket subassembly.   (Courtesy Atomic Power Development Associates.)

TABLE 10-2
Some Design Data of the Enrico Fermi
Fast-Breeder-Reactor Power Plant

Reactor core power	268,000 kw(t)
Radial blanket power	28,000 kw(t)
Axial blanket power	4,000 kw(t)
Total reactor power	300,000 kw(t)
Power-plant output	104,000 kw(e) gross, 94,000kw(e) net
Coolants (Primary and secondary)	Sodium
Moderator	None
Fuel elements	Pin type 0.158 in. OD; 144 per subassembly; 91 subassemblies
Maximum cladding temperature	1017°F
Maximum fuel	1235°F
Core volume	11.65 ft^3
Core heat-transfer surface	1,378 ft^2
Average core heat flux	652,000 Btu/hr ft^2
Core power density	23,000 kw/ft^3
Average core neutron flux	0.5×10^{16} neutrons/sec cm^2
Maximum core neutron flux	0.8×10^{16} neutrons/sec cm^2
Maximum-to-average heat flux in core	1.43
Blanket volume	161 ft^3 radial, 13 ft^3 axial
Blanket heat-transfer surface	9,250 ft^2 radial, 520 axial
Blanket heat flux	8,500 Btu/hr ft^2 radial, 1,300 axial
Maximum-to-average heat flux in blanket	55 radial
Reactor vessel dimensions	35 ft high, 14.5 maximum diameter
Reactor coolant inlet temperature	550°F
Reactor coolant outlet temperature	800°F
Total reactor pressure drop	97 psi
Primary-coolant flow rate (total three loops)	13.2×10^6 lb$_m$/hr
Core-sodium velocity	31.2 ft/sec
Intermediate-sodium temperatures	750°F outlet, 500°F inlet to heat exchanger
Secondary-sodium flow rates	13.2×10^6 lb$_m$/hr
Steam conditions	600 psia and 740°F
Steam flow	960,000 lb$_m$/hr
Feedwater temperature	340°F
Plant net thermal efficiency	31.3%

picked up from the core and blankets by sodium in the primary loop, transferred via heat exchangers to sodium in the intermediate loop, and finally used to generate steam in the working-fluid loop.

The primary system consists of three loops, each divided into two

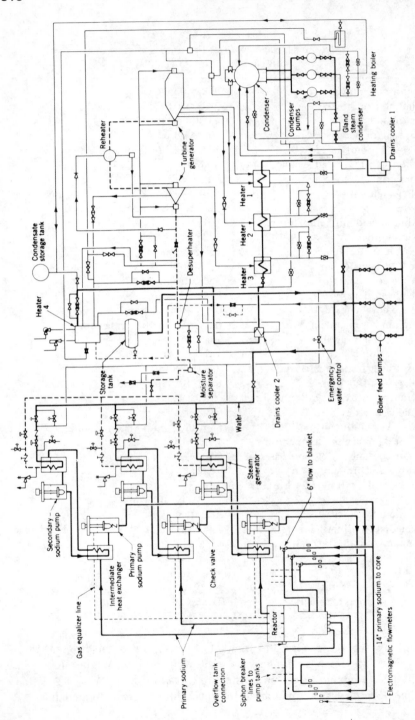

FIG. 10-7. Flow diagram of Enrico Fermi power plant.   (Ref. 94.)

paths before entering the reactor. One path, carrying about 90 percent of the total primary-sodium flow, cools the reactor core and the axial blankets. The other path carrying the remaining 10 percent admits sodium to an annular ring in the reactor via a radial plenum, from which it passes into the radial blanket. This flow is regulated by throttle valves.

The sodium coolant from both paths collects in a 15,000-gal pool whose surface is about 11 ft above the core. This pool reduces the probability of complete loss of coolant and thermal shock in the case of scram and acts as a chimney for natural circulation, removing decay heat in case of power failure. From there, sodium flows via three outlets to the shell sides of three intermediate heat exchangers, placed within the reactor containment vessel since they contain radioactive primary sodium. These heat exchangers are situated at such an elevation with respect to the core to permit the natural circulation mentioned above. From the intermediate heat exchangers, primary sodium goes to three 11,800-gpm vertical-shaft, centrifugal sump-type pumps. In each, sodium fills a tank surrounding the impeller and the discharge pipe. The pumps then send the sodium back to the reactor vessel. A check valve between the pump tank and pump discharge line allows natural circulation of the sodium and prevents backflow. The time required to complete one primary circuit is 55 sec.

The thin dashed line of Fig. 10-7 represents an inert-gas system. The gas used, argon, is maintained at a pressure approximately 1 in. of water below atmospheric. This prevents the highly radioactive argon from escaping out of the system. Argon is used to fill the space above the sodium pool in the reactor, the intermediate heat exchangers and pumps. Argon was chosen because it is inert, available, and has a higher density than air.

The intermediate circuit also consists of three loops. Sodium goes from the tube side of the intermediate heat exchangers to the shell side of once-through-type steam generators and then to 11,800-gpm secondary-sodium pumps. A slightly higher pressure is maintained in the secondary system than in the primary system to prevent radioactive primary sodium from entering the intermediate system in the event of leakage.

The steam generators are vertical, combination cross and counter-flow, shell-and-tube, once-through-type units. Water enters via an annular header at the top, flows down through a cluster of tubes in the center of the shell, then upward through 1,200 seamless, single-wall tubes $\frac{5}{8}$-in. OD and 0.042-in. wall thickness. These tubes are formed into horizontal U-type sections. Sodium enters the shell near the top and flows downward, leaving through a nozzle at the bottom. This arrangement gives 10,800 sq ft of heat-transfer area per heat exchanger. The secondary pumps and steam generators are located outside the main containment vessel.

Superheated steam at 600 psig and 740°F (corresponding to about 254°F of superheat) collects in a common pipe and is fed into the high-pressure end of the steam turbine. The turbine exhausts at 1 in. Hg and the condensate is pumped from the condenser by three parallel condensate pumps into three feedwater heaters. The condensate of the high-pressure heater is led to the shell side of the intermediate heater whose condensate goes to the low-pressure one. The condensate of the latter is used to heat the incoming main condensate and is then fed into the main condenser. There are provisions for bypassing any or all of the heaters. The hot feedwater now goes to the steam generators via storage tanks, shown near the top of Fig. 10-7, entering at 340°F.

The Enrico Fermi reactor has had its share of troubles, however. These included steam-generator weld failures and a partial core meltdown on October 5, 1966 due to coolant flow blockage by loosened zirconium protector plates. It is gratifying to note that this "maximum credible accident" did not endanger personnel or the environment. The reactor has been reloaded with various modifications and restarted in late 1970.

10-4. THE SEFOR PLANT

SEFOR (Southwest Experimental Fast Oxide Reactor), a 20 Mw(t), 50 Mw(t) potential, ceramic-fueled, sodium-cooled, fast reactor [95] is a joint venture of the Southwest Atomic Energy Associates (a group of 17 utilities in the Southern and Southwestern United States), the Karlsruhe Laboratory of West Germany, Euratom, and General Electric Company. Research and Development is supported by the United States Atomic Energy Commission. SEFOR, situated near Fayetteville, Arkansas, is intended to provide data on operating and safety characteristics and economic potential of soft-spectrum, fast-breeder reactors fueled with PuO_2-UO_2, and to obtain physics and engineering data on fuel compositions and structure, temperatures and the Doppler coefficient of reactivity at high fuel temperatures up to near melting. Design for SEFOR was undertaken in March 1964, and construction started in September 1965. Permit for operation at 1 Mw(t) was granted in April 1969.

During the early phases of construction, the plant suffered from rising costs due to increased component costs, delays in delivery times, and added safety considerations. An experimental program, concurrent with design, was undertaken to provide needed data and resulted in some design changes. Table 10-3 lists some design and operating data of SEFOR.

TABLE 10-3
The SEFOR Plant

Power	20 Mw(t), no electric power
Fuel	Pellets 0.875 in. diam, 5/8 in. high, cupped at both ends
Cladding	Type-316 stainless steel, 0.040 in. thick
Design fuel composition	Pu^{239} & Pu^{241} 18.78%, total Pu 20.4%, depleted U 79.6%
Fission reactions in isotopes	Pu^{239} 86%, Pu^{240} 2%, U^{238} 11%, U^{235} 1%.
Core I Composition	Fuel 43.2%, Sodium 29.5%, BeO 5.7%, Steel 21.6%
Heat flux, Btu/hr ft^2	152,000 av., 280,000 max
Specific power kw/kg$_m$ (U + Pu) . . .	10.5 av., 19.1 max
Power Density kw/liter	37.7 av., 69.6 max
Linear power generation, kw/ft . . .	11.3 av., 20.8 max
Fuel temperature	1900°F av., < 5000°F(melting) max
Coolants	Sodium (primary and secondary)
Primary coolant velocity, ft/sec . . .	5.4 av., 8.3 max
Primary coolant temperatures	700°F inlet, 820°F exit, 1050°F max. during prompt transients
Primary coolant flow, gpm	5000 total, 700 core bypass, 250 auxiliary circuit
Secondary coolant temperatures . . .	550°F inlet, 670°F outlet
Heat transfer coefficients, Btu/hr ft^2 °F	1000 and 1060, main and auxiliary intermediate heat exchangers, 9.0 and 7.3 airblast exchangers
Control	Nickel reflector control
Control rods	Ten 6-in.-thick segments surrounding core. Two used for fine and eight for coarse control
Doppler coefficient (temperature) . .	$- 0.0085 \ T^{-1}$
Sodium void coefficient	$-\$4$ & $-\$10$ for total loss from core and vessel, $+17¢$ for loss of 10% of core volume from center of core

The SEFOR reactor building consists of two containments in series (Fig. 10-8). The inner containment, a reinforced concrete structure designed for 10 psig, houses the control drives, the primary sodium system, the irradiated fuel storage (below grade) and the refueling facility (above grade). Leakage is prevented by an inner welded steel liner in intimate contact with the concrete. The outer containment is a welded cylindrical pressure vessel 50 ft in diameter and 114.5 ft high.

The main coolant enters the reactor vessel at 700°F through a single inlet nozzle (Fig. 10-9) flows down an annulus between the vessel wall and a shroud surrounding the core, reverses direction and enters the core

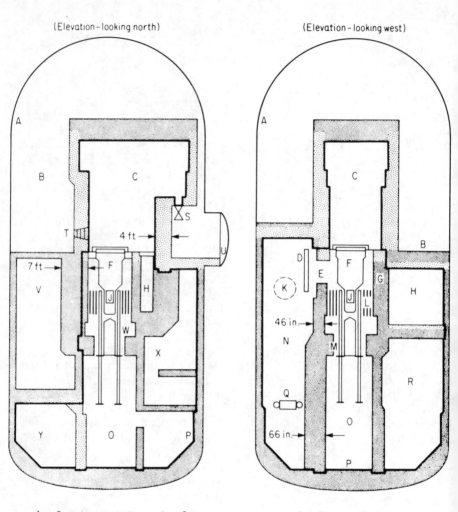

(Elevation – looking north)   (Elevation – looking west)

A – Containment shell   J – Core   S – Fuel transfer valve
B – Operating floor   K – Main IHX   T – Window
C – Refueling cell   L – Blast protection   U – Equipment door
D – Shadow shield   M – Bottom shield plug   V – Air shaft
E – Sodium-pipe tunnel   N – Sodium equipment cell   W – N_2 inlet plenum
F – Reactor   O – Reflector drive cell   X – Nitrogen equipment area
G – Nitrogen outlet ducts   P – Steel liner   Y – Sodium drain tank cell
H – Cold trap cell and   Q – Main pri. pump
    nitrogen duct area   R – Nitrogen equipment cell

▨ Heavy concrete   ▨ Ordinary concrete

FIG. 10-8.   SEFOR reactor building.   (Ref. 95.)

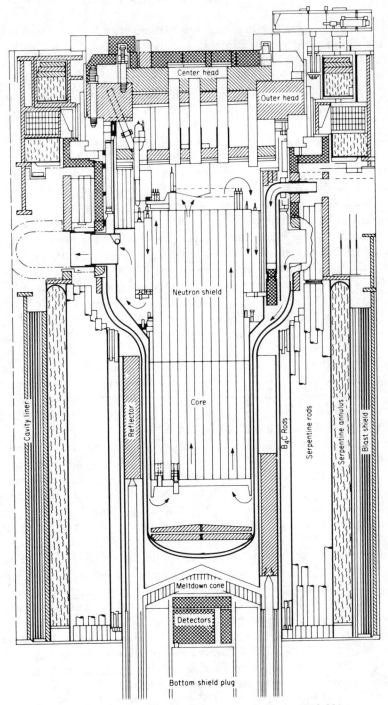

FIG. 10-9.   SEFOR reactor vertical cross section.   (Ref. 95.)

through an orificed grid plate and flows through 109 fuel channels and through a channelized neutron shield. At the top it flows radially, then down short neutron shield channels, up through an annulus plenum, and finally out, at 820°F, through a single exit nozzle. The main coolant design flow rate is 4,300 gpm but with a bypass leakage flow of 700 gpm. An auxiliary coolant system supplies 250 gpm through a single check-valved inlet line and nozzle.

The shielding above the core, made of B_4C and occupying extension handles above the fuel rods, is intended to prevent significant activation of the reactor vessel flanges and head. A thermal shield is provided below the core. The reactor vessel is 70 in. in major diameter, 17.25 ft. long, stainless-steel (Type 304) vessel designed for a pressure of 100 psig at 1050°F. The vessel's main closure flange heads and bolting were designed so that no portion of the vessel or surrounding support structure shall become missiles in the refueling cell during a design basis accident (DBA) of an explosive energy release of 400 Mw-sec (equivalent to 200 lb_m TNT). The vessel has a double wall below the main nozzles. The outer or safety vessel is also designed for 100 psig and 1050°F and will normally contain argon. Four leak detectors are placed between the vessels to monitor for sodium leakage. The core shroud is made of Type-304 stainless steel.

Each fuel subassembly (Fig. 10-10) contains 6 fuel rods and a center tightener rod containing BeO for flux softening. (In the second core the BeO will be replaced by another material to run tests with a harder spectrum but with a loss of reactivity of 1.5–2.0 percent). The core volume fractions of core I are given in Table 10-3.

The fuel rods are divided into two segments, separated by a spacer welded to the cladding. This reduces axial expansion and helps separate the Doppler coefficient from the overall fuel temperature coefficient in the experimental program. Each fuel rod is 0.970 in. diameter and 49 5/8 in. long. The active fuel is 33.812 in. long. There is a 2-in. gap between the two fuel segments and insulator pellets separating the fuel from the axial reflectors. The fuel is PuO_2–UO_2 pellets approximately 0.875 in. diameter and 5/8 in. long. The pellets are dished to 0.015 in. at both ends to decrease axial expansion and are held in place by Inconel-X springs. The fuel material is Pu^{239} and Pu^{241} (18.7 percent), other plutonium (up to 20.4 percent) and depleted uranium. The upper and lower reflectors are made of nickel and are 0.875 in. diameter and 4 in. long each. Cladding is a 0.04 in. thick Type-316 stainless steel. The gaps inside it are helium-filled. Extension rods above the fuel and center-tightening rods contain B_4C powder for shielding, clad in Type-316 stainless steel rods with gas spaces filled with helium.

Figure 10-11 shows a simplified flow diagram of SEFOR. There

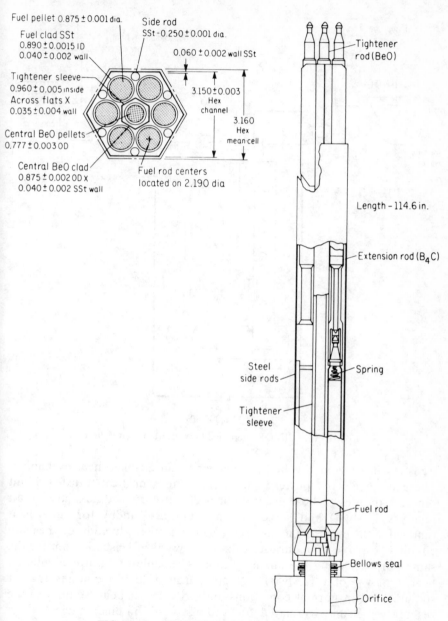

Fuel pellet 0.875 ± 0.001 dia.

Side rod
SSt - 0.250 ± 0.001 dia.

Fuel clad SSt
0.890 ± 0.0015 ID
0.040 ± 0.002 wall

0.060 ± 0.002 wall SSt

Tightener sleeve
0.960 ± 0.005 inside
Across flats X
0.035 ± 0.004 wall

3.150 ± 0.003
Hex channel

3.160
Hex mean cell

Central BeO pellets
0.777 ± 0.003 OD

Central BeO clad
0.875 ± 0.002 OD x
0.040 ± 0.002 SSt wall

Fuel rod centers
located on 2.190 dia

Tightener
rod (BeO)

Length - 114.6 in.

Extension rod (B₄C)

Steel
side rods

Spring

Tightener
sleeve

Fuel rod

Bellows seal

Orifice

FIG. 10-10.   SEFOR fuel subassembly.   (Ref. 95.)

is a main and an auxiliary coolant system in parallel.   Each consists of
one primary loop and one secondary loop.   The primary loop carries
reactor heat to an intermediate heat exchanger.   The secondary loop

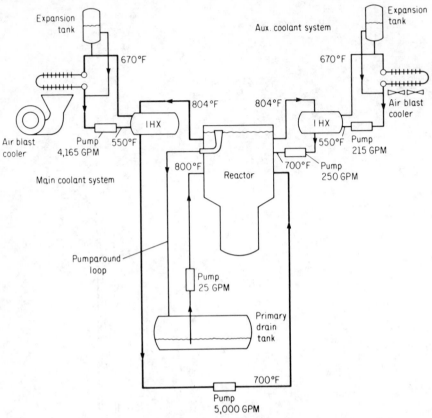

FIG. 10-11.   SEFOR simplified flow diagram.   (Ref. 95.)

transfers heat from the heat exchanger to an air-blast heat exchanger. There is no electric power generation.   The main heat transfer circuit removes the design heat generation of 20 Mw(t).   The auxiliary circuit removes 1 Mw(t) and serves as an operating standby for decay-heat removal in the event the main circuit is inoperative.   In addition the main circuit is capable of natural-circulation flow of at least 250 gpm for the same purpose, while the auxiliary loop is capable of natural circulation heat removal of 0.1 Mw(t).   A pumparound loop maintains reactor sodium level in normal conditions and provides makeup sodium to the reactor vessel and irradiated fuel storage vessel in emergencies.

A primary sodium drain tank serves as the principal expansion volume for the reactor and primary loops.   The secondary loops have expansion tanks with regulated gas pressure to maintain higher pressure than in the primary loops so that leaks may not occur from the radioactive side.

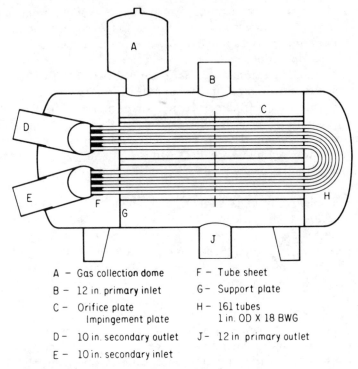

A – Gas collection dome   F – Tube sheet

B – 12 in. primary inlet   G – Support plate

C – Orifice plate   H – 161 tubes
       Impingement plate   1 in. OD X 18 BWG

D – 10 in. secondary outlet   J – 12 in. primary outlet

E – 10 in. secondary inlet

FIG. 10-12.   SEFOR main intermediate heat exchanger.   (Ref. 95.)

The intermediate heat exchangers (Fig. 10-12) are of the shell-and-tube type. Primary sodium flows in a vertical, single pass, cross flow-fashion on the shell side. Secondary sodium flows in a two-pass (6 passes for the auxiliary heat exchanger) tube bundle. Differential thermal expansion between shell and tubes is accommodated by the flexible tube bundle.

The air-blast heat exchangers are of the finned-tube type and use forced air. Steel gratings above and below the tube bundles protect this outdoors facility against flying objects. Also, the bundle and support structure are designed for 300-mph wind loads. The blower speed is regulated by a temperature indicator-controller to provide constant sodium outlet temperature. The blower also has manually operated inlet guide vanes for control at very low load conditions. The tube bundle housing is insulated and electrically heated to maintain temperatures during startup and shutdowns.

Sodium in the main primary and secondary circuits is pumped by two identical linear-induction, electromagnetic pumps, designed for 5,000 gpm of 700°F sodium, and capable of 1,000 gpm of 1050°F sodium each. The pumps are controlled by varying the voltage input and are hermetically sealed and gas-cooled. Power to two stator windings of the main primary pump is supplied by two separate synchronous motor-generator sets operating in parallel for additional safeguard protection. Also, each motor-generator set is equipped with a flywheel which provides a coastdown time of one minute during which flow reduces from 5,000 gpm to 1,000 gpm. Generator excitation during coastdown is provided from a 125-volt battery. Power to the main secondary pump is supplied from an autotransformer, induction voltage regulator combination. Power supplies are designed to provide sinusoidal voltage oscillations at a frequency of about 2 cycles/min for reactor oscillator tests.

The radioactive primary sodium (containing Na^{24}) loops are in nitrogen-filled cells which are concrete shielded within the reactor building. The secondary sodium loops are mainly located outside the reactor containment vessel within the operations building. The air-blast heat exchangers are located outside the operations building. All components in contact with sodium are made of Type-304 stainless steel. All are equipped with tubular electric resistance heaters inside the thermal insulation to keep the sodium molten during filling and extended shutdown. All have sodium-leak detectors with control-room alarms.

Argon is used to cover sodium surfaces and in the refueling cell which is exposed to reactor and irradiated fuel storage sodium during refueling. A circulating nitrogen atmosphere is used in the reactor and other primary component cavities to prevent sodium burning with air in the event of leakage as well as provide cooling for some of the components.

Neutron lifetime and delayed neutron fractions are important in discussing fast reactor control (Sec. 9-7). Recall that θ_0 is a strong function of the neutron energy spectrum and that β is a strong function of fuel composition. In SEFOR core I, the calculated median fission energy is 174 Kev average and 207 Kev at the center of the core and the fuel is a mixture of Pu and depleted U. The prompt neutron lifetime θ_p is 0.66 microsec (compared to about 0.1 microsec for a hard-spectrum fast reactor) and β is 0.0032 (compared to 0.002 for Pu^{239} alone).

SEFOR is reflector-controlled. The control system consists of ten 6-in.-thick nickel segments completely surrounding the core but located outside the reactor vessel. These control "rods" are lowered below the core for reactor shutdown. In addition, a 2-in. annular region of B_4C outside the segments followed by a serpentine moderator, decrease neutron reflection during shutdown. Two of the rods are used for fine control, 8 for coarse control.

The Doppler temperature and power coefficients are negative. The expansion and bowing coefficients are also negative. The sodium coefficient in this relatively small 40-in.-diameter, 37-in.-high core is negative. Reactivity decreases of $4 and $10 occur for complete sodium loss from the core and from core and vessel respectively. A small reactivity increase of 17¢ maximum occurs when sodium is removed from the center 10 percent of the core only.

10-5. LARGE SODIUM-COOLED FAST REACTORS

The AEC has sponsored a program to develop basic fast-breeder concepts for large, 1,000-Mw(e) commercial power plants. The objectives of the program were to develop designs, optimize fuel-cycle costs, attain attractive fuel-doubling times, generate high-temperature steam, all consistent with safety requirements in fast reactors. Four companies participated in this program. The design parameters are shown in Table 10-4, and the main configurations in Fig. 10-13. The studies [96] are sumarized in the following.

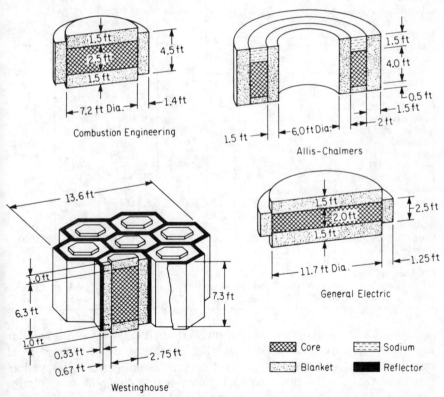

FIG. 10-13. Proposed reactor core arrangements for four 1000-Mw(e) design studies. (Ref. 96.)

TABLE 10-4
Some Design Parameters of Four 1,000-Mw(e) Sodium-Cooled Fast Reactors

	Allis-Chalmers	Combustion Engineering	General Electric	Westing-house
Power, Mw (t), core	2,125	1,950	2,125	2,170
Axial blanket		200	300	35
Radial blanket	375	350	75	295
Total	2,500	2,500	2,500	2,500
Core geometry	Annular	Flat cylinder	Pancake	Modular
Core volume, Liters	7,520	2,895	6,030	7,050
Core and axial blanket composition, percent .				
Fuel	29.4	25.6	34.8	29.4
Sodium	40.0	66.5	46.4	55.1
Steel	30.6	7.9	18.8	15.5
Fuel form, core and axial blanket	Oxide	Carbide	Oxide	Carbide
Fissionable fuel, kg_m	3,690	1,155	2,304	3,686
Core loading, ceramic, kg_m	20,200	9,400	14,600	25,200
Power density, kw(t)/liter	282	695	365	308
Specific power, kw/kg_m				
Fissile material	640	1,680	920	520
U + Pu ceramic material	100	198	150	86
Temperatures °F				
Core inlet	950	850	800	979
Core outlet	1,200	1,120	1,100	1,200
Maximum cladding	1,330	1,400	1,332	1,400
Maximum fuel	4,615	2,600	4,700	2,184
Coolant flow, lb_m/hr $\times 10^{-6}$	114	113.6	95.4	128
Core velocity, ft/sec	10	20	11.1	26.5
Loop pressure drop, psi	90	59.8	40	113
Fuel pin cladding OD, in.	0. 30	0.30	0.25	0.30
Fuel cladding thickness, in. .	0.028	0.011	0.015	0.010
Fuel pellet OD, in.,	0.24	0.259	0.22	0.268
Fuel pellet ID, in.	0.10		0.06	
Fuel cladding bond	He	Na	He	Na
Number of fuel pins per subassembly	123	169	470	120
Total number of cores per subassembly	498	157	225	252
Fuel enrichment, average percent	20.8	13.0	18.0	15.7
Blanket enrichment, average percent	0.3	0.3	0.3	0.3
Core burnup, Mwd/ton	100,000	110,000	110,000	100,000
Breeding ratio	1.32	1.42	1.25	1.57
Fuel cycle costs mills/kwhr	0.69	0.30	0.57	0.26
Doubling time, years	19.5	6.2	15.8	11.7
Na void worth (core completely voided), %Δk .	0	1.0	0.4	0.3
Doppler coefficient $Tdk/dT \times 10^3$	− 2.6	− 5	− 5	− 12

Allis Chalmers reduces the sodium void coefficient by a thin annular high leakage core. Leakage is further increased by having no axial breeder blanket, only sodium space. Fuel pins contain $PuO_2 - UO_2$ pellets distributed axially in three compartments within the pins to minimize reactivity effects due to fuel compaction. Fuel pellets are hollow centered to permit maximum linear power rating of 13 kw/ft at maximum

1	Reactor core	11	Intermediate heat exchanger
2	Refueling cell wall		(shown in raised position)
3	Shield plug (in raised position)	12	Secondary sodium inlet
4	Control rod drives	13	Secondary sodium outlet
5	Shield plug lifting screw and guide	14	Primary sodium pump and
6	Fuel transfer machine		drive motor (3 units)
7	Fuel shuffling machine	15	Primary sodium vessel
8	Fuel decay tank	16	Insulation and outer tank
9	New fuel storage	17	Steel liner on concrete
10	Intermediate heat exchanger	18	Concrete shielding
	(3 units)		

FIG. 10-14.   1000 Mw(e) follow-on Design of LMFBR.   (Courtesy General Electric Co.)

fuel temperatures.   The radial blanket is shuffled in-out to avoid large changes in blanket sodium-exit temperatures due to Pu buildups.   The plant contains six coolant loops.

Combustion Engineering uses a flat cylindrical core and carbide fuel in core and blankets.   Each blanket is rotated 180° halfway throughout life to equalize exposure and Pu buildups.   Fission gas release from the fuel, to the extent of 11 percent over the 100,000 Mwd/ton burnup is accommodated in a 15 in. gas space above the upper axial blanket.

General Electric uses a highly pancaked core to obtain a negative void coefficient. The core consists of three radial zones with different Pu enrichments for power flattening. The linear power rating of 22.7 kw/ft is obtained with hollow-centered oxide fuel. The core contains 7 percent BeO to increase the negative Doppler coefficient. The entire primary system, consisting of reactor core, 6 heat exchangers and 6 pumps, is immersed in a 52-ft-diameter sodium tank to minimize sodium leakage. A shielded inert-gas cell above the reactor is used for refueling. General Electric also proposes no BeO addition for harder spectrum and therefore higher breeding ratios, and fuel burnups up to 200,000 Mwd/ton.

Westinghouse uses a reactor composed of seven hexagonal modules, each a high-leakage core surrounded by 8-in.-thick blankets of UO_2 and a 4-in. layer of graphite to decouple the modules. The fuel is assembled in a controlled fuel expansion assembly to provide a negative reactivity coefficient due to thermal expansion, supplementing the Doppler coefficient. Fission gases are vented to the primary sodium coolant to minimize stresses in the cladding. Adjustments in the sodium flow are made periodically to accommodate the buildup of Pu in the blankets.

All four designs have neutron lifetimes of $3-6 \times 10^{-7}$ sec, effective delayed neutron fraction of 0.004 and sizable negative Doppler coefficients. The presence of oxygen and, in particular, carbon in the fuel softens the spectrum and results in a large negative Doppler coefficient. The power coefficients are always negative, being of the order of -1 to $-3 \times 10^{-6} \Delta k/\text{Mw}$.

The above preliminary program was succeeded by "follow-on" designs, "add-on" studies, and so on. Figure 10-14 shows the follow-on design by General Electric.

10-6.   GAS-COOLED FAST REACTORS

The Gulf General Atomic Corporation has embarked on a program to develop the technology necessary for the construction of helium-cooled fast reactors. A Gas-Cooled Fast-Reactor Experiment (GCFRE) [90] was designed to provide fuel studies as well as data on reactor physics and kinetics of large, low-enriched fast breeder reactors. In this experiment, a two-zone, 1387 liter, 109 Mw(t) core uses 11 percent-enriched uranium-plutonium oxide fuel, clad in stainless steel and clustered into 75 fuel subassemblies. Three phases of testing are contemplated: (a) basic physics and kinetic testing, (b) fuel testing, including the screening of different fuel elements and subassemblies, testing burnup (100,000 Mwd/ton is contemplated) and burnup effects, fuel-cladding interactions, bowing, vibrations and heat-transfer, and (c) testing new fuel of higher enrichment in a small core at power-reactor ratings and flux levels and to develop statistical data on fuel performance and high-neutron irradiation effects.

A typical 1,000-Mw(e) gas-cooled fast-breeder reactor (GCFR) conceptual design was proposed in a joint study by Gulf General Atomic [97, 98] and the East Central Nuclear Group (comprising of 14 investor-owned electric utilities).

The criteria chosen for the GCFR design were:

(a) The reactor has no fertile blanket and must be large, since low leakage is required for good breeding ratios and good use of fuel.

(b) The design retains the trend of using prestressed concrete pressure vessel with gas cooling. Here the tension members (tendons) are well shielded from core irradiations, since they are imbedded several feet inside the concrete walls. It may be recalled that a large number of prestressed concrete pressure vessels have been built in Europe and for the Fort St. Vrain HTGR, built by the same company, (Sec. 8-9).

(c) High coolant pressures are provided to attain high mass-flow rates and consequently high power levels for a given coolant temperature rise. Table 10-5 lists proposed performance characteristics of GCFRE and GCFR. Table 10-6 shows a comparison between GCFR and two thermal gas-cooled reactors. Note in particular the much higher power density but the similar specific power.

GCFR, (Fig. 10-15) uses a horizontal prestressed-concrete pressure vessel (PCRV) which houses the reactor and the primary helium circuit. PCRV serves the dual function of pressure vessel, and because concrete thicknesses required by structural considerations are more than adequate, of biological shield. Boron is added to the inner portions of the concrete.

PCRV is 100 ft long, 33 ft in diameter, with 12 ft thick walls and 15-ft-thick ends. It is divided into three compartments. A center compartment, 20 ft long, contains the reactor core. Two adjacent compartments, each 35 ft long contain the steam generators and blowers. The vessel is designed for a pressure of 1,000 psia and a temperature of about 920°F. It contains a large number of tension members (tendons) as well as some localized ordinary steel reinforcement. There are circumferential tendons to control the hoop stresses and end tendons to transmit the pressure forces exerted on the vessel heads to anchor points in the walls where they are counteracted by the axial tendons. The tendons are tensioned before the vessel is pressurized and are designed to exert compressive forces opposite to and greater than the forces produced by the internal pressure when acting alone. The principal stresses at the inner faces of the vessel are therefore compressive. The vessel is lined intern-

TABLE 10-5
GCFRE and GCFR Design Parameters

	GCFRE	GCFR
Power output, Mw(t)	100	2,633
Power output, Mw(e)		1,000
Fuel	Mixed UO_2 and PuO_2	same
Core average U/Pu ratio	5.75	5.50
Core average fertile/fissle ratio	8.00	8.00
U composition, percent: U^{235}	0.35	0.35
U^{238}	99.6	99.6
Pu composition, percent: Pu^{239}	63.0	66.7
Pu^{240}	24.0	27.6
Pu^{241}	10.0	3.6
Pu^{242}	3.0	2.1
Core volume percent: Fuel	54.0	32.0
Structure	14.0	10.0
Coolant	32.0	58.0
Core height diam, cm	110 × 132.9	126.5 × 316.25
Axial blanket height, cm	30.0	60.0
Radial blanket thickness, cm	23.5	46.0
Fuel subassemblies, in core blanket	75,90	248, 188
Fuel pins per subassembly	91	256
Fuel pin diam, pitch, cm	1,094.	0.76,
	1,236 (triangular)	1.025 (square)
Fuel pin surface	Smooth	Roughened
Maximum fuel temp, °F	1,184	3,018
Maximum cladding temp, °F		1,292
Linear power rating, kw/ft	7.5	13.1
Coolant flow, lb_m/hr		12.744 × 10⁶
Coolant temps, inlet, outlet, °F	392, 1,090	644, 1,184
Coolant pressure, psia	1,250	1,000
Coolant pressure drop, psi		28.3 (core),
		50.0 (total)
Internal conversion ratio	0.97	0.98
Physics		
Doppler coefficient, $T\Delta k/k\Delta t$	0.0068	0.0068
Helium worth	0.0015	0.0042
Median fission energy, kev	170	150
Fraction neutrons captured in core	0.383	0.384
Fraction neutrons in core fission	0.336	0.325
Fraction leaking from core	0.281	0.291

ally with 0.5–1 in. steel sheet to insure leak-tightness. The vessel and liner are kept relatively cool against the 630°F reactor inlet temperature by stainless steel insulation toward the reactor side and cooling on the concrete side. Cooling coils are welded to the outside of the liner and are embedded in the concrete to maintain the latter at no more than 150°F.

The three vessel compartments are separated by 4-ft-thick concrete partitions, water-cooled on both sides. They provide the necessary

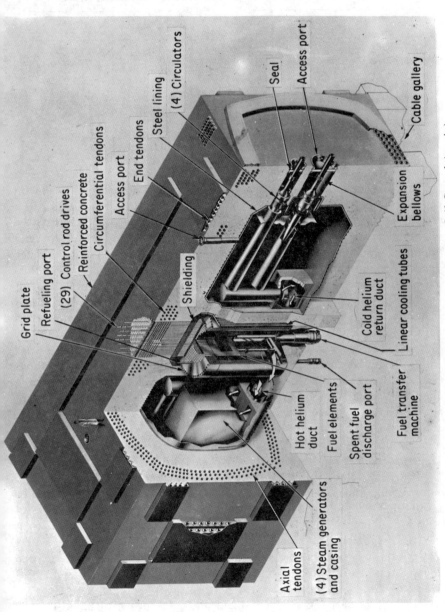

FIG. 10-15. The Gas-cooled Fast Reactor, GCFR. (Courtesy Gulf General Atomic.)

TABLE 10-6
Comparison between GCFR and Two Thermal Gas-Cooled Reactors*

	AGR	HTGR	GCFR
Power, Mw(e)	600	1,000	1,000
Core diam × height, cm	945 × 825	945 × 470	275 × 150
Average power density, watts/cm³	2.5	7.5	275
Coolant fraction in core	0.75–0.8	0.15–0.2	0.51
Specific power kw/kg$_m$ fissile	500	1,000–2,000	1,000
Average core enrichment ...	0.015–0.025	0.04–0.05	0.12

* From Ref. 64.

shielding of the steam generators. The compartment walls are lined with water-cooled, gas-tight mild-steel liners. The helium ducts and casings are insulated so that the voids in the steam-generator compartments are maintained at about 150°F permitting the installation of electrically operated controls and valves. The vessel ends contain 8-ft-diameter penetrations for installation and servicing of the steam generators. Normal maintenance will be possible with the reactor shutdown because of the shielding by the 4-ft partitions. The 8-ft penetrations also contain a helium recirculator. Another is mounted in a smaller penetration above the large one.

The core has a total volume of 10,000 liters and is slightly pancakish with a height-to-diameter ratio of 0.4. It contains 277 fuel subassemblies (Fig. 10-16), each composed of 256 stainless-steel-clad fuel pins 0.8 cm in diameter. The core grid plate is 16 ft in diameter and consists of a square lattice formed by 1/4-in. thick and 40-in. deep cross-members and a pitch of 6.734 in. The core is surrounded by radial and axial blankets containing depleted UO_2 pellets. There are 29 control rods that enter the vessel from the top. The entire core is surrounded by a thermal shield.

Helium coolant flows downward through the core. This keeps the control rods and the grid plate cool and uniform in temperature. Hot helium at 1198°F exits to ducts located in the bottom corners of the vessel partitions (Fig. 10-15). Hot helium discharges into a shielded plenum in the steam-generator compartment, and flows through the reheater, super-heater and the evaporator-economizer in each steam generator. It then goes to the helium circulators and to a cold shielded plenum at the bottom center of the steam-generator partition. It reenters the reactor at 630°F flowing past the thermal shield and back down through the core and blankets.

The primary helium circuit consists of four loops, each containing a separate helium circulator and a steam generator. Each helium circula-

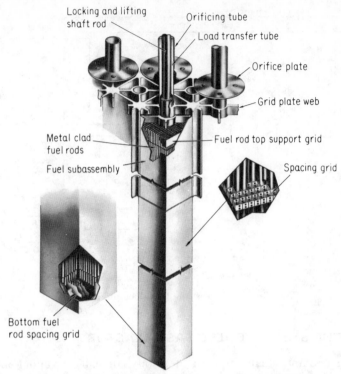

FIG. 10-16. GCFR fuel subassembly and grid plate. (Courtesy
Gulf General Atomic.)

tor is a 40,000 shaft-horsepower two-stage centrifugal unit, driven by a
single-stage axial series steam turbine located in the cold reheat line
(Fig. 10-17). The circulators have oil and helium buffer sealing and are
installed at the outlets of the steam generators.

The entire steam-supply system is enclosed within the PCRV. The
steam generators are once-through units capable of supplying 7.2×10^6 lb_m/
hr of steam at 2,400 psia and 1000°F. The reheat steam from the high-
pressure turbine exhaust is used to drive the helium circulator turbines
before flowing into the reheater sections of the steam generators. Feedwater is at 2,900 psia and 325°F.

A number of steam-generator designs are being contemplated.
These include once-through U-tube designs with vertical or horizontal
tubes and a modular design consisting of vertical modules with radial gas
flow.

The plant contains auxiliary systems for helium purification, storage
and handling, cooling water for PCRV liners, etc. Because of the high
temperatures of the plant, the thermal efficiency is 38.3 percent.

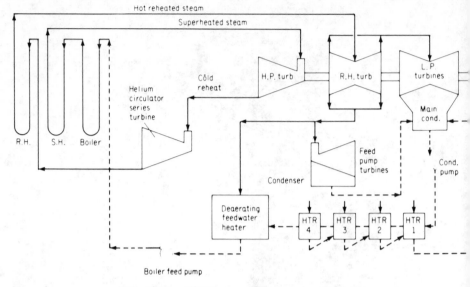

FIG. 10-17.   GCFR steam flow diagram.   (Ref. 98.)

10-7   THE STEAM-COOLED FAST REACTOR

As a coolant, steam (though thermodynamically a vapor) acts as a gas. Consequently, gas-cooled, as well as superheat steam reactor experience can be relied on as background experience in the design of steam-cooled fast reactors. The development program for this type of reactor has thus concentrated on much the same problems as other fast reactors, such as the development of mixed plutonium and uranium oxides, fuels, cladding and structural materials, nuclear data, transient and safety analysis, etc.

The main differences between steam-cooled and other fast reactors are brought about by the nature of the coolant. Steam offers some potential advantages, including: (a) well-known chemical, thermodynamic, and physical characteristics, (b) a wealth of experience in designing and manufacturing steam components so that many off-the-shelf items may be available, resulting in (c) an expected short development program, and (d) the possibility of using a direct cycle (reactor supplying supherheated steam directly to the turbine), with consequent capital-cost economies.

The main disadvantages are: (a) the use of high-pressure steam, required for high mass-flow rates, results in a soft neutron spectrum and a consequent reduction in breeding ratio and, therefore, fuel cycle economics, and (b) shutdown and emergency core-cooling problems that it shares with the gas-cooled fast-breeder reactor.

The market for the steam-cooled fast-breeder reactor is explained as follows. Because of the use of burner reactors at the present stage of nuclear-power development, much plutonium, converted from U^{238} in these reactors, is accumulating. The government buyback policy allowed the so-called plutonium credit to make the fuel-cycle costs attractive in these reactors. This policy, however, is expected to be terminated in the early 1970's. A market for this thermally produced plutonium must therefore be found. It can probably be best used in mixed U-Pu-oxide-fueled, high-breeding-ratio sodium or gas-cooled fast reactors. However, until this reactor type is developed to a point where it has become economically competitive, probably in the early or mid 1980's, the steam-cooled fast-breeder reactor, which, alone among the three types, claims to have no major development problems, can fill the gap. Its introduction date, however, is critical and depends on complex economic matters. Too early an introduction, for example, would disadvantage breeder-reactor economics because it would inflate plutonium prices prematurely, while a late introduction would disadvantage thermal reactor economics because it would deflate them. It is claimed that an introduction date around the mid 1970's is optimum and that in the 10-year interval between the availability of large plutonium reserves from thermal reactors and the introduction of economical high-gain breeders, a 50,000,000 kw(e) plutonium-fueled market will exists. Such is the incentive for the development of the steam-cooled fast breeder reactor.

The above optimistic outlook has, however, been shattered by the inability to develop, at an early stage, of a fuel element that is capable of withstanding the highly corrosive atmosphere of high-temperature steam. Interest in the steam-cooled fast reactor has, therefore, waned considerably by 1970. The principles behind it are, nonetheless, interesting. The next section presents conceptual designs of steam-cooled fast-reactor plants.

10-8. THE EXPERIMENTAL STEAM-COOLED REACTOR (ESCR)

As with the gas-cooled fast reactor (Sec. 10-6) conceptual designs for a small experimental reactor have been prepared, in this case by General Electric Company and a group of 14 electric utilities [99]. This is a 50 Mw(e) Experimental Steam-cooled Fast Reactor (ESCR). In addition, a large, commercial-scale 1,000 Mw(e) Fast Steam-Cooled Reactor (FSCR) has been studied by General Electric. In Europe interest in steam-cooled fast reactors has been shown by Belgium, Sweden, and

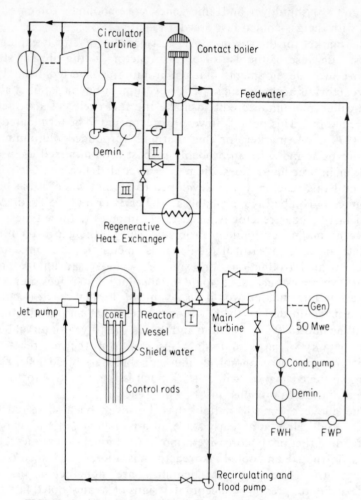

FIG. 10-18. Flow diagram of Experimental Steam-Cooled Reactor, ESCR. (Ref. 99.)

Germany. A major purpose of ESCR was to test fuels and cladding materials under combined fast neutron and superheated steam conditions. Table 10-7 gives conceptual design data of ESCR.

ESCR is a closed-loop direct-cycle plant (Fig. 10-18). Saturated steam at 1,500 psia and 596.2°F is pumped by three steam-driven high-pressure circulators to the reactor. Steam drives are preferred from a control standpoint because of the wider range of flow, (15–100 percent as compared to 30–100 percent for variable-frequency motor-generator sets). Steam enters the reactor via three loops at the top and flows

TABLE 10-7
Conceptual Design Data of ESCR*

Power	128 Mw(t) core, 12 Mw(t) axial blankets, 50 Mw(e)
Core dimensions	40 in. diam, 18 in. high
Core volumetric composition, percent	Fuel 33.3, coolant 30.4, structure (incoloy-800) 27.5, moderator (BeO) 8.1, Control (Ta) 0.7
Fuel material	$PuO_2 + UO_2$
$Pu^{239} + Pu^{241}$ enrichment, percent	27 new core, 25 midcycle core average
Fuel pin pitch/diam ratio	1.2
Cladding	Incoloy-800, 0.228 in. OD, 0.015 in. thick
Maximum clad temperature	1300°F
Heat flux, Btu/hr ft²	390,000 average, 900,000 hot spot
Linear power, kw/ft	6.8 average, 15.7 hot spot
Coolant	Steam
Coolant mass-flow rate	1.73×10^6 lb_m/hr
Coolant temperatures, °F	610 core inlet, 950 outlet
Coolant pressures, psia	1500 core inlet, 1415 outlet
Reactivity worths ($\Delta k/k$)	-0.014 steam loss, $+0.022$ flooding (equil core, no thermal poisoning)
Control requirements (Δk)	Burnup 0.038, cold-to-full power 0.035, safety and shutdown 0.017, 0.090 total
Doppler Coefficient Tdk/dT	-0.01 operating, steam in, -0.006 voided, steam out

* From Ref. 99.

downward through the core where it is superheated to 950°F. It leaves the reactor via twelve pipes.

In this experimental plant, three modes of operation, provided by appropriate valving, were designed. The first is a direct-cycle super-heated mode in which approximately one third of the reactor superheated steam goes directly to the turbine generator to produce 50 Mw(e). The remaining two-thirds mix with the feedwater from the main turbine condenser in three contact boilers to produce saturated steam. This is re-turned back to the reactor by the circulators. The circulator turbines use 1–2 percent of the steam. Their condensate is fed back to the contact boilers. The main turbine inlet is selectively taken from the reactor outlet pipes to use steam of lowest contamination levels.

In the second, a direct-cycle saturated mode, all of the reactor superheated steam is desuperheated and washed in the contact boilers and saturated steam from the contact boilers is used to drive the main

342 Fast-Breeder Reactors and Power Plants

turbine. The entire core steam flow is therefore washed. This mode may also be used to provide wet steam to the turbine at low loads.

The third is a direct-cycle regenerative superheat mode in which all the reactor superheated steam passes through a regenerative heat exchanger before mixing with the feedwater in the contact boilers. Two thirds of the resulting saturated steam goes back to the reactor via the circulators. The remaining one-third is resuperheated in the regenerative heat exchanger, then goes to drive the main turbine. This mode combines the advantages of full-flow washing and the high efficiency of using superheat steam in the main turbine.

Besides acting as decontaminants for the circulated steam, to remove those fission products that can be removed by the scrubbing action of feedwater, the contact boilers also provide a reserve of saturated water to furnish desuperheating for a period of time in the case of feedwater failure. They are also a source of steam and flooding water in case of depressurization. [Tests on a 5 kw(e) turbine driven by direct-cycle superheated steam with fuel defects in the reactor have shown activity deposition in piping and turbine to consist predominantly of the relatively short-lived radioiodines, Te^{132} and Mo^{199} Maintenance was accomplished after appropriate decay periods following shutdown. Further tests with high specific-power fuel will be made.]

A large steam dump tank, not shown, containing unpressurized water, provides an emergency heat sink in the event of the loss of turbine and condenser to the system. In this case reactor steam would be blown into the water.

The 128 Mw(t) core is about 40 in. in diameter and 18 in. high (Fig. 10-19). It is surrounded by 18-in. high upper and lower axial blankets producing 12 Mw(t) and by a 9-in. thick nickel radial reflector. The core contains 6 Y-shaped control rods with tantalum absorber and twelve peripheral control rods with B_4C absorber. In addition, three central fuel subassembly locations are reserved for redundant emergency shutdown control with tantalum balls introduced through a rupture diaphragm. The core shroud is surrounded within the pressure vessel by a large space filled with saturated light water which primarily serves as a radiation shield but also provides a ready supply of flashing steam for core cooling in the event of system depressurization.

Heat flux in the steam-cooled fast-reactor fuel elements is cladding temperature limited. In ESCR the maximum cladding surface temperature has been set at 1300°F to maintain mechanical strength and adequate corrosion resistance. In order not to exceed this temperature and still attain the high power densities of a fast reactor, turbulence promoters [100], used to increase the heat transfer coefficient by about 85 percent are proposed in the lower two-thirds of the core (coolant enters the core

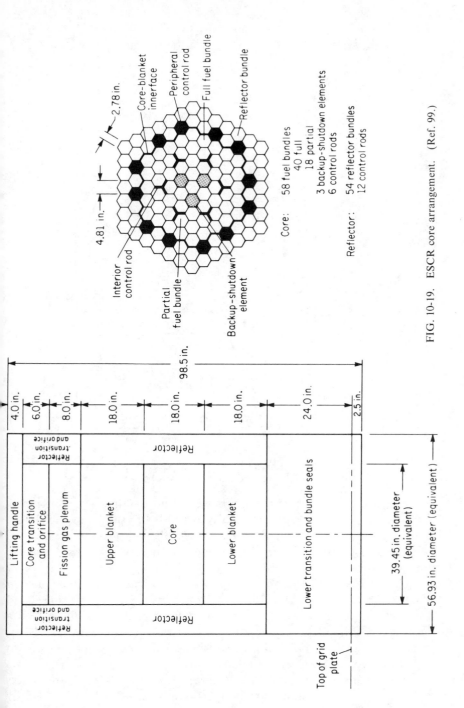

FIG. 10-19. ESCR core arrangement. (Ref. 99.)

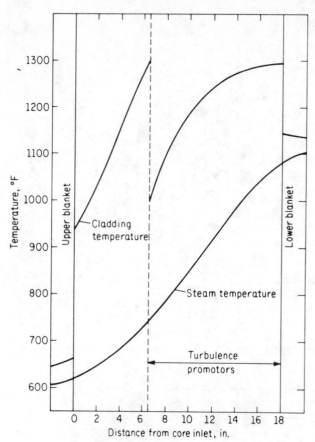

FIG. 10-20. Steam and cladding surface temperatures
with turbulence promoters. (Ref. 100.)

at the top). The expected maximum steam and cladding surface temperatures in a channel at the beginning of core life are shown in Fig. 10-20.
With burnup, the core power decreases and blanket power increases and
temperatures become lower. Turbulence promoters, of course, increase
pressure drops and have other effects.

The problem of shutdown and emergency cooling will be met in
ESCR, depending on steam flow rate, by fog or spray cooling using shield
water. The problem is complicated by the compactness of the core lattice, high heat fluxes, and generally low heat-transfer coefficients. Tests,
therefore, remain to be made to determine emergency operating conditions to prevent excessive clad temperatures.

In normal shutdown, the ESCR core is flooded. In this state it acts
much like a natural-circulation boiling-water reactor (Chap 3). Decay

heat (about 1/2–1 percent full power) causes the flooding water to partially boil and natural circulation flow upward through the core is established. Recirculation water returns to the core via an external line (also used for spray and flooding purposes). The resulting steam is condensed in an external shutdown condenser (not shown) and returned to the core bottom.

chapter 11

Fluid-Fueled Reactors and Power Plants

11-1. INTRODUCTION

In fluid-fueled reactors the fuel, in fluid form, is both the heat-generation medium, when in the core, and the primary coolant carrying its own energy through the primary loop. Figure 11-1 shows a simplified flow diagram of a fluid-fueled-reactor power plant. The reactor core is

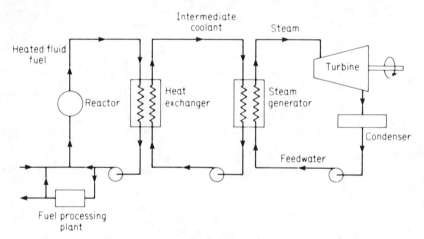

FIG. 11-1. Schematic of a fluid-fueled-reactor power plant.

simply a "bulge" of critical dimensions in the primary-loop piping. The bulk of fission energy is generated within the core, so that the fluid fuel leaves it at a higher temperature than when it entered. (Decay energy continues to be generated within the piping, of course.) The fluid fuel goes to the primary heat exchanger where it gives up its energy to an intermediate coolant. It is then pumped back to the reactor. An intermediate loop is necessary to isolate the highly radioactive fluid fuel from the steam system and to prevent the accidental entry of water into the primary loop and core. Since the fuel is in fluid form, it is relatively

346

easy to provide for the continuous removal, reprocessing, and replenishment of part of the fuel from the primary loop.

While there are several types of fluid-fueled reactors that will be discussed in this chapter, it appears that the most promising type for large economical power plant operation currently is the molten-salt breeder reactor. Molten-salt reactors now being developed use solutions of thorium and uranium tetrafluorides dissolved in lithium and beryllium fluorides. They are thus capable of thermal breeding [recall that breeding in a thermal reactor can be expected only from a $Th^{232} - U^{233}$ cycle (Sec. 9-4)]. In addition, the fluorination of a mixture of thorium and uranium tetrafluorides leads to the formation of UF_6 which is volatile and is therefore readily separated from thorium hexafluoride, which is not. This greatly simplifies the thorium cycle and gives it an economic advantage over the uranium cycle. Molten-salt fuels and materials are readily available, can operate at high temperatures and low pressures, do not undergo violent reactions with air or water, have radiation stability, and are compatible with graphite. They have large specific heats so that very high specific powers are attainable.

11-2. ADVANTAGES AND DISADVANTAGES

Among the advantages of a fluid-fuel system are:

1. Heat is generated within the primary coolant, so that the thermodynamic losses attributed to heat exchange due to a temperature difference between fuel and coolant (Sec. 2-3) are nonexistent, resulting in high thermal efficiencies.
2. Fuel fabrication costs are eliminated.
3. Radiation damage to fuel elements, which limits burnup, is eliminated.
4. The continuous removal of fission products and the addition of fresh makeup fuel during operation are made possible.
5. The excess reactivity, normally built in in solid-fueled reactors, is not necessarily large in this case, since the quantity of fissionable material within the core can easily be regulated.
6. Fluid-fueled reactors have a large negative temperature coefficient of reactivity.

The last two factors make this type of reactor safe from runaway accidents.

Probably the most outstanding advantage of fluid-fueled reactors is the relative ease with which high ratios of conversion and breeding can be built into them. When a fluid breeder blanket is used, the resulting fissionable material can easily be added to the fuel without the need of costly fuel processing and refabrication associated with solid-fueled breeders. Looking at the projected world power demands and the fossil and fissionable fuel reserves makes this development a very desirable goal in any nuclear-energy program.

Some degree of purification of blanket material, however, is necessary before addition to the core. In the U^{238}-Pu^{239} cycle, for example, lack of purification causes a buildup of plutonium isotopes heavier than Pu^{239} in the breeder fluid [101], resulting in a dilution of Pu^{239}, a change in neutron economy, and a worsening of the already acute problem of radioactivity.

Fluid-fueled reactors are, of course, not without their problems. Some of these are:

1. The primary fluid is highly radioactive and is an emitter of neutrons. Piping, heat exchangers, pumps, and other primary-loop components are therefore irradiated. This calls for the development of costly methods and machinery for remote maintenance and expensive plant arrangement.

2. High quality of materials and construction is required to ensure integrity and leak-tightness, especially in a high-pressure system (such as the aqueous reactors, below), and provision must be made for the safe disposition of fuel in the case of accidental leakage.

3. Fluid fuels in general are highly corrosive and therefore require special and costly materials of construction and handling techniques.

4. Radiolytic gases, generated in all fluid-fuel systems (such as H_2 and O_2 in aqueous systems), are a potential hazard and require special provision for handling or recombination by a catalyst.

5. Fluid-fuel-reactor power plants require supporting facilities, such as chemical laboratories and waste-disposal facilities, that are much more extensive than those required for solid-fuel-reactor plants. This, however, is balanced by the reduction in off-site supporting facilities.

6. They are usually limited to low core power densities and require fairly large core tanks, which, because of manufacturing problems, may pose a limitation on total core power.

11-3. FLUID-FUELED-REACTOR TYPES

Fluid-fueled reactors may be subdivided into homogeneous and heterogeneous types (Fig. 11-2). A homogeneous fluid-fuel reactor is

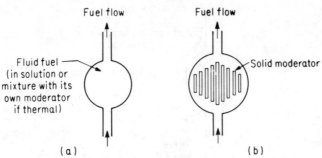

FIG. 11-2. Two types of fluid-fueled reactors. (a) Fast or thermal homogeneous; (b) thermal heterogeneous.

one which may be fast and thus requires no moderator or one in which the fuel carries its own moderator in fluid form, such as uranium salts in water solution. A heterogeneous fluid-fuel reactor is one in which a fluid fuel is passed through a core containing solid moderator elements, such as graphite.

Fluid-fueled reactors are usually classified according to the type of fuel that they use. The most common types are:

1. The *aqueous-fuel* systems. These are systems in which fuel material (such as uranium salts) is in solution or suspension (slurry) in light or heavy water and, if the ratio of water to fissionable material is large enough, are of the thermal homogeneous type.

2. The *liquid-metal-fueled* reactors, LMFR. In this category, nuclear fuel is held in solution or suspension in a liquid metal. An example is uranium metal in solution in molten bismuth.

3. The *molten-salt* reactors. These are reactors using uranium- (or other fuel-) bearing salts of low neutron capture cross sections, such as chlorides, fluorides, and hydroxides. This category of reactors could be of the thermal variety, if the salt contains hydrogen, deuterium, or fluorine, such as Li^7OD, or they could be of the fast type, provided that the salt does not contain any moderating material. Chlorine seems to be feasible in this respect. Such fast reactors have not received much attention, however.

4. *Gaseous-suspension* reactors. These are reactors in which fuel particles are held in suspension in a gas, such as UO_2 dust in helium. If the dust is fine enough, the fuel-laden gas may be used directly in a simple-loop gas-turbine power plant, such as in Fig. 7-3.

A fluid-fueled reactor can also be of the one- or two-region type.

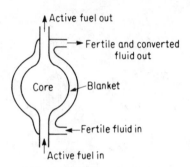

FIG. 11-3.   Schematic of a two-region fluid-fueled reactor.

In the latter (Fig. 11-3), the core contains the active fuel and generates the bulk of reactor power.   It is surrounded by a blanket containing fertile material, where core leakage neutrons are absorbed.   The two-region reactor is believed to be the more promising type as far as breeding gain is concerned.   A one-region reactor (no blanket) can be designed to have a good conversion ratio.   It would, however, have to be large (for low neutron leakage) and thus have a costly reactor vessel.

In the two-region reactor, the core and blanket fuels need not be of the same type.   That is, one or the other, or both, may be of the solution or slurry types.   One of the more promising reactors [102] contains a dilute uranyl sulfate (UO_2SO_4) solution in water in the core and a concentrated ThO_2 slurry in the blanket.   This arrangement is thought to produce a breeding ratio of 1.10 (10-to-15-year doubling time) and relatively high specific power.

The proper choice of fuel for a fluid-fueled reactor is a problem of major proportions.   Much needed information on the physical (including phase relationships of solutions) and chemical properties of the fuels and of their corrosion and erosion properties is still lacking.   Other relatively unknown factors are the extent of solubility of fission products in the original fuel, the effects of these products on the phase relationships, and the effect of nuclear radiations on the fuels.   (There is some evidence that this effect may be significant.)

The following sections will be devoted to those fuel types which show the most promise and which have, therefore, received sufficient attention.

11-4.   AQUEOUS FUEL SOLUTIONS

Several uranium and thorium salts have been considered for core and blanket, including uranyl sulfate, fluoride, nitrate, chromate, phosphate, lithium uranyl carbonate, and thorium nitrate and phosphate. Of these, uranyl sulfate (UO_2SO_4) seems to offer the greatest promise as a core fuel solute. For the blanket, thorium is the preferred constituent. However, no thorium salt solution has been found to be completely satisfactory as far as thermal and radiolytic stability and nuclear characteristics are concerned. Much effort has therefore been devoted to the development of a ThO_2 slurry as blanket material.

Tests on the solubility of UO_2SO_4 in light and heavy water showed [102] that the system behaves as a three-component (rather than binary) system composed of UO_3, SO_3, and H_2O (or D_2O). Solubility relationships (Fig. 11-4) indicate the existence of two miscible-liquid phases, which appear upon heating a stoichiometric UO_2SO_4 solution beyond about 550°F

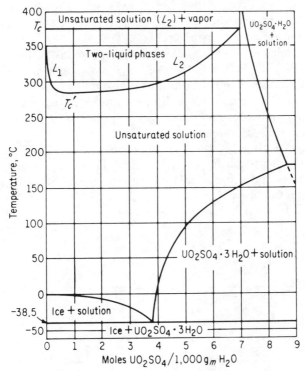

FIG. 11-4. Phase diagram for the system UO_2SO_4-H_2O.
(Ref. 103.)

(288°C). This phenomenon often occurs in organic systems but infrequently in aqueous systems. One of the two liquid phases, L_1, is less concentrated, and the other, L_2, more concentrated in uranium than the original solution. Also, L_1 has been found to be more acidic than L_2, indicating partial separation of SO_3 and UO_3, the former increasing in L_1 and the latter in L_2. A small temperature rise above the phase-separation temperature is sufficient to force a separation of a concentrated uranium solution from an initially dilute one   Phase relationships should be used to ensure that, for given operating temperatures, no undesirable phase separation and precipitation (such as into vapor or solid) take place.

Solubility relationships for the five-component system UO_3-CuO-NiO-SO_3-D_2O are also being studied. This system comes up because of the use of cupric sulfate as catalyst for the recombination of O_2 and D_2 (separated by radiolytic action) and because nickel enters the solution via corrosion of stainless steel.   Here fuel concentrations should be selected so that, if precipitation does occur, the solid phase is limited to non-uranium compounds, such as $NiO_4 \cdot H_2O$ and $3CuO \cdot 2H_2O$. Uranium precipitation, of course, is undesirable because of the changes in reactivity of the system and the possible danger accompanying uranium accumulation in one or more spots.

The solubility of compounds of fission products, such as $La_2(SO_4)_2$, $Y_2(SO_4)_3$, $BaSO_4$, $SrSO_4$, Cs_2SO_4, Ag_2SO_4, and others, in water and different-strength uranyl sulfate solutions is also being studied, with the view to ensuring against separation and precipitation in undesirable locations. Initial studies indicate that these fission products are more soluble in uranyl sulfate than in water.

Some physical properties of uranyl sulfate solutions, necessary for evaluating the heat-transfer behavior (Sec. 11-10) and other parameters of uranyl sulfate, can be calculated with reasonable accuracy from the following expressions [103]:

Density

1. In H_2O

$$\rho = \frac{1}{78.65/C_u - 1.046} + \rho_{H_2O} = \frac{1}{120.9/C_s - 1.046} + \rho_{H_2O} \quad (11\text{-}1)$$

2. In D_2O

$$\rho = \frac{1}{71.0/C_u - 0.944} + \rho_{D_2O} = \frac{1}{109/C_s - 0.944} + \rho_{D_2O} \quad (11\text{-}2)$$

where ρ = density of desired solution, g_m/cm^3

C_u, C_s = mass concentrations in percent of uranium and sulfate

ρ_{H_2O}, ρ_{D_2O} = densities of pure H_2O and D_2O at the same temperatures, g_m/cm^3 (these densities are shown in Fig. 11-5)

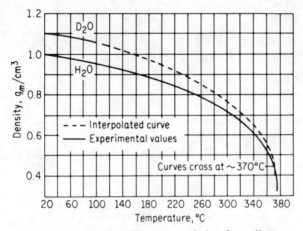

FIG. 11-5. Density-temperature relation for ordinary
and heavy water. Pressure: 1 atm up to normal boiling
point, saturation pressure beyond normal boiling point.
(Ref. 103.)

Viscosity

This is given in Fig. 11-6 in centipoises for various uranyl sulfate
concentrations and for pure H_2O and D_2O. Many of these data are
extrapolated from low-temperature measurements and, while they are the
best available now, may not be very accurate.

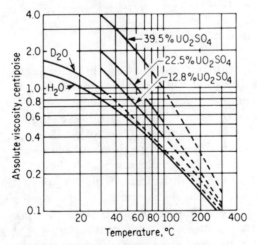

FIG. 11-6. Viscosity versus temperature for
H_2O, D_2O, and aqueous UO_2SO_4 solutions.
(Ref. 103.)

Specific Heat

This is given at 25°C in Table 11-1. At elevated temperatures, it can be estimated from the following relationships:

1. In H_2O $\qquad\qquad$ $c_p = (c_p)_{25} R_{H_2O}$ $\qquad\qquad$ (11-3)
2. In D_2O $\qquad\qquad$ $c_p = (c_p)_{25} R_{D_2O}$ $\qquad\qquad$ (11-4)

where c_p = specific heat at desired temperature, Btu/lb$_m$ °F
$\quad (c_p)_{25}$ = specific heat at 25°C, from Table 19-1
R_{H_2O}, R_{D_2O} = ratio of specific heat at desired temperatures to that at 25°C
$\qquad\qquad$ for pure H_2O and D_2O

TABLE 11-1
Specific Heat of Uranyl Sulfate Solutions

UO_2SO_4, mass Percent	c_p at 25°C, Btu/lb$_m$ °F	
	In H_2O	In D_2O
0	0.998	1.005
10	0.905	0.916
20	0.809	0.830
30	0.714	0.745
40	0.619	0.658
50	0.523	0.568
60	0.428	0.474
70	0.333	0.370
80	0.238	0.250
85.9 (UO_2SO_4-3D_2O)	. . . . .	0.174
87.2 (UO_2SO_4-3H_2O)	0.170	

11-5. AQUEOUS SLURRIES

The problem with slurries, such as ThO_2 slurries (the blanket material), is keeping the solid ThO_2 particles in suspension. These slurries are non-Newtonian in character and turbulent flow is required to ensure their continued suspension. The more non-Newtonian the fluid is, the higher the speed of flow required. This, of course, affects the design of the reactor blanket and the loop components and piping associated with it. For example, the speed of the suspension should be above a certain minimum in all cases. The degree of inclination of the piping has a bearing on settling. Also, it may be necessary to eliminate all sudden changes in directions of flow of the slurry since they may cause the solid particles to separate because of centrifugal action.

The properties of slurries are usually divided into two categories:

1. Intrinsic properties. These are functions only of the properties of the pure components of the slurry. Density, specific heat, and thermal

conductivity are intrinsic properties. They can easily be calculated from the data of the pure components and the concentration of these components.

2. Extrinsic properties. These are functions of the particle size of the solid, as well as the properties of the components. Settling rate and viscosity are such properties.

The following relationships may be used to calculate ThO_2 suspension properties:

Density

$$\rho_s = \epsilon\rho_p + (1 - \epsilon)\rho_f \qquad (11\text{-}5)$$

where ρ_s = desired density of suspension
ϵ = volume fraction of ThO_2 in suspension
ρ_p = density of ThO_2 particles = 10.03 g_m/cm^3 = 626.17 lb_m/ft^3
ρ_f = density of pure fluid suspension carrier, such as H_2O or D_2O, at same temperature as suspension

Specific Heat

$$c_{p_s} = m_p c_{p_p} + (1 - m_p)c_{p_f} \qquad (11\text{-}6)$$

where c_{p_s} = specific heat of slurry, $Btu/lb_m\ ^\circ F$
m_p = mass fraction of ThO_2 particles in suspension
c_{p_f} = specific heat of pure fluid carrier, $Btu/lb_m\ ^\circ F$
c_{p_p} = specific heat of pure ThO_2 particles, $Btu/lb_m\ ^\circ F$, given by

$$c_{p_p} = 0.0623 + 0.0493 \times 10^3 T - \frac{2605}{T^2} \qquad (11\text{-}7)$$

where T = absolute temperature, $^\circ R$

Thermal Conductivity

For the pure (100 percent density) ThO_2, it is approximately equal to 7.26 $Btu/hr\ ft\ ^\circ F$ at 212$^\circ F$ and 3.39 $Btu/hr\ ft\ ^\circ F$ at 572$^\circ F$. Thermal conductivity of ThO_2 slurries in water shows the same temperature dependence as water; that is, they have a maximum around 265$^\circ F$. For the slurry the following empirical relationship may be used.

$$\frac{k_s - k_f}{k_s + k_f} = 1.13\epsilon\,\frac{k_p - k_f}{k_p + 2k_f} \qquad (11\text{-}8)$$

where k_s, k_f, and k_p are the thermal conductivities of the slurry, fluid carrier evaluated at the temperature of the slurry, and the particles, respectively, in $Btu/hr\ ft\ ^\circ F$. Equation 11-8 is believed to yield satisfactory results for ϵ values up to 0.2 and for temperatures up to 572°F (300°C).

Viscosity

For noninteracting solid particles in the suspension, the viscosity is a function only of the volume fraction of these particles in the slurry and therefore is an intrinsic property. The viscosity of the ThO_2 slurry, in such a case, may be obtained from the expression

$$\mu_s = \mu_f(1 + 2.5\epsilon + 7.17\epsilon^2 + 16.2\epsilon^3) \qquad (11\text{-}9)$$

where μ_s and μ_f are the viscosities of the slurry and the carrier fluid, respectively.

However, when the particle size decreases below 5 or 10 μ, the distances between the particles decrease, and the forces between them (colloidal forces) become important. The slurry appears flocculated, the particles joining together in flocs or clusters whose structure affects the viscosity. (The intrinsic value of viscosity given by Eq. 11-9 is equal to the theoretical minimum for a slurry of given mass composition.)

Since flocculated structures are non-Newtonian, the viscosity is not constant but is a function of the rate of shear which is a function of the speed of flow at any radial position. The viscosity can be defined, therefore, only as a point value, true only for a given speed and radial position.

Effective viscosity data for small-diameter tubing have, however, been obtained experimentally [103]. Figure 11-7 shows such viscosities in centipoises for UO_3-H_2O slurries of different concentrations, and for H_2O, versus the reciprocal of the absolute temperature in degrees Kelvin.

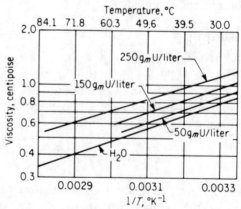

FIG. 11-7. Viscosity of UO_2-H_2O slurries as a function of temperature. (Ref. 103.)

Settling Rate

An important property of slurries is the *settling* rate. This obviously affects the consistency and behavior of the blanket at low speeds. For large particles (no colloidal forces), the settling rate is given by

$$U_s = (1 - \epsilon)^{4.65} U_0 \qquad (11\text{-}10)$$

where U_s = settling rate of slurry, ft/sec
$\quad U_0$ = settling rate of a single spherical particle, ft/sec

For ThO_2 particles less than 10 μ in diameter the settling rate of a single particle in water is given by Stokes' law:

$$U_0 = \frac{g d_p^2 (\rho_p - \rho_f)}{18 \mu_f} \qquad (11\text{-}11)$$

where g is the local gravitational acceleration, in ft/sec^2, and d_p is the particle diameter.

For the flocculated slurries, the settling rate, often referred to as the *hindered* settling rate, is affected by the floc properties, as well as by the parameters included in Eqs. 11-10 and 11-11. In this case, it is some ten to fifty times as great as the single-particle settling rate U_0. In inclined cylindrical containers, the hindered settling rate increases as the angle with the vertical is increased, goes through a maximum, and decreases. In one experiment, the maximum occured at 50° from the vertical and showed a settling rate four times as large. The hindered settling rate also increases with temperature. It has also been found that, for ThO_2 particles larger than 2.4 μ, the settling rate is inversely proportional to the viscosity of the carrier fluid, for temperatures up to 572°F (300°C).

Figure 11-8 shows experimental settling rates of a ThO_2-H_2O slurry of 250 g_m ThO_2/kg$_m$ H_2O strength and different particle sizes versus temperature.

A flocculated suspension is said to go into *compaction* when the settling flocs meet and rest upon each other, thus decreasing the magnitude of the settling rate. Compaction seems to be a function of vessel size, suspension concentration, and yield stress. When the ThO_2 particles accumulate on and cover a surface with a dense rigid layer or form spherical particles 15 to 100 μ in diameter, the suspension is said to have "caked." The ThO_2 *cakes* have a density around 340 lb$_m$/ft^3 and resemble chalk in strength and consistency. They can be deformed only by fracture. Caking is eliminated when the ThO_2 powder is manufactured by calcination at high temperatures (around 2900°F).

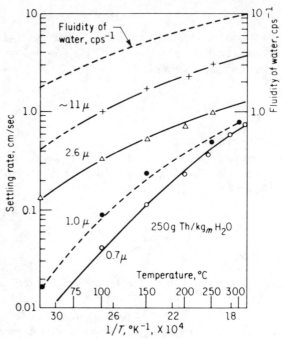

FIG. 11-8.   Temperature-particle-size effects on the
settling rate of thorium oxide slurries, 250 g_m Th/kg
H$_2$O.   (Ref. 102.)

11-6.   LIQUID-METAL SOLUTION FUELS

Liquid-metal fuels are composed of fuel solutions or slurries in a
liquid-metal carrier.   The latter, in general, have elevated neutron cross
sections (Fig. 9-8).   The ratio of liquid metal to fuel, therefore, is limited
by nuclear requirements.   The ratio of the number of neutrons captured
by the liquid-metal carrier to the number of neutrons captured by the
fissionable fuel is given by

$$n_c = \frac{N_f}{N_{ff}} \frac{\sigma_{af}}{\sigma_{aff}} \qquad (11\text{-}12)$$

where N is the number of nuclei in a given volume of the fluid fuel, σ_a is
the neutron absorption cross section, and the subscripts f and ff refer to
the carrier liquid-metal fluid and the fuel, respectively.   Based on a
reasoning similar to that given for Eq. 9-13, n_c in a critical reactor operating at a steady state must fulfill the requirement

$$n_c < \eta_{ff} - 1 - L' \qquad (11\text{-}13)$$

where η_{ff} is the average number of neutrons emitted per neutron absorbed in the fissionable fuel and L' is the number of neutrons lost by parasitic capture in the moderator structure (but not the carrier fluid) and by leakage (Secs. 9-3 and 9-4).

Similarly, for the reactor to be a breeder, i.e., with a conversion ratio greater than unity, the value of n_c must be less than $\eta_{ff} - 2 - L'$. Maximum values of n_c for steady-state and breeder reactors using the three fissionable fuels are given in Table 11-2. These maximum values are based on L' equal to zero. These values must then be reduced further when L' is determined. When n_c and the neutron cross sections are known, the maximum fuel dilution N_f/N_{ff} (or its reciprocal, the minimum fuel concentration) can be calculated. Note that maximum values of n_c are greater for fast than for thermal reactors. The ratio of neutron absorption cross sections in Eq. 11-12 can be determined for a fast reactor, however, only if the actual neutron spectrum for that reactor is known. It is also believed that the inelastic-scattering effects play a part in the case of fast reactors. In general, fast reactors require that dilution be less (or fuel concentration be greater) than in thermal reactors.

TABLE 11-2
Maximum Values of n_c ($L' = 0$)

Fuel	Nonbreeder Reactor		Breeder Reactor	
	Thermal	Fast	Thermal	Fast
U^{233}	1.284	1.490	0.284	0.490
U^{235}	1.070	1.201	0.070	0.201
Pu^{239}	1.103	1.639	0.103	0.639

Example 11-1. Calculate the minimum atomic concentration of 95 percent enriched uranium fuel in solution in molten bismuth in a thermal nonbreeding reactor.

Solution. From Table 11-2,

$$n_c = 1.070$$

From Appendix B of Ref. 2,

for bismuth, $\sigma_a = 0.034$ barns

for U^{235}, $\sigma_a = 678.2$ barns

Thus

$$1.070 = \frac{N_{Bi}}{N_{U^{235}}} \frac{0.034}{678.2}$$

from which

$$\left(\frac{N_{U^{235}}}{N_{Bi}} \right)_{min} = 46.85 \times 10^{-6}$$

For the fuel,

$$\left(\frac{N_U}{N_{Bi}}\right)_{min} = \frac{46.85 \times 10^{-6}}{0.95} = 49.32 \times 10^{-6} \qquad \text{or} \qquad 49.32 \text{ ppm}$$

As with aqueous fuels, liquid-metal fuels may be of the solution or slurry types. The solubility of uranium and thorium metals in a variety of liquid metals at various temperatures can be found in the literature [23, 104, 105]. Only bismuth, lead, and their alloys appear to have significant, though limited, uranium solubility over a wide range of temperatures. Of these, lead has too high neutron absorption cross sections to be considered as a suitable fuel carrier. The solubility of uranium in bismuth is shown in Fig. 11-9. This solubility is reduced

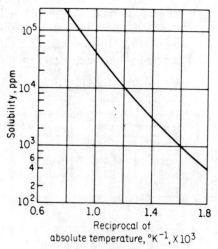

FIG. 11-9. The solubility of uranium
in liquid bismuth. (Ref. 23.)

slightly whenever zirconium or magnesium is present in the system. Since the fuel concentration is rather low, the properties of the uranium-in-bismuth solution are approximately equal to those of pure bismuth, given in Table 11-3. Also, since bismuth (and most other liquid metals) is stable under nuclear radiations, these properties are not expected to change with irradiation. It has also been found that static corrosion characteristics of fuel solutions in bismuth are similar to those of pure bismuth, indicating that those materials that have good corrosion resistance to bismuth can be used safely with the fuel solution.

TABLE 11-3
Some Physical Properties of Pure Bismuth

Melting point, °F	520
Boiling point, °F	2690
Latent heat of fusion, Btu/lb_m	7.22
Latent heat of vaporization, Btu/lb_m .	1113.5
Vapor pressure, lb_f/in.^{2}:	
1683°F	0.01934
1953°F	0.1934
2295°F	1.934
2417°F	3.868
2552°F	7.736
Density, lb_m/ft^3:	
572°F	643
752°F	618.7
1112°F	603
1476°F	586.8
1764°F	574.4
Specific heat, Btu/lb_m °F:	
520°F	0.0340
752°F	0.0354
1112°F	0.0376
1472°F	0.0397
1832°F	0.0419
Viscosity, lb_m/ft hr:	
579°F	4.022
844°F	3.098
1112°F	2.410
Thermal conductivity, Btu/hr ft °F:	
572°F	9.918
752°F	8.950
932°F	8.950
1112°F	8.950
1292°F	8.950

While bismuth has received the greatest attention because of solubility and neutron-cross-section considerations, it has some disadvantages. It is expensive and relatively uncommon. It has a high melting point, requiring outside heating during extended shutdowns, and on freezing expands by 3.3 volume percent. When a bismuth nucleus absorbs a neutron, it is transmuted into Po^{210}, an α emitter and a difficult material to handle. Bismuth is also highly corrosive (Sec. 11-9).

Higher concentrations of fuel in bismuth than are possible by solution can be achieved by the use of slurries. However, not enough data are available on liquid-metal slurries, although it is believed that compounds of uranium and bismuth in suspension in bismuth are sufficiently stable for reactor use [106]. Some work is being done on UO_2-in-Na or -NaK slurries. The intrinsic properties of liquid-metal slurries may be obtained from the properties of the solid and the liquid-metal carrier. Not enough information is available on extrinsic properties such as viscosity and settling rate.

11-7. MOLTEN-FUEL SALTS

This category consists of mixtures of fuel and other salts (to act as diluents and to lower melting temperatures) used in molten form. The salts must fulfill various stringent requirements such as low neutron cross sections (except for the fuel nuclei), thermal and radiolytic stability, low vapor pressure (to minimize pressurization), low corrosiveness, and good heat-transfer-fluid-flow characteristics. They must also remain molten at all primary-loop temperatures and must be capable of containing fission products without appreciable change in properties. Naturally no single salt mixture can meet all these requirements or prove optimum for all reactor types.

The fluorides and chlorides seem to provide the best compromises for molten-fuel choice.* Fluorine has a relatively high scattering cross section for neutrons. The fluorides are therefore more suitable for thermal reactors. They are also more chemically stable than the chlorides; have lower vapor pressures, higher volumetric specific heats, and higher thermal conductivities; are good heat-transfer agents; and require no fluorine-isotope separation. The chlorides in general have lower melting points and higher viscosities. They have relatively high absorption cross sections for thermal neutrons and are therefore more suited to fast reactors.

Of the fluorides, uranium tetrafluoride (UF_4) seems to be the most suitable. Uranium hexafluoride is volatile; UO_2F_2 is an active oxidizer; and the pentavalent uranium fluorides such as UF_5 and U_2F_9 are thermally unstable as well as strong oxidants. Uranium trifluoride (UF_3) is stable in the pure state and when under an inert atmosphere. In molten form, however, it undergoes chemical change to UF_4 and U.

---

* The nonmetals C, N, Si, S, P, and O form high-melting binary compounds and a large number of oxygenated anions. The nitrates, nitrites, sulfates, and sulfites lack thermal stability. The silicates are usually too viscous. The phosphates, borates, and carbonates are corrosive to high-temperature materials (nickel, iron, chromium). It is also doubted that the covalently bonded oxygenated anions posses sufficient stability to radiation.

Not enough information is available on plutonium fluorides. It is believed, however, that PuF_3 is the most suitable of them. Thorium tetrafluoride is the only possible thorium fluoride, since all normal thorium compounds are quadrivalent.

The more suitable fluorides and chlorides have melting temperatures in excess of 1850°F. This means plant temperatures that would tax the available materials of construction. Diluents (also fluorides or chlorides) are therefore used to lower these melting temperatures. A large number of metallic elements of low neutron absorption cross sections are available to form fluorides for that purpose.

In binary systems, low melting temperatures are attained only if the fuel concentration is too high. As an example, the phase diagram of the UF_4-LiF system [103] shows a melting-temperature region below 1100°F, but for UF_4 concentrations of 22 to 38 mole percent.

Some three- and four-component fluoride mixtures, such as UF_4-LiF-KF, UF_4-LiF-RbF, UF_4-LiF-NaF-KF, and UF_4-LiF-NaF-RbF, show melting points below 1100°F at lower fuel concentrations [23]. Low melting over a wide range of concentrations can also be obtained if ZrF_4 is used as a constituent. The three-component system UF_4-ZrF_4-NaF showed characteristics good enough to be used (in 1954) in the Aircraft Reactor Experiment [107], the first experimental molten-salt reactor. Beryllium difluoride, a later diluent, has lower neutron cross sections than the others but slightly higher viscosities and toxicity. It is now thought that the systems UF_4-BeF_2-LiF, UF_4-BeF_2-NaF, and UF_4-BeF_2-LiF-NaF offer the best combinations for use in molten-salt reactors.

Plutonium and thorium fluorides and mixtures of thorium and uranium fluorides are discussed in the literature [23]. Available data on chlorides are much less extensive than on fluorides. It has been determined, however, that UCl_3 and UCl_4 can be used in mixture with many chloride diluents, such as LiCl, NaCl, KCl, and RbCl, giving low melting points with reasonable fuel concentrations. It is also believed that UCl_3 results in improved corrosion characteristics at low temperatures. The ternary systems UCl_4-NaCl-$ZrCl_4$ and UCl_4-KCl-$ZrCl_4$ are also feasible.

Properties of Molten Fluorides

Specific heat, density, viscosity, and melting points of some fluorides are given in Table 11-4. Thermal-conductivity values are given in Table 11-5. For the molten mixtures, these values are believed to be accurate only to ± 25 percent. The values for the solid mixtures are thought to be somewhat more accurate.

The vapor pressures of fluorides are given by the usual equation

$$\log_{10} p_v = A - \frac{B}{T} \tag{11-14}$$

TABLE 11-4

Specific Heats and Melting Points of Some Molten Fluorides.

Composition, Mole Percent					Melt. pt., °F	Specific Heat		Density	Viscosity, lb$_m$/ft hr			
UF$_4$	LiF	NaF	ZrF$_4$	BeF$_2$		Temp. range, °F	Btu/lb$_m$ °F	lb$_m$/ft^3 (t, °F)	600°C	700°C	800°C	900°C
4	50	.....	46	.....	1022	1292	0.28	245.75 − 0.0319t				8.47
5	.....	56	39	.....	986	1053–1638	0.3044 − (3.594 × 10^{-5}t, °F)	244.50 − 0.0319t		0.2517 exp (3798/T, °K)		
12	.....	63	25	.....	1103	1292	0.24	260.15 − 0.033t				
20	.....	65	15	.....	1130	1207–1695	0.20	285.19 − 0.0354t				
25	.....	50	25	.....	1130	1130–1706	0.27	319.53 − 0.0551t	.....	20.57	12.1	
2	.....	47	.....	51	743	1292	0.49	149.65 − 0.0139t	39.93	19.36	10.65	
5	.....	55	.....	40	842	1292	0.42	172.86 − 0.0173t	27.10	13.55	7.74	
12	.....	76	.....	12	957	968–1814	0.32	226-83 − 0.0260t	.....	17.42	10.89	
15	.....	25	.....	60	869	536–1922	0.32	214.91 − 0.0243t	.....	.....	.....	
2.5	67	.....	.....	30.5	867	1292	0.57	149.03 − 0.0139t	20.33	13.31	9.32	
2.5	15	55	.....	27.5	878	1292	0.47	162.82 − 0.0156t	15.97	10.65	8.47	
2.5	34	26.5	.....	37	680	1292	0.51	145.66 − 0.0062t		0.0632 exp (5094/T, °K)		

* From Ref. 23.

TABLE 11-5
Thermal Conductivities of Some Fluoride Mixtures

Composition, Mole Percent						Thermal Conductivity, Btu/hr ft °F	
UF$_4$	LiF	NaF	ZrF$_4$	KF	RbF	Molten	Solid
1.1	44.5	10.9	....	43.5	....	2.3	2.0
4	....	50	46	....	....	1.3	0.5
4	....	....	48	....	48	1.0	
6.5	....	53.5	40	....	....	1.2	
27.5	....	46.5	....	26	....	0.5	
	57	....	....	43	....	2.4	
	46.5	11.5	....	42	....	2.6	2.7
	43	....	....	....	57	1.2	

where p_v is vapor pressure in mm Hg, T is absolute temperature in degrees Kelvin, and A and B are constants, given for some materials and their mixtures in Table 11-6.

TABLE 11-6
Vapor-Pressure Constants for Some Fluorides

Material and composition			Temperature range, °F	A	$B \times 10^{-3}$
	UF$_4$		1898–2165	7.792	9.171
	PuF$_3$		2190	12.468	21.120
	NaF		1857–1967	9.506	12.539
			2545–2835	9.180	12.385
	KF		2464–2737	8.019	9.169
	LiF		1609–1940	9.630	12.420
			2467–2817	9.071	12.057
	RbF		2125–2570	8.124	8.753
	ZrF$_4$		1141–1614	13.433	12.399
UF$_4$	ZrF$_4$	NaF			
4	43	53	1472–1832	7.370	7.105
4	46	50	1472–1832	7.888	7.551
4	50	46	1472–1832	3.281	7.779
20	15	65	1472–1832	5.860	6.944
25	25	50	1742–1832	6.844	6.906

* From Ref. 23.

11-8. GAS-SUSPENSION FUELS

Not much information is available.on the properties of powder-laden gases. It is, however, known that many of these properties are functions of the temperature and pressure of the carrier gases. The density of

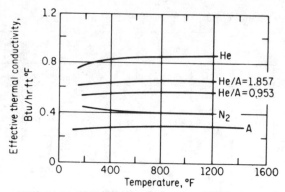

FIG. 11-10.   Effective thermal conductivity of some
uranium-oxide-power-laden gases.   Porosity, 0.405;
mean particle size, 0.00027 ft, about 80 μ.   (Ref. 108.)

the gaseous suspension can be calculated from the density of the pure
gas (a function of pressure and temperature) and the density and volume
fraction of the powder.   Experimental effective thermal conductivity
[108] of UO_2 powder in several gases and gas mixtures is shown in Fig.
11-10 as a function of temperature.   It is interesting to note that the UO_2-
He system shows an effective thermal conductivity some eight times that
of pure He at 200°F and five times that at 1000°F, both at atmospheric
pressure.   Since most of the resistance to heat conduction occurs in the
gas, the effective conductivity is usually a strong function of that of the
gas.   Figure 11-11 shows that UO_2-He suspensions have effective conductivities about three times those of UO_2-argon and vary linearly with

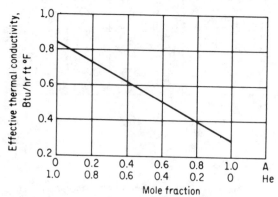

FIG. 11-11.   Effective thermal conductivity of uranium oxide powder in HeA mixtures, independent of
temperature.   (Ref. 108.)

gas composition. Effective conductivities were also found initially to increase with pressure and level off at high pressures, as shown for MgO powders (Fig. 11-12). This effect is explained by the fact that, beyond

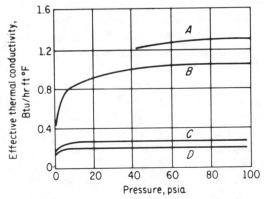

FIG. 11-12. Effect of gas pressure on effective thermal conductivity of some MgO-powder-laden gases. A, MgO in He, 750 900°F; B, MgO in He, 370-450°F; C, Mg in air, 300-900°F, pressures above 10 psi; D, MgO in argon, 350-430°F, pressures above 10 psi. (Ref. 108.)

a certain pressure, the mean free path of the gas molecules becomes small compared with the distances between the solid particles which are effective in transferring heat.

11-9. CORROSION AND EROSION CHARACTERISTICS OF FLUID FUELS

A rather extensive program of studies has been conducted in order to determine the corrosive and erosive characteristics of fluid fuels. The program aimed at determining exact operating conditions under which the more promising materials can be used. Corrosion rates are highly sensitive to these conditions, changing in many cases by several orders of magnitude with small changes in them.

Solution-type aqueous fuels are strongly acidic. Also, radiation was found to increase their corrosion rates. Aqueous slurries have been found to be both corrosive and erosive. Erosion alone occurs when the material of construction is chemically inert to the slurry. Corrosion occurs because of the aqueous phase of the slurry. The most serious damage to materials occurs by a combination of corrosion and erosion, in which the solid particles erode a weak protective oxide film, and the exposed metal reacts with the aqueous phase. Resistance to attack by slurries is greater the smaller the particle size. The extent of damage has also been found to increase with solid concentration and flow speed, and to decrease with temperature and oxygen content.

Materials found to be of primary interest in aqueous systems are (1) zirconium alloys for the reactor core tank, separating the core from the blanket, preferred to stainless steel there because of lower absorption cross sections for thermal neutrons and (2) austenitic stabilized stainless steels (primarily Type 347) for most of the fuel and blanket circulating systems outside the reactor. Titanium and zirconium are also used for special equipment outside the core. Titanium has extremely good corrosion resistance which justifies its high cost. Other materials of interest include the noble metals, Stellite and aluminum oxide (for bearings), and others. The objective here has not been to eliminate corrosion but to aim at a life expectancy of 20 years, corresponding to a corrosion rate of no more than about 2 mills/year.

Fuel solutions in Bi behave as pure Bi, as far as corrosion is concerned. A wide variety of materials have been suggested as suitable containers for such fuels. These include steels containing chromium, molybdenum, and silicon; carbon steels; beryllium; molybdenum; tungsten; and many ceramics. Nickel seems to take uranium out of solution in Bi and precipitate it as UNi_2 and UNi_5. Magnesium is used as a deoxidant although it tends to reduce slightly the solubility of U in Bi. Magnesium can be added to the fuel as a corrosion inhibitor with no effect on solubility.

Most early corrosion information on molten-fuel salts was obtained during early experiments on the Aircraft Reactor Experiment [107]. The reactor vessel was made of Inconel (approximately 16 percent Cr and 5 percent Fe in Ni) and fueled with UF_4-NaF-ZrF_4. This fuel was observed selectively to oxidize Cr in Inconel, especially in regions of high temperature. This resulted in the formation of voids, uniformly distributed over the surface of the metal. The depth of these voids increased with time and with operating temperatures and slightly with uranium concentration. Nickel-molybdenum alloys and alloys including Ni and Mo, such as Hastelloy B (5 percent Fe), have shown excellent stability in UF_4-NaF-ZrF_4. Alkali fluoride fluids such as UF_4-LiF-NaF-KF showed considerably higher rates of corrosion, although it is believed that NiMo alloys are satisfactory for such fuels. Corrosion is greatly accelerated if metals are exposed to fluorides in the presence of air or oxygen. Fluorides at high temperatures must therefore be contained only under an inert or, if possible, a reducing atmosphere.

More data on the behavior, properties, and corrosion characteristics of fluid fuels may be found in Ref. 23.

11-10. REACTOR VESSEL DESIGN

The design of the reactor vessel for a fluid-fueled reactor is influenced by nuclear, flow, and mechanical considerations. A cylindrical vessel would be simple, but unless flow is properly guided, would suffer from a poor flow pattern and consequent accumulation of gases and solids formed during operation, as well as hot spots in the corners. A spherical vessel has a low surface-volume ratio, low neutron and heat losses, and a high power density. In the case of high-vapor-pressure fuels, such as the aqueous fuels, the vessel must be designed for high pressure. Here again a spherical vessel gives maximum structural strength with minimum wall thickness. A true sphere in which the fluid enters at one end and exits at another still suffers, though to a lesser extent, from an unfavorable flow pattern. Variations of the spherical design have been suggested and tried. Figure 11-13 shows four such variations, suggested in conjunction with the aqueous homogeneous reactor program [102].

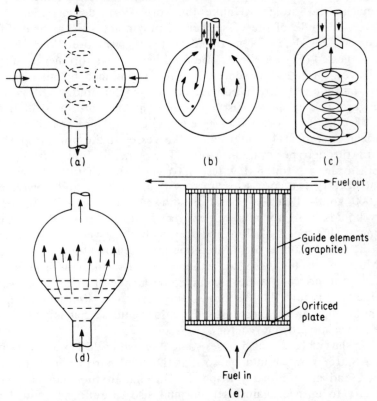

FIG. 11-13. Fuel flow in five fluid-fueled reactor core geometries. (a) Vortex flow, (b) mixed flow, (c) modified vortex flow, (d) straight-through flow, (e) cylindrical with flow guides. (Ref. 102.)

In the vortex or rotational flow design (Fig. 11-13a) used in the Homogeneous Reactor Experiment HRE-I (Sec. 11-13), the fluid enters the vessel tangentially at the equator and leaves at the two poles, thus generating an axial vortex as shown. Fission and other gases are drawn into the vortex and removed. The heated fluid (with longer residence time) moves mainly in the center, so that the vessel walls are in contact with cooler fluid. This design proved satisfactory at power densities up to 32 kw/liter. Flow tests for reactor sizes much larger than that of HRE-I showed, however, that the pressure drops in such a core would be too high.

In the recirculation flow design (Fig. 11-13b), the fluid inlet and outlet are axial, concentric, and at the top of the spherical vessel. The fuel is introduced vertically, flows downward, hits the bottom of the sphere, and recirculates to the annular exit at the top. The flow is highly turbulent so that gases and solids formed within or introduced into the reactor remain in suspension. The single vessel opening simplifies the mechanical design. Unlike the vortex-flow design (Fig. 11-13a) however, the bulk of the vessel here is in contact with the heated fluid. To correct this defect, a modified vortex-recirculation flow vessel has been suggested (Fig. 11-13c). The fluid here enters through the annulus and leaves through the center. Vortex is induced by inlet vanes as shown, and the walls are in contact with the cooler fluid.

Figure 11-13d shows a straight-through flow core, used in HRE-II (Sec. 11-11). The core vessel here is conical in its lower half and semi-spherical on top. The fluid enters the vessel at the bottom and flows upward through perforated plates. The general flow pattern in this vessel is orderly. Initial flow tests have suggested that this design would be satisfactory for core sizes up to 4 ft in diameter and flow rates up to 4,000 gpm. Reactor operation showed, however, that solids and heavy liquids were not readily removed, leading to possible local over-heating. The speed of flow and surface boiling were not sufficient to keep the vessel walls at the desired operating temperatures.

Much work on these and other core vessel designs is now being carried on to arrive at a more completely satisfactory vessel for this type of reactor. Figure 11-13e shows the type of cylindrical core, with graphite moderator elements used as flow guides, that is proposed for large power molten-salt reactors.

The slurry blanket also poses some interesting design and flow prob-lems. Slurry flow should be so that the solid particles remain in sus-pension and are uniformly dispersed throughout the blanket. This is necessary to ensure against settling and also to uniformly and steadily absorb and reflect core leakage neutrons. Some degree of swirl, there-fore, is necessary to sustain mixing. In the HRE-II this was accomplished by introducing the slurry through nozzles located in the bottom of the reactor vessel. These help to create a swirl in the blanket region between the core and pressure-vessel walls.

11-11. SOME AQUEOUS-FUEL REACTORS

The first aqueous homogeneous reactor built was the Los Alamos zero-power, critical experiment, LOPO, fueled with uranyl sulfate in light-water solution. It began operation in May, 1944. This was followed by HYPO, a 5-kw(t) uranyl nitrate reactor (December, 1944), and SUPO, a 45-kw(t) modification (March, 1951). At the higher powers, the fuel solution underwent boiling within the reactor vessel. This type of reactor thus became known as the *water boiler,* not to be confused with the boiling-water reactor (Chapters 3 and 4). Further experimentation at Los Alamos led to the design of nonaqueous (enriched UO_2 dissolved in concentrated phosphoric acid) reactor packages for remote locations. These were the Los Alamos Power Reactor Experiments LAPRE-I and LAPRE-II, initiated in 1955. The first of these two reactors, in particular, has suffered from severe corrosion difficulties.

In 1950 a homogeneous-reactor project was initiated at Oak Ridge National Laboratory, Tennessee, to develop and build power breeder reactors. This effort resulted in the Homogeneous Reactor Experiments HRE-I and HRE-II.

The HRE-I was a 1,000-kw(t) two-region reactor with a stainless-steel core fueled with a solution of uranyl sulfate in distilled light water and having heavy water in the blanket. The design maximum fuel temperature was 482°F. The HRE-I started operation in April 1952, and continued on an intermittent basis for about two years during which it attained a maximum power output of 1,600 kw(t), corresponding to a maximum power density of 58 kw/liter. The HRE-I was primarily constructed to provide information on physics and chemistry and to demonstrate that stable operation was possible with aqueous homogeneous reactors. Such stable operation was in doubt prior to HRE-I because of the production of relatively large volumes of gases in such reactors. Other considerations such as corrosion, fuel reprocessing, and ease of maintenance were ignored. The HRE-I showed that no gross nuclear instabilities occurred during operation. Having fulfilled its main objectives and because there were no provisions for maintaining it, it was finally dismantled in the spring of 1954 to make way for the HRE-II.

The HRE-II started operation in December 1957. Its purpose is to provide further information on materials, components, remote maintenance and repair methods, and procedures for removal of fission and corrosion products from the fluids. The HRE-II is fueled with uranyl sulfate in solution in heavy water. A core tank made of the low-neutron-absorption material Zircaloy 2 was used (instead of the stainless steel used in HRE-I) to provide true breeding in the blanket. Figure 11-14 shows a cross section of the HRE-II reactor. Some of its design and operational data are given in Table 11-7.

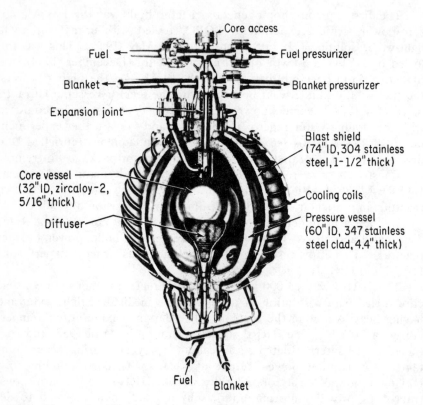

FIG. 11-14. HRE-II reactor vessel assembly. (Courtesy Oak Ridge National Laboratory.)

The HRE-II is a two-region reactor (Fig. 11-14). The inner core is a 32-in.-diameter, 5/16-in.-thick Zircaloy 2 vessel. It is surrounded by a 60-in.-diameter, 4-in.-thick carbon-steel pressure vessel clad on the inside with a 1/2-in. layer of stainless steel. The pressure vessel in turn is surrounded by a 74-in.-diameter, 1.5-in.-thick stainless-steel blast shield. This shield is surrounded on the outside with cooling coils and thus has the added function of cooling the pressure vessel by absorbing radiant heat from it.

TABLE 11-7
Some Design and Operational Data of HRE-II

Reactors heat output	5,200 kw(t)
Reactor overall diameters	32-in. core, 60-in. vessel
Reactor materials	Zircaloy 2 core, carbon steel (4 in. thick) clad with $\frac{1}{2}$-in. stainless steel
Fuel	4 mole $\%$ UO_2SO_4 in D_2O with 2 mole $\%$ $CuSO_4$ as a recombination catalyst
Fuel mass	1.84 kg_m U^{235} in core, 5 kg_m U^{235} in system
Reactor inlet temperatures	493°F core, 532°F blanket
Reactor outlet temperatures	572°F core, 540°F blanket
Reactor pressure	1,700 psig, core and blanket
Reactor pressure drops	2 psi, core; 6 psi, blanket
Fuel flow	180,000 lb_m/hr, core; 97,000 lb_m/hr, blanket (10-sec residence time in core)
Thermal-neutron flux	1.2×10^{14} maximum, 7.5×10^{13} at core wall, 0.9×10^{14} average
Power density	22.5 kw/liter maximum, 15.5 kw/liter at core wall
Type of pumps	Canned-motor, centrifugal
Steam generators	Two; one for core, one for blanket
Steam conditions	471°F, 505 psig, core; 532°F, 885 psig, blanket
Steam flow	16,680 lb_m/hr, core; 740 lb_m/hr, blanket
Feedwater temperature	200°F

As is the case with all reactors containing aqueous material, the reactor pressure must be sufficiently high to restrain boiling at reactor temperatures; hence the 1,700-psig reactor pressure and the thick pressure vessel. An electrically heated pressurizer (Sec. 5-4) is provided at the top of the core outlet pipe. In the pressurizer, steam at 1,700 to 2,000 psig is generated from heavy water.

Figure 11-15 shows a schematic flow diagram of the HRE-II plant. Roughly, the left side of the diagram contains the fuel system, the right side, the blanket system. The fuel solution is pumped upward into the bottom of the core at 493°F. It then goes through nine perforated plates installed in the diffuser section of the core to provide a velocity distribution approximately corresponding to the neutron flux distribution. The fuel solution then leaves the core at the top at 572°F. Note (Fig. 11-14) that no control rods are provided in the core. Reactivity control (as was demonstrated in the HRE-I) can be accomplished by varying the fuel concentration and by the large negative temperature coefficient of the fluid fuel.

The heated fuel flows past the pressurizer to a gas separator. In this separator, vortex flow is induced in the fluid by vanes, and the undissolved gases, such as oxygen and deuterium, are bled together with some fuel from the vortex center and sent to a low-pressure system. The remaining fuel then goes to the fuel heat-exchanger-steam generator, where it flows inside U-tube bundles, transferring its heat to water on the outside, then flows downward, and is pumped back to the core by a canned-motor centrifugal pump.

The fluid from the gas separator goes through a "letdown" heat exchanger and a "letdown" valve and into fuel dump tanks. There, D_2O vapor boils off at low pressure and mixes with gas from the "letdown" system. The gas mixture then flows upward, through an iodine-removal bed and a catalytic recombiner. From there it goes to a condenser where D_2O returns as condensate either to the dump tank or to a holdup tank. Noncondensable fission gases are vented through cold traps (to remove more condensable gases remaining in the mixture) and then to an activated-charcoal adsorber.

The fuel solution is then repressurized by the fuel feed pumps shown and sent to the letdown heat exchanger. It is then fed into the main high-pressure fuel system and returned to the core via the steam generator and main pump.

Fuel concentration is controlled in the dump tank. This is done by varying the amount of D_2O which dilutes the uranium. The inventory of the fuel in the system can be determined at any time by sampling and weighing the dump and condensate tanks.

The blanket material goes through a circuit quite similar to that of the fuel material just described.

Steam produced in the fuel and blanket heat exchangers is sent to two steam drums and then to a turbogenerator (Fig. 11-15). A condenser, condensate pump, deaerator, and two feed-water pumps complete the steam circuit. The turbogenerator used in HRE-II is that originally used for HRE-I and is therefore too small to handle the maximum HRE-II output. An air-cooled steam condenser (not shown) was added to handle the extra output.

The HRE-II achieved high-power operation in February 1958. No materials problems (other than a relatively minor stress corrosion cracking that developed early in the reactor leak-detector system) have been encountered in 6,200 hr of operation at 250 kw(t) or higher [109]. The reactor, however, showed signs of extreme instability, especially at high powers and temperatures, evidenced by sudden surges in power of up to seven times the normal operating level, a condition not encountered with HRE-I. This instability was found to be accompanied by loss of uranium from the fuel solution and its deposition on the core walls, causing local

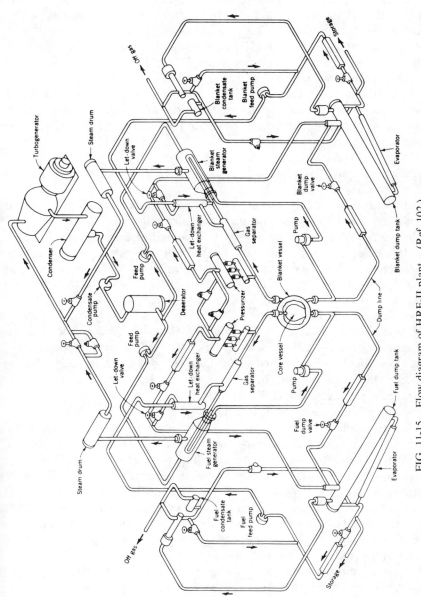

FIG. 11-15. Flow diagram of HRE-II plant. (Ref. 102.)

heating which led to further uranium deposition and so on. (This resulted in the burning of two holes in the core walls.) It is believed that the poor flow pattern of the HRE-II core is responsible for this behavior. The fluid fuel moves upward slowly, with no vigorous scouring action (in contrast to the HRE-I), making it easy for the core and screens to collect sediment. The HRE-II core will therefore be modified by removing some of the screens, reversing the direction of fuel flow, and increasing the flow speed along the walls to 1.5 fps.

The removal of fission and corrosion products from the HRE-II system is done by hydraulic centrifuges (also called hydroclones). Enough information regarding the effectiveness of these centrifuges is not yet available.

11-12. LIQUID-METAL-FUEL REACTORS

Liquid-metal-fuel reactors were suggested as early as 1941. Much pioneering work on this type has been done by Brookhaven National Laboratory (BNL) and the Babcock & Wilcox Company. In general, the liquid-metal-fuel burner (i.e., nonbreeder) reactor uses U^{235} as fuel. As operation continues and the concentration of fission products and higher uranium isotopes builds up, more U^{235} is added to maintain the reactivity of the reactor. This continues until the maximum limit of solubility in bismuth is reached. The bismuth is then cleaned by removing gases, fission products, and uranium. The operating period between cleanups increases with the size of the core. It has been estimated [110] that, for an 8-ft-diameter-core, 70-kw/liter power density, 200-Mw(e) reactor, the period would be between 6 and 12 years, with an average uranium concentration of 1,055 ppm.

In one liquid-metal-fuel breeder reactor concept, it is expected that the reactor will be fueled with U^{233} and blanketed with Th^{232}. Here, in addition to the fuel cleanup system required in the burner reactor, a blanket processing plant is required. The converted U^{233} will be added to the core as needed. The U^{233}-fueled reactor will operate with longer periods between cleanups because of the much lower production of nonfissionable uranium. The time here has been estimated [110] at close to 100 years but conservatively advised to be around 30 years. The average uranium concentration will be about 1,200 ppm. Breeding ratios around 1.0 are expected.

The Liquid-metal-fueled Reactor Experiment (LMFRE-I) designed and constructed by the Babcock & Wilcox Company and located at Brookhaven National Laboratory in Upton, N.Y., is a 5-Mw(t) thermal (graphite-moderated) reactor fueled with 95 percent enriched uranium dissolved in molten bismuth. The fuel enters and leaves a cylindrical core at 775 and 910°F, respectively. Because of the low vapor pressure of the fuel solution, the reactor pressure is only around 75 psig (compared with 1,700 to 2,000 psig for the aqueous HRE-II). The reactor heat is transferred to a secondary bismuth loop and is dissipated into the atmosphere through an air-blast heat exchanger.

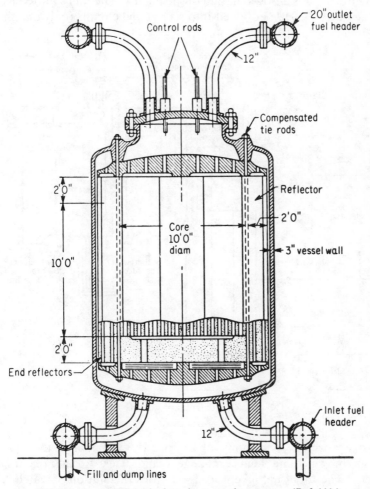

FIG. 11-16.   LMFR with plutonium-recycle reactor.   (Ref. 111.)

Under study at BNL is a liquid-metal-fuel reactor with recycled plutonium [111]. This is to be a single-region, single-fluid thermal reactor fueled by a solution in bismuth of Pu^{239}, U^{238} and Mg and Zr as corrosion

inhibitors.   Figure 11-16 shows a vertical section of the proposed vessel.
The fuel will be pumped upward through perforated graphite moderator
blocks.   The core will be reflected in the radial as well as in the axial
directions by graphite.   Several reactor-design problems arise because of
the larger size of Pu-fueled reactors (larger than U^{233}-or U^{235}-fueled
reactors).

Figure 11-17 is a flow diagram of the plant.   The fuel solution leaving
the reactor at 1020°F is divided into six parallel circuits (only one shown)

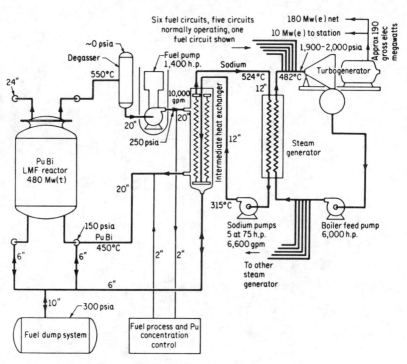

FIG. 11-17.   Flow diagram of LMFR.   (Ref. 111.)

with one normally a standby or in repair.   The fuel is then pumped
through intermediate heat exchangers and back to the reactor inlet
header.   The total fuel inventory is 600 tons of solution of which 72 percent fills the primary loop outside the reactor.   The total time required
by the fuel to complete one full loop is 17 sec, of which 4.8 sec is the
residence time in the reactor itself.   Sodium is used in the intermediate
loop, giving off its heat to a steam generator.   Steam is produced at
900°F.   The plant also includes a degasser, a fuel dump, and fuel-processing and control systems.

11-13. MOLTEN-SALT REACTORS

Much of the developmental work on molten-salt systems has been done at Oak Ridge National Laboratory (ORNL), Tennessee. The motivation for the work, like that for other fluid-fueled systems, is the ability to utilize economically, through breeding and simple fuel processing, all grades of uranium and thorium. Molten salts, like liquid-metal fuels, do not suffer from the high vapor pressure of aqueous fuels and therefore do not require high reactor pressurization. They are capable of operation at higher temperatures than aqueous or liquid-metal fuels.

Molten fluorides were first developed for use in aircraft reactor power plants, which require compactness and therefore high-temperature operation. More recently, much study has been directed toward adapting these salts to civilian power thermal breeder reactors.

By 1960 only one molten-salt reactor experiment had been constructed and operated. This is the high-temperature, 2,600-kw(e) Aircraft Reactor Experiment (ARE) [107]. Its main purposes were to gain experience in molten-fuel-salt handling and to study the kinetics of this type of reactor. The ARE was fueled with UF_4-ZrF_4-NaF and moderated by BeO. The fuel core temperatures were 1210°F inlet and 1650°F outlet. The core material was Inconel. The ARE demonstrated that a molten-salt reactor is extremely stable at high powers and that maintaining all primary-loop lines hot enough to prevent fuel freezing is not a difficult task. It was finally dismantled in 1954.

In October 1960 the United States Atomic Energy Commission announced that it will build at Oak Ridge a single-region thermal reactor that has been under study for some time. This is known as the Molten Salt Reactor Experiment (MSRE) [112]. The major objectives are to demonstrate the dependability, serviceability, and safety of this type of reactor and to develop technology suitable for a thorium breeder. It uses a Li^7F-based fuel which includes U, Th, Be, and Zr. The design calls for a core outlet temperature of 1225°F. MSRE has a cylindrical graphite core 4.5 ft in diameter and 5.5 ft high. Fuel is pumped through 600 channels in graphite columns that extend the full height of the core. The reactor vessel is made of a nickel-base alloy, containing molybdenum and chromium, that was developed under the Aircraft Nuclear Propulsion program. Reactor heat is transferred via a primary heat exchanger to a LiF-BeF_4 secondary fluid. No electrical power is produced. Instead, heat is dumped to the atmosphere via a secondary heat exchanger. MSRE has been operating at various power levels since 1965 with U^{235} fuel. Operation with U^{233} is in the planning stage at this writing.

Original thinking in the design of a large molten-salt breeder reactor (MSBR) envisioned a two-region, two-fluid system with fuel salt separated from the blanket by graphite tubes.   Later, due to developments in fuel-processing and reactor-core configurations, a change to a single-fluid system was prompted primarily by (a) a new evidence that Pa^{233} and possibly the rare-earth fission products can be separated from mixed uranium-thorium salts (by reductive extraction methods employing liquid bismuth), (b) new calculations showing that neutron leakage from a single-fluid core of modest size can be reduced to desirable levels by increasing the ratio of fuel salt to graphite, thus having an undermoderated region in the outer part of the core (core leakage in a two-fluid reactor was used in the blanket), and (c) the greater simplicity and reliability of a single-fluid over a two-fluid breeder where, in the latter, graphite piping in the core was intended to separate fuel and blanket salts.   Graphite suffers from dimensional changes due to irradiation and the large amounts of structural graphite in a two-fluid design would have necessitated the periodic replacing of the entire core and reactor vessel as the most economical way.

In the single-fluid concept, the graphite (in the form of bars) functions only as a moderator, need not have firm connections at top or bottom, and can be easily removed individually or in groups.   To reduce neutron leakage to an acceptable low level, resulting in a desirable low. fuel inventory, the size of the core, and consequently the power output, had to be large, although smaller cores can be made satisfactorily by proper zoning.

The fluid now consists of fissile uranium and fertile thorium in tetrafluoride form, both dissolved in a lithium fluoride-beryllium fluoride carrier salt.   The reference design is for a large 4,444-Mw(t) reactor and power plant in which an intermediate coolant (sodium fluoroborate) transfers heat from the fluid fuel to the steam generators which produce superheated steam.   The steam is used to generate 2,000 Mw(e) for a thermal efficiency of 45 percent.

Because graphite shrinks on irradiation, and to avoid an uneven liquid volume increase through the core, the graphite pieces are located at the top and bottom of the core by a metallic structure that is dimensionally stable.   The considerable load of graphite also poses forces that must be supported.   These forces are gravitational if the core is empty.   Because the salt is twice as dense as the graphite, the forces become buoyant (upward) and equal to the weight of the graphite.   These are in addition to the large upward forces which are caused by the pressure drop across the core when the pumps are running.

The reactor vessel is 18 ft in diameter and 24 ft high with a dished head at the bottom (Fig. 11-18).   Four fuel streams from the heat exchangers mix in the plenum formed by the dished head.   A flat support plate, with perforated webs for reinforcing, locates and supports the graphite

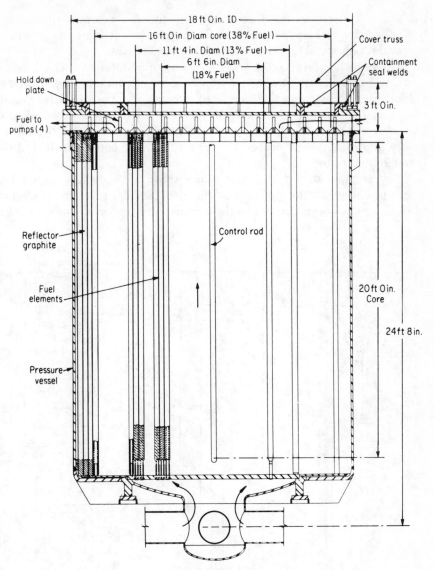

FIG. 11-18.   Elevation of MSBR reactor vessel.   (Ref. 112.)

in the center part of the core, as well as contains the orifices for flow control through the graphite pieces.   The outer core region, mounted the same way as the center region (on thimbles), receives reduced flow.   The top of the vessel contains a 1/4-in.-thick grid welded to it.   Each square in the grid locates nine core graphite pieces.   The graphite is retained in one radial position but other wise supported freely and is free to float in the salt.   A flat reinforced lid which constitutes the top of the vessel, held down by I-beams clamped to the perimeter, restrains the graphite pieces through metal spiders which also act as orifices.

The core has three radial regions containing different fuel salt fractions.   The central region, one-sixth of the core, contains 19 percent by volume salt; the surrounding one-third, region 2, contains 17 percent and the outer half, region 3, contains 44 percent.   The radial power distribution is matched to salt flow so that the temperature rise of the salt is nearly uniform.   This prevents overheating and conserves fuel flow through the reactor.

The graphite core pieces mentioned above are shown in Fig. 11-19. The space outside the pieces will have the same equivalent diameter (4 times the flow area divided by the wetted perimeter) as the inside of the

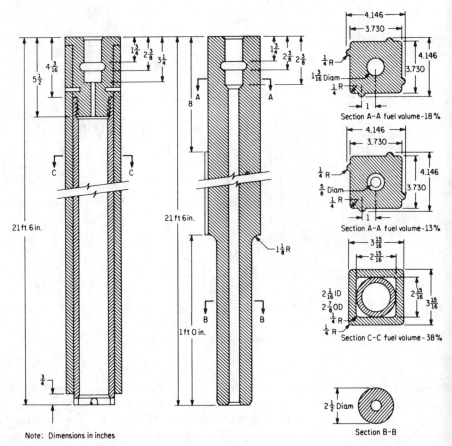

Note: Dimensions in inches

FIG. 11-19.   Configuration of MSBR core elements.   (Ref. 112.)

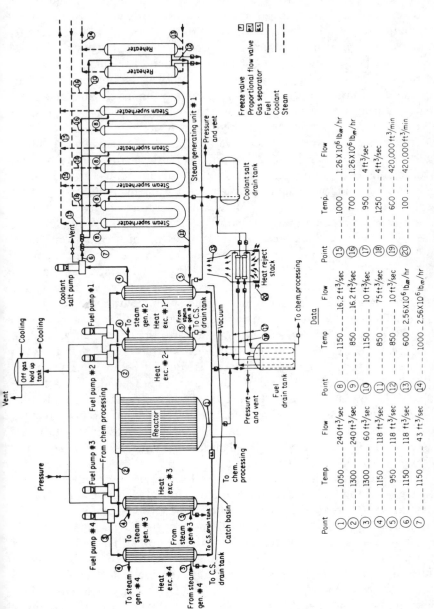

FIG. 11-20. MSBR power-plant flow diagram. 2,000 Mw(e) unit. (Ref. 112.)

Data

Point	Temp.	Flow
①	1050	240 ft³/sec
②	1300	240 ft³/sec
③	1300	60 ft³/sec
④	1150	118 ft³/sec
⑤	950	118 ft³/sec
⑥	1150	118 ft³/sec
⑦	1150	43 ft³/sec

Point	Temp	Flow
⑧	1150	16.2 ft³/sec
⑨	850	16.2 ft³/sec
⑩	1150	10 ft³/sec
⑪	850	75 ft³/sec
⑫	850	10 ft³/sec
⑬	600	2.56×10⁶ lbm/hr
⑭	1000	2.56×10⁶ lbm/hr

Point	Temp.	Flow
⑮	1000	1.26×10⁶ lbm/hr
⑯	700	1.26×10⁶ lbm/hr
⑰	950	4 ft³/sec
⑱	1250	4 ft³/sec
⑲	600	420,000 ft³/min
⑳	100	420,000 ft³/min

pieces themselves in the central region of the core, thus resulting in uniform radial pressure throughout the region. In regions 2 and 3 the outside spaces will be reduced and the pieces increased in cross sections to obtain the desired fuel fractions. Proper orificing will be used to equalize radial pressures, minimizing cross flow between regions.

Figure 11-20 shows a proposed flow diagram of the plant. 200 ft^3/sec of primary coolant (fuel salt) enter the reactor at the bottom at 1000°F, leave at the top at 1300°F, are pumped by four pumps situated above the reactor through four one-pass heat exchangers which are at the same elevation as the reactor. A concrete reactor cell 52 ft in diameter and 47 ft deep will contain the reactor vessel, four heat exchangers, and four fuel pumps (Fig. 11-21). The cell will have a thermal shield, to prevent overheating of the concrete, and double containment. A total of 75 ft^3/sec of secondary coolant enters the heat exchangers at 850°F, leaves at 1150°F, and is pumped to four steam generators and two reheaters. The steam superheaters heat 1.26 × 10^6 lb$_m$/hr of steam from 700°F to 1000°F. The reheaters heat 2.56 × 10^6 lb$_m$/hr of steam from 600°F to 1000°F. The steam generators are located in four steam-generating cells, each 33 ft wide by 46 ft long and 43 ft deep, symmetrically located in relation to the reactor cell (Fig. 11-21). These cells contain only nonradioactive secondary coolant salt and steam, and therefore no thermal shielding or double containment is needed.

A fuel drain tank, Fig. 11-20, connected to the core inlet via a 12-in. line, is housed in a separate cell 28 ft wide, 49 ft long, and 38 ft deep and having double containment. This cell is situated below the reactor cell in order for the fuel salt to drain by gravity. A secondary salt-storage tank with little shielding is also provided. An off-gas cell, doubly contained, receives fission gases and contains water-cooled gas holdup tanks and charcoal absorber beds. Chemical processing and off-gas processing cells are also included in the plant layout, Fig. 11-21. Expansion in all pipes is taken up by bellows.

The fuel drain tank is necessary in the case of pump power failure, since even if the reactor is immediately scrammed, decay-heat generation may raise the core temperatures to intolerable levels because the fuel salt, being its own coolant, is no longer in motion. A single pump failure is not, however, a sufficient reason to drain the core. The volume of the fuel drain tank, 2,247 ft^3, is large enough to hold all the fuel salt in the entire primary system. Once in the tank, the fuel salt is cooled by the fluoroborate secondary coolant salt through U-tubes which extend into the tank from headers located at the top. Maximum cooling requirements are 60 Mw(t) but the system is designed for 300 Mw(t). The fluoroborate coolant is circulated by natural convection (since power failure may have caused the initial emergency). The heat is then dumped

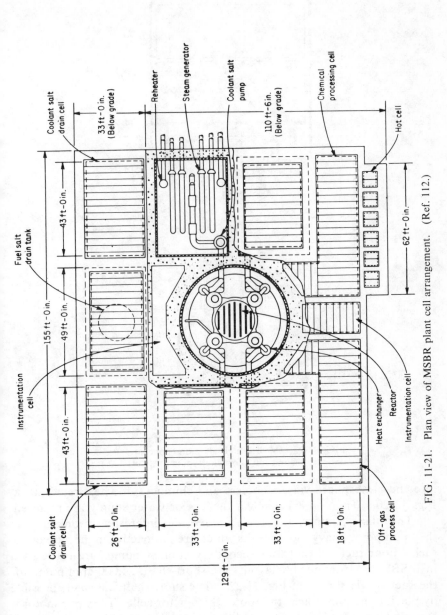

FIG. 11-21. Plan view of MSBR plant cell arrangement. (Ref. 112.)

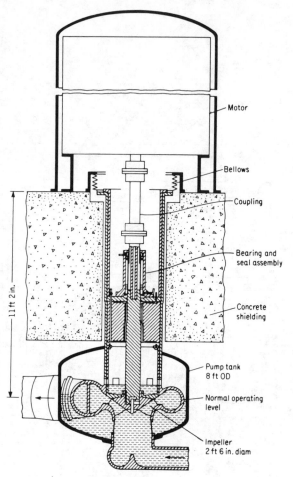

FIG. 11-22.   MSBR fuel-salt pump concept.   Preliminary layout of short-shaft pump.   (Ref. 112.)

into the atmosphere via air coolers and chimneys, the heat-reject stack,
Fig. 11-20.   Thus no mechanical devices or water supply are required.

The plant pumps represent a unique and interesting problem since
they pump hot heavy molten salts which are radioactive in the case of the
fuel.   Four fuel-salt and four secondary coolant pumps are to be used
in MSBR.   A preliminary layout of a short-shaft, 24,000-gpm pump for
the fuel salt is shown in Fig. 11-22.   The short shaft was made possible
by placing the pump near the top of the reactor cell.   The drive motor is
mounted on top of concrete shielding into which the bearing housing is
recessed to reduce shaft overhang.   A one-bearing impeller overhung-
design (about 6.5 ft below the lower bearing) is used.   The pump will be

TABLE 11-8
Some Design Parameters of MSBR Reference Design

Power output	2,000 Mw(e), 4,444 Mw(t) (3,646 in core, 798 in blanket)
Fuel-salt composition	BeF_2, LiF, ThF_4, UF_4; in 20, 67.7, 12, 0.3 mole %, respectively
Fuel properties at 930°F	$c_p = .29$ Btu/lb_m °F, $\rho = 223$ lb_m/ft³, $k = 0.49$ Btu/hr ft °F, $\mu = 34$ lb_m/hr ft
Secondary coolant composition	BF_3NaF; in 48, 52 mole %
Secondary coolant properties at 715°F	$c_p = 0.37$ Btu/lb_m °F, $\rho = 125$ lb_m/ft³, $k = 0.46$ Btu/hr ft °F, $\mu = 34$ lb_m/hr ft
Reactor-vessel dimensions	18.3 ft diam, 24.5 ft high
Core dimensions	16 ft diam, 20 ft high, 4,020 ft³ volume, 19% in region 1 (center), 17% in region 2, 44% in region 3
Number of graphite elements in core	1,760
Core-reflector thickness	12 in.
Fissile and fertile inventories	1,880 and 90,000 kg_m, respectively
Fuel-salt volumes	2,445 ft³ total; 1240 in core, 25 in reflector, 590 in plena, 320 in heat exchangers, 120 in pumps, 150 in piping
Fuel-salt velocity in core	13 fps, maximum
Average core power density	64 kw(t)/liter, 2 peak-to-average core power ratio
Average core specific power	2.34 Mw(t)/kg_m
Fuel-salt pumps	4, 1300°F design temperature, 24,000 gpm, 7,500 hp each, at 200 ft head, 890 rpm
Secondary coolant pumps	4, 1700°F design temperature, 53,000 gpm, 7000 hp each, at 200 ft head, 890 rpm
Isothermal reactivity coefficients	3.6×10^{-5}/°C Doppler, -2.4×10^{-5}/°C prompt salt, -045×10^{-5}/°C overall
Breeding ratio	1.068
Fuel-doubling time	12.6 years (5.6%/year fuel yield)
Fuel-cycle costs	< 0.5 mill/kwhr probable
Maximum neutron flux	1.1×10^{15} (<1.86 w), 7×10^{14} (> 50 kw)

permitted to move freely with thermal expansion of the reactor system. A floating coupling or universal joints will accommodate pump displacements due to such expansions. A molten salt-lubricated bearing made of cermet hard coatings which are plasma-sprayed on Hastelloy-N subtrate will probably be used. The secondary coolant salt pumps will probably be similar to the fuel salt pumps.

(1) Reactor
(2) Primary heat exchangers
(3) Fuel-salt circulating pumps
(4) Coolant-salt circulating pumps
(5) Steam/generator superheaters
(6) Steam reheaters
(7) Fuel-salt drain tank
(8) Double containment structure
(9) Confinement building

FIG. 11-23. Cutaway perspective of reactor cell portion of MSBR plant. (Courtesy Oak Ridge National Laboratory.)

No final design of the heat exchangers has been set at the time of this writing, although a one-pass flow design is being considered.

Reactivity coefficients are of particular interest in fluid-fueled reactor concepts. Isothermal reactivity coefficients for MSBR have been calculated. The total power coefficient is, as expected, negative. A Doppler coefficient of reactivity equal to $-3.5 \times 10^{-5}/°C$, is primarily due to the fertile thorium. A graphite spectral coefficient of $+1.9 \times 10^{-5}/°C$ is positive because of the competition between thermal neutron capture in the fuel, which decreases less rapidly than $1/V$, and thermal capture in thorium, which decreases nearly as $1/V$, with increasing temperature. All effects associated with changes in density of salt and graphite (including reasonance) result in $-0.26 \times 10^{-5}/°C$. A large negative prompt salt coefficient of $-2.4 \times 10^{-5}/°C$ includes all reactivity effects associated with an isothermal core temperature change except graphite spectral. This is the most important coefficient since it largely determines the fast transient response of the system and should afford adequate reactor stability and controllability.

The MSBR is expected to be load following at the expense of modest temperature changes. Calculations (on a two-fluid system) showed that a load reduction of 10 percent/sec results in a heat exchanger inlet temperature increase from 850°F to 1100°F at the rate of 25°F/sec. Thermal lags between heat exchanger and core caused the power to lag behind the load somewhat, but the response appeared satisfactory. Large load changes are believed to be amenable to easy control by a properly designed control system that maintains some desired temperature condition.

Figure 11-23 shows a perspective layout of the reactor cell of the MSBR power plant.

chapter 12

Organic-Cooled
and -Moderated Reactors

12-1. INTRODUCTION

This chapter discusses a reactor concept which involves the use of organic liquids as coolants or coolant-moderators. The advantages of such a system are many:

1. Compared with the use of water as coolant, where the reactor pressure must be high (about 1,000 psia for boiling-water and 2,000 psia for pressurized-water reactors), an organic coolant with low vapor pressure may be chosen. Thus high-temperature–low-pressure operation, resulting in high thermal efficiency as well as a lighter and less costly pressure vessel and other components, is possible.

2. In general, the organics have low corrosion characteristics with most materials and are compatible with most fuels, cladding, and structural materials. A direct result of this is the ability to use conventional materials, such as carbon or low-alloy steels, instead of stainless steels, in the pressure vessel, pumps, piping, etc., resulting in significant savings in capital costs.

3. The induced radioactivity in pure organic liquids is very low, resulting in reduced shielding requirements of the primary loop and therefore further reduction in capital costs.

4. Organic liquids in general have well-known physical, chemical, and handling characteristics.

On the debit side there are these disadvantages:

1. Organic materials are unstable when subjected to high temperatures and to nuclear radiations, resulting in the decomposition of these coolants under power-reactor conditions and possible plugging or fouling of flow channels and heat-transfer surfaces. There is thus a need for purification or cleanup facilities to remove decomposition products from the system. The removed decomposition products must be replaced with fresh coolant. This results in an increase in operating costs, which, under given con-

ditions, may nullify the reduction in capital costs associated with organic coolants. It also limits the freedom of choice of the coolant to those organic liquids that exhibit stability under reactor operating conditions.

2. Organic liquids generally have poorer heat-transfer and pumping characteristics than either water or liquid metals.

These and other characteristics of organic coolants and organic-cooled reactors will be discussed further in the rest of this chapter.

The first reactor using organic liquids was the Organic Moderated Reactor Experiment (OMRE) designed by Atomics International, a division of North American Aviation, Inc., for the U.S. Atomic Energy Commission. The OMRE was an organic-cooled and -moderated reactor of 16 Mw(t) maximum output, designed specifically as a test facility to investigate organic cooling. The heat generated in OMRE was therefore dissipated to the atmosphere and not used to generate useful power. Much experience was gained with OMRE [113, 114]. The first power-producing organic-cooled and -moderated power plant to be built was the Piqua reactor [115], a 45.5 Mw(t) and 11.4 Mw(e) plant. Both OMRE and Piqua have now been decommissioned.

Organic-cooled and -moderated reactor power plants did not prove economically competitive with water- or gas-cooled reactor power plants, and had a rather short history. The concept is still of interest, however, and may someday be revived. Hence this chapter.

12-2. ORGANIC LIQUIDS FOR REACTORS

The early search for organic liquids that are resistant to high temperatures and nuclear radiations included various aliphatic, fused-ring, and synthetic compounds. This search [116] indicated that the aromatics in general and the *polyphenyls* in particular are the most suitable organics in this respect.

The aromatic family of hydrocarbons is based on the benzene ring (Fig. 12-1). This is an unsaturated hexagonal ring depicting the benzene

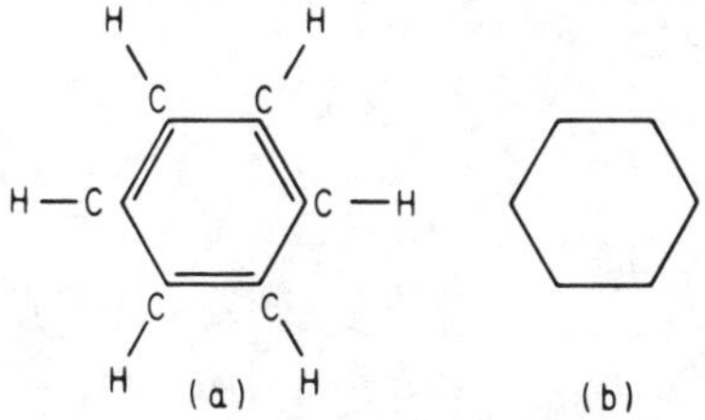

FIG. 12-1. The benzene ring C_6H_6.

molecule C_6H_6. In Fig. 12-1a the full structure of the ring is shown; six carbon atoms are connected by alternate single and double bonds (indicating nonsaturation). Because of the alternate double bonding, only one hydrogen atom branches from each carbon atom. The benzene ring is often represented by a hexagon (Fig. 12-1b) for simplicity.

The phenyl radical is a benzene ring minus one hydrogen atom, that is, C_6H_5 (Fig. 12-2). Again the complete structure and the simplified

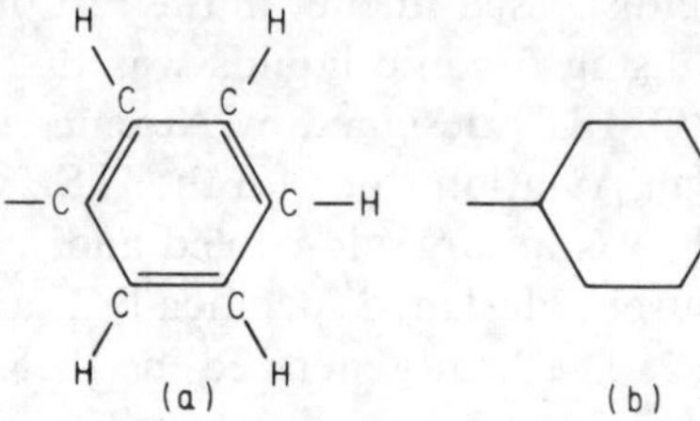

FIG. 12-2. The phenyl radical C_6H_5.

symbol are shown. When a phenyl radical joins a benzene ring, it replaces one of the hydrogen atoms of that ring, so that there are essentially two phenyl radicals, each C_6H_5, joined together. This new molecule, shown in Fig. 12-3, is therefore called the *diphenyl, biphenyl,* or *phenylbenzene,* and has the chemical formula $C_6H_5C_6H_5$.

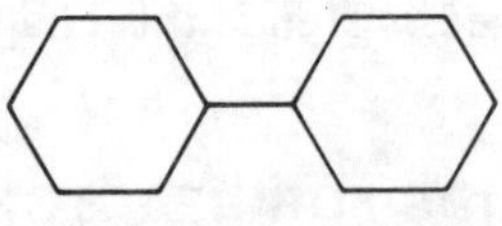

FIG. 12-3. The diphenyl molecule.

When two phenyl radicals join a benzene ring, each replaces one of the hydrogen atoms of the ring. The resulting molecule is called the *terphenyl* or *diphenyl-benzene* and has the chemical formula $(C_6H_5)_2C_6H_4$. The shape of the terphenyl molecule depends upon which two of the hydrogen atoms around the benzene ring are replaced by phenyl radicals. There are three possibilities, i.e., three terphenyl isomers (Fig. 12-4).

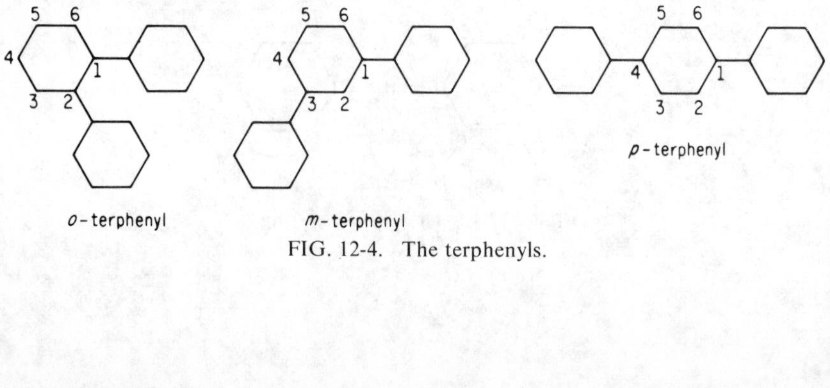

o-terphenyl   *m*-terphenyl   *p*-terphenyl

FIG. 12-4. The terphenyls.

Ortho-terphenyl or simply o-terphenyl, also called 1,2-diphenylbenzene, has the phenyl radicals connected to adjacent corners of the benzene ring; i.e., they replace hydrogen atoms 1 and 2. Similarly, meta- or m-terphenyl is 1,3-diphenylbenzene, and para- or p-terphenyl is 1,4-diphenylbenzene.

The polyphenyls therefore are composed of one or more phenyl radicals connected to a benzene ring or to each other. For example, we may have octaphenyls (a product of polyphenyl decomposition) which are composed of eight rings connected to each other in some fashion. The molecular masses can be approximately calculated from the masses of carbon and hydrogen atoms and their numbers in each molecule. Diphenyl has a molecular mass of 154.20 whereas each of the three terphenyls has a molecular mass of 230.29. Molecules heavier than terphenyl have low vapor pressures (high boiling points), high melting points, and high viscosities and are not suitable as coolants. Organic reactor coolants have so far been limited to diphenyl, the terphenyls, and their mixtures.

Table 12-1 lists some properties of the pure polyphenyls before irradiation and decomposition.

TABLE 12-1
Some Properties of Unirradiated Polyphenyls*

Property	Diphenyl	o-terphenyl	m-terphenyl	p-terphenyl
Melting point, °F	157	135	189	416
Boiling point, °F	492	630	689	709
Density, g_m/cm³:				
200°F	1.01	1.05	1.03	
600°F	0.83	0.90	0.93	0.96
800°F	0.66	0.77	0.83	0.86
Viscosity, lb_m/hr ft:				
200°F	2.47	5.76	8.08	
600°F	0.29	0.53	0.73	0.41
800°F	0.16	0.34	0.39	0.20
Vapor pressure, psia:				
200°F	0.062	0.040	0.020	
500°F	16	3	1	1
600°F	47	16	10	7.5
800°F	110	27	16	12
900°F	400	132	77	60
Molecular mass	154.20	230.29	230.29	230.29
Critical temperature, °F	924	1039	1117	1150
Critical pressure, psia	467.3	349.8	379.8	349.8

* Mainly from Ref. 117.

The OMRE was originally planned to be cooled and moderated by diphenyl. Later experiments under reactor conditions, however, indicated that the terphenyls showed greater stability. Thus commercial mixtures bearing the name Santowax* were used in the OMRE and planned for use in future organic-cooled reactors. Table 12-2 contains the composition and some properties of two such mixtures, Santowax R and Santowax O-M. Santowax R is less costly but has a higher melting point. Because the OMRE coolant loop was originally designed for diphenyl, the lower-melting Santowax O-M was used in it.

TABLE 12-2
Composition and Some Properties of Santowax

	Santowax O-M	Santowax R
Composition, percent:		
Diphenyl	16.0	1.0
o-terphenyl	46.1	13.0
m-terphenyl	31.8	59.0
p-terphenyl	6.1	27.
Properties:		
Molecular mass	218.1	229.5
Specific gravity at 800°F	0.770	0.777
Specific heat, Btu/lb$_m$ °F at 800°F	0.600	0.555
Thermal conductivity, Btu/hr ft °F at 800°F	0.061	0.061
Viscosity, centipoise, at 800°F	0.180	0.180
Final melting point, °F	125	311

12-3. RADIOLYTIC AND PYROLYTIC DAMAGE TO ORGANIC COOLANTS

While organic coolants are practically inert to most reactor materials, they decompose when at high temperatures and when subjected to ionizing nuclear radiations. Since both these conditions are present in a power reactor, a discussion of the damage and its effect on reactor operation is in order. As mentioned previously, early experiments on organic coolants indicated that polyphenyls showed exceptional stability under reactor conditions. The following discussion will therefore be limited to this family of organics and will be largely based on results of early experimentation concerning polyphenyl decomposition which was done on OMRE in the United States and at Harwell, England [114, 116, 118].

Damage or decomposition of organic coolants occurs under the effects of (1) ionizing radiation, called *radiolytic* decomposition, and (2) high temperature, called *thermal* or *pyrolytic* decomposition.

The mechanism of radiolytic damage is a complex one. It is caused by both light particles, such as gamma photons and electrons, and heavy particles, such as neutrons. The energy associated with these particles is absorbed by the coolant, with most of the energy due to absorbed

* Reg. U.S. Patent Office by the Monsanto Chemical Co.

γ-rays and to the thermalization of epithermal neutrons. Since organic materials are covalent-bonded–i.e., made up of molecules whose atoms share electrons, they are consequently held together by weak forces, and the ionization caused by gamma rays or neutrons is sufficient to break this covalent bonding, resulting in the production of atoms and lighter molecules (gases) and in rearrangement into different and heavier molecules and even into different types of chemical bonds.

It is believed that radiolytic decomposition is additive in nature and thus may be represented by the equation

$$D_t = \Sigma x_i D_i \qquad (12\text{-}1)$$

where D_t = total rate of decomposition, percent/hr (by mass)

D_i = rate of decomposition due to particular type of radiation i, percent/hr

x_i = fraction of radiation of type i of total radiation

The second type of damage, pyrolytic damage, results in decomposition products similar to those of radiolytic damage. The rate of decomposition in this case is insignificant at low temperatures but increases exponentially with temperature [116], becoming significant beyond about 800°F. Experiments on many polyphenyls have shown that, at 800°F, the rate of pyrolytic decomposition is approximately 0.03 percent/hr. At 900°F, the rate increases to about 0.7 percent/hr. Pyrolytic damage can be approximately represented by the relationship

$$D_r = e^{(0.031 T - 42.5)} \qquad (12\text{-}2)$$

where D_r = pyrolytic decomposition rate, percent/hr

T = absolute temperature, °R

Decomposition products are classified into (1) gases, (1) low-boiler liquid material, and (3) high-boiler material. The first two, gases and low-boilers, are compounds that are more volatile (usually lower molecular mass) than any of the original material. The high-boiler material, also called simply the *high-boilers,* are compounds that are heavier and less volatile than the original material.

Because there is less recombination at high temperatures, gases and the low-boilers are mostly formed under high-temperature irradiation. The gases formed in OMRE [119] were composed of about 63 percent hydrogen, plus smaller amounts of methane, ethane, ethene, propane, propene, butane, butene, and traces of others. The low-boilers [120] include light aromatics such as benzene, toluene, xylene, and others. Gases comprise about 2 to 3 percent of the total decomposition products while the low-boilers contribute less than 1 percent of the total. The great majority of decomposition products are made up of high-boilers.

The high-boilers are made up of heavy molecules of complex structure, in which the carbon-hydrogen atomic ratio is higher than in the
original material. They are mostly composed of polyphenyls (up to
octaphenyl) and triphenylenes in the approximate ratio of 3:1 [120]. The
average molecular mass of the high-boilers is about 460 (as compared
with about 154 for diphenyl and about 230 for the terphenyls).

When decomposition products are left in the coolant, they cause the
physical properties of the coolant to change. Because these products
are mostly high-boilers, they cause the density and viscosity to increase
and the specific heat to decrease. These effects are shown in Figs. 12-5
to 12-8. The greatest effect is the increase in viscosity. It can be seen

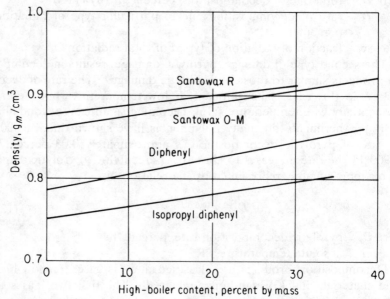

FIG. 12-5. Effect of high-boiler content on density of some organic
coolants.

that, at a high-boiler content of 30 percent by mass, the viscosity is
roughly doubled, while the density is increased and the specific heat
decreased by less than 10 percent. Another change of interest is the
increase in the hydrogen atomic density (hydrogen atoms per cubic
centimeter of coolant) with irradiation. Although the ratio of carbon to
hydrogen atoms is increased, hydrogen atomic density increases because
of the increase in coolant density. This effect, shown in Fig. 12-8, has to
be taken into account in determining the change in nuclear behavior
(moderation, etc.) of the organic with irradiation.

The rate of decomposition of polyphenyls has been experimentally

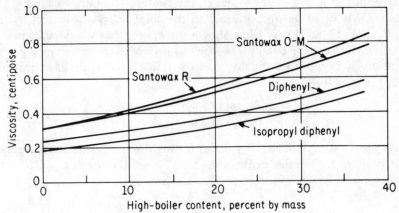

FIG. 12-6. Effect of high-boilers on viscosity.

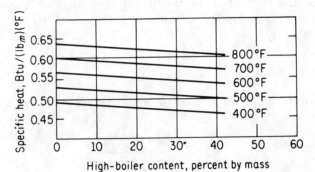

FIG. 12-7. Effect of high-boilers on specific heat.

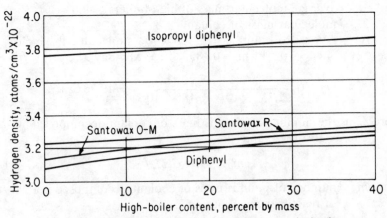

FIG. 12-8. Effect of high-boilers on hydrogen atomic density.

found, both in in-pile loops in MTR, the Material Testing Reactor [121], and in OMRE, to be proportional to the total energy absorbed by the liquid. (Work at Harwell has shown, however, that the dependence of radiation damage on energy absorbed is different for different types of radiation.) The proportionality factors, based on 100 ev of energy absorbed, are as follows:

$$G_p = \frac{\text{molecules of high-boilers formed}}{100 \text{ ev of energy absorbed}} \qquad (12\text{-}3)$$

where G_p is the proportionality factor for high-boiler formation (subscript p for polymer). Similar expressions may be written for gaseous, G_g, and low-boiler, G_b, decomposition products. Similarly,

$$G_c = \frac{\text{molecules of coolant decomposed}}{100 \text{ ev of energy absorbed}} \qquad (12\text{-}4)$$

where G_c is the proportionality factor for coolant decomposition. Because decomposition products are mostly high-boilers, G_p and G_c may be directly related by

$$\frac{G_c}{G_p} = \frac{\bar{M}_p}{M_c} \qquad (12\text{-}5)$$

The mass of decomposed coolant for a given reactor power can be found as follows:

$$m_d'' = G_p \frac{\bar{M}_p}{\text{Av}} = G_c \frac{M_c}{\text{Av}} \qquad (12\text{-}6)$$

where m_d'' = number of pounds mass of coolant decomposed per 100 ev of energy absorbed
$\bar{M}_p$ = average molecular mass of high-boilers $\simeq 460$
M_c = molecular mass of coolant
Av = Avogadro's number = 2.733×10^{26} molecules/lb_m mole
Since 1 ev = 4.44×10^{26} kwhr = 0.185×10^{-29} Mw-day, then

$$m_d' = G_c \frac{M_c}{2.733 \times 10^{26}} \frac{1}{0.185 \times 10^{-27}} = 19.78\, G_c M_c$$

where m_d' is the number of pounds mass of coolant decomposed per Mw-day of energy absorbed, and

$$m_d = 19.78\, G_c M_c f \qquad (12\text{-}7)$$

where m_d = number of pounds mass of coolant decomposed per Mw-day of reactor thermal output
f = fraction of reactor power absorbed by coolant

Conversely, G_c (and subsequently G_p) may be determined if m_d is measured and M_c and f are known.

It has also been experimentally found that G_p (and G_c) varies with the percentage of high-boilers already present in the coolant. Figure 12-9

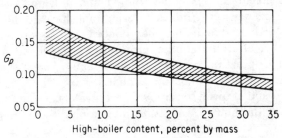

FIG. 12-9. G_p versus high-boiler content for poly-phenyls.

contains the result of such tests [118] and shows that G_p decreases rapidly with high-boiler content. The relationship can be given by the approximate equation

$$G_p = G_{p_0}(1 - C_{HB})^2 \qquad (12\text{-}8)$$

where G_p = desired value of proportionality factor for high-boilers

G_{p_0} = proportionality factor for fresh coolant (zero high-boiler content)

C_{HB} = fraction of high-boilers present in coolant

Thus for a high-boiler content of 30 percent, the value of G_{p_0} is roughly cut in half.

In determining G_p in the OMRE, the fraction of energy absorbed by the coolant-moderator, f, was calculated to be 8.4 percent of the total fission energy, a large percentage because of the exceptionally large moderator-fuel ratio (10.4:1) in the OMRE core and the high enrichment of the fuel (93 percent). Of this energy it is estimated that 28 percent is absorbed in the neutron slowing-down process and 72 percent in γ-ray attenuation. Organic power reactors with more normal moderator-fuel ratios would be expected to lower the total energy absorbed by the coolant-moderator to the normal 4 to 6 percent, with corresponding reduction in decomposition rates. The OMRE has shown a decomposition rate of approximately 75.5 lb_m/Mw-day at 600°F. This corresponds to a $G_p = 0.1$ (for an average high-boiler molecular mass of 460). The gaseous-decomposition proportionality factor G_g was found to be about 0.02. Note that this corresponds to a much smaller mass ratio because of the much lower molecular masses of the gaseous products. It is necessary to know the gaseous decomposition rate accurately in order to design an adequate reactor degasification system.

In order to reduce the cost of replacement of the decomposition products, the organic coolant may be allowed to operate with a certain high-boiler content, where the rate of high-boiler formation is sufficiently reduced. This, however, has to be balanced against the decrease in heat-transfer coefficient and possible fouling and clogging of heat-transfer surfaces and passages. Atomics International have found from experiments in OMRE that a high-boiler content of 30 percent by mass, at which $G_p = 0.1$, offered a good compromise. This level can be attained by suspending operation of the coolant cleanup system (Sec. 12-7) until the high-boiler content builds up to that value. The cleanup system is then restarted and allowed to remove an equivalent amount of high-boilers which form at the new G_p rate, thus maintaining the 30 percent content figure.

12-4. INDUCED RADIOACTIVITY IN ORGANIC COOLANTS

It has been pointed out (Sec. 12-1) that one of the main advantages of organic reactor coolants is the low induced radioactivity of the pure liquid. Barring any accumulation of radioactive material in any one spot, the low induced activity results in reduced requirements for shielding of the primary-coolant loop and consequently in savings in mass and cost of the plant. It is indicated [114] that accesibility for maintenance of the primary-loop components is possible even during reactor operation.

A pure organic is a hydrocarbon, composed only of carbon and hydrogen. Thus only low activity levels may be attributed to it. The main activity in an organic system, however, is due to impurities present in the organic. The main impurities encountered in the OMRE coolant (Santowax O-M) were found to be composed of inorganics such as sodium, manganese, chlorine, and copper in varying amounts. In addition, quantities of rust, metal filings, and other impurities were found to be collected and carried by the coolant from primary-loop surfaces during initial testing. These last impurities must be removed by the organic cleanup system prior to operation. In OMRE experience, the induced activity consisted mainly of short-lived γ emitters listed in Table 12-3. When the

TABLE 12-3
Main Activity of OMRE Coolant*

Isotope	Half-life	Concentration, ppm
Na^{24}	15.0 hr	0.10
Mn^{56}	2.58 hr	0.03
Cl^{38}	37.3 min	2.00
Cu^{64}	12.90 hr	
Cu^{66}	5.10 min	0.20

* From Ref. 114.

reactor was operating at 10 Mw(t), the highest radiation level outside the reactor shelter area was found to be 80 mr/hr (at the surface of a filter removing particulate matter from the coolant).

12-5. OTHER CHARACTERISTICS OF ORGANIC COOLANTS

Safety

Organics are materials with well-known handling properties. They are compatible with reactor materials. There is very little hazard of exothermic chemical reaction between them and these materials. However, they are pyrophoric; i.e., they ignite spontaneously when hot and in contact with air. The flame-point temperatures are 255°F for diphenyl and 375°F for Santowax R. The dust-cloud ignition temperature is, however, 1200°F for both coolants [116]. When organics burn, they form dense black smoke clouds, making fire fighting difficult. It is recommended that automatic fire-extinguishing equipment be installed in critical closed areas, such as in the vicinity of the reactor, and that a blanket of an inert gas be used to separate or shield the organic from possible exposure to air. The OMRE, for example, has a pressurized-nitrogen blanket above the coolant pool in the reactor vessel; this doubles as a primary-loop pressurizer (Sec. 5-4.).

Corrosion

Chemical interaction between polyphenyl coolants and most reactor materials is very small. Test runs to determine the degree of corrosiveness of reactor materials subjected to polyphenyls indicate that only magnesium and zirconium are susceptible to slight chemical reaction, magnesium by oxidation and zirconium by hydride formation. Other materials such as aluminum, stainless steel, and mild steel showed insignificant mass change and varying degrees of discoloration, with mild steel exhibiting the most discoloration in the form of surface blackening.

12-6. HEAT TRANSFER OF POLYPHENYL COOLANTS

Among the disadvantages of organic coolants are the relatively low (compared with water and liquid metals) heat-transfer characteristics and high pumping-power–heat-transfer ratio [2].

Forced-convection heat-transfer tests were made on diphenyl, Santowax R, and Santowax O-M by Atomics International [122]. The ranges of operating conditions were 481 to 772°F bulk temperature, 3.4 to 33.4 ft/sec velocity, and heat fluxes between 39,200 and 291,000 Btu/hr ft². The following correlation was recommended for the pure coolant:

$$Nu = 0.015\ Re^{0.85}\ Pr^{0.30} \qquad (12\text{-}9)$$

where Nu, Re, and Pr are the Nusselt, Reynolds, and Prandtl numbers, respectively.

The forced-convection heat-transfer coefficient decreases gradually with high-boiler formation. At the 30 percent high-boiler-content operating level, the heat-transfer coefficient has been found to decrease by about 10 percent [123]. It has also been noted that the decrease in heat-transfer coefficient observed is completely recoverable on reduction of high-boiler concentration. This seems to indicate that, in the tests cited, no significant fouling (i.e., deposition of decomposition products) of the heat-transfer surfaces occurred. Only a small fraction of 1 mill of hard dark film appeared to adhere to the heat-transfer surface, with this film apparently not interfering with the heat-transfer mechanism. Some instability in heat-transfer coefficient, particularly at low coolant velocities and low surface temperatures [122], was observed, however. A representative value of forced-convection heat-transfer coefficients is 1300 Btu/hr ft² °F for clean santowax O-M, with 600°F bulk temperature and 14.6 ft/sec velocity. Because of the low heat-transfer coefficients of organics, clad finning may be used.

More recently, in order to attain higher heat-transfer coefficients, some attention has been given to subcooled nucleate boiling of organic coolants [124]. This has the added effects of lowering the operating pressure of the coolant at the high temperatures desired (since surface boiling is permitted), lowering the pumping requirements, and increasing the coolant exit temperature (because of increased heat transfer). Experiments showed that the boiling heat-transfer coefficient decreased with time and stabilized after about 700 hr. It is believed that this is due to the smoothing out (i.e., reduction in nucleation centers) of the boiling surface because of the buildup of smooth film on the surface. Some attention has also been given to the use of polyphenyls in a boiling-type reactor (similar to the boiling-water reactor).

Jordan and Leppert [125] conducted experiments on nucleate saturated boiling of various polyphenyls at atmospheric pressure. The results of these experiments are shown in Fig. 12-10, which presents mainly the nucleate-boiling regime of some polyphenyls in which the heat flux is given as a function of the difference between the heating wall temperature t_w and the saturation temperature of the coolant, t_{sat}. During runs with diphenyl, the authors observed that, when the critical heat flux was reached and the heating surface was at about 1900°F and was allowed to remain at that temperature for 15 to 30 sec, surface fouling in the form of a thin film deposition occurred. This fouling was then found

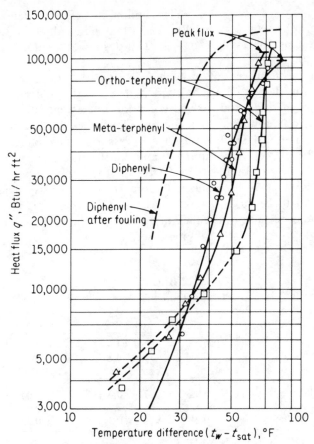

FIG. 12-10.   Nucleate-boiling heat flux of polyphenyls at
atmospheric pressure.   (Ref. 125.)

to improve the boiling heat-transfer coefficient, shown by the dotted line
in Fig. 12-10.

For subcooled boiling, the authors suggest the use of the method of
Forster and Greif [126] in order to take into account the effects of the
degree of subcooling of the coolant (saturation temperature t_{sat} minus
bulk temperature t_f) and the coolant velocity on the transition point.
This is the point where heat transfer changes from pure convection to
nucleate boiling [2]. [In this method, the heat flux is first plotted as a
function of the difference between the wall temperature and the coolant-
fluid bulk temperature, that is, $t_w - t_f$.   This assumes that heat transfer
is by convection only.   Equation 12-9, which contains velocity in Re,
may then be used to calculate the convective heat-transfer coefficient
h which appears in Nu.   The heat flux q'' due to convection is then

calculated from $q'' = h\,(t_w - t_f)$. The boiling heat flux is then plotted on the same graph. Values from Fig. 12-10 may be used. The two curves will intersect at some value of q''. The desired transition heat flux is greater by 40 percent.]

Boiling heat fluxes of polyphenyls may be obtained from the Rohsenow correlation [2]. The coefficient C_{sg} in that correlation, which is a function of the liquid-surface condition but essentially independent of pressure, was obtained for some polyphenyls by Jordan and Leppert from experimental data at atmospheric pressure. The value of C_{sg} was found to be 0.0086 for diphenyl and 0.0105 for m-terphenyl. For o-terphenyl, it was found to be 0.018 but the exponent on the right-hand-side term must be changed from 0.33 to 0.135. The heating surface in all three cases was stainless steel. Other polyphenyls were not correlated because of the lack of physical-property data. Properties of diphenyl, m-, and o-terphenyl used by the authors are given in Table 12-4.

TABLE 12-4
Physical Properties of Some Polyphenyls at Atmospheric Pressure*

Property	Diphenyl	m-terphenyl	o-terphenyl
t_{sat}, °F	946	1164	1084
c_p, Btu/lb$_m$ °F	0.611	0.584	0.569
h_{fg}, Btu/lb$_m$	136.5	122	117
μ, lb$_m$/ft hr	0.67	0.570	0.825
σ, lb$_f$/ft	0.001296	0.00113	0.00124
ρ_f, lb$_m$/ft³	52.59	50.70	52.0
ρ_g, lb$_m$/ft³	0.210	0.248	0.236

* From Ref. 125.

Experimental burnout or critical heat fluxes for two unirradiated organic coolants have been obtained by Core and Sato [127]. The empirical equations defining these limits include the effects of subcooling and velocity. They are as follows:

For diphenyl:

$$q_c'' = 454\,\Delta t_{sub}\,V^{0.63} + 116{,}000 \tag{12-10}$$

For Santowax R:

$$q_c'' = 552\,\Delta t_{sub}\,V^{2/3} + 152{,}000 \tag{12-11}$$

where q_c'' = critical heat flux, Btu/hr ft²
Δt_{sub} = degree of subcooling of coolant
$\quad = t_{sat} - t_f$, °F
V = coolant velocity, ft/sec

These two correlations have limits of uncertainty of $\pm$ 106,000 and $\pm$ 160,000 Btu/hr ft^2, respectively. They are based on the following operating conditions: For diphenyl, $\Delta t_{sub} = 0$ to 328°F (also low-quality tests), $V = 0.5$ to 17 ft/sec, pressure 23 to 406 psia, and coolant bulk temperature 510 to 831°F; for Santowax R, $V = 5$ to 15 ft/sec, pressure $= 100$ psia, and coolant bulk temperature 595 to 771°F. All tests were made on coolant flowing in an annular passage formed by a 0.46-in.-ID electrically heated tube containing an unheated 0.188-in.-diameter coaxial rod.

The authors point out that burnout heat fluxes of polyphenyls are several times lower than those found in nonboiling-water reactors at their characteristic operating pressure and that they are appreciably affected by impurities, especially those that affect the saturation temperature and consequently the degree of subcooling in Eqs. 12-10 and 12-11.

12-7.   POLYPHENYL CLEANUP AND OPERATING COSTS

It has been pointed out that a sizable reduction in capital costs is made possible by the use of polyphenyl coolants. The coolant, however, must be purged of some of its decomposition products so as to allow its high-boiler content to remain at a fixed level. This level, found to be 30 percent by experiments in OMRE, is a compromise mainly between the decomposition rate and the heat-transfer mechanism (Sec. 12-3). Removal of high-boilers also decontaminates the system, since most of the coolant radioactivity is associated with the high-boilers.

The cost of reprocessing or cleanup and replenishment is composed of two items: (1) capital cost of the cleanup system and (2) cost of the polyphenyl replacement.

It has been found that the capital cost of the cleanup system is rather low. In OMRE, for example, the cleanup system consists of a single-stage vacuum distillation unit plus associated equipment (Fig. 12-11). This system will now be described. A 300-gal batch of coolant is bled from the OMRE system each day (out of a total 9,200-gpm coolant flow) and led to a still operating at 30 mm Hg. About 100 gpd of residue is taken from the still bottom and led to waste storage. Then 200 gpd of vaporized and purified organic is condensed by a spray of cooled recirculated organic. Next 100 gpd of fresh makeup is added to the purified organic, and the mixture is fed back to the coolant system. About 20 gpm of coolant in a different OMRE loop is filtered to remove particulate matter.

While a batch cleanup system was used in OMRE when decomposition-rate data were sought, continuous cleanup systems should be designed for organic power reactors. The capital cost of the cleanup

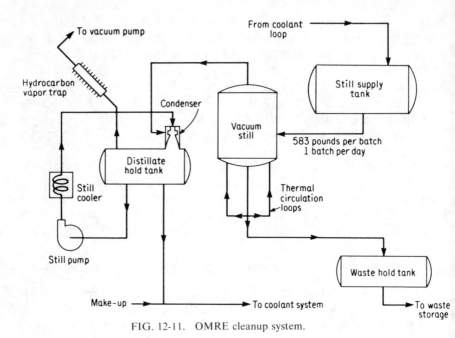

FIG. 12-11.   OMRE cleanup system.

system, in any case, is a minor portion of the total cost of cleanup.   The cost of replacement of the organic therefore constitutes the major portion.

If the proportionality factor G_c or G_p is known, the cost of coolant makeup in mills per kilowatthour of plant net output may be easily obtained, by using Eq. 12-7, to give

$$\text{Makeup cost} = 0.824 \; \frac{G_c M f C}{\eta} \; \text{mill/kwhr(e)} \qquad (12\text{-}12)$$

where C = cost of coolant $/lb_m$

η = net thermal efficiency of plant

Thus for a plant using Santowax R as coolant, at $0.17/lb_m$, and having a net efficiency of 30 percent and $G_p = 0.10$, so that $G_c = (460/229.5) \times 0.1 = 0.20$ (where 460 is the average molecular mass of high-boilers and 229.5 is the molecular mass of Santowax R) and where $f = 0.04$ (a value attainable in large reactors where the ratio of organic to fuel is lower than in OMRE), the cost of makeup would be 0.85 mill/kwhr(e).   Lower values around 0.60 mill/kwhr(e) have been indicated for a large, 300-Mw(e), organic-cooled design [128].

This replacement cost represents a fairly large portion of the cost of power.   It is, however, believed to be low compared with the savings in capital cost attained through the use of conventional materials and low pressures, so that net economic gain is achieved in organic-type reactors

In the 300-Mw(e) reactor-design study cited above [128], the power costs are compared with a conventional 305-Mw(e) coal-fired power plant (Table 12-5). The comparison is based on a coolant makeup price of

TABLE 12-5
Power Cost Comparison of Organic-cooled and Coal-fired Plants*

Costs, mills/kwhr	300 Mw(e), organic-cooled	305 Mw(e), coal-fired
Fixed charges	4.14	3.30
Fuel costs	2.23	3.25
Coolant makeup	0.60	
Operating and maintenance	0.34	0.22
Total	7.31	6.77

* From Ref. 128.

$0.17/lb$_m$, nuclear-fuel average burnup of 15,000 Mw-day/ton, and $0.35/million Btu for coal. It is indicated that further savings in nuclear-fuel costs due to reductions in fabrication costs, as well as in coolant makeup price because of expected volume production, are to be expected.

Two more approaches to reducing the costs of coolant makeup are under investigation. These are:

1. The addition of radiation-damage inhibitors to the organic coolant. Some 9 or 10 inhibitors have been found to reduce both gas and high-boiler formation by 20 percent or more [124]. The most effective of these is 3,4-benzpyrene which reduces both G_p and G_g by 50 percent. The effects of these inhibitors on the heat-transfer, fluid-flow, and other characteristics of the coolants are yet to be determined.

2. Chemical regeneration of the high-boilers with hydrogen, so that they can be reused in the coolant loop. Again more information on this system must be obtained before adopting it in organic-cooled plants.

12-8. SOME OTHER CHARACTERISTICS OF ORGANIC-COOLED REACTORS

Fluid Flow

Not mentioned in connection with the heat-transfer mechanism in organic-cooled reactors is the instability of the heat-transfer surface [122]. This instability manifests itself in reductions in the value of the convective heat-transfer coefficient of several percent per 100 hr of operation. The instability is more pronounced at low coolant velocities but is also a function of heating (fuel-element) surface temperatures and, to a lesser degree, of coolant bulk temperatures and heating-surface conditions. A low limit is thus imposed on coolant velocity which increases pressure losses and pumping work. When fuel-channel orificing is added to equalize coolant temperature rise across the reactor, the pressure drop of the coolant becomes particularly high. Besides increased pumping work, the increased pressure drop imposes a penalty on neutron economy because of the necessity of increased structural material to cope with it. Power flattening is a cure for such a situation, since it would reduce or eliminate orificing and thus reduce the pressure drop. Of course, it also increases the average specific power of the reactor core.

Separate Moderator

In large stationary plants where mass and size are not critical, a separate moderator such as heavy water or graphite may be used along with an organic coolant. Such an arrangement has two main advantages:
1. It improves the neutron economy of the reactor, since either of the moderators has lower absorption cross sections than the organic.
2. It reduces the rate of coolant decomposition since the amount of gamma and neutron energy absorbed by the coolant will now be materially reduced.

As indicated before, a similar effect may be obtained in an organic-cooled and -moderated reactor by reducing the moderator-fuel ratio in the core. In this case, however, only the gamma energy absorbed by the coolant-moderator is reduced (because of increased gamma absorption by the fuel), while the neutron energy absorbed by it remains practically unchanged. In the case of a separate moderator, the bulk of neutron slowing down falls on the moderator and the coolant receives a reduced dose of both gamma and neutron energies.

In heavy-water-moderator, organic-cooled reactors [129, 130], as in D_2O-moderated, gas-cooled reactors, there is the problem of keeping the moderator from boiling. It can be solved by cooling and pressurizing the moderator, with further mass and cost penalty on the normally large heavy-water reactor vessel, or by cooling the moderator to low temperatures and containing the coolant and fuel in pressure tubes.

Fuel and Cladding

No unusual problems regarding the choice of fuel material arise in organic-cooled reactors. Either uranium metal or uranium dioxide may be chosen, provided that the enrichment is compatible with reactor physics. Natural uranium may be used with heavy-water moderator and organic cooling [130]. Either plate, thin-cylindrical, rod, or rod-with-internal-cooling fuel-element shapes may be used. The design of fuel elements must take into account the low heat-transfer characteristics of organic coolants. Depending upon the magnitude of the heat-transfer coefficient designed into the system, finned cladding may be used.

The choice of cladding is more of a problem. Magnesium and zirconium are not suitable with organic coolants because of magnesium oxidation and zirconium hydride formation (Sec. 12-5). Stainless steel, although not required on the basis of corrosion, has been used in organic reactors on the basis of the wide experience already gained regarding its manufacture and operation under reactor conditions. It, however, has a low heat conductivity and high neutron absorption cross sections, necessitating the use of enriched fuels. Beryllium and aluminum are good organic-reactor cladding materials. Beryllium has the advantage, over aluminum, of near-zero neutron absorption—a clear advantage in the case of reactors using low neutron-absorption coolants and moderators. This advantage, however, is not so clear-cut in the case of organic-cooled reactors because of the large neutron absorption by the organic coolant.

Aluminum is compatible with organic coolants but has poor structural strength at power-reactor temperatures. If used, the main structural support of the fuel element must therefore fall on the fuel itself. In this respect, alloyed metallic fuel is necessary. The cladding in this case serves only to separate the fuel from the coolant and provides the necessary heat-transfer surface. Such a combination is employed in the Piqua organic-cooled reactor (Sec. 12-10). When a structurally weak fuel such as UO_2, which has excellent burnup characteristics, is preferred, a strong aluminum-based material called SAP (sintered aluminum power), or APM (aluminum powdered metal), is used [131]. This material consists of aluminum particles with about 6 to 8 percent Al_2O_3 dispersed through it. It is believed that these oxide particles contribute to the structural strength of SAP, in much the same way as carbon particles do in steel. The tensile strength of SAP is about five times that of aluminum, and it maintains its structural strength up to about 1000°F, about 300°F higher than any other aluminum alloy. It can easily be shaped into finned tubing. Its disadvantages are the reduced ductility with oxide content and the difficulty of making joints (welding). A UO_2-SAP fuel-element combination is suggested in an OMCRE design study [128].

Organic-Loop Heating

Because of the high melting points of polyphenyls (Tables 12-1 and 12-2), any organic loop must be heated prior to reactor operation, during low-power testing, or during shutdown. Either electric resistance or induction heating may be used, with the latter having the advantage of simplicity and ease of installation. The OMRE employed an induction heating system capable of maintaining the primary loop at about 300°F.

12-9. THE ORGANIC MODERATED REACTOR EXPERIMENT (OMRE)

The OMRE was designed, constructed, and operated by Atomics International for the U.S. AEC. It is located at the National Reactor Testing Station (NRTS), Arco, Idaho. It had a maximum thermal output of 16 Mw(t). The plant was conceived for the sole purpose of studying the feasibility and characteristics of organic-cooled and -moderated reactors and for testing organic liquids and reactor materials associated with this type. No conversion of heat to electricity was therefore attempted. The reactor energy is instead dissipated into the atmosphere by a forced-air heat exchanger. Some of OMRE plant data are given in Table 12-6. Figure 12-12 shows a cross section of the OMRE reactor, Fig. 12-13 is a flow diagram, and Fig. 12-14 is a perspective of the OMRE facility.

TABLE 12-6

Some OMRE Design and Operational Data

Thermal output	5–16 Mw(t)
Net electrical output	None
Fuel type	93% enriched, 25% UO_2, 75% stainless-steel alloy
Fuel geometry	Plate type, 2.5 in. by 36 in. active dimensions; 16 per subassembly in a channel; 2.9 in. by 2.8 in. plates separated by 0.134-in. gap, 4.5-in. square lattice spacing
Cladding	Type 304 stainless steel, 0.005 in. thick
Core dimensions	22.5 in. square, 36 in. high
Maximum fuel temperature	824°F
Core heat flux	313,000 Btu/hr ft² max., 99,200 average
Maximum neutron flux (thermal)	2.1×10^{13} at 700°F and 35% high-boiler content
Coolant-moderator	Santowax O-M
Coolant velocity	15 fps, single pass through core
Coolant temperatures	500–700°F inlet, 530–710°F outlet
Coolant pressure	200 psia
Coolant pressure drop in reactor	55 psi
Coolant flow	7,200 gpm
Coolant pumps	Two; each, 150-hp single stage centrifugal
Steam conditions	None
Control rods	Twelve, 1.25-in. steel tubes filled with boron carbide powder

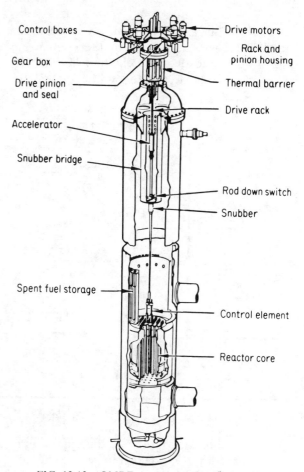

Control boxes

Gear box

Drive pinion
and seal

Accelerator

Snubber bridge

Spent fuel storage

Drive motors

Rack and
pinion housing

Thermal barrier

Drive rack

Rod down switch

Snubber

Control element

Reactor core

FIG. 12-12.   OMRE reactor vessel.   (Ref. 114.)

The OMRE reactor vessel, Fig. 12-12, is made of 1-in.-thick mild steel, $4\frac{1}{2}$ ft in diameter and 28 ft high.   The core, located in the lower part of the vessel, is covered with a pool of organic coolant 14 ft high. This pool acts as a biological shield during fuel handling and other operations.   A nitrogen blanket at 200 psig is provided above the free surface of the pool.   It acts as a seal, preventing air-coolant chemical reactions, and maintains coolant pressure.   The OMRE coolant saturation temperature corresponding to this pressure is 920°F, much higher than the 710°F maximum coolant temperature.   The nitrogen blanket becomes contaminated by the gases (hydrogen, methane, etc.) formed by coolant decomposition.   Therefore a continuous purge of the nitrogen blanket is undertaken, at a rate of about 30 ft³/hr, to prevent gaseous buildup.

The coolant loop (Fig. 12-13) contains an expansion tank (Sec. 5-4) which provides for changes in volume of the coolant in case of large changes in temperature and thus prevents much change in the coolant pool level. Coolant leaves the reactor at the rate of 9,200 gpm and

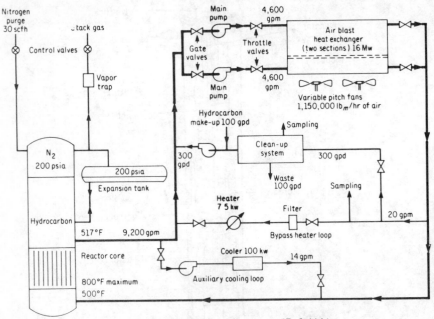

FIG. 12-13. OMRE flow diagram. (Ref. 114.)

mixes with 20 gpm of coolant from the cleanup and filter systems. The coolant is then pumped by two 4,600-gpm centrifugal hot-oil pumps, of the type commonly used in petroleum refineries, and enters an air-blast heat exchanger which dissipates the reactor energy to the atmosphere. The coolant then returns to the reactor after 20 gpm is extracted for cleanup and filtration. The OMRE cleanup system has been described in Sec. 12-7.

The filter loop contains a bypass heater which is used to provide heat-transfer data on the coolant during changes in its properties as a result of irradiation. A centrifuge (not shown) has been added to provide information on its ability to remove particulate matter from the coolant.

The auxiliary cooling loop, shown in Fig. 12-13, removes decay energy at the rate of 100 kw(t) from the reactor after shutdowns, when the main air-blast heat exchanger is inoperative. Heat removal here is by a water spray.

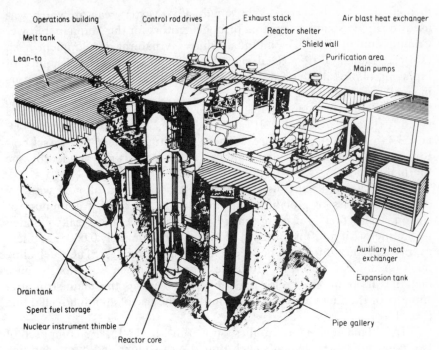

FIG. 12-14. Perspective of OMRE facility, NRTS, Idaho. (Courtesy Atomics International, Division of North American Aviation Corporation.)

The entire coolant loop is made of carbon steel with the exception of the reactor vessel, which is mde of mild steel. The fuel elements are in plate form 2.5 in. wide and 60 in. overall length, but the active length is 36 in. The meat consists of 0.002-in.-thick alloy of 93 percent enriched UO_2 in stainless steel. Cladding is unfinned stainless steel 0.005 in. thick on each side.

The OMRE was originally designed to use the lower-melting diphenyl. When it was found that the terphenyls showed greater resistance to radiation damage, Santowax R, made up of 99 percent of the three terphenyls and 1 percent diphenyl and lower priced than either diphenyl or Santowax O-M, was selected for use in future organic reactors. However, because of the design of the OMRE coolant system, the lower-melting mixture, Santowax O-M, which contains more diphenyl than Santowax R, was used.

12-10. THE PIQUA OMR POWER PLANT

The Piqua OMR, now decommissioned, is the first organic-cooled and -moderated reactor power plant to generate electricity for commercial use. It is a relatively small, 45.5 Mw(t), reactor which was used to supply steam to existing turbogenerators for the Municipal Power Commission of the City of Piqua, Ohio. A natural development from OMRE, Piqua OMR was designed to demonstrate the feasibility of replacing old fossil-fueled steam generators with nuclear steam generators in small power stations.

The Piqua reactor vessel is made of 2-in.-thick low-carbon steel and is 7.6 ft ID and 27 ft high. It is designed to meet ASME codes for 750°F and 300-psia internal pressure. The core, situated near the vessel bottom, consists of 85 fuel subassemblies. The core is surrounded by one thermal shield.

The fuel is 1.9 percent enriched metallic uranium alloyed with 3.5 percent by mass of molybdenum and about 1 percent of aluminum or silicon. The fuel is clad in finned aluminum, metallurgically bonded to the fuel with a diffusion layer of nickel approximately 0.0005 in. thick, to prevent interdiffusion between fuel and cladding. The fuel alloy mentioned has creep-rate properties that are much superior to pure metallic uranium, thus allowing the fuel to act as the main structural support of the fuel element and permitting the use of aluminum cladding (Sec. 12-8).

The Piqua fuel subassemblies, shown in Fig. 12-15, are composed of two concentric fuel elements each finned on its inner and outer cladding surface. The fins are slightly spiral to improve further the heat-transfer characteristics. Note that the two central finned surfaces, facing each other, have their fins inclined in opposite directions. One outer and one inner concentric steel-tube cover the exposed sides of the fuel elements and are fastened to them at the upper and lower ends. (The OMCR [128] fuel elements are made of UO_2 and SAP cladding, are flat-plate-shaped, and have finned cladding on both sides of each plate.)

In the Piqua reactor, the coolant enters the core at the top and flows downward through the fuel elements to a plenum below the bottom grid plate. It then flows upward in the annular space between the core and the thermal shield to an outer plenum of the vessel. The organic thus serves as core reflector as well as coolant and moderator. The entrance to each fuel subassembly is properly orificed to maintain equal coolant-temperature rise throughout the core.

As shown in Fig. 12-16, the Piqua flow diagram, the coolant enters the reactor vessel at the top, at a rate of 5.5×10^6 lb_m/hr (12,000 gpm) and 120 psia and 525°F. It receives 45.5 Mw(t) from the core and is thereby heated to 575°F. It leaves the reactor vessel via two 6,000-gpm pumps and enters the steam superheater, where 150,000 lb_m/hr of steam at 450 psia is superheated to 550°F. It then enters the boiler where feedwater at 268°F is converted to steam at 450 psia. From the boiler it

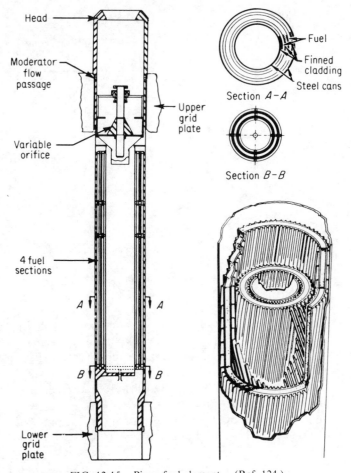

FIG. 12-15.   Piqua fuel element.   (Ref. 124.)

goes back to the reactor.   The steam formed is supplied to the steam header of the Piqua municipal power plant.

The boiler is constructed of carbon steel and is in the form of a vertical shell-and-tube heat exchanger and contains an internal moisture separator.   The organic coolant flows in the tubes, and the water flows on the shell side of the boiler.   The superheater is also made of carbon steel and is in the form of a shell-and-tube heat exchanger but with the organic coolant on the shell side and steam in the tubes.

The organic-purification system contains degasification and high-boiler removal systems.   For gas removal, 200 gpm of hot coolant from the reactor is sprayed into the degasifier, which is in the form of a tank maintained under vaccum.   Decomposition gases (hydrogen, methane,

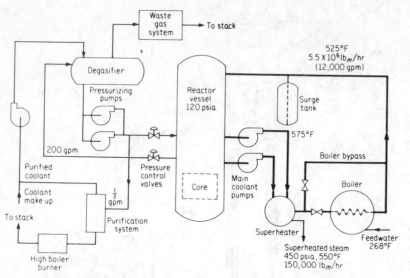

FIG. 12-16.   Piqua OMR flow diagram. (Ref. 125.)

etc.) and water vapor (from the steam loop) are separated and vented to the stack.   The rest of the coolant is pumped back to the reactor. To remove the high-boilers (beyond the 30 percent content), part of the coolant is first withdrawn from the core, preheated to 700°F, and fed into a vacuum distillation column.   The high-boiler residue is transferred to a waste-disposal system which consists of storage, hydrocarbon-burning, and gas-filtering facilities.

Provision for decay-energy removal (not shown) is made by pumping the coolant through a heat exchanger where heat is removed by evaporating water.

The plant is automatically controlled so that it is load-following between 15 and 100 percent of full load.   It can also be manually controlled during start-up.   Load-following is accomplished by varying the reactor power level according to steam demand.   This is done by varying the coolant flow in the boiler by a bypass (Fig. 12-16) while maintaining constant coolant flow, coolant reactor-outlet temperature, and constant steam pressure.

chapter 13

Thermionic
Energy Conversion

13-1. INTRODUCTION

This and the next two chapters discuss methods of converting thermal energy and some nuclear radiations to electricity without the use of a thermodynamic cycle–for example, the Rankine or Brayton cycles. These rely on a working fluid (steam) and rotating machinery (turbine-generator) to produce electricity. While such cycles are highly developed both technically and economically and have high thermodynamic efficiencies, they are most suitable to large power outputs.

There has long been a fascination with the possibility of directly converting into electricity, the energy of thermal sources such as chemical or nuclear reactions, or of radiation sources such as photons, beta particles, or fission fragments without resort to an intermediate cycle. Such devices would have the advantages of simplicity, reliability, quietness, and long life (no moving parts).

The first of such devices to be discussed in this book is the thermionic energy converter. A thermionic converter is basically a vacuum or gas-filled device composed of a heated metal emitter or cathode from which electrons "flow" or are "boiled off," a cooled metal collector or anode which collects the electrons, and suitable electrical leads. The heat source could be the fuel in a nuclear reactor.

Because a thermionic converter converts heat to electricity, it is, like a Rankine cycle, a heat engine which obeys the second law of thermodynamics and is therefore Carnot efficiency-limited. Its efficiency at present is, however, much lower than that of a Rankine cycle, but it has features that make it useful for special applications. The thermionic converter operates best at high temperatures with optimum cathode temperatures between 1000 and 2200°C. There are no great advantages, with present day thermionic materials, to be had with too low (below 400°C) anode temperatures. Present rotary-turbine-blade materials, on the other hand, are not capable of operating at such high temperatures. The thermionic converter could therefore be used as a *topping* device with

417

high temperature nuclear sources. (A topping device is one that utilizes heat at very high temperatures and rejects heat to another device, such as a Rankine cycle, which utilizes the heat at lower temperatures). Also, where there may be a need for an electric power source with no moving parts for quietness or long life, a nuclear reactor with natural convection cooling and thermionic conversion only, offers interesting possibilities.

The release of electrons from a hot body (thermionic emission) was first discovered by Edison in 1883. The thermionic diode rectifier was invented by Fleming in 1904. Thermionic conversion itself was first proposed by Schlichter [132] in 1915, but serious development of the idea began only in the early 1950's.

13-2. COMPONENTS OF A THERMIONIC CONVERTER

Figure 13-1 is a schematic of a thermionic converter. It consists of a heated metal *emitter* or *cathode* receiving heat Q_A and a cooled metal *collector* or *anode,* rejecting heat Q_R. The cathode and anode are sepa-

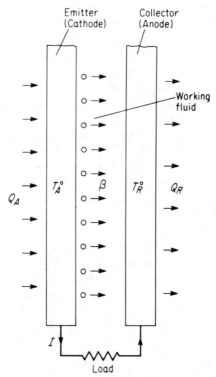

FIG. 13-1. Schematic of a thermionic energy converter.

rated by a working fluid. Electrons are emitted or boiled off the hot cathode and migrate toward the anode. There are, however, retarding forces at the surface of the emitter (below), and sufficient energy must be imparted to the electrons if they are to overcome these retarding forces, escape from the emitter and migrate to the collector surface. Part of the energy of the electrons reaching the collector is lost in the form of heat which must be removed from the collector and which makes up part of Q_R. Q_R is further made up of thermal radiation and conduction through the working fluid from the emitter. The balance of the energy of the electrons drives them through the external load and back to the emitter.

The working fluid separating the emitter and collector, or the electrodes, may simply be the electron gas, the so-called *vacuum converter* or *diode*; or it may be a partially ionized plasma—i.e., a mixture of neutral atoms and an almost-equal number of electrons and positive ions, the so-called *vapor-filled converter* or a *plasma diode*.

To understand the operation of a thermionic converter, it is necessary to consider several solid and surface phenomena, such as conduction electron energies, thermionic emission, and surface ionization and recombination; as well as space (or volume) phenomena, such as space-charge fields and plasma transport properties. Some of these phenomena are discussed in the next three sections. Others are discussed where appropriate.

13-3. SOLID PHENOMENA

The interior and surface of a metallic crystalline solid contain atoms that have nuclei surrounded by both electrons that are tightly bound to them (the shell model) and *valence* electrons that are weakly bound to them, are free to migrate, and are considered to belong to the entire crystalline solid rather than to any one particular atom. It is the valence electrons that concern us here. They are also called *conduction* electrons since they constitute the primary mechanism of electric and heat conduction in a metal. The conduction electrons in the *interior* of the metal are considered to be free, since, there are no *net* forces on them due to the other free electrons or the ionized nuclei and their bound electrons. They move in an essentially equipotential field, each having a constant electrostatic *potential energy* E_i which is independent of its location inside the crystal. At the surface boundary, however, there exists a potential energy *barrier*, since there are no positive ions on one side of the boundary to give the free electrons equal attractive forces. In other words, the free electrons are easily moved by electric fields within a metal, but it takes considerably more energy to boil them out of the metal into free space.

One therefore has an *electron gas*, which is confined within the metal [133]. It is not, however, equivalent to an ordinary gas whose energy distribution is given by the Maxwell-Boltzmann law. Electrons, instead, exist in states restricted by the so-called Pauli exclusion principle (which specifies that no two electrons in the same atom can exist in the same state at the same time) and their energy distribution is given by the *Fermi-Dirac* law. In temperatures that are not too high (not greater than 3000°K) the Fermi-Dirac distribution is given by

$$n(E)\,dE = \left[\frac{4\pi}{h^3}(2m_e)^{3/2}\right]\frac{E^{1/2}}{1 + e^{(E - E_F)/kT}}\,dE \qquad (13\text{-}1)$$

where $n(E)\,dE$ is the number of electrons per unit volume in the energy range dE. The quantities between brackets are constants: h is Planck's constant (6.625×10^{-34} Joule-sec), m_e is the mass of the electron (0.00055 amu). The balance of the right-hand side is energy dependent. k is the Boltzmann constant (1.38×10^{-23} Joule/molecule °K). The quantity E_F is called the *Fermi energy*. It is a constant for many cases of interest, being nearly independent of temperature.

A plot of the energy distribution of an electron gas, Eq. 13-1, is shown in Fig. 13-2 at different temperatures. The quantity $E^{1/2}$ contributes the parabolic rise of the curves from $E = 0$. The quantity

$$P(E) = \frac{1}{1 + e^{(E - E_F)/kT}} \qquad (13\text{-}2)$$

is called the *Fermi-Dirac probability distribution function*. It is interesting to note that, unlike a classical gas, free electrons do *not* all have zero energy at absolute zero, but that they have finite energies *up to a maxi-*

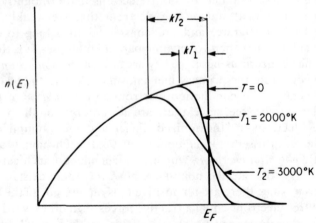

FIG. 13-2. Energy distribution of an electron gas at different temperatures.

mum given by E_F, the Fermi energy. At $T = 0$, the probability distribution function is 1 from $E = 0$ to $E = E_F$, meaning the probability that any state between energies 0 and E_F is occupied by an electron is 1, and is zero for $E > E_F$.

At higher temperature only the high-energy portion of the distribution is different from that at $T = 0$. The probability that states much less than E_F are occupied at higher temperatures is still 1, the extent depending upon the value of T. For $E = E_F$ the probability is exactly 1/2 (independent of temperature). For energies much greater that E_F, the exponential term is much greater than the 1 in the denominator of Eq. 13-2, and the probability reduces to the Maxwell-Boltzmann distribution probability (given by $Ae^{-E/kT}$, where A is a constant).

The Fermi energy can be computed by evaluating the total number of free electrons per unit volume, n, as

$$n = \int_0^{E_F} n(E)\, dE = \frac{4\pi}{h^3}\, (2m_e)^{3/2} \int_0^{E_F} E^{1/2}\, dE$$

$$= \frac{8\pi}{3h^3}\, (2m_e)^{3/2} E_F^{3/2}$$

from which

$$E_F = \frac{h^2}{2m_e} \left(\frac{3n}{8\pi} \right)^{2/3} \tag{13-3}$$

The values of E_F for metals are typically of the order of a few electron volts, being about 7 ev for copper and 3.1 ev for sodium. It can also be easily shown that the *average* energy of a free electron at $T = 0$ is equal to 3/5 E_F, or a few ev, much higher than the average kinetic energy of a particle in a classical gas even at high temperatures, which is given by 3/2 kT, thus being zero at $T = 0$ and only 0.03 ev at room temperature.

The behavior of free electrons at temperatures higher than 0 is most important to our discussion. At moderate temperatures, the corners of the zero temperature distribution, Fig. 13-2, are only slightly rounded. The difference increases as the temperature increases, becoming significant only within the energy range

$$|E - E_F| \simeq kT \tag{13-4}$$

and, as indicated previously, the probability that the state $E = E_F$ is occupied is 1/2. The free electrons whose energies are much less than the Fermi energy remain locked in the same energy states they occupied at $T = 0$ irrespective of temperature. A fraction of the most energetic electrons, those having energies within kT of the Fermi energy occupy higher energy states than the Fermi energy. Such electrons can be elevated to these energies by thermal excitation (heating).

13-4. SURFACE PHENOMENA

We talked earlier about the potential energy barrier at the surface of a metal. An electron leaving a solid surface is acted upon by a net positive charge in the interior, the so-called *image* force at the surface,

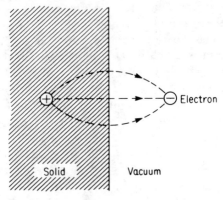

FIG. 13-3. Electrostatic field barrier between electron outside metal and its image inside.

due to the lack of positive ions outside the surface, Fig. 13-3. In order for the electron to overcome this force and not return to the surface, it must have sufficient energy, equal to the work required to remove it from the Fermi energy level to a point outside the metal (a short distance beyond the last ion). This energy, per unit charge, is called the *work*

TABLE 13-1
Thermionic Properties of some Metals

Metal	A amp/$m^{2o} K^2$ × 10^{-6}	φ, volts	Metal	A amp/$m^{2o} K^2$ × 10^{-6}	φ, volts
Copper, Cu		4.48	Platinum, Pt	0.32	5.36
Calcium, Ca		3.20	Potassium, K		2.25
Cesium, Cs	0.5	1.89	Rhenium, Re		5.00
Cromium, Cr		4.45	Selenium, Se		4.87
Galbium, Ga		4.16	Silver, Ag	0.60	4.08
Germanium, Ge		4.62	Sodium, Na		2.28
Iron, Fe		4.63	Tantalum, Ta	0.55	4.20
Manganese, Mn		3.95	Thorium	0.60	3.35
Molybdenum, Mo ...	0.55	4.20	Tungsten, W	0.60	4.53
Nickel, Ni	0.30	5.00	W + Cs	0.03	1.5
Niobium, Nb		4.00	Vanadium, V		4.11
Osmium, Os		5.50	Zinc, Zn		4.27

function of the metal, φ, and has the unit of volts. The work functions of some metals of interest are given in Table 13-1.

Electrons within a metal obtain kinetic energies by thermal excitation, and therefore only a few electrons possess the necessary escape energies when the metal is at low temperature. As the metal is made hotter and hotter, more and more electrons will have sufficient kinetic energies to overcome the electrostatic barrier and leave the surface. The rate of electron emission from a surface is therefore a function of the metal temperature, as well as the energy barrier at the surface. More electrons are emitted the higher the temperature and the lower the energy barrier. The energy barrier in electron volts, is equal to the product of the electron charge e and the work function φ in volts. The rate of electron emission is given by the *Richardson-Dushman* equation,

$$I = A\,T^2\,e^{-\left(\frac{e\varphi}{kT}\right)} \tag{13-5a}$$

where I = rate of electrons emitted in amp/m^2

A = constant thought initially to be a universal one with a value of 1.2×10^6, but actually varies considerably from metal to metal, given in Table 12-1, amp/$m^2\,{}^\circ K^2$

T = absolute temperature, ${}^\circ K$

e = electron charge = 1.6×10^{-19} coulomb

φ = work function, Table 13-1, vólts

K = Boltzman constant = 1.38×10^{-23} joules/${}^\circ$K

= 8.62×10^{-5} eV/${}^\circ$K

Introducing the numerical values of e and k, Eq. 13-5a can be written in the form*

$$I = A\,T^2\,e^{-\left(\frac{11,600\varphi}{T}\right)} \tag{13-5b}$$

Equations 13-5 show that for a high value of I, a material of low work function must be chosen and operated at as high a temperature as possible. Figure 13-4 shows a plot of I versus temperature for pure and treated tungsten.

The internal energy E_i that must, as a minimum, be possessed by an electron in order to escape from a metal surface, must be equal to the sum of the mean energy level of the electrons within the metal, the Fermi level, and the surface energy barrier $e\varphi$, Fig. 13-5, or

$$E_i = E_F + e\varphi \tag{13-6}$$

* The voltage equivalent of temperature is given by $V = \dfrac{kT}{e} = \dfrac{T}{11,600}$, where V is in volts and T in degrees Kelvin.

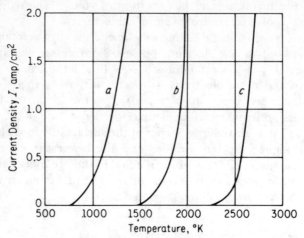

FIG. 13-4. Electron current density-temperature emission characteristics for tungsten, (a) oxide coated, (b)
thoriated, and (c) pure tungsten.

Electrons which possess energy barriers in excess of $e\varphi$ above the
Fermi level will escape. All others will return. The measure of the

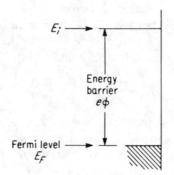

FIG. 13-5. Energy-level
diagram of electron attemp-
ting to leave metal surface.

barrier is of course the work function φ in volts (since e is constant) and
is seen to vary between 1.89 for cesium to 5.5 for osmium in Table 13-1.
Suitable surfaces with low work functions are often obtained by combining metals, such as by using a thin coating of cesium on tungsten.

If a metal is heated sufficiently so that kT becomes comparable to
its work function, some of the free electrons will have energies equal

to or exceeding the internal potential energy and thus escape the surface, Fig. 13-6. This is thermionic emission.

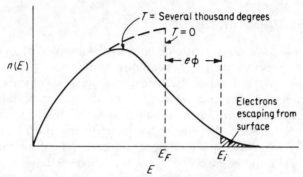

FIG. 13-6. Electron energy distribution at high temperatures showing electrons with sufficient energies to escape from surface.

13-5. SPACE PHENOMENA

Now that some free electrons are out of the metal, they encounter a new problem in the space between the emitter and the collector. As they leave the emitter, they form a "cloud" of negative charges near the surface of the emitter. This cloud repels electrons subsequently emitted back to the emitter surface. This is the so-called *space-charge effect*. The space-charge effect is counteracted by (a) reducing the gap between the electrodes resulting in the so-called vacuum diode, (b) by filling the space between the electrodes with a plasma, resulting in the so-called plasma diode, or (c) by using crossed electric or magnetic fields. The first two methods are discussed in the following sections. The third has not yet resulted in practical thermionic conversion.

When the electrons are finally captured by the collector they give up an amount of energy equal to the work function φ of the collector. They then must flow through an external load back to the emitter. To do this a voltage difference must exist across the two surfaces. In thermo-electricity, Sec. 14-3, the voltage difference is caused by the different electrical properties of p and n semiconductors. In thermionics, both the emitter and collector are good metallic conductors, rather than semiconductors. The voltage difference is thus obtained by choosing the emitter and collector such that the emitter would have a higher work function than the collector. Characteristically, the difference is about 3/4 volt. The hotter emitter thus has a larger population of electrons in the higher energy levels than the collector. This difference in electron density is the driving force that produces the useful voltage. It is, therefore, the difference in work functions that makes the thermionic converter a power device.

13-6. THE VACUUM DIODE

A *vacuum diode* is a thermionic converter composed of an emitter and collector, henceforth called the cathode and anode, separated by the electron gas only but otherwise vacuum. Such a converter uses very close spacing between the electrodes, of the order of microns (it is sometimes called the *close-spaced diode*), to overcome the space charge of the electrons between the electrodes by reducing the number of such electrons. Such close spacings pose severe practical limitations. Nevertheless, an analysis based on a vacuum diode is basic to thermionic conversion and gives an insight into the operation of other types.

Figure 13-7 shows a typical potential diagram between cathode and anode of a vacuum diode. The ordinates of the diagram are in volts.

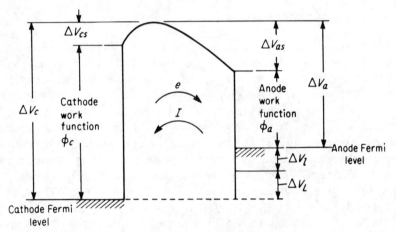

FIG. 13-7. Potential diagram of a thermionic vacuum diode.

(Corresponding energies in electron volts are obtained by multiplying the ordinates by e, the constant charge on the electron, so that the diagram could also be thought of as an energy-level diagram.) The potential of emitted electrons, above the cathode Fermi level, must, at a minimum, be equal to the cathode work function φ_c. However because of the presence of a cloud of electrons near the cathode surface electrons need an extra potential, ΔV_{cs}, called the *space-charge potential barrier* to overcome the space-charge barrier so that they may not be repelled and returned back to the cathode. There, therefore, is an *effective* cathode work function ΔV_c given by

$$\Delta V_c = \varphi_c + \Delta V_{cs} \tag{13-7}$$

The current leaving the cathode is now obtained by using ΔV_c in place of φ in Eqs. 13-5. An electron possessing ΔV_c succeeds in overcoming the hump or *potential peak* in the diagram and is accelerated towards the anode. There it liberates an effective anode potential ΔV_a (and an energy $e\Delta V_a$) until it reaches the Fermi level of the anode. This is given by

$$\Delta V_a = \varphi_a + \Delta V_{as} \tag{13-8}$$

If the anode work function is, as shown, smaller than the cathode work function, there will be a net potential difference which can be connected to a useful load, ΔV_L between the two electrodes. Each electron thus going through a load liberates useful energy $e\Delta V_L$ but also energy loss through electrical leads due to their electrical resistance, ΔV_l, where ΔV_l is the voltage drop through the leads, before returning to the cathode Fermi level.

The *net* current I_n given by the diode is now given by

$$I_n = A T_c^2 e^{-\left[\frac{e(\phi_c + \Delta V_{cs})}{kT_c}\right]} \tag{13-9a}$$

$$= A T_c^2 e^{-\left[\frac{e(\varphi_a + \Delta V_{as} + \Delta V_l + \Delta V_L)}{kT_c}\right]} \tag{13-9b}$$

This assumes that the reverse current, from anode to cathode, is negligible due to the lower temperature of the anode. Equation 13-9b shows that for a given output voltage and space charge, the net current is only a function of the anode work function and independent of that of the cathode.

The load voltage ΔV_L is dependent upon the difference between the work functions of the electrodes, and therefore an output voltage of only a few volts at best can be expected as can be seen from Table 13-1. This points to one of the difficulties encountered with thermionic conversion, namely the conflicting requirements of large output current and large output voltages. As can be seem from Eqs. 13-5, the former requires a low cathode work function while the latter a large one.

Another and greater difficulty is that due to the space charge and its potentials ΔV_{cs} and ΔV_{as} for they reduce both the output voltage and current as can be seen by examining Fig. 13-7 and Eqs. 13-5 and 13-7. Vacuum diodes attempt to reduce this by using extremely close spacings between the electrodes.

The Interelectrode Potential Profile

Figure 13-7 showed that the hump of the interelectrode potential profile occurs near the cathode. The exact position of this hump (which really corresponds to a potential minimum that may occur at either or between the electrods) would materially affect the choice of work functions which would result in reasonable output voltages. It can be seen, by examining Fig. 13-8, that a greater difference in work functions must be chosen if the hump occurred at the cathode than at the anode, for the same diode output voltage $\Delta V_l + \Delta V_L$ and the same ΔV_c and ΔV_a.

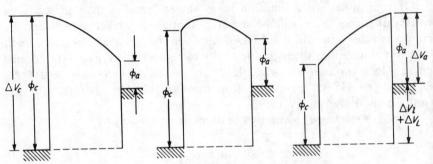

FIG. 13-8. Three interelectrode potential profiles for the same ΔV_c, ΔV_a and $\Delta V_l + \Delta V_L$.

Langmuir [134] presented a solution of the interelectrode potential-distance relationship, based on a Poisson distribution. The solution is given by the normalized plot of Fig. 13-9, in which the dimensionless ordinate ψ and abscissa ζ are related to potential and distance by

$$\psi = \frac{eV(x)}{kT_c} = \frac{11,600}{T_c} V(x) \tag{13-10}$$

and

$$\zeta = 9.18 \times 10^5 \frac{I_n^{1/2}}{T_c^{3/4}} x \tag{13-11}$$

where $V(x)$ is in volts, T_c in $°K$, I_n in amp/cm^2 and x in cm, measured from the point of minimum potential. If the work function and the output voltage are known, the difference between the space-charge potentials at the cathode and anode will be known. Values of these potentials can then be assumed and the distances to the electrodes obtained from Fig. 13-9. For space charges of the order of half a volt, the total spacing is very small, being of the order of microns.

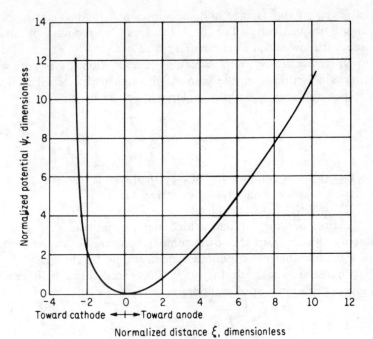

FIG. 13-9. Normalized interelectrode potential vs. distance for a vacuum diode.

13-7. EFFICIENCY OF A VACUUM DIODE

The primary factors entering into the design of diodes are the spacing between electrodes, already discussed, and the temperature of the cathode. The latter should be as high as possible for maximum current, but limited by the metallurgical limitations of the cathode, as well as the heat losses by radiation. The efficiency is limited by power losses. These are:

(a) *Radiation heat losses.* These are the energy losses, just mentioned, from cathode to anode due to the large temperature difference between them. If the cathode and anode are to be considered as two infinite parallel plates (a good assumption because of the close spacing), radiation heat flux loss P_r (energy per unit area) would be given by the familiar equation

$$P_r = \sigma \frac{(T_c^4 - T_a^4)}{\dfrac{1}{\epsilon_c} + \dfrac{1}{\epsilon_a} - 1} \tag{13-12}$$

where σ is the Stephan-Boltzmann constant (5.67×10^{-8} watt/m²°k⁴), ϵ

the emissivity of the surface and the subscripts c and a refer to cathode and anode respectively. The heat loss may be reduced by choosing materials with low surface emissivities.

(b) *Heat conduction and I^2R losses.* Conduction losses are the heat losses by conduction from the cathode through its electrical lead to the load, P_k per unit area of cathode. They are given by

$$P_k = \frac{1}{A_c} k_l A_l \frac{T_c - T_L}{l} \tag{13-13}$$

where A_c = area of cathode
    k_l = thermal conductivity of electrical lead
    A_l = cross-sectional area of electrical lead
    l = length of electrical lead
    T_L = temperature of load, which may be ambient
Conduction losses from the lower-temperature anode are ignored. To reduce P_k, a low thermal conductivity lead should be chosen.

I^2R losses are the Joule losses, P_j, dissipated through the electrical lead, given, per unit area of cathode, as

$$P_j = \frac{1}{A_c} (I_n A_c)^2 R_l \tag{13-14}$$

where R_l is the electrical resistance of the lead. Half this loss is assumed to flow toward the cathode and the other half towards the load.

The combined conduction and I^2R losses $P_{k,j}$, from the cathode, per unit area of cathode, are therefore given by

$$P_{k,j} = \frac{1}{A_c} \left[k_l \frac{A_l}{l} (T_c - T_L) - \frac{1}{2} (I_n A_c)^2 R_l \right] \tag{13-15a}$$

Recognizing that

$$R_l = \rho_l \frac{l}{A_l}$$

where ρ_l is the electrical *resistivity* of the lead, Eq. 13-13a may be written in the form

$$P_{k,j} = \frac{1}{A_c} \left[\rho_l k_l (T_c - T_L) \frac{1}{R_l} - \frac{1}{2} (I_n A_c)^2 R_l \right] \tag{13-15b}$$

The electrical and thermal conductivities of metals go hand in hand. They are related by the *Wideman-Franz* law [133] which states that the product of the electrical resistivity and thermal conductivity are a direct function of temperature:

$$\rho_l k_l = \frac{\pi^2}{3} \left(\frac{k}{e}\right)^2 T_l \tag{13-16}$$

where k is the Boltzmann constant and e the charge on the electron. T_l is the lead temperature which may be taken as the arithmetic average of the cathode and load temperatures, or

$$T_l = \frac{T_c + T_L}{2} \tag{13-17}$$

The *efficiency* of the diode η can now be formulated as

$$\eta = \frac{P_L}{P_e + \text{losses}} = \frac{P_L}{P_e + P_r + P_{k,j}} \tag{13-18}$$

where P_L is the useful load power per unit cathode area, given by

$$P_L = I_n \Delta V_L \tag{13-19}$$

P_e is the potential energy imparted to the electrons plus their average kinetic energy in terms of the cathode temperature. P_e, per unit area of cathode, is given by

$$P_e = I_n \left(\Delta V_c + \frac{2kT_c}{e}\right) = I_n \left(\Delta V_L + \Delta V_l + \Delta V_a + \frac{2kT_c}{e}\right) \tag{13-20}$$

in which

$$\Delta V_l = I_n A_c R_l \tag{13-21}$$

The efficiency is now written by combining Eqs. 13-18 and 13-21 to give

$$\eta = I_n \Delta V_L \bigg/ \left[I_n \left(\Delta V_L + I_n A_c R_l + \Delta V_a + \frac{2kT_c}{e}\right) \right.$$

$$\left. + P_r + \frac{\pi^2}{6} \left(\frac{k}{e}\right)^2 \frac{(T_c^2 - T_L^2)}{A_c} \frac{1}{R_l} - \frac{1}{2} I_n^2 A_c R_l \right] \tag{13-22}$$

It can be seen that R_l (which includes the lead geometry factor l/A_l) affects the denominator of Eq. 13-22 in different ways. R_l also enters into the expression for the current (via ΔV_l in Eq. 13-9b). The efficiency is maximized with respect to R_l and R_L (or ΔV_l and ΔV_L) by obtaining the derivatives of $1/\eta$ with respect to these quantities and equating to zero. The result [135] gives the maximum efficiency as

$$\eta_{\text{max}} = \frac{1}{1 + \Gamma} \tag{13-23}$$

and the corresponding lead resistance R_l^*, load resistance R_L^* and current I_n^* for operation at maximum efficiency as

$$R_l^* = \frac{1}{P_r A_c} \left(\frac{kT_c}{e}\right)^2 \left[\frac{\pi\Gamma}{\left(\frac{3}{2} + 3\Gamma\right)^{1/2}}\right] \tag{13-24}$$

$$R_L^* = \frac{1}{P_r A_c} \left(\frac{kT_c}{e}\right)^2 \left[2 + \frac{\Delta V_a e}{kT_c} + \Gamma + \frac{\pi(1 + \Gamma)}{\left(\frac{3}{2} + 3\Gamma\right)^{1/2}}\right] \tag{13-25}$$

and

$$I_n^* = \frac{P_r e}{kT_c} \cdot \frac{1}{\Gamma} \tag{13-26}$$

where

$$\Gamma = \frac{\dfrac{\Delta V_a e}{T_c k} + 2 + \pi\sqrt{\dfrac{2}{3}(1 + 2\Gamma)}}{\ln\left(\dfrac{A T_c^3 k}{P_r e}\right) + \ln\Gamma - \dfrac{\Delta V_a e}{T_c k} - 1} \tag{13-27}$$

Example 13-1. An optimized thermionic diode produces 150 amp with a cathode area of 10 cm². The cathode and anode operate at 1800 and 700°K and have readiant emissivities of 0.333 and 0.667 respectively. The cathode material has $A = 1.2 \times 10^6$ amp/m²°K². The electric lead is 3 cm long and made of copper with electrical resistivity of 1.72×10^{-8} ohm-m. Find (a) the electric lead diameter, (b) the load resistance, (c) the power output, (d) the efficiency of the diode, and (e) the number and arrangement of such diodes necessary for a 1 Kw load at 34 volts.

Solution. The radiant loss, from Eq. 13-12.

$$P_r = 5.67 \times 10^{-8} \frac{1}{\dfrac{1}{0.333} + \dfrac{1}{0.667} - 1} (1800^4 - 700^4)$$

$$= 1.662 \times 10^5 \text{ watt/m}^2$$

$$I_n = \frac{150}{10 \times 10^{-4}} = 1.5 \times 10^5 \text{ amp/m}^2$$

Using Eq. 13-26;

$$\Gamma = \frac{1.662 \times 10^5 \times 11,600}{1.5 \times 10^5 \times 1800} = 7.139$$

(a) From Eq. 13-24,

$$R_l^* = \frac{1}{1.662 \times 10^5 \times 10^{-3}} \left(\frac{1800}{11,600}\right)^2 \left[\frac{\pi \times 7.139}{\left(\frac{3}{2} + 3 \times 7.139\right)^{1/2}}\right]$$

$$= 0.677\,5\, \textrm{¡}\, 0^{-3}\ \textrm{ohm}$$

Since $R_l = \rho_l \dfrac{l}{A_l}$, then

$$A_l = \frac{1.72 \times 10^{-8} \times 3 \times 10^{-2}}{0.667 \times 10^{-3}} = 7.62 \times 10^{-7}\ \textrm{m}^2$$

$$\textrm{lead diameter} = \sqrt{7.62 \times 10^{-7} \times \frac{4}{\pi}} = 9.85 \times 10^{-4}\ \textrm{m}$$

$$= 0.985\ \textrm{mm}$$

(b) From Eq. 13-27,

$$7.139 = \frac{\left(\dfrac{\Delta V_a \times 11,600}{1800}\right) + 2 + \pi \sqrt{\dfrac{2}{3}(1 + 2 \times 7.139)}}{\ln \dfrac{1.2 \times 10^6 (1800)^3}{1.662 \times 10^5 \times 11,600} + \ln 7.139 - \left(\dfrac{\Delta V_a \times 11,600}{1800}\right) - 1}$$

from which $\Delta V_a = 3.834$ volt.
From Eq. 13-25

$$R_L^* = \frac{1}{1.662 \times 10^5 \times 10^{-3}} \left(\frac{1800}{11,600}\right)^2 \left[2 + \frac{3.834 \times 11,600}{1800}\right.$$

$$\left. + 7.139 + \frac{\pi(1 + 7.139)}{\left(\frac{3}{2} + 3 \times 7.139\right)^{1/2}}\right] = 5.66 \times 10^{-3}\ \textrm{ohm}$$

(c) Power output $= I_n^2 R_L = (150)^2 \times 5.66 \times 10^{-3}$
$$= 127.35\ \textrm{watt}$$

(d) efficiency $= \dfrac{1}{1 + \Gamma}$ \qquad Eq. 12-23

$$= \frac{1}{1 + 7.119} = 0.123 = 12.3 \textrm{percent}$$

(e) Number of diodes to produce 1 kw

$$= \frac{1000}{127.35} = 7.85$$

or 8 diodes would be needed. Since each produces a voltage equal to $I_n R_L = 8.5$

volts; and since $34/8.5 = 4$, a network composed of two parallel circuits, each containing 4 diodes connected in series, must be used.

13-8. THE PLASMA DIODE

In vacuum diodes, the space-charge problem was reduced by using very narrow spacings between the electrodes of the order of microns. Such close spacings, coupled with high temperatures, raise problems of operational stability during the lifetime of the diode. The vacuum converter generates low current densities, and is not considered seriously for power production. Instead, *gas-filled* or *plasma diodes* are used.

In the plasma diode, the space-charge problem is attacked by neutralizing the cloud of electrons between the electrodes by positive ions. A gas is introduced and made into a partially-ionized plasma (a mixture of neutral atoms and an equal number of electrons and positive ions) so that positive ions are continuously generated with sufficient density to neutralize the electron cloud electrostatic field. An electron, leaving the cathode, will then not be retarded by a space charge and will travel freely to the anode. Plasma diodes generate much larger current densities and are built with wider spacings than vacuum diodes.

The gas must have a low ionization potential (lower than the cathode work function). The "gas" found most suitable is *cesium vapor* which has the lowest ionization potential of 3.87 volt. (Others are the vapors of *Rb* and *K* with 4.16 and 4.32 volt.) It is obtained by placing a small quantity of cesium in a reservoir, Fig. 13-10. Cesium will vaporize and the vapor will fill the interelectrode spacing. The vapor pressure, and consequently the concentration of cesium atoms, depend upon the temperature of cesium in the reservoir, and can be regulated by changing that temperature. Another very important function, besides supplying positive ions, cesium, which has a work function of only 1.89 volts, Table 13-1, will *cesiate* or condense on the cooler anode surface (since its vapor pressure corresponds to a saturation temperature higher than the temperature of the anode), and therefore contributes a desirable low work function to the anode.

Complete neutralization of the space charge requires that the densities, not currents, of ions and electrons in the spacing be equal. The currents are proportional to the average speeds according to

$$\frac{I_e}{I_i} = \frac{n_e e \, \bar{u}_e}{n_i e \, \bar{u}_i} \tag{13-28a}$$

where n, e, and u are the density, particle charge, and mean speed, respectively, and the subscripts e and i are for the electrons and ions re-

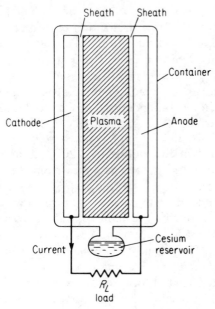

FIG. 13-10. A cesium vapor-filled
plasma diode.

spectively, and since n_e and n_i should be equal, we have the requirement

$$\frac{I_e}{I_i} = \frac{\bar{u}_e}{\bar{u}_i} \qquad (13\text{-}28b)$$

Considering the ions and electrons to be in thermal equilibrium, their kinetic energies will be equal, so that

$$\frac{1}{2}\, m_e \bar{u}_e^2 = \frac{1}{2}\, m_i \bar{u}_i^2 \qquad (13\text{-}29)$$

where m_i and m_e are the masses of ions and electrons respectively. Combining Eqs. 13-28 and 13-29 gives the ratio of currents required for complete neutralization as

$$\frac{I_e}{I_i} = \sqrt{\frac{m_i}{m_e}} \qquad (13\text{-}30)$$

For cesium, the above ratio is about 490, so that for complete space-charge neutralization, one ion must be generated for each 490 electrons leaving the cathode, and the electrons experience no retarding electrostatic fields. The ratio

$$\beta = \frac{I_i}{I_e} \sqrt{\frac{m_i}{m_e}} \qquad (13\text{-}31)$$

is called the *ion-richness ratio*. $\beta = 1$ means complete neutralization while $\beta > 1$ means complete neutralization but with an excess of ions, and $\beta < 1$ means incomplete neutralization. The method of generation of the positive cesium ions is of interest. There are actually two mechanisms: *contact* (or *resonance*) *ionization,* and *volume ionization.*

Contact ionization takes place when an atom comes into contact with a hot surface if the ionization potential of the atom is lower than the work function of the surface. (Cs, Rb and K vapors which have ionization potentials of 3.87, 4.16 and 4.32 volts can therefore be used with Ta, Mo and W which have work functions of 4.20, 4.20 and 4.53 volts). The valence electron of the gas atom detaches from the atom and attaches instead to the surface material. If the surface is hot enough, the electron is then emitted, and an electron-ion pair are produced at the surface. The rate of production is dependent upon the cesium vapor pressure, which in turn is dependent upon the cesium reservoir temperature. It has been found that for the most effective rate of production the cathode temperature must at least be 3.6 times the reservoir temperature. Operation with contact ionization only is said to occur in the *unignited mode.*

Volume ionization occurs when the vapor pressure of cesium is high enough so that the electron mean free path is small compared to the spacing between the electrodes. In this case, the electrons will experience collisions as they travel from cathode to anode. By contrast, pure contact ionization occurs when the pressure is so low that the mean free path is large enough compared to the spacing, that no collisions take place. Usually the rate of current output obtained with contact ionization alone is not attractively high, except at cathode temperatures higher than 1900°K. Such temperatures pose metallurgical problems, and the unignited mode of operation is not attractive for power production.

With "high-pressure" plasmas, at about 1 mm Hg (torr), both contact and volume ionization occur and operation is said to be in the *ignited* or *arc mode.*

13-9. PLASMA DIODE CHARACTERISTICS

The potential profiles of a plasma diode show that only a small potential gradient occurs across the plasma itself to maintain charge flow. There are two narrow gaps, however, between the plasma and each electrode, called *sheaths,* Fig. 13-10, which contain either electrons or ions and consequently large potential gradients. The magnitude and sign of the gradients depend upon the current flow as well as the ion richness ratio β, Fig. 13-11.

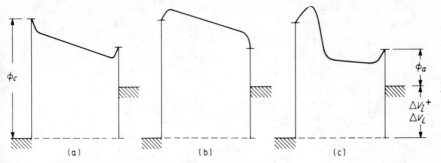

FIG. 13-11. Potential profiles of plasma diodes when the space charge is (a) completely neutralized, $\beta \geqslant 1$, with ion-rich sheaths, (b) partially neutralized, $\beta < 1$, with electron-rich sheaths, and (c) partially neutralized with operation in the obstructed regime.

While no satisfactory theoretical model has yet been developed for the ignited mode and it is not as well understood as the unignited mode (mainly because of lack of understanding of the exact ionization mechanisms), it remains the only acceptable means of obtaining adequate power densities at reasonable cathode temperatures, about and below 1800°K.

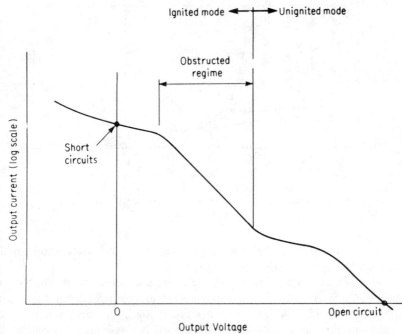

FIG. 13-12. Output current-voltage characteristics of a cesium plasma diode.

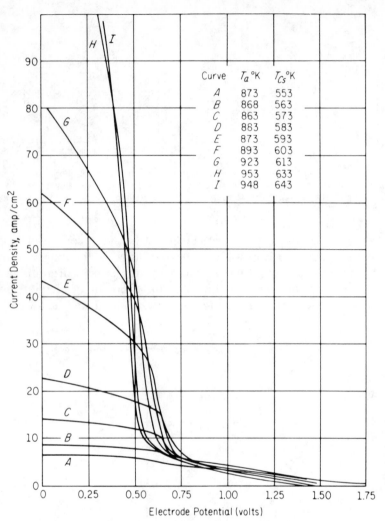

FIG. 13-13. Output current density-voltage curves for a thermionic converter at various collector temperatures. (After Lawrence and Perdew, Ref. 135.)

At these temperatures, the output current of an unignited mode is insignificant compared to that of an ignited mode.

The output current-voltage characteristics of a plasma diode is composed of the two modes as well as several regimes depending upon loading, Fig. 13-12. At zero (open-circuit) loading, operation is in the unignited mode, since no volume ionization is possible. As load is applied and current is increased, voltage drops, first slowly then more rapidly, until the ignited mode is reached. The most important regime is called the *obstructed regime*. In it the product of current and voltage is maximum. The obstructed regime is characterized by a potential profile with an electron-rich double sheath near the cathode, Fig. 13-11c. The electrons crossing the double sheath near the cathode are acted upon by a large accelerating potential and thus have sufficient energy for volume ionization. The resulting ions are accelerated toward the double sheath causing liberation of more electrons.

With further loading (more current), the double sheath disappears, the space charge is completely neutralized and the potential profile is similar to that of Fig. 13-11a. Operation, to the left of the obstructed regime, called the *quasi-saturation regime,* is characterized by a much slower rate of current increase with voltage drop.

The extent of the obstructed regime depends on many factors. Figure 13-13 shows results of experiments by Lawrence and Perdew on a cesium plasma diode with a tungsten cathode and nickel anode, at one

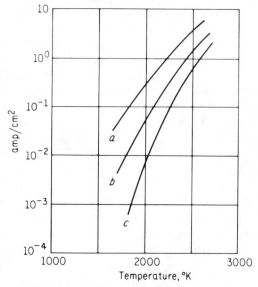

FIG. 13-14. Effect of temperature and cesium vapor pressure on current in a plasma diode with tantalum cathode. Cesium pressures: (a) 0.4, (b) 10^{-2}, (c) 10^{-5} mm Hg.

cathode temperature of 1855°K and various anode and cesium pool temperatures as indicated. It can be seen that current densities of 10–100 amp/cm² are obtained at voltages ranging from 0.3 to 0.7 volt, corresponding to power densities of 7–30 watt/cm². Higher currents

and voltages at efficiencies approaching 30 percent are now being reported. Figure 13-14 shows the effect of cathode temperature and cesium
vapor pressure on current density of a cesium diode using a tantalum
cathode.

While the spacing between electrodes in a plasma diode is not as
critical as for a vacuum diode, it should neither be so large as to materially exceed the mean free path of the electrons, nor too small such that
the electrons will not engage in sufficient ionization collisions. Indeed
it has experimentally been found that there is an optimum electrode
spacing, around 50 microns (0.05 mm), compared to a few microns for
a vacuum diode. The diode of Fig. 13-13 has a spacing of 0.002 in.
(0.0508 mm).

13-10. HEAT SOURCES FOR THERMIONIC CONVERTERS

Present-day practical converters are achieving efficiencies less than
one-half Carnot. Despite this, the advantageous features of thermionics,
including the absence of rotary machinery, compactness, high-temperature operation, quietness, etc., are attracting interest for potential applications with various energy sources. They show the most promise for
space applications where high-temperature heat addition and rejection
are necessary, and where higher outputs than those obtainable with
thermoelectric devices are desired.

The heat sources for thermionic converters can be chemical, solar
or nuclear. All must provide temperatures in the range 1000 to 2000°C.
Chemical (fossil fuel) sources are of course common and easily handled,
but have the disadvantages of bulk and mass, and therefore limited life,
especially for space applications. They also require regular maintenance
to avoid poisoning converter elements by their products. The solar
source is attractive for space applications since it is not life-limited.
It however requires parabolic reflectors to concentrate the energy to
obtain the required operating temperatures. On earth, solar energy is
less intense than in space and is affected by periodicity (night and day),
weather limitations, and sun-tracking requirements.

Nuclear sources using either reactor or isotopic fuels are receiving
the most serious considerations. They have the advantages of relatively
long life, low maintenance requirements, and small size. The main
disadvantage is in the radioactivity associated with them, which gives rise
to heavy masses because of the necessity for shielding. While both
reactor and isotopic fuels are being considered, the latter are the less
attractive of the two because their constantly decaying energies pose

temperature-control problems. With reactors, there are three general concepts for combining thermionic converters and reactor heat, all based on the placement of the thermionic cells relative to the reactor core. They are described below.

(a) The *in-pile* concept, in which the cells are placed inside the core and their emitters (cathodes) are heated directly by fissioning fuel within each cell. This concept allows higher power densities over a broad range of cell diameters. Figure 13-15 shows a schematic of an in-pile system for space applications. The in-pile cathodes are heated directly by the fuel and the anodes are cooled by liquid metal which flows from

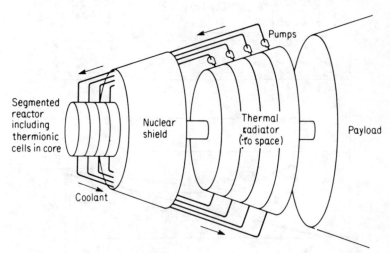

FIG. 13-15. Schematic drawing of an in-pile nuclear thermionic power system for space applications.

the anodes in the core to a space radiator for heat rejection and then pumped back to the anodes. A shield protects the radiator and payload. It may be added that radiators of thermionic systems are smaller than for other space power systems because of higher heat rejection temperatures.

(b) The *surface* concept, in which the thermionic cells are placed in contact with the outside surface of a compact core and heated by conduction from the core fuel. Heat is rejected directly from the surrounding anode to space. This concept results in simple geometry but a greater mass-to-power ratio than the in-pile concept.

(c) The *out-of-pile* concept, where the cells are placed external to the core with the emitters heated by a liquid-metal coolant loop or by heat pipes* from the core, Fig. 13-16. Heat is rejected through an additional liquid metal loop to the radiator. This concept results in a

> * A heat pipe is a simple, highly efficient heat-transfer device that can transfer several hundred times as much heat as metal conductors of the same cross section. It consists of a closed tubular chamber with a capillary-wick structure on the inner wall and a working fluid. Heat is transferred by evaporating the fluid in the heating zone and condensing it in the cooling zone. Circulation is then completed by return flow of the condensate to the heating zone through the capillaty structure. Because of this two-phase action, the heat pipe is nearly isothermal with little loss along its length, and is therefore highly efficient. Heat pipes can be straight, curved, circular, or corkscrew in. shape, with round, square, elliptical, or rectangular cross sections. They can be made

Schematic and Pressure Characteristics of a Heat Pipe

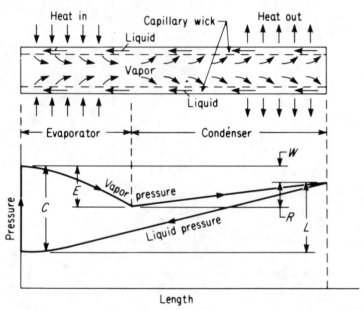

C = Capillary driving pressure E = Evaporator pressure drop
R = Vapor pressure recovery W = Wall friction pressure drop
L = Liquid pressure drop

to go over, under or around components. They can be built to operate over several temperature ranges, from cryogenic to about 3600°F, although no single heat pipe operates efficiently over this entire range. The working fluid can be liquid oxygen, nitrogen or helium in the cryogenic range, water ($\simeq$ 170°F), potassium ($\simeq$ 1100°F), sodium ($\simeq$ 1800°F) and lithium, lead or silver ($\simeq$ 3000°F). A heat pipe less than 1 in. in diameter using water can transport 1 Kw(t). With a metallic fluid it can transport up to 10 Kw(t). In the high-temperature ranges, the wick material must be compatible with the fluid, and refractory metals are often used. The heat-pipe principle was first conceived by R. S. Gaugler of General Motors in 1942 and applied by G. M. Groven of Los Alamos Scientific Laboratory in 1963 as a method of transferring reactor heat in space power systems. There is now widespread interest in heat pipes, especially for space, reactor, and thermionic applications.

reactor core of "conventional" design. The difficulty, common to many space systems, is the temperature limit imposed by the liquid-metal corrosion limit as well as the high-temperature strength characteristics of materials. The intermediate loop, thus results in lower cathode temperatures and therefore lower thermionic power densities than the in-pile

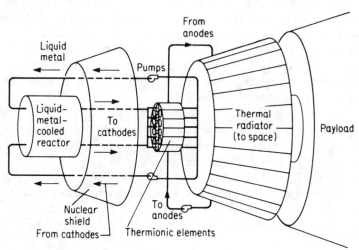

FIG. 13-16. Schematic drawing of an out-of-pile nuclear thermionic power system for space applications.

concept where fuel and cathode are in contact. This also results in higher overall mass-to-power ratios than the in-pile concept.

The in-pile concept has been more fully studied than the others and will be discussed further in the next section.

13-11. THE IN-PILE THERMIONIC CONVERTER

The incorporation of thermionic cells within a reactor must of course take into account the design problems of the cells themselves, the coolant, and the nuclear characteristics of the reactor. The reactor type most suitable for in-pile thermionics is the compact fast reactor. This is because the cells are best cooled by a coolant at 700-1000°C, which is a temperature level much higher than the temperature limitations of any efficient moderators of would be thermal reactors. Refractory materials can also be used without severe neutron resonance absorption in the fast neutron spectrum. The fuel materials UC, UC-ZrC, UO_2 and UO_2-W are used. Critical reactor diameters for these materials (with 25 percent fuel volume fraction, 3-in. Be reflector and height/diameter ratio of 2) range from 20 to 25 in. [137]. These diameters are relatively insensitive to materials other than the fuel.

UC can be used as an unclad emitter. It has a low work function and high electrical and thermal conductivity. It gives about 15 amp/cm^2 at 2000°C surface temperature. The addition of zirconium carbide ZrC to UC improves its resistance to thermal stress without appreciable effect on its electron emission characteristics. Unclad UC, however, is life-limited by fission-gas retention problems which cause distortion, and by evaporation at high temperatures. These limitations are overcome by cladding it in refractory metals, but with the disadvantage of increasing the work function and thus lowering electron emission. This is partially offset by the action of cesium on the surface.

A conceptual design of an in-pile thermionic reactor [136] is made of segmented reactor elements coupled to segmented radiators as suggested by Fig. 13-15. Each reactor segment, Fig. 13-17, may be further subdivided into a double row of unit thermionic cells. In this design, the cell anodes are cooled by liquid metal in crossflow heat-exchanger fashion. Each cell contains an encapsulated UO$_2$ fuel element in the center, with the cladding surface acting as cathode. The fuel is electrically isolated from the anode by an insulator. The anode is a trilayer tube made of a

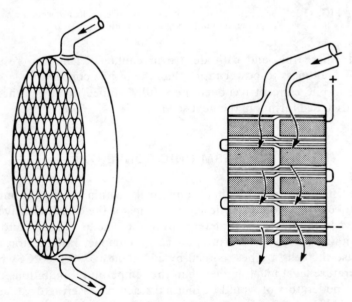

FIG. 13-17. Reactor segment showing individual thermionic cells in crossflow heat exchanger in in-pile thermionic reactor concept. (Ref. 136.)

thin layer of aluminum oxide sandwiched between layers of niobium. The anode is cooled by the liquid metal loop. Cesium is introduced to the interelectrode space from a cesium reservoir or cesium dispenser material. Each thermionic cell develops about 3/4 volt. Each coaxial pair is connected in series forming a module developing about 1.5 volts. The modules are then connected in a series-parallel network to provide the voltage-current characteristics required by the load.

There are two possible in-pile thermionic cell arrangements: (a) The radial internal fuel, and (b) the radial external fuel arrangements. In the internal arrangement (Fig. 13-18a) a cylindrical fuel element and cathode

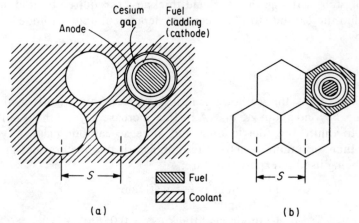

FIG. 13-18. In-pile thermionic cell arrangements: (a) internal fuel, (b) external fuel. (Ref. 137.)

cladding are surrounded by an annular cesium gap and a hollow cylindrical anode. The coolant fills the void between such cells. In the external arrangement (Fig. 13-18b) the fuel has a hexagonal cross section and surrounds an annular cesium gap, a hollow cylindrical anode, and a central circular coolant channel. An analysis of the two geometries [137] shows that the internal arrangement of Fig. 13-18a yields higher output, in watts per unit axial length, for the same fuel volume fraction, and that the difference increases with cell diameter.

Perry and Homeyer [137] derived the following relationship, based on a simplified model of an internal arrangement, between the total reactor electrical power, the power density and the cell and reactor critical dimensions:

$$P = \left[\frac{\pi^2}{2\sqrt{3}} \cdot \frac{D_c}{S^2} \cdot \frac{l_c}{(l_c + s)} HD^2\sigma \right] \tag{13-32}$$

where P = power, watt

$\quad D_c$ = cathode diameter, cm

$\quad S$ = spacing between thermionic cell centers, cm

$\quad l_c$ = cathode axial length, cm

$\quad s$ = axial spacing between cells, cm

$\quad H$ = critical reactor height, cm

$\quad D$ = critical reactor diameter, cm

$\quad \sigma$ = net average thermionic power density, watt/cm²

Taking into account the fact that the power density, cathode length, and material and thickness of the cathode cladding are related by the joule electrical losses, it was shown, based on a power density of 10 watt/cm² and tungsten cladding, that the clad thickness t_c required to hold the losses within the cathode to 10 percent are dependent upon cathode length l_c as follows:

$$t_c, \text{ in.} \quad 0.02 \quad 0.04 \quad 0.06 \quad 0.08$$
$$l_c, \text{ in.} \quad 1.70 \quad 2.20 \quad 2.70 \quad 3.00$$

This shows that the cathode length increases only slightly with clad thickness beyond about 2.5 in. A further increase beyond that point only serves to dilute the fuel loading and increase cathode mass. Typical, rather than optimum, reactor thermionic cell parameters that are currently being used in experiments are:

Power density = 5 watt/cm²

Cell length = 2.5 in.

Cathode clad thickness = 0.05 in.

Axial spacing = 1.0 in.

Cesium gap = 0.01 in.

Anode thickness = 0.03 in.

Shroud thickness = 0.03 in.

Calculations based on these parameters showed variations of reactor power and specific masses of the cells and reactor with cell diameter as shown in Fig. 13-19. The power output may be increased, and specific mass of cells decreased, by increasing the power density as shown in the following tabulation.

Power density, watt/cm²	5.0	7.0	10.0
Cell specific mass, lb_m/Kw(e)	7.1	5.1	3.5
Reactor power, Kw(e)	66.0	92.0	132.0

The in-pile, internal arrangement thermionic reactor concept, then, results in a low-mass compact reactor. The performance can be materially improved by increasing fuel enrichment and power density, and by increased compaction of the cells. The cell length is limited to a

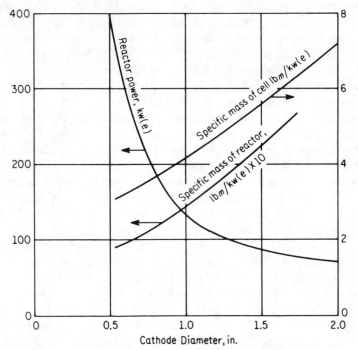

FIG. 13-19. Calculated parameters of an in-pile internal-arrangement thermionic reactor, based on 5 watt/cm² power density, and fixed dimensions; see text. (Ref. 137.)

narrow range due to electrical losses. The cell diameter is an important parameter that must be selected with care to meet both power and nuclear criticality requirements.

The performance goals of in-pile thermionics are *net* power exceeding 5 watt/cm² at 2600°F and 10 watt/cm² at 3000°F, with efficiencies approaching or exceeding 30 percent of Carnot.

13-12. SOME THERMIONIC STUDIES

1. The JPL Studies

Studies of thermionic converters for space applications at the Jet Propulsion Laboratory, JPL, [138] showed a maximum power density of 22 watt/cm² with lifetimes of more that 7,400 hr under high-temperature operating conditions. The problems that still exist are chiefly of a design and metallurgical nature. In these studies tantalum was initially used as emitter material because of the ease of machining and welding, relatively satisfactory high-temperature properties and the availability of high-purity bar stock. It is however, susceptible to chemical reactions and hydrogen embrittlement, and was discontinued in favor of rhenium, which has a higher work function. Another material, molybdenum, resulted in unstable performance with time due to bulk-oxygen contamination. Because of metallurgical considerations, polycrystalline emitters have been used in all of their converters. Grain growth and plastic deformation of an emitter, more pronounced in tantalum than in rhenium, cause distortion of the surface and disturbance of interelectrode spacing.

The JPL collectors were mostly made of molybdenum because of its low cost, good heat-transfer properties, and low work function when cesiated. Molybdenum was occasionally capped by tantalum, rhenium, or nickel, all showing low work functions when cesiated. Some converters showed power degradation. In these, deposits ranging in thickness between several hundred angstroms to several microns were found to have occurred on collector surfaces, resulting in increased work functions by as much as 0.3 ev. All converters employed high-purity alumina for ceramic bodies. Ceramic to metal seals are limited in temperature for long term operation to 600-700°C if the seals were made by a metal powder technique, and 700-800°C for seals made by active metal brazing. Present ceramic-to-metal seals must not experience thermal shocks greater than 0.7°C/sec to ensure against overstressing. The metal component of the ceramic-to-metal seals mostly used by JPL was niobium because of the close match in thermal expansion.

2. The Harwell in-Pile Diodes

The United Kingdom Atomic Energy Authority (UKAEA) has supported a research and development program at its Research Establishment at Harwell to study fission-heated thermionic diodes [139]. Several rigs were studied. Of these the Mark III, a multielement type, was intended as a first approach to the design of the core of a thermionic reactor. It is capable of replacing a reactor fuel element. The others were intended to provide research information. Two designs of the Mark III have been produced, using clad and unclad UC-ZrC fuel material for the emitter, Fig. 13-20. Each design contained four cylindrical internal-fuel cells in series. The collector in each cell acts as a catchpot in the event of fracture of the fuel and is made of a refractory metal. The successive collectors are insulated by spraying alumina onto the refractory collectors, which are then ground and assembled in a honed Zircaloy tube. This also provides good mechanical support and containment and requires no vaccum seals between the successive cells. The interelectrode spaces are 0.5 mm for the top cell and 0.25 mm for the remainder. The emitter, collector, and cesium reservoir operate in the ranges 1500-1800°C, 550-650°C, and 200-400°C respectively.

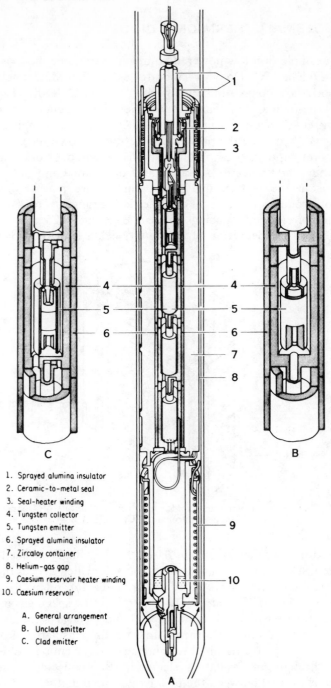

1. Sprayed alumina insulator
2. Ceramic-to-metal seal
3. Seal-heater winding
4. Tungsten collector
5. Tungsten emitter
6. Sprayed alumina insulator
7. Zircaloy container
8. Helium-gas gap
9. Caesium reservoir heater winding
10. Caesium reservoir

    A. General arrangement
    B. Unclad emitter
    C. Clad emitter

FIG. 13-20. Mark III series thermionic rig. (Ref. 139.)

13-13.   A SMALL THERMIONIC DIODE

An example of a low-powered compact thermionic diode is one built by the Donald W. Douglas laboratories of the McDonnel Douglas Astronautics Company and trademarked "Isomite" [140]. Isomite is a radioisotope-fueled, low-temperature thermionic generator, operating with emitter temperatures in the 700-1400°K range. It operates in the vacuum mode, despite the presence of cesium. Low-pressure (10^{-3}-10^{-2} torr) cesium vapor is used only to establish favorable work functions on the emitter and collector surfaces and does not contribute transport effects as in the conventional high temperature plasma diode. The generator produces low current densities in the range 0.1 to 400 milliamp/cm^2 with interelectrode spacings between 0.025 and 0.25 mm.

A general configuration of an Isomite cell is shown in Figure 13-21.

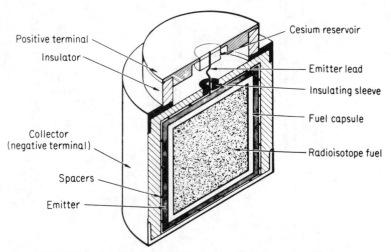

FIG. 13-21.   Isomite thermionic power cell.   (Courtesy McDonnel
Douglas Astronautics Company.)

The radioisotope fuel is encapsulated in a primary welded capsule, completely surrounded by a secondary capsule which acts as the emitter. The emitter is in turn surrounded by the collector. This 4π enclosure design improves the efficiency. The interelectrode spacing is maintained by small spacers which are both thermal and electrical insulators. The positive electric terminal is attached to a ceramic insulator brazed to the collector, the negative terminal is connected to the emitter by a wire lead passing through an insulating sleeve in the collecter. The cesium reservoir is located between the positive terminal and the top of the collector.

Four optimized Isomite sizes covering a four-order-of-magnitude power range have been fabricated. Their characteristics are given in Table 13-3. The output voltage of the cells is increased by the use of

TABLE 13-3
Characteristics of Isomite Power Cells

	Isomite No.			
	1	2	3	4
Power range, mw(e)	0.1-1	1-10	10-100	100-1000
Diameter, cm	1.0	1.3	2.0	2.8
Height, cm	1.3	1.7	2.5	3.4
Mass, g_m	11	23	87	150
Fuel	$^{147}Pm_2O_3$	$^{238}PuO_2$	$^{147}Pm_2O_3$	Electrical Simulation
Power input, w(t)	0.27	1.3	3.5	12.4
Demonstrated output, mw(e)	0.8	4	19	430
at, volt	0.1	0.18	0.15	0.23
Efficiency, percent	0.3	0.3	0.55	3.4

miniaturized dc-dc voltage converters. Figure 13-22 shows a photograph of Isomites No. 2 and 3 with the converter shown on No. 2.

Both promethium-147 and plutonium-238 in oxide form are used as fuel. They have half-lives of 2.50 and 86.0 years respectively. Because of the inevitable decay in thermal power of the radioisotopes, the ther-

FIG. 13-22. Isomite batteries. (Courtesy McDonnel Douglas Astronautics Company.)

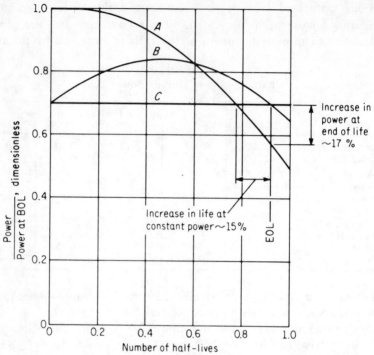

FIG. 13-23. Typical power-decay characteristics of an Isomite battery.
(Ref. 140.)

mionic diode characteristics have been optimized at other than beginning-of-life (BOL) conditions.   Figure 13-23 shows the resulting power-time characteristics.   Curve *A* represents the normal decay curve of the thermal power of the radioisotope.   Because maximum efficiency is made to occur late in life, the power output, shown by curve *B*, peaks later in life.   The result, based on a rated power given by line *C* is an increase in life by about 15 percent as shown.   Improvements in per-formance are expected to result in efficiencies of the order of 10 percent and power outputs as high as 1 w(e).

Isomites belong to the general class of *nuclear batteries*, which include compact isotopic thermoelectric devices (Sec. 14-10).   Nuclear batteries fulfill special requirements where compact, low power, remote, reliable, maintenance-free service is desired, and where high efficiency and low mass/power ratios are not of prime importance.   Potential uses of the Isomites include cardiac pacemakers (50 μw-250 μw), medical and ecological telemetry units (1 mw to 100 mw), transmitters for inaccessible monitors (10 mw to 1 w) and special purpose terrestial, space and hydro-space power supplies (1 w to 100 w).   The No. 1 Isomite, for example, can be used to power cardiac pacemakers, where the 2-7-volt requirements can be obtained by the use of miniature dc-dc voltage converters. Such a battery would be capable to power a cardiac Pacemaker for more than 10 years. For biomedical applications, in particular, the radiation-dose rates of nuclear batteries are of importance. Isomite 1 and 2 show dose rates *at the surface* of about 20 and 150 mrem/hr respectively, with these doses rapidly decreasing with distance.

PROBLEMS

13-1. A thermionic converter is to be used in an outer space application where heat must be rejected to space by thermal radiation. To reduce the mass of the radiator, the anode is to be operated at no less than 1000°C. The maximum temperature of the cathode is 2000°C. What is the maximum theoretical amount of electrical power in watts that can be obtained per Btu/hr input to the cathode?

13-2. Calculate the Fermi-Dirac probability distribution function at 0, 1000, and 2000°K for electron energies of 0, 90, 100, and 110 percent of the Fermi energy if that energy is 3.8 ev.

13-3. Calculate the electron density of a material that has a Fermi energy of 3.8 ev (as in the above problem).

13-4. What is the maximum possible current in amp, that can be emitted from 1 cm² of surface area of (a) tungsten, and (b) cesium-coated tungsten, at 2000°K?

13-5. Select suitable anode work function and material for a vacuum diode that generates 200 amp/m² of the tungsten cathode operating at 2000°K if the load and lead resistances are 0.02 ohm each and the space charge potential at the anode is 0.4 volt.

13-6. A vacuum diode is composed of a cesiated tungsten cathode and a cesium anode. It produces 1.0 amp/cm² net current with the cathode at 1840°K. The load and lead voltage drops total 0.2 volt. Using Langmuir's solution, obtain (a) the interelectrode spacing in microns, and (b) the position of the maximum potential within that spacing.

13-7. A nuclear reactor-thermionic conversion system generates 12 kw(e). The basic thermionic element is a vacuum diode that has an efficiency of 10.82 percent and generates 100 amp with a cathode temperature of 2000°K. The cathode work function is 4 volt. The total losses per diode are 100 watt. The load resistance is 1.2 ohm. Calculate (a) the number of diodes, and (b) the reactor power is kw(t).

13-8. A vacuum diode generating 120 amp has a cathode area of 8 cm² and operates with 2000°K (cathode), 1000°K (anode), and 400°K (load) temperatures. The cathode work function and space charge potential are 3 and 0.2 volt respectively. The cathode and anode have radiant emissivities of 0.3 and 0.7 respectively. The cathode electrical lead is 3 cm long and 1 mm in diameter and has a thermal conductivity of 1.5 watt/cm°C and electrical resistivity of 2×10^{-8} ohm-m. The load resistance is 0.005 ohm. Calculate (a) the power losses in watts, (b) the power output in watts, and (c) the efficiency of the diode.

13-9. It is desired to design a vacuum diode for optimum efficiency. The power output is 380 watt and the efficiency (optimum) is 10 percent. The cathode and anode operate at 1700°K and 600°K and have radiant emissivities of 0.35 and 0.65 respectively. The effective anode work function is 3 volt. Calculate (a) the net current in amp/m², (b) the cathode area in cm², and (c) the load resistance in ohm.

13-10. Calculate the minimum cesium ion concentration and partial pressure that would cause complete neutralization of the space charge in a plasma diode that generates 3.6×10^5 amp/m² of cathode area at 0.5 volt between the electrodes. Assume the ion temperature to be 2000°K. Note: Cesium vapor behaves as a perfect gas at the temperatures and pressures encountered in the diode. The electron charge $= 1.60209 \times 10^{-19}$ coulomb (amp-sec).

13-11. Determine the cell cathode length and diameter and reactor core overall critical dimensions for a 500 kw(e) in-pile internal arrangement thermionic reactor system if the core is to be a right cylinder ($H = D$) and uses cells capable of 10 watt/cm² average power density. The average volumetric thermal source strength in the fuel is 4×10^7 Btu/hr ft³. Cooling requirements necessitate a 1 in. spacing between cell centers. The overall conversion efficiency is 12 percent. Make and state any reasonable assumptions you wish.

chapter 14

Thermoelectric Energy Conversion

14-1. INTRODUCTION

Thermoelectricity, like thermionics, is a mechanism by which thermal energy is directly converted to electrical energy. Unlike thermionics, however, which depends upon electrons crossing a *gap* between different hot and cold metals because of the differences of their Fermi levels, thermoelectricity depends upon the *Seebeck effect,* in which dissimilar materials *joined* in a circuit at two hot and cold junctions will generate a voltage and, when connected through a suitable electrical load, will result in the generation of current. While unit thermionic cells produce current densities of several amp/cm² at about 0.7 volt, thermoelectric couples produce milliamperes at very small fractions of a volt. Both type unit cells can be combined in series-parallel circuits to produce desired outputs, but thermoelectric devices remain less powerful than thermionic devices. Both thermionic and thermoelectric devices convert heat to electricity and are therefore heat engines that obey the second law of thermodynamics and are Carnot-efficiency limited. While thermionic devices work best only at high temperatures, thermoelectric devices, depending upon materials, work at low and high temperatures. In fact, the *reverse* thermoelectric effect, refrigeration, has already found commercial applications and isotope-powered cooling units are now in the development stage.

Present thermoelectric converters are low in efficiency (up to one-sixth Carnot), and costly. They, however have the advantages of simplicity and reliability (no moving parts) and have therefore found applications such as electric generators for remote areas (weather stations, buoys, space satellites, lunar and instrument packages), medical research and others. They have been conveniently adapted to many heat sources, including fossil fuels, radioisotopic and reactor heat sources.

The history of thermoelectricity dates back to the early discoveries of Seebeck (1821), Peltier (1834), and Lord Kelvin (1854) among others. The practicality of early thermoelectricity was limited to temperature and energy-flux measurements by thermocouples and thermopiles. Serious attempts at generating power occurred only in recent times with the discovery of superior thermoelectric materials. Telkes studied PbS-ZnSb and other elements beginning in the late 1930's. Joffee, in 1956, after years of study, discovered the semiconductors PbTe and PbSe and developed several thermoelectric generators, including a well-known kerosene-lantern-heated thermoelectric generator that powered a small radio. This discovery of tellurides and selenides enabled devices to be built with up to 10 percent efficiency. Highest efficiencies are now obtained with n-type PbTe and p-type BiTe-SbTe. The tellurides can only be used up to about 1250°F. Higher temperature (up to 1950°F) though less efficient (lower figure-of-merit) materials include silicon-germanium alloys. It is expected that future improvements in efficiency and thermal stability of thermoelectric materials will broaden thermoelectric applications on earth and in space.

This chapter is mainly devoted to thermoelectric energy conversion. Other devices which convert nuclear radiations, rather than heat, directly to electricity, such as the betavoltaic cell which uses beta sources, and the fission electric cell which utilizes the positive fission fragments emitted from a fissioning fuel layer, are discussed in the next chapter. The first part of this chapter is devoted to some solid-state principles necessary for the study of thermoelectric and these other generators.

14-2. THE BAND THEORY OF SOLIDS

In order to study thermoelectricity a knowledge of the distinction between conductors, insulators and semiconductors is necessary. Electrical materials differ in their electrical conductivities by factors as high as 10^{30}. The *band theory* of solids helps us understand this distinction as well as the phenomena of electrical and thermal conductivities. A quantitative treatment of the band theory involves detailed wave mechanics and is beyond the scope of this book. The following is an adequate (for our purposes) qualitative presentation of some aspects of that theory.

As is known, nuclei of atoms in a crystalline lattice are located at fixed positions, called *sites*. The atoms contain positive nuclei and electrons such that the whole crystal is electrically neutral. Most of the electrons of the atoms, the inner ones, are tightly bound to the nuclei of the atoms and remain at all times with them. The outer electron of each atom, called the *valence* or *conduction* electron, on the other hand, is weakly bound to the nucleus, may wander from one nucleus to another, and is considered, according to the Bloch approach to the band theory, to belong to the entire crystal rather than to any one nucleus. Since the nuclei and their bound electrons form a geometrically ordered array that is repeated throughout the crystal, the valence electron sees a periodic potential. This is represented by Fig. 14-1.

In a system of noninteractive (or isolated) atoms of the same species, as in a gas, the valence electron of each atom may occupy a set of single-

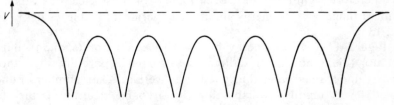

FIG. 14-1. Periodic potential experienced by an electron in a crystalline solid.

energy levels, corresponding to permitted states of that species. This is shown for noninteracting lithium atoms in Fig. 14-2a. When these atoms are brought together such as in a crystalline solid, they interact stongly. This interaction causes the energy levels of the atoms to no longer be coincident with each other, but to spread, essentially continuously,

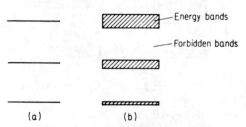

FIG. 14-2. (a) Energy levels in a noninteracting lithium atom, and (b) energy bands in interacting lithium atoms.

throughout a set of *energy bands*. This is shown for a set of interacting lithium atoms in Fig. 14-2b. Much as the regions between the energy levels of the noninteracting atoms, the regions between the energy bands of the interacting atoms cannot be occupied by electrons, and are called *forbidden bands*. The widths and spacings of the energy and forbidden bands depend upon the crystalline material in question. Energy bands may be occupied, be only partially filled, or completely unoccupied. The shading in Figs. 14-2 to 14-4 indicates which bands are occupied.

High conductivity in a metallic crystal can be accounted for by the fact that its uppermost band is *not* completely filled with electrons. An external electric field causes the electrons in this band to gain small amounts of energy that are sufficient to promote them to the continuum of available states immediately above this band. The effect of temperature here is to vary the energy distribution of electrons among the available states only slightly by shifting of their energies. The shift is controlled, as in the case of therminonics, by the Fermi-Dirac statistics, and therefore significant shift occurs only for those electrons which are within a range $\pm kT$ of energy about the uppermost filled level at $T = 0°K$, Sec. 13-3.

Figure 14-3 shows energy bands of a conductor (sodium) and an insulator (diamond). For sodium, the $3s$ band is partially filled, the $3p$ band is unoccupied, and the two bands overlap. The number of unoccupied levels readily available for the $3s$ band electrons is therefore large, resulting in high electrical conductivity. Diamond, on the other

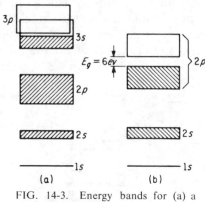

FIG. 14-3. Energy bands for (a) a conductor, sodium, and (b) an insulator, diamond.

hand, has two $2p$ energy bands, one filled, one unfilled, separated by a forbidden band, 6 ev wide. This *energy gap*, E_g, is much larger than kT at room temperature ($\simeq 0.03$ ev) so that virtually no electrons reach the upper $2p$ band even at higher temperatures. Electric fields will not impart sufficient energy to the electrons to promote them to the unoccupied $2p$ band and create electron flow, or an electric current. (The same applies to photons–the reason why diamond is transparent to visible light.)

The uppermost filled, or partially filled, energy band is called the *valence band,* and the next empty, but available, energy band is called the *conduction band.* A substance is a conductor if these two bands are separated by a vary narrow forbidden gap, or if they overlap. A substance is an insulator if they are separated by a large gap.

14-3. SEMICONDUCTORS

There is a class of crystalline solids where the forbidden band between the valence and conduction bands is relatively small, about 1 ev. Silicon and germanium are examples, having gaps of 1.1 and 0.7 ev respectively. For such solids, thermal excitation at low temperature cannot cause a measurable number of electrons to be promoted to the conduction band and they behave as insulators at low temperatures. At high temperatures, an appreciable number of electrons receive sufficient thermal energy to be promoted into the conduction band, the number being a function of both the temperature and the gap width—and the material becomes a conductor.   Such a material (a pure material) is called an *intrinsic semiconductor*.

Semiconductor materials usually have controlled amounts of impurity added to them.   In that case they are called *extrinsic semiconductors*. They are also referred to as *doped* semiconductors.   An impurity adds an allowed energy level (not band) in the forbidden band between the valence and conduction bands.   There are two types, *n*-type semiconductors, and *p*-type semiconductors, Fig. 14-4.

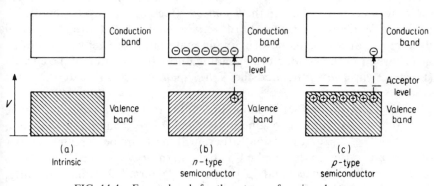

FIG. 14-4.   Energy bands for three types of semiconductors.

n-type semiconductors are characterized by an allowed energy level near the bottom of the conduction band, called the *donor level*.   At room temperatures, almost all the electrons from the donor atoms are thermally excited into the conduction band (since the gap is very narrow). By contrast only a few electrons (one shown) are intrinsically (due to the pure material) elevated from the valence to the conduction bands.   The electrical properties are therefore largely determined by the amount and form of the impurity.

p-type semiconductors are characterized by an allowed energy level near the top of the valence band, called the *acceptor level*.   The impurity has an electron deficiency. At room temperatures, electrons are easily thermally excited from the valence band to the acceptor level, leaving a deficiency of electrons in the valence band. The absence of an electron is called a *hole*, and may be considered a positively charged particle. Again intrinsic motion of electrons is small by comparison and conduction in a *p*-type semiconductor, therefore is largely due to the movement of holes to the valence band. As in *n*-type semiconductors, the conductivity of *p*-type semiconductors is largely determined by the impurities.

14-4. THERMOELECTRIC EFFECTS

There are three important effects in thermoelectricity. They are: (a) the Seebeck, (b) the Peltier, and (c) the Thomson effects.

In the *Seebeck* effect, discovered by the German phycisist Thomas Johann Seebeck in 1821, a voltage is produced in a loop of two different materials whose two junctions are at two different temperatures. The Seebeck effect, as is well known, is widely used in thermocouples. The *Peltier* effect, discovered by the French phycisist Jean Charles Peltier in 1834, is the opposite of the Seebeck effect. When a current is passed through a loop of two different materials, one junction becomes hot while the other becomes cold. In the *Thomson* effect, discovered by the British Phycisist William Thomson (Lord Kelvin) in 1854, if current is passed through a conductor made of a single material, heat is either generated or absorbed in the conductor depending upon the direction of the current (Lord Kelvin also formulated the *Kelvin relations* which relate the three effects.)

Let us now consider a loop, Fig. 14-5, made of *p*-type and *n*-type semiconductors whose two junctions 1 and 2 are maintained at T_1 and T_2 where $T_1 > T_2$. At junction 1, there will therefore be a greater number of electrons raised to the conduction band in the *n*-type material, and holes moved to the valence band in the *p*-type material, than at junction 2. This excess of charges (of both kinds) in junction 1 over junction 2 causes a diffusion of the charges from 1 to 2 in the directions shown by the arrows. At 2 the electrons and holes recombine. The flow of electrons and holes around the loop constitutes the current I in the direction shown, which is conventionally the same as that of the positive holes and opposite to that of the negative electrons. Heat Q_A, must be added at junction 1 to supply the necessary thermal excitation energy to raise the charges to the respective bands. Heat, Q_R, must be removed from junction 2 to carry away the reverse or recombination energy. Current will continue to flow as long as heat is supplied and removed in this fashion.

If the loop is open-circuited, Fig. 14-6, a voltage, ΔV, called the *Seebeck voltage*, will develop. It is proportional to the temperature

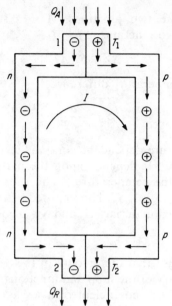

FIG. 14-5. Flow of charges
in a closed thermoelectric
loop.

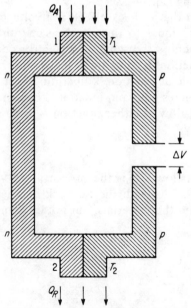

FIG. 14-6. An open-circulated thermoelectric loop.

difference between the two junctions. For a small temperature difference between the two junctions, $\Delta T = T_1 - T_2$,

$$\Delta V = S_{pn} \Delta T \tag{14-1a}$$

or for a differential temperature difference,

$$S_{pn} = \frac{dV}{dT} \tag{14-1b}$$

where S is a temperature-dependent constant of proportionality, called the *Seebeck coefficient*. It depends upon the particular combination of materials as well as the temperature. The subscript pn indicates that current flows from p to n. S_{pn} is composed of *absolute* Seebeck coefficients of the individual materials, S_p and S_n, according to

$$S_{pn} = S_p - S_n \tag{14-2}$$

It should be noted here that the amount of the Seebeck voltage is independent of the geometry of the loop and junctions.

The absolute Seebeck coefficient of a single material is obtained by joining it to another material at a temperature in its superconducting region and therefore has a Seebeck coefficient of zero, or to a material with well-known absolute coefficients over a wide range of temperature, such as lead, and using Eq. 14-2.

By contrast with the Seebeck effect, if a voltage is applied such that the same current continues to flow in the same direction as before, the flow of charge carriers is the same, and electrons and holes continue to be generated at junction 1, absorbing heat from the surroundings, and continue to recombine at junction 2, liberating heat to the surroundings. Thus junction 1 becomes cold and junction 2 becomes hot.

The rate of heat flow at either junction is proportional to the current and is represented by

$$Q = \pi_{pn} I \tag{14-3}$$

where Q is the heat flow. π is the proportionality factor and is called the *Peltier coefficient*. As with the Seebeck coefficient, it is dependent upon temperature and the materials, but not on geometry, and

$$\pi_{pn} = \pi_p - \pi_n \tag{14-4}$$

and

$$\pi_{np} = - \pi_{pn} \tag{14-5}$$

The Thomson effect may be evaluated by considering only one leg of the loop, say that of the p-type material, Fig. 14-7. If current flows from junction 1 to junction 2, holes will be moving in the same direction and, as above, will cool junction 1 and heat junction 2. A temperature-gradient profile will be established along the length of the material.

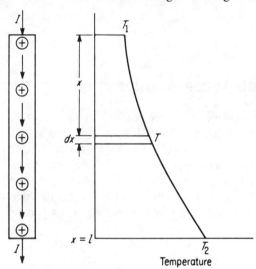

FIG. 14-7. The Thomson effect in a p-type conductor.

Heat, called the *Thomson heat*, Q_τ, will be absorbed along the path by the holes. The heat absorbed, considered positive, in a differential path length dx per unit length, dQ_τ/dx, is proportional to the current and to the temperature gradient at the same location. Thus

$$\frac{dQ_\tau}{dx} = \tau I \frac{dT}{dx} \tag{14-6a}$$

where τ is the proportionality factor, called the *Thomson coefficient*. τ is also temperature dependent. The total Thomson heat absorbed is

$$Q_\tau = I \int_{T_1}^{T_2} \tau \, dT \tag{14-6b}$$

If the direction of current is reversed in the p-type material, heat will be liberated (negative) instead of absorbed. The opposite picture is true for n-type materials. The Thomson coefficient is positive if heat is absorbed when current flows from low-to high temperature. It is therefore positive for all p-type materials and negative for all n-type materials. It should be noted that Q_τ is reversible and is different from Joule heating ($I^2 R$). The total heat liberated in a given conductor is therefore less or greater than $I^2 R$, depending upon whether Q_τ is absorbed (p-type material) or liberated (n-type material) respectively.

The Seebeck and Peltier coefficients are related by the *Second Kelvin relation,*

$$S = \frac{\pi}{T} \tag{14-7}$$

14-5. THERMOELECTRIC CONVERSION PRINCIPLES

If the open loop of Fig. 14-6 is connected to a load, it will produce load current I at voltage ΔV_L. It is easily seen that a thermoelectric conversion device is a heat engine that converts heat directly to electricity and is therefore Carnot-cycle-limited. The losses in such a device include Joule heating and heat conduction losses between junctions 1 and 2. Ignoring these losses for the moment (i.e., 100 percent efficiency), the electric power generated would be equal to the *net* heat input, or

$$\Delta V_L I = Q_{\pi_1} - \dot{Q}_{\pi_2} + Q_{\tau_p} + Q_{\tau_n} \tag{14-8}$$

where the Q_{π_1} and Q_{π_2} are the Peltier heats at junctions 1 and 2, and Q_{τ_p} and Q_{τ_n} are the Thomson heats for the p- and n-type legs respectively. Using the relationships in Eqs. 14-1 through 14-5 and noting the proper signs for the coefficients, we have:

The Peltier heats:

$$
\left.
\begin{array}{c}
-Q_{\pi_1} = \pi_{np_1} I = -\pi_{pn_1} I \\[4pt]
Q_{\pi_2} = \pi_{pn_2} I \\[12pt]
\text{The Thomson heats:} \\[12pt]
Q_{\tau_p} = I \int_{T_1}^{T_2} \tau_p \, dT \\[12pt]
Q_{\tau_n} = -I \int_{T_1}^{T_2} \tau_n \, dT
\end{array}
\right\} \tag{14-9}
$$

where the subscripts np and pn on the Peltier coefficients indicate current flow from n to p and p to n respectively. The subscripts 1 and 2 indicate evaluation at temperatures T_1 and T_2 respectively. Substituting in Eq. 14-8 and dividing by I give

$$\Delta V_L = \pi_{pn_1} - \pi_{pn_2} + \int_{T_1}^{T_2} (\tau_p - \tau_n) \, dT \tag{14-10}$$

Using the absolute Seebeck and Peltier coefficients from Eqs. 14-2 and 14-4, Eqs. 14-1b and 14-7 can be written in the forms

$$S_p = \frac{dV_p}{dT} \tag{14-11a}$$

$$S_n = \frac{dV_n}{dT} \tag{14-11b}$$

$$S_p = \frac{\pi_p}{T} \tag{14-12a}$$

and
$$S_n = \frac{\pi_n}{T} \tag{14-12b}$$

where V_p is the Seebeck voltage associated with the p-material only, and V_n is the Seebeck voltage associated with the n-material only.

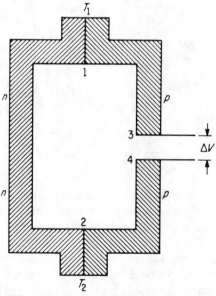

FIG. 14-8. Diagram for the derivation of open-circuited voltage.

The *open-circuited* voltage ΔV may now be obtained, with reference to Fig. 14-8, from

$$\Delta V = V_4 - V_3 = (V_4 - V_2) + (V_2 - V_1) + (V_1 - V_3) \tag{14-13}$$

where

$$V_4 - V_2 = \int_{T_2}^{T_4} S_p \, dT$$

$$V_2 - V_1 = \int_{T_1}^{T_2} S_n \, dT \qquad\qquad\qquad \right\} \qquad (14\text{-}14)$$

and $$V_1 - V_3 = \int_{T_3}^{T_1} S_p \, dT$$

Combining Eqs. 14-13 and 14-14 gives

$$\Delta V = -\int_{T_2}^{T_1} S_n \, dT + \int_{T_2}^{T_4} S_p \, dT + \int_{T_3}^{T_1} S_p \, dT \qquad (14\text{-}15)$$

Assuming $T_3 = T_4$ gives finally

$$\Delta V = \int_{T_2}^{T_1} (S_p - S_n) \, dT \qquad (14\text{-}16)$$

ΔV can also be evaluated in terms of an *average Seebeck coefficient*, $\overline{S}$, over the temperature range $T_2 \rightarrow T_1$, so that

$$\Delta V = \overline{S}(T_1 - T_2) \qquad (14\text{-}17)$$

where

$$\overline{S} = \frac{1}{T_1 - T_2} \int_{T_2}^{T_1} S_{pn}(T) \, dT \qquad (14\text{-}18)$$

It is convenient to introduce at the hot and cold junctions of the thermoelectric loop intermediate conductors that are large enough to provide a large surface through which heat is added and rejected with relative ease. Being relatively large, they would be at the uniform temperatures T_1 and T_2. In that case, it can be shown [135] that these conductors do *not* alter the Seebeck effect in the loop. With a conductor now at the cold junction, it would also be convenient to place the electrical connections in that conductor. Such a loop is shown in Fig. 14-9.

14-6. CONVERTER-PERFORMANCE PARAMETERS

The above discussion pertained to an ideal thermoelectric converter.

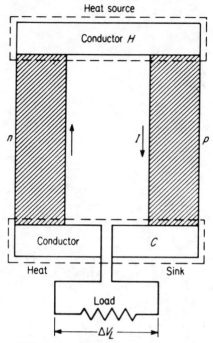

FIG. 14-9. Thermoelectric loop with intermediate conductors and electrical terminals at the cold conductor.

We shall now take into account some of the practical aspects of a converter such as that shown in Fig. 14-9. The converter supplies a load L with current I at a voltage ΔV_L, Fig. 14-10. The value of this voltage is equal to the open-circuit voltage ΔV minus the total internal losses in the loop, or

$$\Delta V_L = \Delta V - IR_C \qquad (14\text{-}19)$$

where R_C is the internal resistance of the converter given by

$$R_C = \rho_p \frac{l_p}{A_p} + \rho_n \frac{l_n}{A_n} \qquad (14\text{-}20)$$

and ρ_p, ρ_n = electrical resistivities of p and n materials
   l_p, l_n = length of p and n members
   A_p, A_n = cross sectional areas of p and n members

Using ΔV from Eq. 14-17 gives

$$\Delta V_L = \overline{S}\Delta T - IR_C \qquad (14\text{-}21)$$

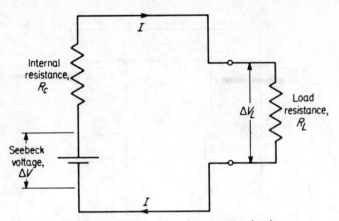

FIG. 14-10.   Thermoelectric converter circuit.

and since

$$\Delta V_L = IR_L \qquad (14\text{-}22)$$

then

$$I = \frac{\overline{S}\Delta T}{R_C + R_L} \qquad (14\text{-}23)$$

Using the average Seebeck coefficient $\overline{S}$, which is independent of temperature within the temperature range $T_2 \rightarrow T_1$, means that, $dS_p/dT = dS_n/dT = 0$, and therefore the effective average Thomson coefficient is zero.* As a consequence, a heat balance on the loop shows that.

Heat added at source $= \begin{cases} \text{Peltier heating at source + heat} \\ \text{conducted through materials toward} \\ \text{heat sink} - \dfrac{1}{2} \text{ Joule heating} \end{cases}$

Heat rejected at sink $= \begin{cases} \text{Peltier heating at sink + heat} \\ \text{conducted through materials toward} \\ \text{heat sink} + \dfrac{1}{2} \text{ Joule heating} \end{cases}$

where it is assumed that half the Joule heating reaches the heat source and the other half the heat sink. The Peltier heats are πI, and since $S = \pi/T$, they are given by $\overline{S}TI$ in the range $T_2 \rightarrow T_1$. Thus

* Since, as can be shown, $\tau = TdS/dT$.

$$Q_A = \bar{S}T_1 I + Q_k - \frac{1}{2} I^2 R_C \qquad (14\text{-}24)$$

and

$$Q_R = \bar{S}T_2 I + Q_k + \frac{1}{2} I^2 R_C \qquad (14\text{-}25)$$

where

$$Q_k = K_C(T_1 - T_2) = K_C \Delta T \qquad (14\text{-}26)$$

and K_C is the total thermal conductance of the converter given by

$$K_C = \left(k_p \frac{A_p}{l_p} + k_n \frac{A_n}{l_n} \right) \qquad (14\text{-}27)$$

where k_p, k_n = thermal conductivities of p and n materials

The thermoelectric converter is a heat engine that has a power output given by

$$\left. \begin{array}{c} P = \text{Power output} = I^2 R_L \\ \text{or} \\ P = I\Delta V_L = I(\Delta V - IR_C) = I\bar{S}\Delta T - I^2 R_C \end{array} \right\} \qquad (14\text{-}28)$$

and a heat input, at the hot junction, given by Eq. 14-24. It therefore has a thermal efficiency given by

$$\eta = \frac{P}{Q_A} = \frac{I^2 R_L}{\bar{S}T_1 I + K_C \Delta T - \frac{1}{2} I^2 R_C} \qquad (14\text{-}29)$$

It is common to alter the above equation to a more convenient form, by multiplying both numerator and denomenator by $\Delta T / T_1 I^2 R_C$, combining it with Eq. 14-23 and rearranging to give

$$\eta = \frac{M}{(1 + M) + \frac{K_C R_C}{\bar{S}^2} \frac{(1 + M)^2}{T_1} - \frac{\Delta T}{2T_1}} \cdot \frac{\Delta T}{T_1} \qquad (14\text{-}30)$$

where M = resistance ratio = $\dfrac{R_L}{R_C}$. $\qquad (14\text{-}31)$

Apart from the temperatures and resistance ratio M, Eq. 14-30 contains one other variable, called the *figure of merit* which has the dimension $^\circ K^{-1}$ and is given the symbol Z. It is given by

$$Z = \frac{\bar{S}^2}{K_C R_C} \qquad (14\text{-}32)$$

14-7. CONVERTER EFFICIENCY OPTIMIZATION

The efficiency of a thermoelectric converter operating in a given temperature range is optimized by optimizing both geometry and the resistance ratio M. Considering M fixed initially, the efficiency is optimized if the figure of merit, Eq. 14-32, is made as large as possible. For a given temperature range, $\overline{S}$ is fixed and Z is optimized by minimizing the product of conductance and resistance $K_C R_C$. This is given by

$$K_C R_C = \left[\left(k\frac{A}{l}\right)_p + \left(k\frac{A}{l}\right)_n\right]\left[\left(\rho\frac{l}{A}\right)_p + \left(\rho\frac{l}{A}\right)_n\right] \quad (14\text{-}33a)$$

For fixed materials the thermal conductivities k and electrical resistivities ρ are fixed, and the product $K_C R_C$ depends upon the ratios of the lengths l to cross sectional areas A in opposite manners. The *geometry* of the converter is then optimized to give the minimum product. In the general case the ratio l/A for the n and p materials is different. If the ratio of the ratios is called g,

$$g = \frac{(l/A)_n}{(l/A)_p} \quad (14\text{-}34)$$

Equation 14-33a becomes

$$K_C R_C = (k\rho)_p + (k\rho)_n + gk_p\rho_n + \frac{1}{2}\,k_n\rho_p \quad (14\text{-}33b)$$

$K_C R_C$ will now be minimized with respect to g by equating the derivative of $K_C R_C$ with respect to g to zero. This gives a value of g_{opt}, given by

$$g_{opt} = \sqrt{\frac{k_n\rho_p}{k_p\rho_n}} \quad (14\text{-}35)$$

The corresponding $K_C R_C$ is

$$(K_C R_C)_{opt} = \left[\sqrt{(k\rho)_p} + \sqrt{(k\rho)_n}\right]^2 \quad (14\text{-}36)$$

and the *thermal efficiency corresponding to optimum geometry* is

$$\eta\,(\text{opt geom}) = \frac{M}{(1 + M) + \dfrac{1}{Z_{opt}}\dfrac{(1 + M)^2}{T_1}\dfrac{\Delta T}{2T_1}} \cdot \frac{\Delta T}{T_1} \quad (14\text{-}37)$$

where Z_{opt} is an *optimum figure of merit* given by

$$Z_{opt} = \frac{\overline{S}^2}{(K_C R_C)_{opt}} = \frac{\overline{S}^2}{\left[\sqrt{(k\rho)_p} + \sqrt{(k\rho)_n}\right]^2} \quad (14\text{-}38)$$

The geometry was optimized, above, by considering the resistance ratio $M(=R_L/R_C)$ initially constant. With the geometry optimized, there is now an *optimum resistance ratio* that results in maximum efficiency, M_{opt}. It is obtained by differentiating the efficiency in Eq. 14-37 with respect to M and equating the derivative to zero resulting in

$$M_{opt} = \sqrt{1 + Z_{opt} T_m} \qquad (14\text{-}39)$$

where

$$T_m = \frac{T_1 + T_2}{2} \qquad (14\text{-}40)$$

and the *maximum thermal efficiency with optimized geometry and resistance ratio* is given by

$$\eta_{max} = \frac{M_{opt}}{(1 + M_{opt}) + \dfrac{(1 + M_{opt})^2}{Z_{opt} T_1} - \dfrac{\Delta T}{2T_1}} \cdot \frac{\Delta T}{T_1} \qquad (14\text{-}41)$$

This equation can, with the help of Eq. 14-39, be reduced to the more convenient form

$$\eta_{max} = \frac{M_{opt} - 1}{M_{opt} + \dfrac{T_2}{T_1}} \cdot \frac{\Delta T}{T_1} \qquad (14\text{-}42)$$

Since the thermoelectric converter is a heat engine, it is Carnot cycle efficiency limited and the ratio of the thermal efficiency of the converter to that of the corresponding Carnot cycle (operating between the same temperature limits) is called the *material efficiency** ϵ or

$$\eta = \epsilon \eta_c \qquad (14\text{-}43)$$

where η_c is the Carnot cycle efficiency given by

$$\eta_c = \frac{T_1 - T_2}{T_1} = \frac{\Delta T}{T_1} \qquad (14\text{-}44)$$

and is the last term on the right-hand side of Eqs. 14-30, 14-37, 14-41, and 14-42. The first terms on the right-hand side of these equations are obviously the material efficiency corresponding to the different efficiencies. The *maximum material efficiency,* from Eq. 14-42, is given by

$$\epsilon_{max} = \frac{M_{opt} - 1}{M_{opt} + \dfrac{T_2}{T_1}} = \frac{\sqrt{1 + Z_{opt} T_m} - 1}{\sqrt{1 + Z_{opt} T_m} + \dfrac{T_2}{T_1}} \qquad (14\text{-}45)$$

---

* In engines, this ratio is called the *engine efficiency.*

Both the material and maximum thermal efficiencies are functions of Z_{opt}, T_1 and T_2.* The material efficiency is plotted in Fig. 14-11 and the thermal efficiency in Fig. 14-12. Both are plotted vs T_1 for various

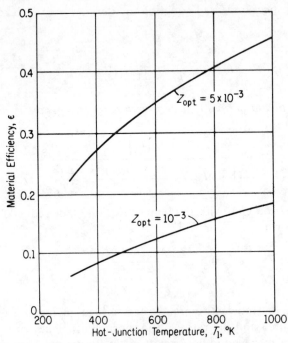

FIG. 14-11. Material efficiency as a function of T_1
for the case $T_2 = 300°$K.

values of Z_{opt}, and for a fixed $T_2 = 300°$K. Also plotted is the Carnot efficiency η_c, which is a function of T_1 and T_2, and since T_2 is fixed, there is only one curve for η_c. It can be seen that the thermal efficiency of the converter increases with an increase in hot junction temperature and with an increase in the figure of merit.

---

* Often in the literature, the maximum material efficiency is given, in the limit when T_1/T_2 approaches unity, as

$$\epsilon_{max} = \frac{\sqrt{1 + Z_{opt}T_m} - 1}{\sqrt{1 + Z_{opt}T_m} + 1}$$

showing that ϵ_{max} is a sole function of the dimensionless variable $Z_{opt}T_m$. This is rather meaningless, since as T_1 approaches T_2, the Carnot efficiency approaches zero. ϵ, as Eq. 14-45 and Fig. 14-11 show, is a function of Z_{opt}, T_1 and T_2.

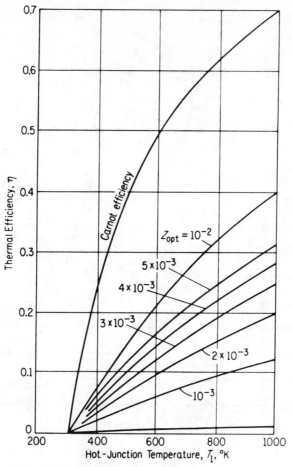

FIG. 14-12. Variation of thermal efficiency with T_1 and Z_{opt} for a thermoelectric converter and a Carnot cycle, for the case $T_2 = 300°K$.

It is often stated that efficiency is higher, the higher the dimensionless product ZT. Applied here, this means that the larger the product $Z_{opt}T_m$ the closer is the thermal efficiency to the Carnot efficiency, or the closer ϵ_{max} becomes to unity, see Eq. 14-45. The material efficiency at any value of T_1 and Z_{opt} can be obtained by dividing the thermal efficiency at that point by the corresponding Carnot efficiency at T_1 (T_2 is fixed for both). Figures 14-13 and 14-14 show values of the product ZT for some p- and n-type semiconductors at various temperatures. Figure 14-15 shows some thermoelectric properties of some semiconductors.

The *optimum current, voltage,* and *power,* meaning these parameters evaluated for operation at maximum efficiency, are given by

Optimum current $\qquad I_{opt} = \dfrac{\bar{S}\Delta T}{R_C(1 + M_{opt})}$ $\qquad\qquad$ (14-46)

Optimum voltage $\qquad \Delta V_{L_{opt}} = \dfrac{\bar{S}\Delta T M_{opt}}{(1 + M_{opt})}$ $\qquad\qquad$ (14-47)

Optimum power $\qquad P_{opt} = \dfrac{M_{opt}}{R_C}\left(\dfrac{\bar{S}\Delta T}{1 + M_{opt}}\right)^2$ $\qquad\qquad$ (14-48)

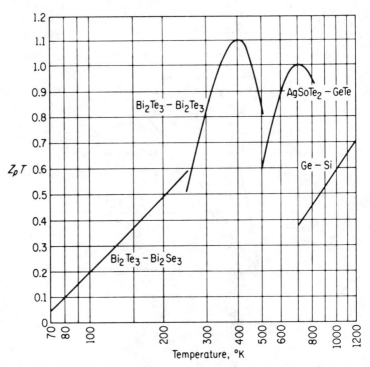

FIG. 14-13.   The product ZT vs. temperature for some p-type semiconductors. (Ref. 135.)

Example 14-1. Design a thermoelectric generator that would operate between hot and cold junctions at 800°K and 600°K, and that would produce 450 watts at 45 volts with maximum efficiency. Also calculate the efficiency.

Solution. In the above operating temperature range the semiconductors with the best figures of merit, according to Figs. 14-13 and 14-14, are the p-type AgSbTe$_2$-GeTe and the n-type SnTe-PbTe.

For these materials the following data is found in the literature for the temperature range in question:

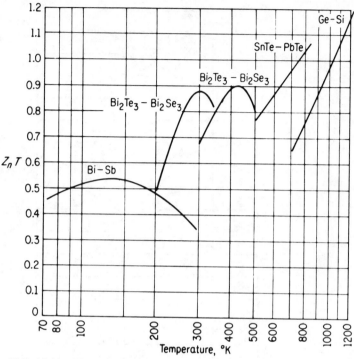

FIG. 14-14. The product ZT vs. temperature for some n-type semiconductors. (Ref. 135.)

Property	p-type AgSbTe$_2$-GeTe	n-type SnTe-PbTe
S volt/°K	240×10^{-6}	140×10^{-6}
$\bar{Z}$ °K^{-1}	1.8×10^{-6}	1.3×10^{-3}
$\bar{\rho}$ ohm-m	4×10^{-5}	1.2×10^{-5}
$\bar{k}$ watt/m°K*	0.8	1.25

* Calculated from $Z = \bar{S}^2/\bar{\rho}\bar{k}$.

Noting that $\bar{S}_p$ and $\bar{S}_n$ have opposite signs,

$$\bar{S} = (240 + 140)\,10^{-6} = 380 \times 10^{-6}\ \text{volt}/°\text{K}$$

The open-circuited (or internally generated) voltage ΔV can now be obtained with the help of Eq. 14-17 as

$$\Delta V = 380 \times 10^{-6}(800 - 600) = 0.076\ \text{volt}$$

From Eq. 14-38,

$$Z_{\text{opt}} = \frac{[(240 + 140)\,10^{-6}]^2}{\left[\sqrt{0.8 \times 4 \times 10^{-5}} + \sqrt{1.25 \times 1.2 \times 10^{-5}}\,\right]^2} = 1.59 \times 10^{-3}\ °\text{K}^{-1}$$

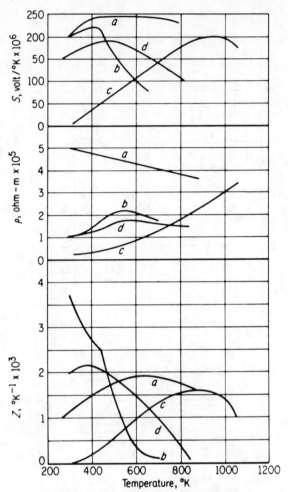

FIG. 14-15. Seebeck coefficient, electrical resistivity, and figure of merit for (a) p-type $AgSbTe_2$, (b) p-type $75\%Sb_2Te_3 - 25\%Bi_2Te_3$, (c) n-type $75\%PbTe - 25\%SnTe$, and (d) n-type $75\%Bi_2Te_3 - 25\%Bi_2Se_3$. (Ref. 135.)

From Eqs. 14-39 and 14-40,

$$M_{opt} = \sqrt{1 + 1.59 \times 10^{-3} \times 700} = 1.45$$

The output load voltage ΔV_L is obtained from ΔV with the help of Eqs. 14-19 and 14-22 as

$$\Delta V_L = \Delta V \left(\frac{R_L}{R_L + R_C} \right) = \Delta V \left(\frac{M_{opt}}{1 + M_{opt}} \right) = 0.076 \left(\frac{1.45}{1 + 1.45} \right) = 0.045 \text{ volt}$$

It thus takes 1,000 thermoelectric converters, connected in series to develop the required 45-volt output.

Since the total current required is $450/45 = 10$ amp, the total load resistance must be $45/10 = 4.5$ ohm, and the individual converter resistance is $4.5/1000 = 4.5 \times 10^{-3}$ ohm. The internal converter resistance is now obtained with the help of Eq. 14-31. At maximum efficiency,

$$R_C = \frac{R_L}{M_{opt}} = \frac{4.5 \times 10^{-3}}{1.45} = 3.1 \times 10^{-3} \text{ ohm}$$

From Eq. 14-35,

$$g_{opt} = \sqrt{\frac{1.25 \times 5 \times 10^{-5}}{0.8 \times 1.2 \times 10^{-5}}} = 2.28$$

From Eq. 14-20 and the definition of g_{opt}:

$$\left(\frac{l}{A}\right)_p = \frac{R_C}{\rho_p + g_{opt}\,\rho_n} = \frac{3.1 \times 10^{-3}}{4 \times 10^{-5} + 2.28 \times 1.2 \times 10^{-5}} = 46\,\text{m}^{-1}$$

also,

$$\left(\frac{l}{A}\right)_p = \left(\frac{l}{A}\right)_p g_{opt} = 46 \times 2.28 = 104.9\,\text{m}^{-1}$$

At this point the cross sectional areas are chosen. The choice is usually based on maximum permissible current flow density. Assuming this to be 10^5 amp/m^2 for both materials, the minimum cross sectional area is $10/10^5 = 10^{-4}\,$m^2. Choosing $l_p = l_n$, then

$$A_p = g_{opt}A_n$$

and since the minimum area is 10^{-4}m$^2 = 1$ cm^2,

$A_n = 1$ cm^2 corresponding to diameter $d_n = 1.128$ cm
$A_p = 2.28$ cm^2, corresponding to diameter $d_p = 1.703$ cm

and

$$l_p = l_n = \frac{46 \times 2.28}{100} = 1.049 \text{ cm}$$

The material efficiency, from Eq. 14-45, is

$$\epsilon_{max} = \frac{1.45 - 1}{1.45 + \dfrac{600}{800}} = 0.205$$

The Carnot efficiency is $(800 - 600)/800 = 0.25$, so that the efficiency of the generator is

$$\eta = 0.25 \times 0.205 = 0.051 = 5.1 \text{ percent}$$

14-8. CONVERTER POWER OPTIMIZATION

A thermoelectric converter can be described by the circuit shown in Fig. 14-10, which is that of a battery with an internal resistance R_C, connected to a load resistance R_L. Maximum efficiency, above, occurs when the impedance ratio R_L/R_C is given by Eq. 14-39. Often, as in some space applications, it is the *maximum power output per unit volume* (and therefore mass) of the converter, that is of interest. This is obtained, instead, by the usual impedance matching–i.e., when the external and internal resistances are made equal, $R_L = R_C$, or

$$M = 1 \qquad (14\text{-}49)$$

It can be shown that this results in

$$\frac{A_p}{A_n} = \sqrt{\frac{\rho_p}{\rho_n}} \qquad (14\text{-}50)$$

and a maximum power per unit volume (where volume $= l_p A_p + l_n A_n$), called $P_{\max}$, given by

$$P_{\max} = \left[\frac{\overline{S}\Delta T}{2(l_p\sqrt{\rho_p} + l_n\sqrt{\rho_n})}\right]^2 \qquad (14\text{-}51)$$

For materials with the highest figure of merit, the load resistance at maximum efficiency is some 30 to 40 percent larger than at maximum power and the current at maximum efficiency, the optimum current, is always lower than at maximum power. The *thermal efficiency at maximum* power is given by

$$\eta_{P_{\max}} = \frac{1}{2 + \dfrac{4}{\overline{S}^2 T_1}(K_C R_C) - \dfrac{\Delta T}{2T_1}} \cdot \frac{\Delta T}{T_1} \qquad (14\text{-}52)$$

where $K_C R_C$ is determined with the help of Eq. 14-50. It is to be noted that the material properties entering into the expression for $P_{\max}$ take the form

$$\frac{\overline{S}^2}{(\sqrt{\rho_p} + \sqrt{\rho_n})^2} \qquad (14\text{-}53)$$

(assuming $l_p = l_n$) instead of the figure of merit given by Eq. 14-38. The figure in the expression 14-53 is the one to be used, therefore, when selecting materials for maximum power rather than maximum efficiency. It and the expression for $P_{\max}$ lack a thermal conductivity term which reflects the fact that since thermal efficiency is unimportant, the heat leakage by conduction has no bearing on the problem. It may be added here that while the total mass of a converter is proportional to the product of the cross-sectional area A and length l of the legs, the heat flux and power capacity are a function of the A/l ratio. Thus shorter legs would have larger specific powers. The shortness of the legs, however, is limited by the contact resistances (the thermal and electrical resistances at the contact points) which degrade the performance of the converter. Commercial elements have ranged in size from 1/2 in. diameter by 1 in. long to $1/4 \times 1/4$ in.

The above performance analyses assumed constant properties for simplicity. In general the materials used in thermoelectric converters have properties that are strong functions of temperature. An analysis using temperature-dependent properties is complicated by the need for determining the exact temperature distribution in the thermoelectric elements and, as expected, involves solutions of nonlinear equations [142, 143]. Such an analysis is beyond the scope of this text.

A word on performance with nonoptimum dimensions is in order. Such deviations are often due to the uncertainty in material properties, but also to manufacturing tolerances. It was found [142] that a deviation from optimum dimensions by as much as 20 percent changes the apparent figure of merit by as little as 1 percent and that the thermal efficiency is insensitive to changes in the ratio of electrical to thermal losses. It was also found that best overall converter performance is obtained when the two materials have properties that are not too dissimilar.

14-9. THERMOELECTRIC MATERIALS

A large number of materials are suitable for thermoelectric operation [144]. The suitability is largely determined by the value of the dimensionless product ZT where T is the operating temperature, °K, and Z is the *figure of merit for a material,* °K^{-1}, given by

$$Z = \frac{S^2}{k\rho} \qquad (14\text{-}54)$$

where S is the absolute Seebeck coefficient for the single material, volt/°K (which may be positive or negative), k its thermal conductivity, watt/cm °K and ρ its electrical resistivity, ohm cm. Recall that Z was previously defined for a junction, Eq. 14-38.

If one were to strive to obtain efficiencies near those obtained in modern steam generating plants, ZT would have to be of the order of 5 or 6 as might be ascertained from Eq. 14-41. The best pure metals have low figures of merit resulting in ZT values of the order of 10^{-2} to 10^{-3} and are obviously unsuitable for practical thermoelectric power conversion (although suitable for use in thermocouples to measure temperatures). Insulators are also unsuitable because of their high resistivities. Suitable materials must have high figures of merit, meaning high Seebeck coefficients and low thermal conductivities and electrical resistivities.

Seebeck Coefficient

The most promising materials are the extrinsic (doped) semiconductors. Tables 14-1 and 14-2 show properties of some metals and semiconductors. Semiconductors have large Seebeck coefficients, in the neighborhood of 200 microvolt/°K.

TABLE 14-1
Thermoelectric Properties of Some Metals and Semiconductors

	Material	Electrical Resistivity ρ, ohm-cm	Thermal Conductivity k, watt/cm°K	Seebeck Coefficient S, volt/°K	Figure of Merit $Z = S^2/k\rho$, °K^{-1}
Metals	Copper	1.69×10^{-6}	3.96	$+2.5 \times 10^{-6}$	$9.3i \times 10^{-7}$
	Nickel	6.67×10^{-6}	0.87	-18×10^{-6}	5.6×10^{-5}
	Bismuth	1.16×10^{-4}	0.08	-75×10^{-6}	6.0×10^{-4}
Semiconductors	Germanium	0.0010	0.636	2.0×10^{-4}	6.3×10^{-7}
	Silicon	0.0020	1.133	2.0×10^{-4}	1.8×10^{-5}
	Indium antimonide	0.0005	0.170	2.0×10^{-4}	4.7×10^{-4}
	Indium arsenide	0.0003	0.315	2.0×10^{-4}	3.8×10^{-4}
	Bismuth telluride	0.0010	0.020	2.2×10^{-4}	2.3×10^{-3}

Thermal conductivity

Heat conduction in metals is due to both lattice vibrations and electron motion, but primarily due to electron motion (the reason for the high thermal conductivity of liquid metals which suffer little conductivity loss upon the near complete destruction of the orderly lattice upon melting). By contrast, most semiconductors have a low density of charge carriers and their heat conduction is primarily due to lattice vibration. The thermal conductivity k may be thought of as composed of thermal conductivities due to electron motion k_e and due to lattice vibration k_l. The value of k_e for doped semiconductors depends upon the doping level. The value of k_l can be lowered by introducing lattice disorder. For example, germanium and silicon by themselves have high values of k_l and are unsuitable for thermoelectric conversion. On the other hand, their alloys exhibit minimum k_l near 50 percent Ge. Also, bismuth telluride Bi_2Te_3 has a minimum k_l when half the bismuth is replaced by antimony, Sb; and lead telluride, PbTe, when half Te is replaced by selenium, Se. k_l is further reduced by mixing semiconductors in solution, hence the use of Bi_2Te_3-Bi_2Se_3, Bi_2Te_3-$SbTe_3$ and others. It may also be added that k_l is lower the larger the atomic mass, hence the use of the heavy-atom materials Bi, Te, Pb, etc.

The doping level in semiconductors can then be optimized to obtain a maximum figure of merit. For most materials this doping level is about 10^{25} atom/m^3 and corresponds to a Seebeck coefficient of about 200 μvolt/°K. With optimum doping and the use of large atom material solutions, values of the product ZT in the neighborhood of unity have been achieved. (A value of 1.8 was obtained for a mixed crystal of 0.4 PbTe-0.6 AgSbTe$_2$ at 900°K). Still, this is of course much too low than the 5 or 6 needed to make thermoelectricity as efficient as modern steam plants. It is often said that this will not happen until thermoelectric materials are found for which the lattice thermal conductivity is reduced considerably below the electron thermal conductivity. Despite their low efficiency, thermoelectric devices, of course, have advantages that highly recommend them for special applications, Sec. 14-1.

A word on dopants might be appropriate here. There are probably 30 different dopants that have been considered. Lead telluride is doped with PbT$_2$ and Na to produce n- and p-type materials respectively. Alkali metal dopants in general create holes because they have one less outer electron than lead. On the other hand halide dopants increase the electron density. Other n-type dopants include Bi_2Te_3, TaTe$_3$, MnTe$_3$ and others.

Figures 14-13 and 14-14 show ZT for some p- and n-type materials respectively. As expected ZT is a strong function of temperature. It can also be seen that different materials are suitable for different temperature ranges. Some materials, including the binary compounds of bismuth, antimony, tellurium, and selenium have large ZT at low temperatures and are therefore suitable for *low-temperature* power production and for *thermoelectric refrigeration*. Recall that the Peltier effect causes cooling at one junction and heat rejection at the other. That cooling can be used for refrigeration. Thermoelectric refrigeration is costly but is simple, compact (small devices are used for local freezing in some surgical operations), silent, has long life (no moving parts), easy to control, and can be reversed to heating by reversing the current.

In the *intermediate temperature* range, the materials of most interest are the tellurides and selenides, especially those of lead, bismuth and antimony. In the 1300-1900°F temperature range, the material with probably the best ZT is an alloy of silicon and germanium, SiGe with Ge content between 20 and 30 percent. It has good mechanical strength and low coefficient of expansion in that range. Its ZT reaches about 1.0 for the n-type and 0.8 for the p-type at about 1200°K. Other materials with good ZT in the same range include 95 percent GeTe-5 percent

Bi_2Te_3, 90 percent GeTe-10 percent AgSbTe, and the compounds InAs, InSb, and GaAs.

At *high temperatures,* the silicides and sulfides are used. At very high temperatures, the oxides are used. Over 3000°F, few materials are insulators since thermal excitation makes most ordinary insulators intrinsic conductors. At such temperatures, thermionic generators, Chapter 13, are more efficient. (Thermionic conversion, on the other hand, ceases to be practical below 2200°F.)

Another factor to be considered is that high-temperature operation, required for efficient power generation, requires relatively high energy gaps E_g (Sec. 14-2) to avoid intrinsic behavior. Both Be_2Te_3 and PbTe have narrow energy gaps, and despite their high atomic masses are therefore not suitable for very high-temperature operation. Unfortunately as k_l decreases with an increase in atomic mass, E_g also decreases (both k_l and E_g depend upon the bonding between atoms). Exceptions are indium telluride and selenide which have large energy gaps despite their low values of k_l, Table 14-2.

TABLE 14-2
Properties of Some Doped Semiconductors*

Material	Type	Maximum Operating Temperatures, °C	Melting Point, °C	Energy Gap, E_g ev	Thermoelectric Properties				
					at °K	$\alpha_0^2 \sigma^*$	k_e^*, w/cm°C	k_l, w/cm°C	Z_{max}
Bi_2Te_3	n or p	180	575	0.15	300	4×10^{-5}	0.004	0.016	2×10^{-3}
$BiSb_4Te_{7.5}$	p	180	—	—	300	4.6×10^{-5}	0.004	0.010	3.3×10^{-3}
Bi_2Te_2Se	n	330	—	0.3	300	3.6×10^{-5}	0.003	0.013	2.3×10^{-3}
PbTe	n or p	630	904	0.3	300	2.6×10^{-5}	0.003	0.02	1.2×10^{-3}
GeTe(+Bi)	p	630	725	—	800	3.2×10^{-5}	—	0.17	1.6×10^{-3}
ZnSb	p	330	546	0.6	500	2×10^{-5}	—	0.17	1.2×10^{-3}
$AgSbTe_2$	p	630	576	0.6	700	1.4×10^{-5}	0.002	0.006	1.8×10^{-3}
InAs(+P)	n	830	940	0.45	900	4×10^{-5}	—	0.07	6×10^{-4}
CeS(+Ba)	n	1030	—	—	1200	8×10^{-5}	—	0.01	8×10^{-4}
Cu_8Te_3S	—	—	930	—	1100	1.8×10^{5}	—	0.033	1.5×10^{-3}
Ge-Si	n	930	—	0.8	900	3×10^{-5}	—	0.033	9×10^{-4}
Ge-Si	p	930	—	0.8	900	2×10^{-5}	—	0.033	6×10^{-4}

* Ref. 142.

Other material factors to be considered are mechanical and metallurgical. As an example, lead telluride PbTe must at high temperatures be handled in a slightly reducing atmosphere because it oxidizes easily. It melts at 904°C. Both lead and tellurium are soluble in PbTe to a certain extent with excess lead resulting in a *n*-type and excess tellurium in a p-type material, though not to the degree required for optimum operation. The semiconductors and their contacts must also tolerate thermal cycling and shock, a difficult task since they have low thermal conductivities. They are known to have better mechanical properties when produced by powder metallurgy rather than by casting. Thermoelement materials must be chemically stable. If intended for long service, they must be annealed so that their metallurgical structure does not change with time. Most low- and moderate-temperature materials, such as tellurium, are volatile and susceptible to easy oxidation. The preparation of such elements must therefore be done in an oxygen-free atmosphere. At high temperatures, the vapor pressures of the material and dopants are high, so that encapsulation in a pressurized inert atmosphere becomes necessary. High-molecular-mass rare gases are preferred because of their low thermal conductivities.

It is obvious that the thermoelectric material picture is a complex one. Research is continuing to find materials that are suitable at different temperatures.

14-10. THERMOPILES, CASCADED AND SEGMENTED CONVERTERS

Thermopile

Since a thermoelectric converter element is a low-voltage, high-current device (a few tenths of a volt, but a few tens of amperes) it is often necessary to construct a *thermopile* (a stage in a cascade, below), which is several converter elements in series, but operating within the same temperature range. The p and n legs of the elements are alternately connected at the hot and cold junctions, Fig. 14-16. The elements in a thermopile are thus connected electrically in series and thermally in parallel. The materials connecting the legs are electrical and thermal conductors. The materials at the top and bottom of the pile are thermal conductors but electrical insulators.

As already shown, figures of merit and other material properties vary considerably with temperature, the figures of merit of suitable materials are maximum over fairly narrow temperature ranges. When a thermoelectric converter is required to operate over a large temperature difference, therefore, different materials should be used that would most efficiently span that temperature range. This can be done by either cascading or segmenting the converter.

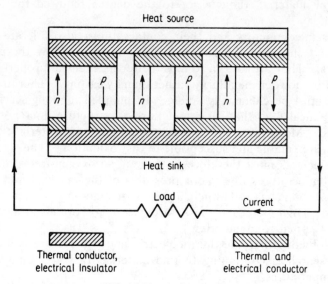

FIG. 14-16.   A thermopile.

Cascaded Converter

A *cascaded* converter (Fig. 14-17) is one in which the heat rejection from one converter stage (thermopile) supplies heat input to the next stage in the cascade.   The stages are insulated from each other electrically but are connected thermally.   Two and three stages in cascade are not uncommon, and it is not necessary to have the same number of thermoelements in each stage.   The stages can be connected electrically in series but only if the current outputs are the same.   The load impedance in that case would be the sum of the impedances that match the individual stages.   Otherwise the stages will be electrically independent, Fig. 14-17. A cascaded converter can attain high theoretical efficiencies.   However, the actual efficiency is less, due to contact resistances (contact resistances dissipate electrical energy, oppose heat flow and have zero Seebeck coefficients), heat transfer from and temperature drops in the materials between stages which result in thermodynamic irreversibilities.

Segmented Converter

A *segmented* converter (Fig. 14-18) is one built from legs made from different materials which are electrically and thermally connected and therefore act in series.   Here at each junction of two dissimilar materials there will be a Peltier heat that must be accounted for in the analysis.

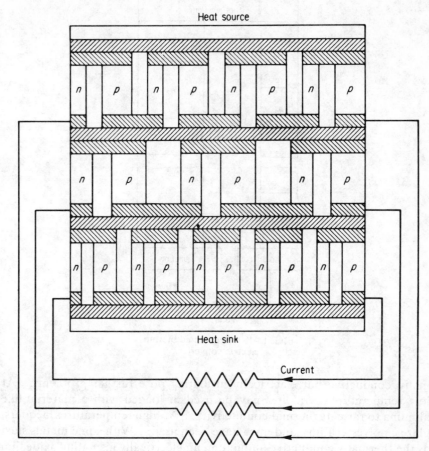

FIG. 14-17.   Cascaded thermoelectric converter.

Contacts

Contacts at low temperatures, for Bi_2Te_3 and $BiSbTe_3$, are made by plating a thin layer of nickel to the semiconductors and soft-soldering the contacts, or by using a bismuth antinmony alloy as solder. For high temperatures with Bi_2SeT_2, GeTe, and PbTe, contacts are usually made by heating the semiconductor to near melting and bringing it into contact with either nickel, iron, or stainless steel.   Great care should be taken to avoid oxide films, copper diffusing into Bi_2Te_3, or diffusion of these materials into the semiconductors to an extent that might affect their thermoelectric properties.   The contacts should be as ductile as possible and should have thermal expansion coefficients similar to those of the semiconductors.   The materials between the stages that need be ther-

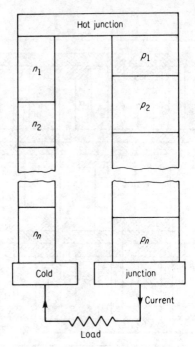

FIG. 14-18. Segmented thermoelectric converter.

mally conductive but electrically insulating pose further problems. At low temperatures Araldite may be used, if loaded with a material like alumina to raise its thermal conductivity. At high temperatures lacquers, glazes, polyester films, and mica have been used. With some metals such as the thermally conductive aluminum, an electrically insulating oxide film may be used provided that local short-circuiting is carefully avoided. The electrode material should not interact with the thermoelements or doping materials. Copper forms a Cu-PbTe eutectic at 500°C and must be excluded from operation with PbTe at high temperatures. Similarly, zinc and cadmium react with PbTe. Solders at moderate, and spring-loaded iron at high temperatures have been found satisfactory. Silicon-germanium alloys have been bonded successfully to tungsten which has a thermal expansion closely matching that of the alloy. Tungsten is used only for the cold junction because it oxidizes rapidly at high temperatures.

14-11.   THE SNAP 10A SYSTEM

An example of actual thermoelectric power devices will be presented in this Section. SNAP (Systems for Nuclear Auxiliary Power) are systems which use as heat sources radioisotopes (the odd-numbered ones) or reactors (the even-numbered ones) to generate relatively small amounts of power for special applications, such as in remote earth locations or in space. Power is converted from the heat source by conventional Rankine or Brayton cycles or by thermoelectric means.

SNAP 10A is composed of a nuclear reactor coupled to a thermoelectric converter and designed to produce a minimum of 500 watts for one year in space. The SNAP 10A reactor was developed by Atomics International Division of North American Aviation, Inc., and the conversion unit by RCA. The SNAP 10A program was initiated by the U.S. Atomic Energy Commission in early 1961. The goals were (a) to demonstrate that a reactor power plant can be handled and launched within acceptable safety criteria, (b) to show that it can survive a launch and begin power generation upon remote command from the ground, (c) to prove that a reactor-thermoelectric power plant is simple and reliable in space, and (d) to provide in-space operating data useful for designing systems with higher performance.

The SNAP 10A reactor is slightly epithermal and has a compact core, 9 in. in diameter and 12 in. long, that contains 37 zirconium-hydride fuel-moderator elements of 10 percent enrichment. The hydrogen density of 6.5×10^{22} atoms/cm^3 is approximately that of cold water and results in the small size and low fuel inventory of the core. Zirconium hydride fuels are also sufficiently stable to operate at the elevated temperatures required for space. The core is reflected by a $2\,^1/_4$-in.-thick beryllium reflector. Startup and short-term control are effected by varying the neutron core leakage by varying reflection. This is done with the help of four external rotatable semicylindrical beryllium drums located in the reflector region. Long-term control is achieved through the inherently large negative temperature coefficient of the zirconium hydride fuel. The reactor is shielded by a lithium hydride radiation shield.

Figure 14-19 shows a cutaway view of the SNAP 10A satellite. The reactor is coupled to a thermoelectric converter by a eutectic sodium-potassium (NaK-78) coolant loop. NaK is pumped by a dc Faraday electromagnetic pump. It flows from the reactor to a circular inlet header on top of the truncated conical structure. From there it flows through 40 tubes of roughly semicircular cross section, placed axially on the inner conical surface. It then collects in the circular lower header from which it returns back to the reactor. A bellows-type expansion compensator in the coolant loop accommodates NaK volume changes and provides coolant pressurization to a minimum of 5 psia.

Figure 14-20 shows a schematic of the SNAP 10A power system. The reactor and pump are shown on the left. On the right, a thermoelectric conversion module with a section of NaK tube is shown. Cylin-

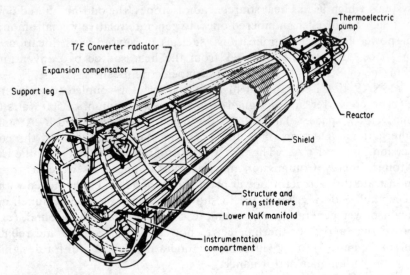

FIG. 14-19.   The SNAP 10A satellite.   (Courtesy Atomics International, Division of North American Rockwell Corporation.)

drical p- and n-type pellets of SiGe are alternately spaced along the length of the tube which constitutes the heat source.   SiGe was preferred to PbTe because it is stable above 1250°K while PbTe sublimes above 700°K and would have required encapsulation.   SiGe can also be bonded to low-resistance electrical contacts and has better mechanical properties. Its figure of merit is, however, lower than that of PbTe.   The pellets are electrically insulated from the NaK tube by thin alumina disks.   The pellets are connected in series-parallel fashion resulting in a rated output of 580 watts at 2.6 volts on a matched load of 1.6 ohm.   Electrical connection is by copper straps at the hot junctions and by aluminum straps at the cold junctions.   Since in outer space heat can be rejected only by thermal radiation, the aluminum straps are made to form the outer surface of the conical structure and act as the radiator.   The straps are made as large as possible and are insulated electrically from each other by a small clearance gap.   The total surface area of the radiator is 5.8 in.^{2} All contact interfaces from the NaK tube to the aluminum radiator are metallurgically bonded to reduce thermal and electrical conductive resistances.

Power to the electromagnetic NaK pump is from a number of thermoelements that are connected in parallel to supply the low-voltage, high current required by this type of pump.   The pump's magnetic field is supplied by a permanent Alnico-V magnet.   The entire system is heat shielded to prevent coolant freeze up in space prior to operation.

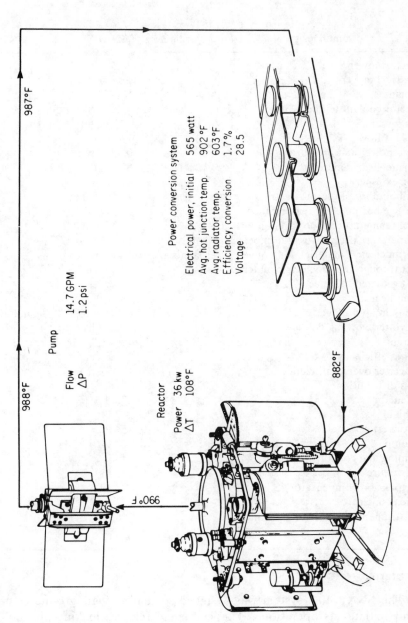

FIG. 14-20. Schematic of the SNAP 10A system. (Courtesy Atomics International.)

987°F

988°F

990°F

882°F

Pump

Flow   14.7 GPM
ΔP   1.2 psi

Reactor

Power   36 kw
ΔT   108°F

Power conversion system

Electrical power, initial   565 watt
Avg. hot junction temp.   902°F
Avg. radiator temp.   603°F
Efficiency, conversion   1.7%
Voltage   28.5

Table 14-3 contains some of the operating parameters of the SNAP 10A System.

TABLE 14-3
Some Design and Operating Parameters of SNAP 10A

Reactor power	33.5 kw(t)
Electrical power	0.58 kw(e)
Coolant	NaK
Reactor coolant temperatures	472°C, 881°F inlet, 533°C, 991°F outlet
Coolant flow	13 gpm
Pump power and pressure rise	0.625 kw, 1.1 psi
Thermoelectric material	SiGe alloy
Figure of merit, Z	$0.58 \times 10^{-3}/°K$
Total number of thermoelectric couples	1440 (720 in series)
Average hot-junction temperature	502°C, 935°F
Total temperature drop (average)	187°C, 337°F
Temperature drop across thermoelectric material . .	170°C, 306°F
Thermoelectric material and converter resistances . .	1.41 and 1.6 ohms
Heat through thermoelectric material	30 kw(t)
Heat radiated to space	33 kw(t)
Shunt heat loss	3 kw(t)
Open-circuit voltage	61 volts (0.084 v/couple)
Converter working voltage	28.5 volts
Current	19 amp
Carnot efficiency	24 percent
Converter overall efficiency	1.43 percent
Mass distribution	
Reactor	275 lb_m
Shield	217 lb_m
Converter	154 lb_m
Pump	20 lb_m
Piping and coolant	45 lb_m
Structure	83 lb_m
Expansion compensator	28 lb_m
Heat shield	32 lb_m
Instrumentation and compartment	106 lb_m
Total	960 lb_m

Startup

The SNAP 10A system was designed so that a single ground command initiates its operation. This command releases locking pins holding the form control drums, applies power to a startup controller, and activates a circuit that will later eject the heat shield. Two control drums immediately are rotated to the full in position. The other two are rotated inward by the control system at the rate of 0.2 degrees per minute, requiring about 7 hours to reach criticality. Shortly after criticality, the as yet cold reactor experiences an initial positive reactivity and corresponding power insertion which is terminated by its own negative temperature coefficient. With continued reactivity insertion, the reactor gradually heats up until, at 275°F NaK temperature, the heat-shield halves surrounding the entire unit are released and then ejected by preloaded springs. The electromagnetic pump works immediately upon an increase in reactor temperature. At full operating temperature, a temperature switch terminates ramp insertion. The same switch functions like a thermostat, allowing occasional reactivity insertion to maintain reactor-coolant outlet temperature above the switch setpoint. When equilibrium is reached, the control system is turned off and the reactor, because of its large negative temperature coefficient of reactivity, behaves as a constant temperature heat source.

The SNAP 10A Flight

The SNAP 10A power plant was attached to the forward end of an Atlas-Agena rocket (Fig. 14-21). The launch, called the *Snapshot Flight*, took place at 1:24 p.m. on April 3, 1965 from Point Arguello, California, on a 700 n.m. (nautical mile) target circular orbit, and achieved a 717 n.m. apogee and 699 n.m. perigee. A start command was given at 5:05 p.m., and the reactor reached criticality at 11:15 p.m. that evening. It was

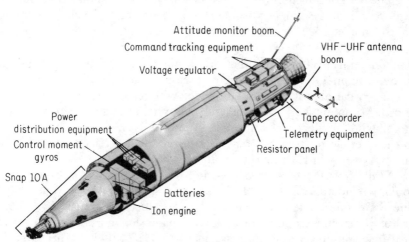

FIG. 14-21. In-orbit configuration of the *Snapshot Flight* spacecraft using the SNAP 10A system. (Courtesy Atomics International.)

operating at full power at 1:45 a.m. of the next day. Maximum power
of 0.65 kw(e) with reactor thermal power of 42 kw(t) and NaK flow of
14.3 gpm, all above design, was reached (the space heat sink was slightly
more efficient than predicted). The onboard control system was deacti-
vated by ground command approximately 6 days after launch. The
power plant was then solely controlled by the negative temperature coefficient of the reactor, and the system continued operation without electronic or mechanical moving parts. Hydrogen losses, fuel burnup and
other causes resulted in gradual decrease in power and average system
temperature, until at the end of about 43 days the system was producing
0.535 kw(e). It was predicted, on the basis of ground-test performance,
that the system would produce 0.510 kw(e) at the end of 90 days and
0.475 kw(e) at one year.

On May 16, 1965, 43 days and 553 orbits after startup, contact with
the spacecraft was lost abruptly, but was picked again by telemetry
transmitter on orbit 574, 40 hours later. The spacecraft was then operating on reserve battery supply, the command system was inoperative,
the reactor reflectors were ejected, and SNAP 10A was producing zero
power. While the loss of data during 40 hours made it virtually impossible to pinpoint the precise cause of shutdown, the available data indicated that the system itself did not fail but had responded normally to
spurious signals from electronic components onboard that had failed.

14-12. SECOND-GENERATION SYSTEMS

The data obtained during 43 days of SNAP 10A operation encouraged
exploration of system performance improvements. Second-generation
improvements focus particularly on temperature, the single most important parameter. Since any amount of heat is available as a source,
heat rejection by thermal radiation (a fourth-power function of radiator
absolute temperature), and radiator area are crucial to the performance.
The electrical output must be optimized with respect to the radiator area,
in order to minimize size and mass of the system. It has been shown that
the maximum electrical output per unit radiator area is proportional
to the average heat source temperature of the converter, a relationship
that has been verified experimentally.

While SNAP 10A reactor hydride fuel has operated with maximum
coolant temperature of 991°F, it has been successfully tested at 1300°F.
The SiGe thermoelectric material has also been tested at 1800°F and
prototype modules operating at 1300°F have shown a factor of four improvement in power per unit mass over SNAP 10A [145]. This and other
design improvements are reflected in the two advanced system perfor-
mances in Table 14-4.

TABLE 14-4
Advanced Reactor Thermoelectric Data*

Electric power, min kw(e)	0.5	2.0
Reactor power, kw(t)	25	96
Overall diameter × length, in.	50 × 100	63 × 160
Mass, lb_m	750	1150
Radiator area, ft²	25	110
NaK temperatures, average, max, °F .	1225, 1300	1210, 1300
Design life, yr	3	3
Payload separation distance, ft	35	50

* From Ref. [145].

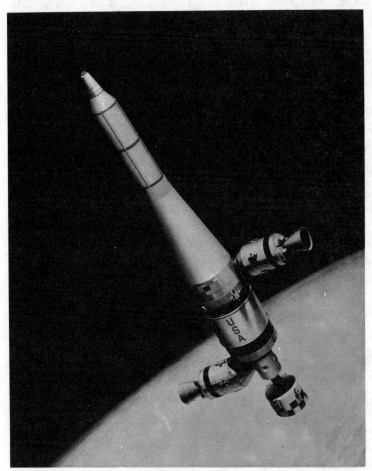

FIG. 14-22. Artist's conception of an orbiting manned space station, powered by a compact reactor-thermoelectric system at the tapered end. Crew quarters are between the Apollo-type shuttle spacecraft near the bottom. (Courtesy Atomics International.)

Figure 14-22 is an artist's conception of an orbiting manned space station using an advanced reactor-thermoelectric system to furnish electrical power. In this concept the reactor is located at the narrow tapered end of the space station. Astronauts work and live in the bottom third of the spacecraft, between the docking points of two Apollo-type shuttle spacecraft. At the bottom of the space station is an unmanned research satellite.

14-13. ISOTOPIC POWER DEVICES

In isotopic power devices, heat is generated due to the radioactive decay of a radioisotope (see Ref. 2, Sec. 4-11), and converted to electricity by a suitable conversion system. The most suitable conversion system at present is thermoelectricity.

Of the more than 1,300 radioisotopes available, those with the proper characteristics for use as isotopic power device fuels number only eight at present (1971). The proper characteristics are favorable combinations of half-life, power density, gamma emission, physical and manufacturing properties, and cost. Table 14-5 lists the eight isotopes, the fuel material form in which they are used, their half-lives and other characteristics. They are divided into two types. Type 1 isotopes are β emitters, separated from spent reactor fuel elements. Type 2 are α emitters, prepared by neutron irradiation in a nuclear reactor. Type 1 fuels generally produce lower power densities, but are lower in cost than Type 2 fuels. They also present a greater hazard of X and γ radiation and must be encapsulated more carefully in suitable shielding material.

TABLE 14-5
Radioisotopic Fuels

Type	Radioisotope	Activity	Half-life	Fuel Material	Fuel Material density, g_m/cm³	Initial power density, w(t)/g
1	Strontium-90	β, few γ	28 y	$SrTiO_3$	4.8	0.54
	Cesium-137	β, few γ	33 y	CsCl	3.9	1.27
	Cerium-144	β, many γ	285 d	CeO_2	6.4	12.5
	Promethium-147	β, few γ	.5 y	PmO_2	6.6	1.1
2	Polonium-210	α	138.4 d	Po	9.3	1320
	Plutonium-238	α	86 y	PuC	12.5	6.9
	Curium-242	α	163 d	CmC	11.75	1170
	Curium-244	α	17.6 y	CmC	11.75	

Strontium-90, as an example, is one of the most widely used β emitters. It has a relatively low cost and is plentiful in reactor waste storage tanks. Its fuel material form, strontium titanate $SrTiO_3$ was selected (by the Martin Company) because of its fire resistance (1910°C melting point), shock resistance, and low solubility in water, which helps keep the bone-seeking strontium from getting into biological systems. $SrTiO_3$ is prepared by first isolating the strontium isotopes from other elements by solvent extraction resulting in strontium carbonate $SrCO_3$. Sr^{89}, an undesirable short-lived isotope is then separated by aging (since its 52-day half-life is much shorter than the 28 years of Sr^{90}). The carbonate is then converted to the titanate. The latter is usually prepared in small cylindrical pellet form.

Alpha emitters (Type 2), while not fission products, are produced in reactors and must still be separated from other elements. They involve costly and complex chemical processing with high radiation levels.

All isotopic fuels when used in conversion devices must be encapsulated in suitable containers. The capsules serve the functions of structural strength, shock absorption, barrier against oxidation, and they must transfer heat from the fuel material to the converter elements efficiently. They also act as radiation shields for β emitters and as reservoirs to collect the helium (α particles) formed in the decay process of α emitters.

The advantages of isotopic power devices are that they are compact, reliable, have no moving parts and generate power continuously. They are, however, costly (fuel prices currently range between one hundred to several thousand dollars per watt), produce relatively low power densities which decay with time at a rate depending upon the half-life of the isotope used. They are therefore life limited.

Substantial reduction in the cost of isotopic fuels may be obtained by using the fission products without separation. Such *mixed fission products* (MFP) do not have a constant half-life, but decay with continually increasing half-lives, and therefore with activity that decreases much faster with time than if it had a single average half-life. At least one demonstration device was built with mixed fission products as fuel.

In the United States, isotopic power devices are designated SNAP *(Systems for Nuclear Auxiliary Power)* and given odd numbers. (Even-numbered SNAP systems utilize reactors as heat sources, such as SNAP-10A, Sec. 14-11). The first efforts to develop SNAP isotopic power devices began in 1956. An early generator was SNAP-1, a 500 watt, 2-month life generator which used Cerium-144 as fuel and a small turboelectric generator for energy conversion. The project was abandoned in favor of thermoelectric conversion. Table 14-6 lists SNAP systems that have been developed since, together with their functions, fuels, and other characteristics.

TABLE 14-6
SNAP Isotopic Power Generators

SNAP No	Function	Fuel	Power, w(e)	Height × Diam. in.	Mass, lb_m	Life
3	Demonstration	Po^{210}	2.5	4.75 × 5.5	4	90 d
–	Satellite power	Pu^{238}	2.7	4.75 × 5.5	4.6	5 y
–	Weather station	Sr^{90}	5	18 × 20	1680	2 y min.
7A	Navigational buoy	Sr^{90}	10	20 × 21	1870	2 y min.
7B	Fixed navigational light	Sr^{90}	60	22 × 34.5	4600	2 y min.
7C	Weather station	Sr^{90}	10	20 × 21	1870	2 y min.
7D	Floating weather station	Sr^{90}	60	22 × 34.5	4600	2 y min.
7E	Ocean bottom beacon	Sr^{90}	7.5	20 × 21	1870	2 y min.
9A	Satellite power	Pu^{238}	25	20 × 9.5	27	5 y
11	Moon probe	Cm^{242}	21-25	20 × 12	30	90 d
13	Demonstration	Cm^{242}	12	2.5 × 4	4	90 d
–	Demonstration	MFP	5-10	–	–	up to 10 y
15	Military	Pu^{238}	.001	–	–	5 y
17	Communications satellite	Sr^{90}	30	–	30	3-5 y
19	Nimbus B weather station	Pu^{238}	25	–	22	1-3 y
21	Undersea application	Sr^{90}	10	–	506	5 y
27	Apollo Lunar Surface Experiment Package (ALSEP)	Pu^{238}	63	–	66	1 y

14-14. SNAP-27

The latest SNAP isotopic power device to be built and deployed at the time of this writing is SNAP-27, a plutonium-238-fueled isotopic thermoelectric generator that was deployed on the lunar surface by the Apollo 12 and 14 astronauts in November 1969 and February 1971. It is slated to be similarly used on subsequent Apollo missions. SNAP-27 powers the Apollo Lunar Surface Experiments Package (ALSEP) and furnishes power to transmit uninterrupted data to earth during both the lunar day and the long lunar night (350 hr). The SNAP-27 program began in late 1965 and the first unit was tested in late 1966. Initial electrical power requirements of 56 watts were exceeded and the generator produces more than 67 watts after one year of operation.

The SNAP-27 generator and fuel are separated during the earth to lunar surface transportation phase and assembled on the lunar surface. There are, therefore, four hardware elements: (a) The generator consists of the thermopile, structure, and heat rejection system. It is housed in an argon-pressurized container and stowed in one of the lunar module (LM) scientific equipment bays during the trip to the moon. (b) The fuel capsule contains the isotopic fuel in a separate hermetically sealed structure to prevent release of the fuel. (c) A graphite LM fuel cask which supports the fuel capsule and provides thermal and blast protection to the fuel capsule in the event of a pad abort or a space abort resulting in re-entry. It is mounted on the outside of the LM descent stage during the trip to the moon. (d) Auxiliary support equipment provides storage, ground and flight handling and testing. On the lunar surface an astronaut removes the fuel capsule from the graphite cask and inserts it into the generator. Initially the generator power leads are short circuited to minimize the hot junction temperature during generator warm-up. When placed on the line, the astronaut removes the short circuit at the main connector and the generator begins to operate at 16 volt dc.

The SNAP-27 generator shown in Fig. 14-23, uses lead telluride thermoelectric material manufactured by 3M Company (also used in

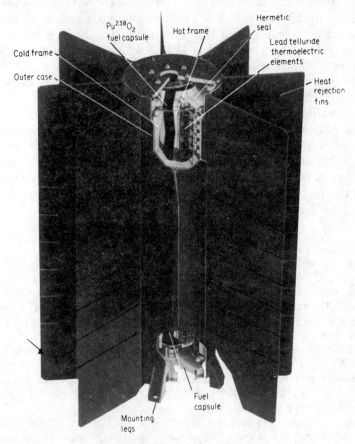

FIG. 14-23. The SNAP-27 generator. Overall dimensions: 18 in. high, 16 in. diameter. Masses: 28 lb_m unfueled, 15.5 lb_m fuel capsule, including 8.6 lb_m fuel. (General Electric photo, Courtesy Space Nuclear Systems, U.S. Atomic Energy Commission.)

SNAP-17, 21, and 23). There are 104 thermoelectric couples in the generator. The fuel material is $Pu^{238}O_2$ in microsphere form, 50 to 250 microns in size. The fuel capsule is 2.5 in. in diameter and 16.5 in. long. It comprises two fuel compartments each containing half the total nominal loading of 1480 w(t). The fuel is held in an annular configuration with the resulting void adding life to the capsule and providing containment for the helium generated in the α decay.

The fuel capsule transmits heat to the generator hot frame by thermal radiation. Normal operation temperatures are 1350°F for the capsule and 1125°F for the hot frame. The hot frame is made of Inconel 102 alloy. Both surfaces are coated with high-emissivity iron-titanate coating (minimum emissivities 0.85 for the capsule and 0.8 for the frame). Heat received by the hot frame, ∼ 1250 w(t), is transmitted to the thermoelectric element hot side through an electrical insulator, an electrode, and a hot button in that order. The normal hot and cold junction temperatures are 1080-1100°F and 525°F respectively. A high-density insulation packing of powdered Min-K insulation with thermal conductivity of 0.04 Btu/ft°F hr is used to minimize heat losses. The generator is filled with argon cover gas having a pressure of 25 psia at the operating temperatures.

The heat leaves the elements via beryllium oxide followers (which interface on the cold caps) to a massive beryllium cold frame which equalizes the temperature of the elements. Heat then goes to the outer case which is a 0.06 in. thick beryllium cylinder with eight beryllium fins brazed to it for heat rejection.

PROBLEMS

14-1. A material has a seebeck coefficient (volt/°K) which varies with temperature T (°K) according to $S(T) = 10^{-3} + 3 \times 10^{-6}T - 1.2 \times 10^{-8}T^2$. It is combined with a second material at a junction through which a current of 1 amp is passed. The junction absorbs 0.2 watt of heat at 300°K. Find (a) the Seebeck coefficient of the second material, and (b) the Seebeck coefficient of the junction at 300°K.

14-2. A material has a Seebeck coefficient (volt/°K) that varies with temperature as $S(T) = 10^{-3} + 2 \times 10^{-6}T - 10^{-8}T^2$ where T is in °K. A 10 amp current is passed through a length of the material having 0.1 ohm resistance, causing the ends of the material to be at 400°K and 380°K. Calculate (a) the total Thomson heat, and (b) the total heat liberated by the material in watts.

14-3. A single thermoelectric converter operates between 1000°K and 500°K. The Seebeck coefficients (volt/°K) of the p- and n-type materials are given by $10^{-3} + 10^{-6}T - 10^{-8}T^2$ and $8 \times 10^{-4} + 10^{-6}T - 1.5 \times 10^{-8}T^2$ respectively where T is in °K. Calculate (a) the open-circuited (Seebeck) voltage, and (b) the average Seebeck coefficient of the converter.

14-4. A single thermoelectric loop operates between 1500°K and 1000°K. The average Seebeck coefficient is 0.0003 volt/°K. The p- and n-type elements are 1.1 cm long, and 2.24 and 1 cm^2 in cross sectional areas, and have electrical resistivities of 5×10^{-5} and 10^{-5} ohm-m respectively. The current density is limited to 10^5 amp/m^2. Calculate (a) the open-circuited voltage, (b) the load voltage at maximum current, and (c) the power output in watts.

14-5. A 1-watt thermoelectric loop operates between 1400°K and 900°K with 1-cm long p- and n-type elements having electrical resistivities of 5×10^{-5} and 10^{-5} ohm-m. The load resistance is 0.01 ohm. The average Seebeck coefficient is 0.00025 volt/°K. Select the necessary element diameters if the p-element diameter is 35 percent larger than that of the n-element.

14-6. A thermoelectric loop operating between 950°K and 550°K uses 1.1 cm long p- and n-type materials having diameters of 2.0 and 1.4 cm, electrical resistivities of 4.5×10^{-5} and 1.25×10^{-5} ohm-m, and thermal conductivities of 0.75 and 1 watt/m°K respectively. The average Seebeck coefficient is 0.00035 volt/°K. The load resistance is 0.008 ohm. Calculate the power output, heat added, heat rejected, efficiency, and figure of merit of the loop.

14-7. Bismuth telluride is used in the p and n elements of a 0.8 watt, 10 amp thermoelectric converter. The elements are 1 cm long and 1 cm in diameter. The temperatures of the hot and cold junctions are 450°K and 320°K respectively. Estimate (a) the loop figure of merit, (b) the efficiency, and (c) the heat input (watt).

14-8. Determine whether the efficiency of the converter in Prob. 14-7 is optimized with respect to (a) geometry, and (b) resistance ratio. Calculate the current, power output, and material and thermal efficiencies if the converter is completely optimized for maximum efficiency.

14-9. p-Type AgSbTe$_2$ and n-type 75 percent Bi$_2$Te$_3$–25 percent Bi$_2$Se$_3$ are used in a converter that is required to generate 0.6 watt when operating between 800°K and 600°K. Find the element diameters and length that would result in maximum efficiency if current density is limited to 10^5 amp/m^2 of element cross sectional area.

14-10. What are the material, Carnot, and thermal efficiencies of the converter in Prob. 14-9?

14-11. p-Type AgSbTe$_2$ and n-type 75 percent Bi$_2$Te$_3$–25 percent Bi$_2$Se$_3$ are used in a converter that is required to generate 0.6 watt with junction temperatures of 800°K and 600°K. Note that these are the same conditions as in Prob. 14-9. This time, however, find the element diameters that would result in the minimum volume of material if the length of the legs is 0.89 cm. Current density is not limited here.

14-12. It is desired to design a system that produces 5 kw(e) using a nuclear reactor as a heat source and thermoelectric converters. p-Type AgSbTe$_2$ and n-type 75 percent PbTe–25 percent SnTe are used between 900°K and 700°K. Elements with diameters not exceeding 3.35 cm are available. The converter is to be designed for maximum efficiency. Find (a) the number and electrical arrangement of the converters, (b) the diameters of the elements, and (c) the necessary reactor thermal output, kw(t).

chapter 15

Direct Energy
Conversion of
Nuclear Radiations

15-1. INTRODUCTION

The preceding two chapters discussed two methods, thermionics and thermoelectricity, by which direct energy conversion from *heat to electricity* is accomplished. Thermionic and thermoelectric devices are therefore heat engines which are Carnot efficiency limited. An intriguing concept is the direct conversion of *nuclear radiations to electricity*. Devices doing this are not heat engines and are not Carnot efficiency limited, although their efficiencies are still quite low. They are usually low-power devices, producing in the micro- to milliwatt range, with currents in the micro- and milliamp range but with voltages ranging from a fraction of a volt to many thousands of volts.

The devices are, however, compact, trouble-free, and have many special applications in the medical, space and other fields.

15-2. NUCLEAR RADIATION CONVERTER TYPES

Devices directly converting nuclear radiations into electricity (and other energy forms) are many and varied. They can be classified into three broad categories:

(a) Radiation-induced ionization.
(b) Radiation excitation of semiconductors.
(c) Direct collection of charged particles.

In the first category, radiation energy is converted to electrical energy by producing ionization, such as in a gas confined between two plates, and separating the ion pairs into positive ions and negative electrons in a manner similar to that in an ionization chamber. In order to produce useful electrical power, however, the plates have to possess different work functions (Sec. 13-3) to produce a voltage difference [146], or be at different temperatures so that the plasma potential relative to each plate will be different, the so-called *Klein effect* [147]. (An ion chamber, on the other hand, uses a voltage difference between the plates from an external source to separate the ion pairs and cannot produce useful power.) Ionizing radiation can also be used in nonelectric production to promote chemical reactions and derive chemical energy in the so-called *chemonuclear reactor* [148], to act as a catalyst, to pump lasers, to produce nonequilibrium ionization in magnetohydrodynamic (MHD) devices [149], to neutralize the space charge in a thermionic diode, and other applications.

It was shown in the preceding chapter on thermoelectricity that semiconductors can be used to produce electrical power when subjected to a source of heat. In the second category, semiconductors are used to produce electrical power when subjected to a source of radiation. Such radiation can be in the form of photons (light), neutrons, protons, α-particles, β-particles, etc. Devices using these radiations may be generally called *radiation-voltaic* energy-conversion devices. In particular, those using photons are called *photovoltaic* devices. Those using β particles are called *betavoltaic* energy conversion devices. The latter have been the subject of extensive study and are discussed in the next section.

In the third category, the charged particles emanating from nuclear sources, such as radioisotopes or fissioning fuel, are used. Examples of this are (a) the *alpha cell,* (b) the *beta cell,* both involving the direct collection of α and β particles from emitters of such particles, (c) the *fission electric cell,* which involves the direct collection of charged fission fragments escaping from a fissioning fuel surface, and (d) the *gamma electric cell,* which involves the collection of γ-photon-induced compton electrons.

In this chapter only two of the currently more promising power devices are discussed, one belonging to each of the last two categories above. They are the beta cell and the fission electric cell.

15-3. THE BETAVOLTAIC CELL

Betavoltaic cells have been made with the semiconductors Si, Ge, Se, and GaAs, and with the beta sources Pm^{147}, and Sr^{90}-Y^{90}. Pm^{147}, with a 2.5-year half-life, is presently the most promising of the sources because of recent advances in Pm^{147} purification and source preparation.

Figure 15-1 shows a schematic of a planar, single-cell betavoltaic generator, composed of a plate beta emitter on a support plate, and *n-p* plates joined together as shown, and called an *n-p* diode, with the *n*-type material facing the emitter. The radiation causes electron-hole production in the *n-p* diode and current is generated in the direction shown.

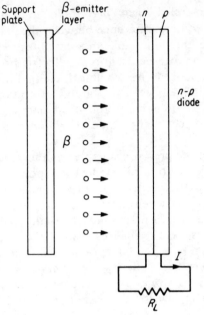

FIG. 15-1. A planar betavoltaic cell.

Figure 15-2 shows an equivalent circuit of a betavoltaic cell.   The ideal
current generated by a betavoltaic cell, $I_{g_{max}}$, is given by

$$I_{g_{max}} = e\varphi_{\beta_0} \frac{\overline{E}_\beta}{\epsilon} \qquad (15\text{-}1)$$

where e is the charge on the electron, φ_{β_0} the incident β flux, $\overline{E}_\beta$ the

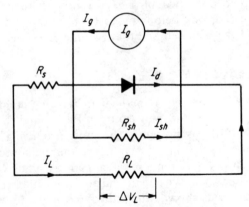

FIG. 15-2.   Equivalent circuit of a betavoltaic cell.

average β particle energy, and ϵ the average energy expended to create an electron-hole pair.

In general the generation current I_g is less than $I_{g_{max}}$ by a factor called the *collection efficiency*, η_{col}, or

$$I_g = I_{g_{max}} \eta_{col} \tag{15-2}$$

η_{col} is the fraction of $I_{g_{max}}$ that can be collected under short-circuit conditions. Typical values of η_{col} for silicon junctions are 0.6 for a p-type resistivity of 0.3 ohm-cm, and 0.9 for 3.0 ohm-cm.

When a load R_L (Fig. 15-2) is applied, a forward-bias voltage develops across the n-p junction, and part of the current flows back through the diode instead of the load. This part is called the diode current I_d and is subtracted from the generation current. I_d is usually given [150] by

$$I_d = I_0(e^{eV/AkT} - 1) \tag{15-3}$$

where I_0 is a *reverse saturation* current of the junction, V the voltage across the diode, k the Boltzmann constant $= 1.3805 \times 10^{-23}$ joules/°K $= 8.62 \times 10^{-5}$ eV/°K and T the absolute temperature. I_0 is given by

$$I_0 = A \frac{kT}{e} \cdot \frac{1}{R_0} \tag{15-4}$$

where R_0 is the junction resistance. A is a constant characterizing the particular junction. For an ideal junction $A = 1$. An ideal defect-free silicon junction has $I_0 = 10^{-11}$ amp/cm² of junction area. These ideal parameters are approached only if the forward bias voltage increases to 0.5 volt. At lower values the defects cause recombinations of the electron-hole pairs, increasing the values of both I_0 and A, and thus contributing to I_d. The range of I_0 in real devices is 1 to 10 μamp/cm²; the range of A is 2 to 3.

A further reduction in current is due to a shunt resistance R_{sh}, Fig. 15-2, across the p-n junction shunting paths, which exist in all real junctions. The shunt current is given by

$$I_{sh} = \frac{V}{R_{sh}} \tag{15-5}$$

Typical values for R_{sh} range between 10 and 1,000 kΩ for I_{sh} in the tens of μamp range. Such values are high for betavoltaic cells which, depending upon the β-emitter layer thickness, deliver between 10 and 100 μamp. R_{sh} should therefore be as high as possible.

Another loss in betavoltaic devices may arise due to a series resistance R_s to the load. Since such resistances are typically of the order of 1 Ω, and V is typically between 0.1 and 0.4 volt, such a loss is not significant. R_s is therefore usually considered to be zero, and I_g the same as the short-circuit current I_{sc}, or

$$I_{sc} = I_g$$

The load current I_L is now given by

$$I_L = I_g - I_d - I_{sh} \tag{15-6a}$$

or, combining with Eqs. 15-3 and 15-5

$$I_L = I_{sc} - I_0(e^{\,eV/AkT} - 1) - \frac{V}{R_{sh}} \tag{15-6b}$$

Other parameters of interest are the ratio of junction-to-load resistance resulting in maximum power $(R_0/R_L)_{P_{max}}$ (Prob. 15-3), the load voltage at maximum power $\Delta V_{P_{max}}$, and the maximum power P_{max}. These are given by

$$\left(\frac{R_0}{R_L}\right)_{P_{max}} = e^{\,eV/AkT} \tag{15-7}$$

$$\Delta V_{P_{max}} = \frac{AkT}{e} \ln \frac{R_0}{R_L} \tag{15-8}$$

$$P_{max} = \frac{[AkT/e \ln(R_0/R_L)s]^2}{R_L} \tag{15-9}$$

The p material electrical resistivity ρ_p has a large effect on performance. As ρ_p increases, I_{sc} and R_{sh} increase, which causes $V_{P_{max}}$ (and the open circuit voltage ΔV_{oc}) to increase. It also causes I_0 to increase, which in turn causes them to decrease. There, therefore, is an optimum value of ρ_p, found to be about 1 ohm-cm for silicon p-n diodes.

For a given betavoltaic cell of fixed design, the current- and power-voltage characteristics have the shape given in Fig. 15-3 for a commercial hetavoltaic cell using a planar Pm_2O_3 (promethia) source of 2.5 mg_m/cm^2 thickness deposited on a metal subtrate with an activity of 678 ci/g_m, and a circular area of 2.38 cm^2. It can be seen that the cell has a maximum power of 8.91 $\mu watts/cm^2$, open-circuit voltage $\Delta V_{oc} = 0.398$ volt, and a short-circuit current $(\Delta V_L = 0)$ $I_{sc} = 31.9$ $\mu amp/cm^2$. The maximum efficiency of the cell is 1.47 percent.

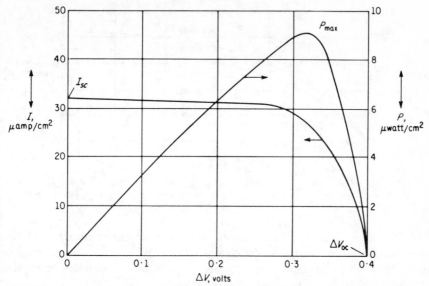

FIG. 15-3. Current and power-voltage characteristics of a betavoltaic cell.   Radioisotopic source: Promethia, Pm$_2$O$_3$, thickness $l_t = 2.5$ mg$_m$/cm^2, semiconductor
material: silicon. (Ref. 150.)

The maximum efficiency of a betavoltaic cell is given by

$$\eta_{max} = \frac{P_{max}}{P_{avail}} \tag{15-10}$$

where P_{max} = maximum electrical power, as in Fig. 15-3, usually in μwatt/
   cm^2

   P_{avail} = available total power provided by the β-emitting radiosotopic layer, μwatts/cm^2, computed from

$$P_{avail} = \frac{l_t A_c}{2.788} \ \mu watt/cm^2 \tag{15-11}$$

where l_t = thickness of radiosotopic layer, mg$_m$/cm^2
   A_c = Activity of radiosotopic material ci/g$_m$

   Maximum efficiency and power vary with the beta-emitting radiosotopic layer thickness, l_t, usually measured in milligrams (mass) per
square centimeter, mg$_m$/cm^2.   This is shown in Fig. 15-4 for a commercial silicon-promethia cell. As l_t increases P_{max} incrèases, while
η_{max} increases rapidly to an optimum value which occurs for fairly thin
layers, then decreases as l_t increases further. P_{max} also increases with an
increase in activity level, but η_{max} is independent of it.   An increase in
l_t and in activity also increase I_{sc}, and the voltages ΔV_{oc} and $\Delta V_{P_{max}}$, though
the latter are fairly independent of l_t beyond $l_t = 1$ or 2 mg$_m$/cm^2.

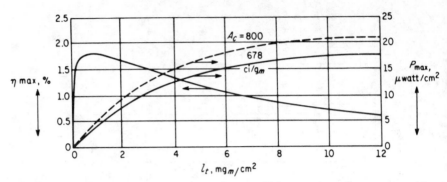

FIG. 15-4. Variation of betavoltaic maximum efficiency and maximum power with radioisotopic layer thickness l_t. Source material: Promethia, semiconductor material: silicon. (Ref. 150.)

The above pertained to planar sources and semiconductors of the unidirectional type. A bidirectional source in which the β flux emerging from both sides of radiosotopic wafer is utilized for conversion would result in nearly doubling the overall efficiency.

15-4.　BETAVOLTAIC BATTERIES

Betavoltaic batteries can be constructed with a wide range of current- and power-voltage characteristics by combining the semiconductor junctions in series-parallel networks. For example, 100 μwatts could be produced from 10 cells of the type with characteristics as shown in Fig. 15-4 with a Pm_2O_3 thickness of 2.2 mg_m/cm^2 and an efficiency of about 1.65 percent, or from only 5 cells with a Pm_2O_3 thickness of 8 mg_m/cm^2 but with a reduced efficiency of 0.85 percent. Thus a savings in volume (and mass) maybe obtained at the expense of efficiency.

As with all radiosotopic energy sources, the power decays with time because of the decay of the β-emitting radiosotopic source. The end-of-life (EOL) characteristics must therefore satisfy the requirements of the particular application the battery is used for. This means that the beginning-of-life (BOL) characteristics are such that the battery is initially overpowered.

The optimum radiosotopic material used with silicon semiconductors for betavoltaic batteries is Pm^{147} in promethia (Pm_2O_3) form. This is because the threshold for permanent radiation damage of silicon semiconductors resulting from β particles is about 0.2 Mev. Pm^{147} has a maximum particle energy of 0.23 Mev with only 1 percent of the particles exceeding 0.2 Mev. Promethium-147 has a half-life of 2.50 years, its β particles of 0.23 Mev energy are accompanied by 0.12 Mev gamma. It usually contains small amounts of the isotopic impurity Pm^{146}. Promethium-146 is dirtier, since it produces higher levels of activity (0.78 Mev beta and 0.75 Mev gamma), and is the chief contributor to activity dose rates outside the battery. Such activities can measure several hundred mrem/hr and should be shielded against, especially for medical uses. Tantalum is usually used as the shield material. The Pm^{146} impurity can be reduced by reirradiation in a high thermal neutron flux, where Pm^{146} has a large resonance absorption cross section for neutrons, and is therefore partly converted to Pm^{147}.

Other candidates for beta-emitting source materials are Sr^{90}-Y^{90} which give particles with energies significantly higher than 0.2, 0.54, and 2.26 Mev respectively. They are in secular equilibrium and cannot be separated, and consequently would have a rather short half-life. It is stated that it seems that isotopes with β particle energy greater than about 0.2 Mev are not suitable for betavoltaic use [150]. For long half-life the β sources H^3 and Ni^{63}, with β energies of 18.6 Kev and 67 Kev and half lives of 12.3 and 92 years respectively appear suitable. They result, however, in lower beta fluxes than promethia.

Betavoltaic batteries, showing efficiencies of the order of 1 percent are the most efficient and economical of all nuclear direct energy conversion sources (thermionic, thermoelectric, etc.) in the microwatt-to-few milliwatt power range. They have several-fold lifetime and power-density advantages over chemical cells (such as mercury oxide cells) in the same range.

Betavoltaic batteries have applications in the biomedical field, such as in the cardiac Pacemaker for heart patients. Such a device could be humanly implantable provided a relatively thick tantalum shield is used (2 or 3 mm) to reduce the surface activity dose rates (to about 5 mr/hr). Other biomedical applications are in telemetry and monitor devices. They can also be used for timing and switching devices, for remote instruments such as lunar instrument packages, and any applications where long-life power devices in the above low power range are needed.

TABLE 15-1
Some Operational Data of Betacels

Characteristic	Type A	Type B
Maximum power, P_{max}, μwatts	43	212
Voltage at P_{max}, $\Delta V_{P_{max}}$, volts	1.35	3.35
Open-circuit voltage, ΔV_{oc}, volts ...	1.79	4.75
Short-circuit current, I_{sc}, μamperes ..	44.0	77.0
Power density, μwatts/cm³	48.7	82.8
Battery efficiency, percent	1.04	0.84
Dimensions, length L, cm	0.78	1.32
diameter D, cm	1.20	1.57

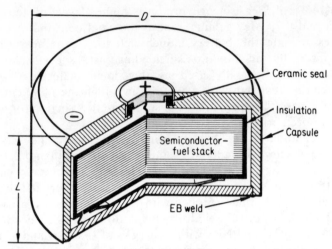

FIG. 15-5. General arrangement of Betacel nuclear batteries.
(Courtesy McDonnell-Douglas Astronautics Company.)

The McDonnell Douglas Astronautics Company has developed beta-voltaic batteries, trademarked "Betacel." Figure 15-5 shows a cross section of a Betacel. Two main prototype devices, labeled A and B, of the same general arrangement, Fig. 15-6, with characteristics shown in Fig. 15-7, have been constructed. Some of their operational data are given in Table 15-1.

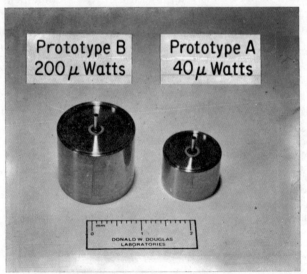

FIG. 15-6. Size B (left) and size A Betacels manufactured by McDonnell Douglas Astronautics Company.

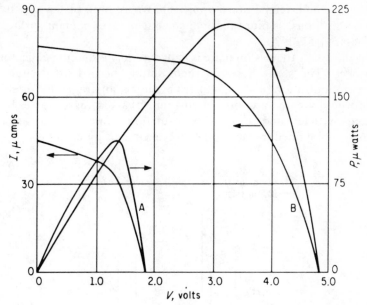

FIG. 15-7. Current-voltage characteristics of prototypes A and B Betacels manufactured by McDonnell Douglas Astronautics Company.

15-5. THE FISSION ELECTRIC CELL

The *fission electric cell* (FEC) is a direct collection device. In principle it is a simple device (Fig. 15-8). It is composed of a fission plate which is a uranium fuel layer, with support plate on one side, and a collector plate on the other.

Fission takes place in the fuel when subjected to neutrons from a reactor or other neutron source. Several radiations as well as fission fragments emerge from the fission plate. It is the *charged* fission fragments that are of interest. Their kinetic energy is directly converted to electricity in the FEC.

The FEC concept was first proposed by Wigner in 1944. In 1957 Safonov [151], during work on the *cavity* reactor (one with a low-density region in the center) suggested the use of it in conjunction with the FEC. Because the FEC, like all other collection devices, is *not* a heat engine and therefore not Carnot-efficiency limited, Safonov concentrated on low-temperature structural materials with optimum nuclear characteristics, such as aluminum and magnesium. For space applications, however, where cooling of materials is exclusively done by thermal radiation, a high-temperature approach, where both plates may be close to material limits, is more feasible since it reduces coolant surface areas and masses [152, 153].

In the FEC, the fission plate is called a *cathode* and the collector plate an *anode*. The space between them should be in hard vacuum to (a) ensure no ionization of a residual gas by the fission fragments or electrons which would create an opposing current, (b) avoid energy losses from the charged particles, and (c) provide insulation. Such a hard vacuum can be easily maintained in space but poses problems on earth.

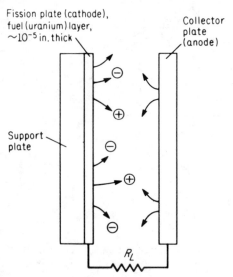

FIG. 15-8. Schematic of a planar fission electric cell (FEC).

A fraction of the fission fragments penetrate the fission layer, cross the vacuum gap, and are collected on the anode. The fuel layer must be thin enough to allow the heavy, highly charged fission fragments to escape. Each fission fragment on the average carries 20 positive charges and has a kinetic energy of 80 Mev. Thus open-circuit operation would, ideally, result in the buildup of about 4 million volts of electric potential. Voltages below this value would of course be obtained under load by matching the load resistance to the internal fission fragment current. The current-voltage characteristics of an idealized planar FEC is shown in Fig. 15-9. The FEC is inherently a high-voltage power source which may have certain applications, such as electrical propulsion in outer space.

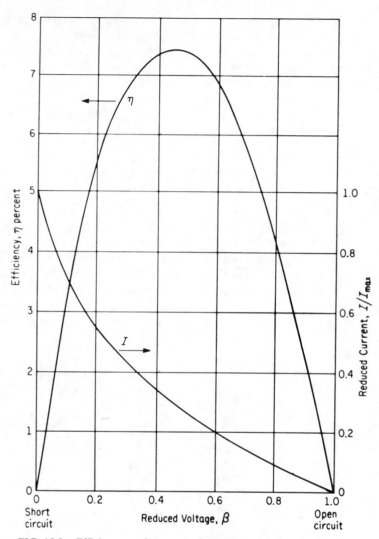

FIG. 15-9.   Efficiency- and current-voltage characteristics of an idealized planar fission electric cell.

15-6.   PERFORMANCE PARAMETERS OF AN FEC

The efficiency of an FEC is defined as

$$\eta = \frac{I \Delta E}{N E_0} \qquad (15\text{-}12)$$

where I = fission fragment "current" = number of particles reaching
anode per unit time and area of fuel, $\sec^{-1} cm^{-2}$
ΔE = kinetic energy change of fission fragments between cathode
and anode, electron volts
N = total number of fragments emitted per unit time and area fuel,
$\sec^{-1} cm^{-2}$
E_0 = fission energy per fragment as released in fission ($\simeq$ 80 Mev),
electron volts
ΔE is given by

$$\Delta E = Ve_f \qquad (15\text{-}13)$$

where V = voltage between electrodes, volts
e_f = the charge per fission fragment that is transferred between
cathode and anode, electron charges
The current I is obtained by considering the angular distribution of
the fission fragments reaching the anode. These fragments are produced
at the cathode with an isotropic distribution. In a planar geometry,
Fig. 15-10, half of them go back toward the support plate and are lost
there, and the other half go in the general direction of the anode. Of

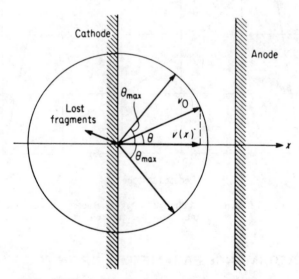

FIG. 15-10. An idealized planar fission electric cell.

these, the particles with too large an angle θ will not possess a sufficiently
large energy component in the x direction to reach the anode at high
potential. There is therefore a *maximum allowable angle of emission,*

θ_{max}. For a single fragment, the x-component of velocity is equal to the emitted velocity times $\cos \theta$. Thus

$$v(x) = v_0 \cos \theta$$

and
$$E(x) = E_0 \cos^2 \theta \tag{15-14}$$

where v_0 and E_0 are the velocity and energy of the fragments as produced from the cathode, considered independent of θ, $v(x)$ is the x component of v_0, and $E(x)$ the corresponding energy. $E(x)$ must, as a minimum, be equal to ΔE, or

$$E(x) = E_0 \cos^2 \theta \geqslant \Delta E = Ve_f$$

$$\cos^2 \theta \geqslant \frac{Ve_f}{E_0} \tag{15-15}$$

The energy ratio Ve_f/E_0 is given the symbol β. It is equal to a reduced voltage, or the ratio between voltage and the open-circuit voltage $V/(E_0/e_f)$. Thus

$$\theta_{max} = \cos^{-1}\sqrt{\beta} \tag{15-16}$$

The current I may now be obtained as

$$I = N \,(\text{solid angle between } \theta = 0 \text{ and } \theta_{max})$$

$$= N \,\frac{\int_0^{\theta_{max}} 2\pi (r \sin \theta)(r\, d\theta)}{4\pi r^2}$$

$$= \frac{1}{2}\,(1 - \sqrt{\beta})N \tag{15-17}$$

The efficiency may now be obtained by combining Eqs. 15-12, 15-13, and 15-17 to give

$$\eta = \frac{\frac{1}{2}\,(1 - \sqrt{\beta})NVe_f}{NE_0}$$

thus

$$\eta = \frac{1}{2}\,(1 - \sqrt{\beta})\beta \tag{15-18}$$

This is the idealized efficiency of an idealized planar FEC. This efficiency is plotted vs reduced voltage β in Fig. 15-9 together with the reduced current-voltage characteristic for such a cell. The low values of efficiency are due to the facts, already stated, that half the fragments are lost in the support plate, and that those produced at the cathode with angles less than θ_{max} are just not energetic enough to reach the anode. If all fragments were emitted perpendicularly to the plates, the idealized efficiency would have been 100 percent. This is reached with a point source surrounded by a spherical collector. (Remember that a FEC is *not* a heat engine and therefore not Carnot-limited.) The idealized efficiency can be increased by going to concentric cylindrical and spherical geometries and by utilizing both sides of the emitter. For concentric cylindrical and spherical geometries, however, the expressions for efficiency are much more complex because the electric field lines between cathode and anode are no longer parallel. The efficiency is dependent upon the fission-layer thickness, and the ratio of the radii of anode and cathode. For thin layers and high voltages, the efficiencies of fission electric cells compare favorably with those of thermionic and thermoelectric converters. Design ingenuity will help improve their efficiencies further.

There are three other major problems with the FEC which are common to all direct-collection devices. One has to do with the losses and energy drops of the fission fragments within the fuel layers themselves. The fragments have a range of the order of 10 microns in uranium. Very thin fuel coatings are therefore required to keep these losses from being significant. The second problem involves *sputtering*. This is the flaking off of neutral and ionized pieces of a surface when rammed into by a fission fragment. Such pieces will cause leakage currents and voltage breakdowns. The third problem has to do with secondary electron emissions from the fuel surface. As the fragments give up energy in passing through the fuel layer, they liberate electrons, some of which also leave the fuel surface. In fact, more electrons leave than positive fragments, and the electrons would create a countercurrent that exceeds the useful fission-fragment current. This countercurrent must be suppressed if the cell is to operate at all. Electrostatic grids or magnetic fields are used to accomplish this suppression.

15-7. SOME PRACTICAL FECs

A cylindrical *triode cell,* using a grid, was studied at Battelle Memorial Institute [154] and is shown in Fig. 15-11. The grid carries a bias of -20 to -30 kv. This particular cell did not prove successful, resulting in very low voltages (of the order of 100 volts out of a possible maximum of 4 million volts) and a reversal of net current flow at grid biases beyond $-1,800$ v.

A design employing a magnet to turn around the secondary electrons was used at the Jet Propulsion Laboratory [155] and is shown in Fig. 15-12.

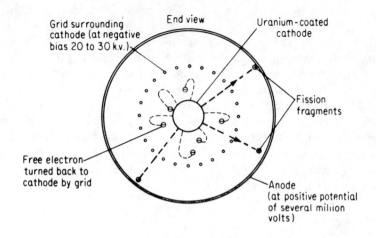

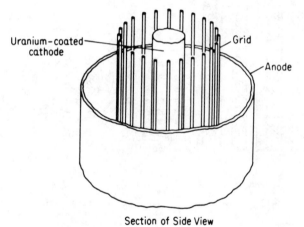

Section of Side View

FIG. 15-11.   Cylindrical triode cell.   (Ref. 154.)

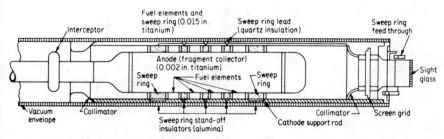

FIG. 15-12.   Modified FEC capsule.   (Ref. 155.)

In this design the positions of the cathode and anode are inverted so that the cathode surrounds the anode. This minimizes electron paths and reduces their collision probability with sputtered material. Those electrons suffering energy losses by collision cannot return to the emitter and create further plasma in the vacuum space resulting in leakage currents. The cell contains additional electrodes called *sweep rings*, which provide positive potential to sweep out the trapped electrons. The cell uses a 15-100-amp magnetic-field current and produces an open-circuit potential, varying exponentially with magnetic-field current, of 1-30 kv respectively with a working potential of 20 kv at 60-amp field current.

A *fission electric cell reactor* may contain an array of fuel subassemblies, as shown in Fig. 15-13. Because of the vacuum gap, an FEC reactor would be large and heavy.

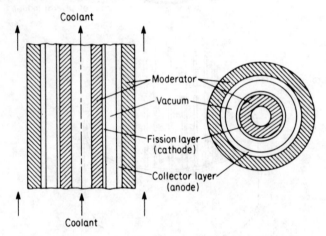

FIG. 15-13. Cross sections of an FEC fuel subassembly.

PROBLEMS

15-1. Estimate the junction resistance and the diode (short-circuit) current per cm² of area of an ideal, defect-free betavoltaic cell that operates at 400 °K and produces 0.15 μamp/cm² at 0.05 volt if the reverse saturation current is 5×10^{-9} amp/cm².

15-2. Consider a single betavoltaic cell that has an area of 2.5 cm², junction and shunt resistances of 10^6 ohm each and operates at 420 °K. What should the β flux (sec⁻¹ cm⁻²) be if the cell were to produce 0.5 μamp at 0.05 volt? Assume the average β energy is 100 kev, $\epsilon = 500$ ev, $\eta_{col} = 0.75$, and $A = 1$.

15-3. Derive Eqs. 15-7 to 15-9 which give the maximum power parameters of a betavoltaic cell.

15-4. A planar betavoltaic cell, 2 cm² in area, operates at 500 °K and receives

β flux from a 2.6 mg_m/cm^2 thick promethia source. The short-circuit current is 10 $\mu\text{amp}/\text{cm}^2$. The junction and shunt resistances are 10^6 ohm each. Calculate the power, load current, voltage, and efficiency at maximum power conditions. Take $A = 1$.

15-5. Pm^{147}, H^3 and Ni^{63} are three β emitters considered for betavoltaic cell operation. They have half-lives of 2.50, 12.26, and 92 years respectively. It is required to design a betavoltaic battery that produces 740 μwatt at maximum power. Calculate the minimum necessary number of diodes in each case if all have diameters of 1.5 cm, are 2 mg_m/cm^2 thick, and all diodes have 1.5 percent efficiency at maximum power.

15-6. An ideal planar fission electric cell generates 200 kvolt. Estimate the cell efficiency and the fraction of the fragments emitted by the cathode that are able to reach the anode.

15-7. Estimate the current and power of an ideal fission electric cell if the maximum allowable angle of emission from the cathode is 80° and the fission layer generates 7.55×10^{14} fragments/sec cm².

15-8. A 2.4 cm² ideal planar fission electric cell has a 10 micron-thick natural uranium layer situated in a 10^{14} thermal neutron flux field. It produces 220 kvolt. Estimate the power, cell efficiency, and overall efficiency (taking into account the fission process). Take the U^{235} nuclear density as 3.4×10^{21} cm^{-3} and the effective fission cross section as 500 barn.

chapter **16**

Fusion Power

16-1. INTRODUCTION

The main incentive for developing the fusion-power reactor is the enormous amounts of energy resources fusion would make available to mankind. The present reserves of fossil fuels are limited and they are plagued by air-pollution problems. The resources of fissile fuels are more abundant, Table 16-1, but also limited, though the successful development of the fast breeder reactor (Chapters 9 and 10) would stretch those resources to perhaps centuries. Besides fossil and fissile resources, those due to solar, hydraulic or geothermal energies are either very erratic or very limited. Fusile fuels, mainly deuterium that exists stably in

TABLE 16-1
World Energy Reserves

	Fuel	Energy in Q Units
Fossil	Coal	32
	Oil and gas	6
Fissile	Uranium and thorium	600
Fusile	Deuterium in the oceans	10^{10}

all natural waters to the extent of 1 part in 6,666, make for an almost unlimited resource that would last man essentially indefinitely (millions of years). Fusion reactions, like fission reactions, produce almost no air pollution. They produce a minimum of manageable radioactive wastes, and, being highly efficient because of the high temperatures of the working fluid, they produce much less thermal pollution per Mw(e) than either fossil or fissile energy.

Fusion reactors, also called *controlled thermonuclear reactors* because the fusion reaction requires ultrahigh temperatures (hundreds of millions or billions of degrees Kelvin) to sustain their reactions, are therefore a high-priority developmental item in many national power programs.

This chapter presents some of the principles of fusion, the attempts at achieving controlled thermonuclear reactions, and conceptual designs of what an operating fusion-power device might look like. It should be stressed here that fusion technology is much more complex than fission technology. While new optimism has been recently generated in the world (1969-70) regarding research on fusion, we are not yet even at the same stage reached on December 2, 1942, when a sustained fission reaction was initiated in Chicago. Since it took fission power plants some 25 years to reach economic maturity after that happening, we should not expect equally competitive fusion-power plants for some time. A demonstration-type fusion reactor, however, might be a reality in 15-25 years.

16-2. FUSILE FUELS AND REACTIONS

Energy from the sun and the stars is thermonuclear, produced by the fusion of four nuclei of hydrogen (protons) into one helium nucleus and two positrons:

$$4H^1 \longrightarrow He^4 + 2_{+1}e^0 + 26.7 \text{ Mev} \tag{16-1}$$

On earth, fusion is accomplished by having two light nuclei fuse into a heavier nucleus. There is a much greater probability of two particles colliding than of four, and there thus is a much greater reaction rate in a necessarily small-sized, man-made fusion reactor. There are several possible fusion reactions between two nuclei:

$$D + D \underset{\longrightarrow}{\overset{\longrightarrow}{\bigg\langle}} \quad \begin{array}{ll} T + n + 3.2 \text{ Mev} & \tag{16-2a} \\ T + p + 4.0 \text{ Mev} & \tag{16-2b} \end{array}$$

$$D + T \longrightarrow He^4 + n + 17.6 \text{ Mev} \tag{16-3}$$

and $\quad D + He^3 \longrightarrow He^4 + p + 18.3 \text{ Mev} \tag{16-4}$

where D = deuteron, deuterium nucleus or $_1H^2$

p = proton or $_1H^1$

T = triton, tritium nucleus or $_1H^3$

The first two reactions (Eqs. 16-2a and 16-2b) occur when all deuterium fuel is used and have about equal probabilities. The third reaction (Eq. 16-3), the most important of all, is shown schematically in Fig. 16-1.

The above fusion reactions are accomplished by accelerating the positively charged nuclei on the left (the reactants) to very high velocities, so that their kinetic energies would overcome the Coulomb repulsive

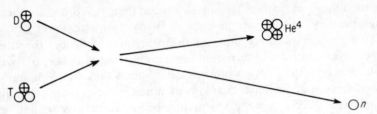

FIG. 16-1.   Schematic illustration of the D-T fusion reaction, Eq. 16-3
⊕ proton, O neutron.

forces between them and they are then able to react.   Such kinetic
energies are equivalent to temperatures of hundreds of millions of degrees.
This results in the reactants becoming a completely ionized gas, or a
completely ionized *plasma*.   The plasma in an operating reactor must
also be held in a configuration that would prevent its quenching by the
walls of the container, at a sufficiently high density (about 10^{15} ion/cm^3)
and for a sufficiently long confinement time (tenths of seconds).   There
are thus three important parameters in an operating fusion reactor:
plasma temperature, confinement time, and density [156].

The plasma will be defined in the next section and the three parameters, discussed in Secs. 16-4 through 16-11.

16-3.   THE PLASMA

It is often stated that the universe contains at least *four* states of
matter: *solid, liquid, gaseous,* and *plasma* states.   The conversion from
one state to the next, in the above order, is dependent upon the kinetic
energy of the fundamental particles of the state, Fig. 16-2.   A solid
becomes liquid when the kinetic energy of its molecules exceeds the
binding energy of the solid crystals.   When the kinetic energy of the
molecules in a liquid increases sufficiently to overcome the van der Waals
forces, the liquid changes to a gas.

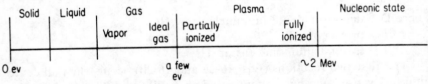

FIG. 16-2.   The states of matter vs. particle energy (nonuniform scale.)

When the kinetic energy exceeds the ionizing potential of the atoms,
the particles become ionized and we have a mixture of free negatively
charged electrons and positively-charged ions.   This is the *plasma*.   A

plasma may be *weakly* or *partially ionized,* in which case it is a mixture of electrons, ions and neutral (nonionized) particles; or *completely ionized,* i.e., a collection of electrons and ions only. In either case, and except for extremely low densities, coulomb forces dictate that the number of negative and positive particles must be equal, i.e., the plasma as a whole is neutral.

The energy at which transition from gas to plasma begins is generally a few ev. In terms of temperature, this is equivalent to 10^4 to 10^5 °K. Above about 2 Mev the nuclear bonds are broken, and matter becomes a collection of free electrons and nucleons (neutrons and protons). This state is essentially a *fifth* state of matter, but one that can only be made to exist in quantities too small for serious study on earth. One of course may visualize other states of matter in which nucleons are broken, etc., as energy is raised beyond the 2 Mev or so level. Of all the states of matter the first three (solid, liquid, and gas) occupy a very narrow range of energies below a few ev, compared to the few-ev-to-2-Mev range for plasma. It is therefore usually estimated that 99.9 percent of all matter in the universe is at least in the plasma state. It is therefore more "normal" for matter in the universe to exist as plasma than as any of the first three states.

Plasma, as seen, is naturally occurring but can also be artificially produced. Many effects due to naturally occurring plasma are observable on earth. These include effects of sunspots and coronas; effects of interstellar matter, which is known to be ionized, on cosmic-ray motion; effects of the van Allen belts, which are charged particles trapped in the earth's magnetic field; effects of the ionosphere (the ionized upper layer of the earth's atmosphere) on radio waves; and effects of plasma beams emanating from the sun which cause magnetic storms disturbing the earth's magnetic field and which are associated with the aurora.

Artificial plasma was first produced by passing electrical current in low-pressure gases before the turn of the century. In 1897 Crookes was first to speculate on the existence of a "... new world, a world where matter may exist in a fourth state ..." from such work on evacuated tubes. Studies on highly ionized plasma, however, did not begin until the early 1920's. Langmuir, about the same time, gave that state of matter its name, *plasma,* * and is credited with developing the basic theory of ionized gases and with many of the experimental techniques still in use today. Other researchers include Tonks, von Engel, Townsend, and Ragowski.

---

* *Plasma:* in biology, the simplest form of organized matter in vegetables and animals of which tissues are formed; the liquid part of the blood after the red and white corpuscles have been removed. In physiology, a liquid expressible from muscular tissue. In pharmacy, a glycerite of starch used in preparing ointment. In pathology, plasma cells are found in mucous membranes and in the lymphanoid tissue. Otherwise plasma is a variety of quartz that was highly valued for engraved ornaments in ancient times.

Since World War II research on plasma has picked up considerably. Research for the purpose of controlling the energy release of fusion reactions is mainly carried on in the USSR, the United States, and the United Kingdom, but also by EURATOM (The European Atomic Energy Community), France, West Germany, Japan and others. In the United States research is carried on under *Project Sherwood* [157].

As contrasted to its nearest accessible state, gas, which is an insulator, a plasma can carry electrical current. Also while a gas is dominated by short-range molecular forces, a plasma is dominated by long-range Coulomb forces. Consequently a plasma behaves characteristically in a magnetic field.

16-4. PLASMA TEMPERATURES

The fusion reaction cross sections for the four fusion reactions of Eqs. 16-2 through 16-4 are shown in Fig. 16-3. The reaction with by far the most favorable cross sections, and consequently the one receiving the greatest attention, is the third reaction:

$$D + T \longrightarrow He^4 + n + 17.6 \text{ Mev} \qquad [16\text{-}3]$$

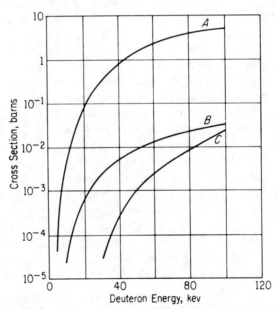

FIG. 16-3. Microscopic fusion cross sections at various deuteron energies. A: D-T reaction, Eq. 16-3, B: D-D reactions, Eqs. 16-2, and C: D-He³ reaction, Eq. 16-4.

It has *favorable* cross sections–that is, it proceeds at an observable rate, only at energies above a few tens of kev. Since 1 ev corresponds to 1.16×10^4 °K,* this corresponds to a few hundred million °K. Even at these energies the cross sections of this reaction are smaller than the cross sections for ionizing collisions (and for cumulative small-angle Coulomb scattering). The energy must therefore be of the order of several tens of kev, corresponding to several hundred million degrees K. At such energies the plasma is *fully ionized.* It may be noted also that the above temperatures are much higher than the approximately 15 million degrees K that exists in the interior of the sun. It should also be noted that, while these temperatures are extremely high by conventional standards, they are actually fairly low by high-energy physics standards. This can be seen if the energy equivalent of temperature is considered. For example, a 30-billion-volt accelerator at Brookhaven produces particles with energies corresponding to a temperature of almost ten trillion degrees.

Ignition Temperatures

The degree to which a plasma must be heated in order to produce useful energy in a fusion reactor will now be discussed. The plasma, as any hot body, loses energy by radiation. Plasma-radiation loss, however, is due to two processes called *bremsstrahlung* and *synchrotron radiation.* Bremsstrahlung is radiation resulting from the interactions of energetic charged particles, which in the plasma is mainly due to the interactions between electrons and ions. Synchrotron radiation is important only at very high temperatures. It is emitted by energetic charged particles, in particular electrons, that are moving in a magnetic field.

Some of the plasma energy is also utilized in heating the incoming gas which fuels the reactor.

While fusion reactions may be accomplished at relatively low temperatures, a fusion reactor becomes self-sustaining at much higher temperatures where the cross sections and consequently the rate of reaction or rate of energy (power) generated are high enough to overcome the losses by radiation and to the incoming gas.

The power (rate of energy) per unit volume radiated by bremsstrahlung is shown by wave mechanics to be given by

$$\Delta Q_r = 1.42 \times 10^{-27} Z^2 n_i n_e T^{1/2} \tag{16-5}$$

where ΔQ_r = radiated power, erg/sec cm^3

Z = atomic number of nuclei present

n_i, n_e = density of ions and electrons respectively, cm^{-3}

T = absolute temperature, °K

* Energy = kT where k is Boltzmann's constant, equal to 8.62×10^{-5} ev/ °K (Appendix E), and T is absolute temperature in °K. Thus 1 ev corresponds to 1.16×10^4 °K.

The loss to the incoming gas is difficult to evaluate since it depends upon the reactor design and fueling scheme.

The rate of energy generated by the plasma is evaluated by first considering the probability that an ion will undergo a fusion reaction per unit time. It is given by

$$P_f = n\sigma v \qquad (16\text{-}6)$$

where P_f = probability of fusion reaction by a nucleus per unit time, sec^{-1}
n = density of the ion species, ions/cm^3
σ = cross section of fusion reaction (Fig. 16-3), cm^2
v = velocity of the ion, cm/sec
Taking into consideration all ions that might interact by fusion in a plasma (such as D and T, Eq. 16-3), the total reaction rate is given by

$$R = n_1 n_2 (\overline{\sigma v}) \qquad (16\text{-}7)$$

where R = reaction rate, reactions/sec cm^3
n_1, n_2 = densities of interacting ions 1 and 2 (such as D and T), ions/cm^3; for the D-T reaction, $n_1 = n_2 = \dfrac{n}{2}$ and $n_1 n_2 = \dfrac{n^2}{4}$, and for the D-D reaction $n_1 = n_2 = n$, but R is $\dfrac{1}{2} n^2 (\overline{\sigma v})$ to avoid counting the same reaction twice

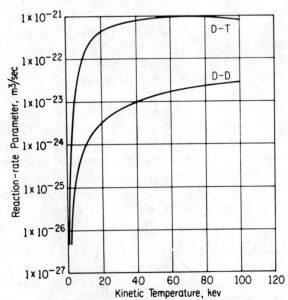

FIG. 16-4. The reaction rate parameter $\overline{\sigma v}$ for the D-T and D-D reactions, assuming a Maxwellian particle distribution. (Ref. 158.)

$\overline{\sigma v}$ = average product of the cross section times velocity, cm³ / sec; the velocity distribution is, in general, Maxwellian, and $\overline{\sigma v}$ is plotted for the D-T and D-D reactions in Fig. 16-4 (see also Fig. 16-22).

The rate of energy, or power, generated per unit volume, ΔQ_g, is now simply equal to $R\Delta E$ where ΔE is the energy generated per fusion reaction, or

$$\Delta Q_g = n_1 n_2 (\overline{\sigma v}) \Delta E \qquad (16\text{-}8)$$

Equations 16-5 for the bremsstrahlung radiation loss and 16-8 for the energy generated by the D-T and D-D reactions, are plotted in Fig. 16-5 for plasma densities of about 10^{15} particles/cm³ [158]. Both ΔQ_r and ΔQ_g increase as the (density)². ΔQ_g, however increases rapidly with temperature because $\overline{\sigma v}$ increases very rapidly with temperature, Fig. 16-4. ΔQ_r on the other hand, increases as $T^{1/2}$. As seen in Fig. 16-5, the energy generated increases more rapidly with temperature than the radiation

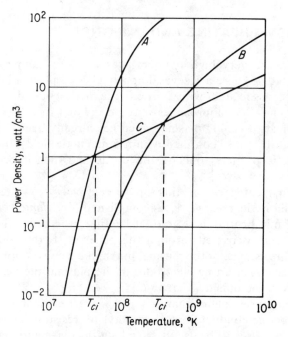

FIG. 16-5. Energy production and losses by radiation (bremsstrahlung) at various temperatures. (A) Energy production by the D-T reaction, (B) energy production by the D-D reaction, (C) radiation loss; T_{ci} = critical ignition temperatures. (Ref. 134.)

loss. When the curves cross–that is, when the generation energy rate equals the radiation loss for either reaction–just enough energy is generated to balance the bremsstrahlung radiation loss. The temperature at which this takes place is called the *critical ignition temperature, T_{ci}*.

Above the critical ignition temperature, more energy is produced by fusion than lost by bremsstrahlung radiation, and net energy production is possible. It is seen that the D-T reaction produces more energy at all temperatures, and has a lower critical ignition temperature (about 45 million °K) than the D-D reaction (about 400 million °K).

The *ignition temperature* at which a self-sustaining reactor may operate, is much higher than the critical ignition temperature, being several hundred million °K for the D-T reaction. This is because of (a) the necessity to produce further energy to heat the incoming gas, (b) the presence of heavier nuclei, such as helium (produced in the D-T reaction) and impurities, in the plasma which materially increase bremsstrahlung losses because of their high Z (see Eq. 16-5), and (c) the emergence of synchrotron radiation losses at higher temperatures.

16-5. PLASMA HEATING AND INJECTION

To attain the very high temperatures indicated above, the plasma must be heated. There are several approaches to plasma heating, some of which are associated with particular experimental devices used in fusion research. The approaches may be divided into two broad categories; one heats a "cold" plasma which has already been injected into a magnetic "bottle" (i.e., confined within a magnetic field), the other produces energetic particles outside the bottle and then injects them into the bottle.

In the first category, heating up to relatively "low" temperatures, about a million degrees K, is accomplished by *ohmic*, or *resistance*, heating. Ohmic heating is a gradual heating process in which dc or ac electrical current is passed through the plasma. It thus depends upon the electrical resistance of the plasma, much like a conventional resistance heater. To avoid inserting electrodes in the plasma, the heating current is induced from the outside, Fig. 16-6.

At temperatures higher than a million degrees K or so, the plasma becomes a better conductor of electricity, its resistance decreases, and ohmic heating ceases to be useful. To heat the plasma to higher temperatures, gradual ohmic preheating is followed by sudden *magnetic compression*. Since the confining magnetic field exerts a pressure on the charged particles of the plasma, a sudden increase in the strength of the magnetic field causes the confined plasma to be compressed and heated.

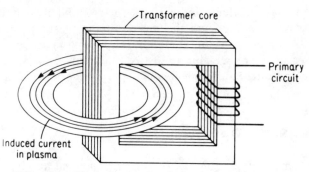

FIG. 16-6. Ohmic heating in plasma by induced current.

This is analogous to a gas becoming heated when compressed by a mechanical device (pump or compressor). Very high plasma temperatures can be attained by two or more successive stages of magnetic compression. Ohmic heating and magnetic compression have been demonstrated in the so-called stellarator (Sec. 16-9) and theta-pinch machines (Sec. 16-7).

In the second category of plasma heating methods, that of preparing energetic particles outside the magnetic bottle and then injecting them into the bottle, two main schemes are used. The first is called *plasma injection* and the second *particle injection.* The first uses a *plasma gun* to produce the high-temperature plasma, which is then injected into the magnetic bottle. Injection is done either by forcing the plasma through the "walls" of the magnetic bottle, or by passing it through a *magnetic valve.* A magnetic valve is formed by some part of the magnetic field whose strength is increased after the plasma has entered the bottle. The second scheme of filling the magnetic bottle from the outside injects a beam of particles. In an early experiment at Oak Ridge, for example, ionized deuterium *molecules* were accelerated to 600 kev and injected as a beam into the bottle. There, each molecule broke up into a deuteron (deuterium ion) and a deuterium atom. The neutral atom escaped but the positive deuteron was trapped in the bottle.

In a variation of the injection method, being studied in the United States, the United Kingdom, and the Soviet Union, a beam of high energy deuterons is first neutralized to a beam of neutral deuterium atoms. The latter easily penetrate the magnetic bottle in which some of them break up into electrons and deuterons which are trapped. This neutral injection scheme involves a two-stage process and may be just as difficult as molecular ion injection. It appears, however, to offer advantages in building up a stable high-temperature plasma.

Another and entirely different method of plasma heating involves the use of laser beams to heat plasmas. The method is receiving some attention in the United States, Italy, and the United Kingdom.

There is some evidence indicating a correlation between the geometry of the plasma container and the most suitable plasma heating method. While all the above-mentioned methods are being investigated in various laboratories, it is conceivable then that the method ultimately used in a fusion reactor might differ from that used in the most successful laboratory experiment.

16-6. PLASMA CONFINEMENT

Plasma at a temperature of the order of several hundred million degrees K is necessary for an operating fusion reactor (Sec. 16-4). The plasma must of course be confined within a container called a *vacuum chamber*. Confinement must keep the plasma away from the walls of the chamber. The problem here then is not the inability of any known materials of construction to withstand the plasma temperatures (though the energy radiated by the plasma in a fusion power reactor may, if not properly removed, damage the walls). Rather the main problem arises because the energetic plasma particles travel in straight but random directions (Fig. 16-7) at speeds averaging tens of thousands of kilometers per second. At such speeds all the particles in a plasma are certain to bombard the walls within a fraction of a microsecond. The bombard-

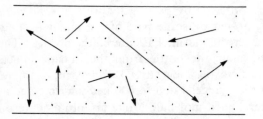 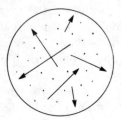

FIG. 16-7. Random plasma particle motion in the absence of a magnetic field.

ment results in the loss of essentially all their kinetic energies and quenches the reaction. In other words, the problem is not that the hot plasma would vaporize the container walls, but rather that the container walls would cool the plasma. Thus even if plasma at the required temperature could be generated, it could not last long enough to generate a practicable amount of energy.

In the sun and stars and in thermonuclear explosions, confinement is accomplished by gravitational or inertial forces. The use of these forces is not practical for a fusion reactor. Instead, the plasma is confined in a *magnetic bottle* away from the physical walls of the container. Advantage is taken of the fact that the plasma is an excellent conductor of electricity and is composed of charged ions and electrons. Charged particles cannot easily cross the lines of a magnetic field, but, to a first approximation, follow helical paths along those lines (Fig. 16-8). Positive and negative particles spiral in opposite directions. A strong enough magnetic field can therefore, in principle, restrict the motion of any charged particle and confine the plasma.

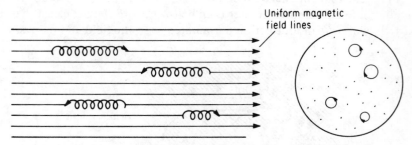

FIG. 16-8.   Helical motion of charged plasma particles along the lines of a
uniform magnetic field.

The radius of a helical path depends among other things upon the mass of the charged particle and the magnetic field strength, being larger for heavier particles. It follows that radii of the helical paths are larger for plasma ions than for plasma electrons.

Unfortunately confinement suffers from the fact that particles, even of the same species and in a uniform magnetic field, move at different velocities in their helical paths. The velocities can be resolved into two components–along the magnetic field lines, and perpendicular to them. The components along the lines determine how fast particles move in the axial direction. The components perpendicular to the lines determine the radii of the helical paths. Extreme cases are for a particle to move rapidly parallel to the magnetic field lines with almost zero radius, or to move essentially in a circle of large radius and not in an axial direction. The result is that charged particles, even of the same species, move with a wide range of radii and axial velocities. Particles, therefore, collide with each other, and may therefore shift from one magnetic line of force to another. Consequently, it is possible for charged particles to *diffuse* in the radial direction and eventually strike the containment walls. The radii of the helical paths, also depend upon the magnetic field strength, being inversely proportional to it. Consequently, a strong magnetic field results in smaller radii, or tighter helical paths, and reduces the tendency of the particles to move radially across the magnetic field.

Another mechanism that causes charged particles to diffuse radially is due to nonuniform magnetic fields. Since the radius of helical motion is inversely proportional to the magnetic field strength, as stated above, a charged particle moving in a nonuniform magnetic field will not have a constant radius as in a uniform magnetic field. This can be illustrated by Fig. 16-9, where a hypothetical magnetic field, perpendicular to and out of the plane of the paper, is assumed to have only two field strengths as shown. The radius of the helical path is smaller in the stronger portion of the field than in the weaker portion. The helical radius above the dividing line is therefore greater than below it, with the net effect that the particle will *drift* in a direction perpendicular to the field and its gradient.

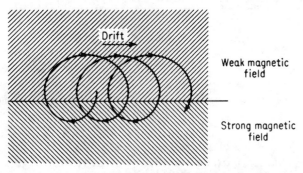

FIG. 16-9. Effect of magnetic field containing hypothetical weak and strong regions on positive particles. Magnetic field lines are perpendicular and outward from plane of paper.

It may be added that drift of positive particles is in a direction opposite to that of negative particles. This causes separation of charges and sets up electrical fields. The combination of electrical and magnetic fields forces the plasma to drift toward weaker magnetic fields, which are often near the container walls.

The above defects in confinement, those of diffusion and drift, can be minimized by building strong uniform magnetic wells. The migration of charged particles along the magnetic lines of force, however, poses a problem of escape from the ends of the reactor.

Research on confinement is proceeding in many countries and on several fronts. The differences in approach in these researches are primarily due to differences in arrangement and geometry of the magnetic field. It was shown above that the plasma particles move both *across* and *along* magnetic lines of force. The various approaches to the problem have therefore been directed at limiting motion across the lines and means of coping with motion along the lines.

There has been essentially a three-pronged attack on the problem, translated in three main methods on confinement. These are: (a) the *pinch concept*, (b) the *open-ended* or *magnetic minor* concept, and (c) the *closed geometry* or *stellarator* concept.

16-7. MAGNETIC PINCH CONFINEMENT

Advantage here is taken of the fact that the plasma, composed of electrically charged ions and electrons, is an excellent conductor of electricity. The electrical conductivity of a plasma is proportional to $T^{1.5}$, where T is the absolute temperature. (At ignition temperatures, the plasma conductivity is some 100 times that of copper at room temper-temperature.) It is possible, therefore, to pass large electrical currents within a plasma.

An electrical current in a conductor (such as a plasma) generates its own magnetic lines of force, that encircle the current, Fig. 16-10. As in simple electromagnetic devices where the right-hand rule applies, if the current is downward, the lines of force are clockwise. The magnetic field, in turn, interacts with the fast-moving plasma particles, exerting

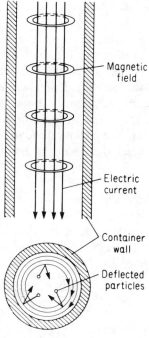

FIG. 16-10. Plasma con-finement by the pinch effect.

radial inward forces upon them, and therefore acts to constrict or confine the plasma to the center of the tube.   This is the so-called *pinch* effect.

To eliminate the need for electrodes, which would quench the plasma, the plasma is confined in a *torus,* Fig. 16-11, and the current is induced by making the plasma essentially a one-turn secondary of a large transformer, as shown in Fig. 16-6, in which the plasma current is regulated by the primary circuit, or by discharging capacitors in the primary.   The plasma current and the constriction would also be relied upon to provide the necessary plasma heating.

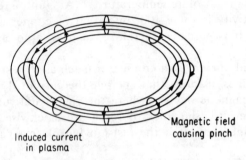

Induced current
in plasma

Magnetic field
causing pinch

FIG. 16-11.   Pinch effect in a torus.

The amount of current required to achieve thermonuclear conditions, however, is enormous.   This can be seen by balancing the gas pressure of the plasma, imagined to be exerted on the magnetic wall and hence called the *outward pressure,* to the confining pressure exerted by the magnetic field lines and called the *inward pressure* (hence the term "magnetic bottle").   Pressurewise, the plasma acts as a normal gas, whose pressure is proportional to the absolute temperature and to its particle density.   The plasma (outward) pressure, p_p, dyne/cm², is therefore given by the gas law

$$p_p = nkT \tag{16-9}$$

where n = plasma particle density, particles/cm³
k = Boltzmann's constant = 1.3805×10^{-16} erg/°K
T = absolute temperature, °K
The confining (inward) magnetic pressure p_B exerted on the plasma by the magnetic field is given by

$$p_B = \frac{B^2}{8\pi} \tag{16-10}$$

where B is the magnetic field strength, in gauss, given by

$$B = \frac{I}{5r} \tag{16-11}$$

where I = current, amperes

r = radius of the pinch, cm

Equating p_p to p_B and combining the above equations give

$$I = \sqrt{200\pi r^2 nkT} \qquad (16\text{-}12)$$

If r and R are the radii of the pinched plasma and the confining tube respectively (so that their cross-sectional areas are πr^2 and πR^2), and N and n are the plasma densities before and after pinch, then

$$nr^2 = NR^2 \qquad (16\text{-}13)$$

and

$$I = \sqrt{2.76 \times 10^{-14}\pi R^2 NT} \qquad (16\text{-}14)$$

It can be seen that for reasonable thermonuclear conditions of $T = 10^8$ °K and $N = 10^{15}$ particles/cm³, and for a confining tube of 10-cm radius only, a current of almost a million amperes would be required. Such currents would be extremely difficult, though not impossible, to generate continuously. *Pulsed* systems in which very large currents are produced intermittently for minute fractions of a second have been considered.

Despite the above difficulty, pinch systems were attractively simple and seemed to be a reasonable approach to the problem. Work proceeded in the early 1950's both in the United Kingdom and the United States on pinch systems. The first U.S. experimental device, at Los Alamos, was dubbed the *Perhapsatron* (at one time also called the "Impossibilitron.") It was in the form of a torus, several feet in diameter, similar to that in Fig. 16-6. Several tens of thousands of amperes were induced in it by discharging a capacitor bank through a coil coupled to the torus.

Pinch Instabilities

The Perhapsatron and independent theoretical work at Princeton showed that pinched discharges are subject to violent and inherent instabilities. The main instabilities are due to *kinks* and *sausages*, Fig. 16-12. When a small kink develops, the magnetic field lines crowd on the concave side with consequent buildup of magnetic pressure. This in turn makes the kink grow until the plasma is pushed to the container wall and quenched. Sausage instability occurs when sausagelike defects occur, also causing the magnetic field lines to crowd at the constriction and the plasma column to tear into pieces at different points. Experiments on the Perhapsatron indeed showed a brilliant discharge which contracted to a finely pinched filament upon applying the voltage of the capacitor bank, followed rapidly by a disappearance of the pinch and a glow throughout the volume of the torus tube.

Several lines of attack were pursued or proposed as solutions for the instability problem. One was to prevent the instability, possibly by creating a weak axial magnetic field in the plasma which would oppose the defects by the resistance of the axial magnetic lines to crowd together. The axial magnetic field would be generated by passing electric currents through coils wound around the tube. Another suggestion was the concept of "radio-frequency pinch," in which the pinch would be produced

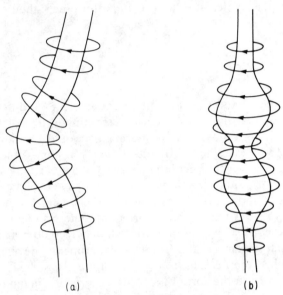

(a)   (b)

FIG. 16-12. Instabilities in pinched plasma: (a) kinks; (b) sausages.

by applying an alternating high-frequency voltage, instead of dc voltage as in the early experiments. A third suggestion was to accept instabilities as facts of life, heat a dense plasma to thermonuclear conditions, extract power from it, all prior to the onset of quenching by the instabilities.

None of the above approaches became really workable, mainly because of a lack of understanding of the true mechanism of instability. Various attempts, such as those that would induce "image currents," etc., and theories, such as the M-theory, the Columbus Concept [156], etc., were made or formulated. While they suggested ways to understanding and even overcoming the effects of the instabilities, the suggestions were very impractical (currents of several 10^8 amperes, pinches several meters in diameter, and extremely large voltage gradients). An experiment called *Columbus-I* was designed and constructed to test some of these solutions.

The Theta Pinch

The *theta pinch* apparatus is a pinch machine of the magnetic mirror type (next section) which combines heating and confinement. The deuterium gas is first ionized into a plasma. The plasma is then heated by a very rapid compression with a magnetic field pulse. The compressed state lasts only a very short time (a few microseconds in current experiments). The magnetic field pulse is produced by discharging a high-voltage low-inductance capacitor bank through a single-turn coil. In General Electric laboratory experiments, 60 kv capacitor banks are discharged in the coil with spark gap switches (operating with a timing accuracy of 10^{-8} sec.). The magnetic field rises from zero to 100 kilogauss in a few millionths of a second, resulting in pressures on the plasma of the order of 70 atmospheres. Similar schemes have been proposed for toroidal configurations.

16-8. CONFINEMENT BY OPEN-ENDED OR MAGNETIC MIRROR MACHINES

The *open-ended* machine is illustrated by the simple *magnetic mirror* shown in Fig. 16-13. The magnetic field lines are allowed to leave the confinement region. The particles are, however, prevented from escaping from the ends of the machine along these lines by making the magnetic field stronger at the ends. As the charged particles approach

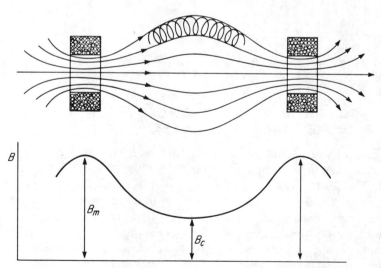

FIG. 16-13. Simple magnetic mirror. Field lines and approximate particle trajectory shown in upper diagram. Axial magnetic field strength B shown below.

the ends, the strong magnetic fields there make them spiral faster in smaller radii so that their velocity components along the lines of forces are reduced. If all the motion energy of the particles along the lines is converted to energy of spiral motion, the particles are reflected; hence the term "magnetic mirror". This, of course, is an ideal situation, and magnetic mirror machines tend to be somewhat leaky. Their success depends upon making the leakage small enough to achieve necessary plasma confinement time, Sec. 16-11, as well as on solving the important problem of plasma instability.

There are more complicated forms of open-ended traps than shown in Fig. 16-13. In one, shown in Fig. 16-14, the mirror field is combined with a multipole field from a set of linear conductors running parallel to the axis of the mirror coils. (A sextupole set is shown in Fig. 16-14.) Such a system, first experimented upon by Ioffe in the Soviet Union is called a *minimum-B trap*. It eliminates one form of instability associated with the plasma.

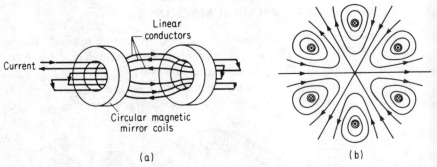

FIG. 16-14.   Minimum-*B* trap magnetic mirror multipole system, (a).   The resultant combined magnetic field is shown in end view (b).

16-9. CONFINEMENT IN CLOSED-GEOMETRY OR STELLARATOR MACHINES

In the *closed*-geometry concept, the magnetic field produced by a toroidal solenoid in a doughnut-shaped (round or oval) configuration is used, Fig. 16-15. The magnetic lines of force are produced by passing current in a coil, or in separate turns, wound around the torus tube and called the *confining* coil or windings. The resulting magnetic lines become closed, thus preventing the axial escape of charged particles from the machine. Because the windings are closer on the inside than the outside, toroidal solenoids have magnetic fields that are weaker near the outer boundary of the solenoid than near the inner boundary. As shown previously, this causes the plasma to drift radially outward because of

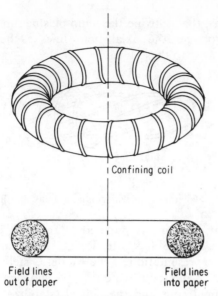

FIG. 16-15. Torus with external confining coil producing internal axial magnetic field.

electric fields developed by the separation of charges due to the nonuniform magnetic fields.

The radial drift in toroidal configurations can be avoided by several methods of more complex field configurations. In one such method the magnetic field lines are made to twist as they move around the torus such that each does not close upon itself after one complete round around the torus. This causes the electric field to essentially short out. This configuration, previously noted, is referred to as the *stellarator*.

In the early models of the stellarator, the twist of the magnetic field lines was achieved by twisting the toroidal tube itself into a figure 8, or a pretzel, Fig. 16-16. In later models the same effect was achieved by adding, to a simple planar torus, helical or corkscrew windings, called

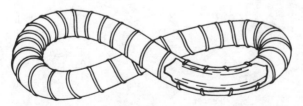

FIG. 16-16. Figure-8 toroidal tube with confining coil. Charged particle drifting upward at one end will drift downward at opposite end.

stabilizing windings, that entwine the main plasma tube, Fig. 16-17.   The confining coil produces the axial field lines as before.   The helical windings provide twist to these lines.

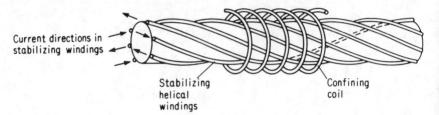

FIG. 16-17.   Stellarator tube with confining and stabilizing windings.

The stellarator principle is pursued in research in many countries, in particular the United States, the Soviet Union, and West Germany. Figure 16-18 is a representation of the Model C stellarator at the James Forrestal Research Center, Princeton, N. J., showing the torus (vacuum tube), confining field coils and the helical (stabilizing) windings.   Also shown are ohmic heating coils which preheat the plasma to about one million degrees K.   The *divertor* is used to remove the impurities from the plasma before it is heated further.   The impurities magnify radiation losses.   The divertor uses a magnetic field to divert the impurity ions from the main tube into a side chamber from which they are removed by vacuum pumps.   One or more ion-heating sections heat the plasma to thermonuclear temperatures by magnetic pumping or compression (Sec. 16-5).   The ion-heating sections contain magnetic coils that surround the

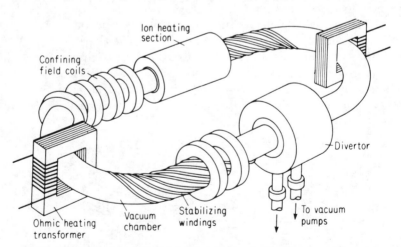

FIG. 16-18.   Sketch of the Model C stellarator built at the James Forrestal
Research Center, Princeton, N.J.

main tube and alternately increase and decrease the local magnetic field strength at high frequency. This alternately, compresses (and thus heats) the plasma, and expands (and thus cools) it. At the proper frequency heating exceeds cooling and a net heating to thermonuclear temperatures is attained.

Stellaratorlike machines, though they still have many problems, show the greatest promise in attaining controlled thermonuclear reactions. One of the characteristics of the stellarator is that the twist of the lines of force due to the combination of confining coils and stabilizing windings is not a simple twist where all lines of force undergo the same angular twist as they move along the main tube. Rather the twist has *shear*. Lines of force occupying the same angular position at a given cross section, rotate at different angular velocities as these lines move along the tube. This is shown in Fig. 16-19. The lines of force farther from the center of the tube rotate through a larger angle than those closer to the center. It has been shown theoretically that such shear fields are capable of preventing the growth of instabilities (provided that the ratio of the plasma "outward" pressure, to the "inward" pressure, given by Eqs. 16-1 and 16-2, called $\beta = 8\pi nkT/B^2$, does not exceed a limiting value–about 20 percent in the model C stellarator).

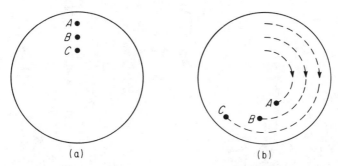

FIG. 16-19. Shear fields in a stellarator. A, B, and C are lines of force, perpendicular to paper. Their positions are shown at an arbitrary starting cross section of the plasma tube in (a), and at a cross section farther downstream in (b).

Unfortunately, model C stellarator was built before the phenomenon was recognized, and had weak shear. The *Tokamak* machines in the Soviet Union, which are stellaratorlike machines, using confining and stabilizing windings *and* an axial current, have, on the other hand, been built with high shear, and have become the first devices to reach parameters (density, temperature, and confinement time) within one order of magnitude of those required in an operating fusion reactor.

16-10. OTHER CONFINEMENT SCHEMES

The above three confinement schemes have received the greatest attention. Other schemes however, have been proposed and were the subject of lesser research efforts. Some of these are the *cusped-geometry* concept, the *astron* concept, and the *rotating-plasma* concept.

The *cusped geometry* program was investigated at Los Alamos and New York University. The basis of the cusped geometry (cusped means pointed, as when two curved lines of force meet at a sharp angle) was the theoretical conclusion that a plasma confined by magnetic lines of force that curve away from it is inherently stable, while a plasma confined by magnetic lines that curve toward it tends, on the other hand, to be unstable. A geometry in which the magnetic lines of force curve away from the plasma everywhere was thus offered in the hope of confining

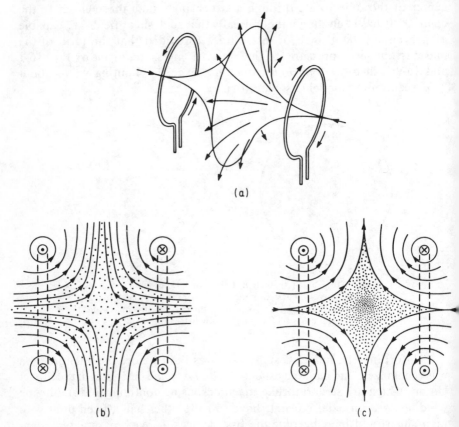

FIG. 16-20.   The cusped geometry.   (a) The development of magnetic lines of force.   (b) The escape of charged particles between lines of force in low-density plasma.   (c) In ideal high-density plasma, sharply cusped geometry is obtained and plasma is confined.

plasma in a stable manner. This is a cusped magnetic field obtained by passing currents in opposite directions in two parallel field coils, Fig. 16-20a. (If the currents were in the same direction, the magnetic field would resemble that of the magnetic mirror, Fig. 16-13.) The system suffers from severe leakage of the charged particles for low-density plasmas, Fig. 16-20b. On the other hand a plasma of high density (actually high β) is well confined in a cusped manner with a well-defined boundary between the plasma and the magnetic field, Fig. 16-20c. At actual thermonuclear conditions it was felt that adequate confinement would be obtained. The main problem was to avoid the high leakage while the plasma density was being built up. The cusped geometry was preceded by the so-called *picket fence* concept which used an array of conductors with current moving alternately in opposite directions. This would produce the desired magnetic field configuration which bends convexly everywhere from the plasma. It was shown, however, that the picket fence would suffer materially from leakage. Work on cusped geometry was discontinued around 1956 in the United States when other concepts proved more promising.

The *Astron* concept is pursued in the United States at the University of California Radiation Laboratory, Livermore. A hybrid concept, it envisions the production of a toroidal magnetic field within an open-ended geometry. This is done by establishing an axial magnetic field in an evacuated, long open-ended tube by passing a current through a coil wound on the outside of the tube. The coils are wound in such a way that they produce one magnetic field "bump" at each end. In this the Astron is somewhat similar to the magnetic mirror. A beam of high-energy electrons (several Mev energy) is injected from one end of the tube. The magnetic field causes the electrons to gather in a cylindrical layer of current, called the E-layer, about the tube axis, Fig. 16-21. The current in the E-layer in turn produces its own magnetic field which

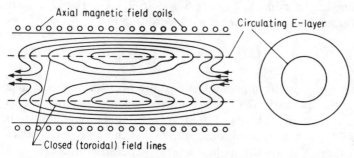

FIG. 16-21. The Astron concept. Interaction of axial magnetic field produced by outside coils with field due to circulating E-layer electrons produces toroidal magnetic field.

interacts with the axial field. This results in a magnetic field configuration in which the lines of force close completely on themselves in the shape of a somewhat distorted torus. If a D and T gas mixture is injected, it will be ionized by the electrons, trapped by the magnetic field and gain energy by collisions with the E-layer. One of the problems of the Astron principle is the instability of the electron layer. The first Astron model was begun in 1957 to experimentally verify some of the principles. Work on the Astron is continuing and includes a proposal to look into proton layers.

The principle behind the *rotating plasma* concept was that the gyroscopic action of a rotating plasma would render the plasma stable. Initial work on rotating plasma was done at UCRL, Berkeley, in a device called the *Homopolar*.

16-11. PLASMA DENSITY AND CONFINEMENT TIME

Practically allowable plasma pressures are limited by the maximum available magnetic field strength, since from Eqs. 16-9 and 16-10,

$$p_p = p_B = \frac{B^2}{8\pi} \qquad (16\text{-}15)$$

Since p_p is thus limited, and since T is extremely high, it follows from Eq. 16-9, $p_p = nkT$, that it is only possible to confine plasmas of low density n. Practical plasmas are believed to fall in a density range between 10^{14} (low-density plasma) to 10^{18} (high-density plasma) particles/cm³, with an average value of the order of 10^{15} particles/cm³ in most cases. This is approximately 1/10,000th the density of a gas at atmospheric temperature and pressure which is about 3×10^{19} molecules/cm³, and is therefore considered an extremely high vacuum.

The maximum energy generated by a pressure-limited fusion reactor is obtained by combining Eqs. 16-8 and 16-9. For the D-T system with equal D and T densities, $n_1 n_2 = n^2/4$, resulting in

$$(\Delta Q_g)_{\max} = \frac{p_{\max}^2}{4k^2} \left(\frac{\overline{\sigma v}}{T^2} \right) \Delta E \qquad (16\text{-}16)$$

where $p_{\max}$ is the maximum allowable pressure. For a given value of $p_{\max}$, the quantity $(\overline{\sigma v}/T^2)$ is a measure of the maximum energy that can be generated. This is plotted against kinetic temperature in kev in Fig. 16-22. It can be seen, once more, that the D-T reaction is superior to the others. It has the highest energy generation in a pressure-limited fusion reactor. Its critical ignition temperature at about 4 kev is below the energy optimum point at 14 kev, so that a D-T reactor may be ignited

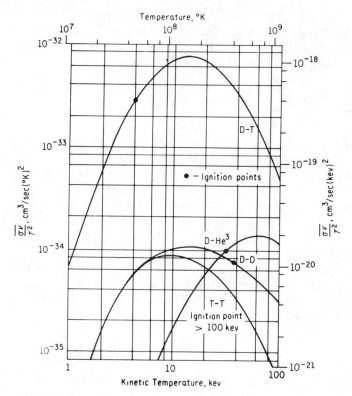

FIG. 16-22.   Variation of the parameter ($\sigma v / T^2$) with temperature for four fusion reactions.

below the optimum, then operated at optimum conditions.   This is not the case for most other reactions.

Additional factors other than magnetic pressure enter into the consideration for plasma density.   The most important is that of *confinement time*.   It should be apparent now, that even under the best confinement scheme, high-energy plasma particles will sooner or later escape the confinement of a magnetic field.   The hot plasma must nevertheless be held in confinement for a sufficiently long period of time so that an appreciable number of fusion reactions takes place.

The confinement time should not be too short such that new fuel is introduced at a rate faster than it can be heated by the existing plasma. It need not be infinitely long, that is, it is not necessary to operate as a continuous reactor.   In fact, it should be short enough such that the temperature of the electrons in the plasma does not rise to a value where the radiation losses become excessive.   There thus seems to be an optimum confinement time for each given set of conditions (density, temperature, etc.). An expression for such an optimum confinement time, derived by Mills [159], was obtained by equating the fraction of the power generated that is retained in the plasma (and not released to the power plant) to the power required to heat the incoming gas:

$$\frac{n^2_i}{4} (\overline{\sigma v}) \Delta E f = \frac{3k}{2\tau} (n_i T_i + n_e T_e) \qquad (16\text{-}17)$$

where the subscripts i and e are for the ion and electron respectively, and f is the fraction of power retained. τ, the confinement time, would then be given by

$$\tau n_i = \frac{6k(T_i + T_e)}{(\overline{\sigma v}) \Delta E f} \qquad (16\text{-}18)$$

A useful figure of merit is the product $n\tau$ where n is the plasma density and τ the confinement time. It has been shown that for any practical fusion reactor that product must be greater than a minimum value. This is known as the *Lawson criterion*. For a fusion reactor using deuterium only as fuel,

$$n\tau > 10^{16} \qquad (16\text{-}19)$$

while a fusion reactor using deuterium and tritium would have

$$n\tau > 10^{14} \qquad (16\text{-}20)$$

where n is in particles/cm^3 and τ is in seconds. Thus for a D-T reactor with 10^{15} particle/cm^3 density, the confinement time need only be greater than 0.1 sec; and there is a trade-off between density and confinement time, a 10^{16} density requires 0.01 sec and so on. The D-D reactor, on the other hand, requires confinement times, or densities, one hundred times greater.

Early plasma experiments failed to meet the Lawson criterion by several orders of magnitude. Almost without exception the experiments suffered from faulty confinement, and much of the research has been aimed at understanding the reasons for the faults. More recently (1968-69) the Tokamak-3, a late model of the stellarator-type machines in the Soviet Union [160, 161], showed particle densities and temperatures both about one tenth of the values required for a fusion reactor. Much optimism has been therefore generated in the early 1970's that a practical fusion reactor will be a reality in 15 to 25 years. Several countries are now building various versions of the Tokamak.

16-12. THE D-T FUSION REACTOR POWER PLANT

Many arguments have been advanced why a D-T-fueled fusion reactor is preferable to all other types. This section will describe a conceptual fusion reactor power plant (also referred to as the *controlled thermonuclear power reactor,* CPTR), using the D-T fuel cycle. Such a plant is shown schematically in Fig. 16-23.

The *fusion reactor* is shown in cross section to the left of the diagram, and could be of either the closed (stellarator) or open (magnetic mirror) type. The plasma is confined magnetically within a container *(vacuum chamber).* A *divertor* is included in the container complex. The divertor, Sec. 16-9, removes impurities from the plasma. The vacuum chamber wall receives large quantities of heat flux from the plasma (Sec. 16-14) in the form of bremsstrahlung and synchrotron radiations (though part of the latter is reflected back to the plasma). The wall must therefore be cooled, probably with the help of a sandwich-type construction. It must also be made of a material that resists radiation damage, resists corrosion by the coolant, and has low vapor pressure. Niobium and molybdenum are likely candidates for the wall material. Molybdenum contributes to neutron multiplication (n, $2n$ reactions) needed for tritium breeding (Sec. 16-13).

The vacuum-chamber-wall *coolant* poses interesting problems. The requirements for such a coolant are that it (a) must operate at high temperature, (b) have good heat transport capabilities, (c) should not be a neutron sink (but if possible contribute to neutron multiplication), (d) should be easily pumped, and (e) must be metallurgically compatible with reactor materials. Ordinary water and organic coolants are excluded because of (a) and (c) and heavy water because of (a). Liquid metals, particularly lithium, are likely candidates. However, they suffer from the difficulty of circulating them in the vicinity of the strong magnetic field of the plasma. There will be excessive ohmic losses in a conducting coolant, and the power required to pump such a coolant in a magnetic field is greater than that required to pump a nonmetallic coolant. Molten salts have therefore been suggested as container wall coolants, with the LiF-BeF_2 eutectic (called *flibe*) receiving the most attention. The beryllium in it contributes to neutron multiplication and moderates the neutrons to tritium-breeding energies (Sec. 16-13). It has also shown satisfactory corrosion characteristics with the alloy INOR-8. Other molten salts receiving attention are the lithium nitrates and nitrites, though the nitrogen and oxygen compete for neutrons [156]. Mixtures of liquid metals and molten salts such as lithium and flibe are also being considered.

The *lithium blanket* surrounding the vacuum chamber has multiple

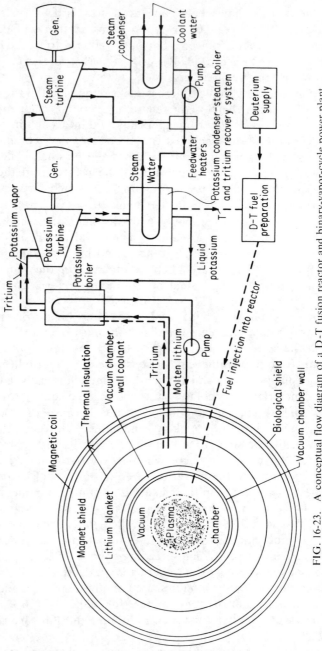

FIG. 16-23.   A conceptual flow diagram of a D-T fusion reactor and binary-vapor-cycle power plant.

functions. (a) It acts as a moderator and neutron absorber that receives the high-energy neutrons of the D-T reaction and converts their kinetic energy to heat. That reaction, Eq. 16-3, is here repeated with a breakdown of energy as

$$D + T \longrightarrow He^4(3.52 \text{ Mev}) + n^1(14.06 \text{ Mev}) \qquad (16\text{-}21)$$

(b) It acts as the heat source for the power plant. The power density in it is a strong function of the radius [estimated to decrease by several orders of magnitude in a 2 to 3 meter-thick blanket in a 6 to 7 Gw(t) Tokamak-type reactor] posing interesting heat removal problems. (c) It acts as the tritium breeder (Sec. 16-13). (d) It absorbs the γ from the neutron inelastic scattering reactions, and (e) it is a first shield for the magnetic coils.

The material of the blanket could be pure lithium metal, a lithium-bearing fused salt such as flibe, or a mixture of the two. Lithium is lower in cost than flibe and has superior heat-transfer characteristics. It suffers, however, from the difficult pumping problem in a strong magnetic field. The electromagnetic resistance to flow of lithium could be minimized by proper design, such as by making the coolant channels parallel to the confining magnetic field lines.

The *magnetic coils* are placed on the outside after further shielding. The coils are placed there, instead of between the plasma and blanket, to protect them against neutron bombardment which would cause structural damage and increased electrical resistivity. Their presence there also would have adversely affected the neutron economy of the reactor (necessary, among other things, for tritium breeding). Normally, the power required to operate the coils would be a sizable fraction of the gross electrical output of the power plant. This is the reason why *superconducting* coils are chosen. Superconducting coils are cooled to cryogenic temperatures (another reason for placing them on the outside of the reactor), and only a small fraction of the electrical output would be needed for refrigeration. Furthermore the helium generated in the D-T reaction would be available as refrigerant.

Table 16-2 compares the costs of producing 100 kilogauss in a 6-in. bore coil using three materials. It shows the superiority of superconductors as coil materials. Because of their low temperatures, the superconducting coils must be shielded against thermal radiation and conduction from the lithium blanket.

Finally, the reactor is surrounded by suitable biological shielding.

In the conceptual CPTR plant of Fig. 16-23, the blanket material (lithium, or lithium salt) gives up heat to liquid potassium in a heat exchanger producing potassium vapor which generates power in a potassium-vapor turbine. The turbine potassium exhaust condenses in a

TABLE 16-2
Relative Costs of Producing 100 Kilogauss in a 6-in. Bore Coil*

Material	Temper- ature, °K	Current Density, amp/cm²	Relative Annual Cost		
			Fixed	Operating	Total
Cu	328	600	51	192	243
Na	8.5	1000	85-155	35	120-190
Nb₃Sn	4	10⁵	100	0	100

* From Ref. 162.

potassium-condenser-steam-boiler which generates steam for a conventional steam cycle. Thus we have a binary-vapor cycle. Such cycles, it is recalled (Sec. 2-10), are used to take advantage of a particularly high-temperature heat source and a low-temperature heat sink (the environment).

In variations on the conceptual design of Fig. 16-23, the potassium, rather than the blanket material is pumped, or gas coolant helium, or heat pipes (Sec. 12-8) transport heat between the lithium and potassium.

Some of the tritium generated in the lithium blanket is given up to the potassium system and is recovered in the potassium-condenser-steam-boiler. The recovered tritium then is combined with fresh deuterium to make new fuel, and the mixture prepared for injection into the reactor.

Because of the high-temperature source of the binary cycle, its thermal efficiency is expected to exceed 60 percent. This is about 50 percent higher than the best efficiencies (42 percent) obtained from modern fossil-fueled power plants and some 100 percent above ordinary-water-reactor power plants. This of course means conservation of fuel, but also very low thermal pollution per Kw(e) installed, as compared with other power-plant types. This is an advantage to be added to the almost complete elimination of radioactive waste products and the absence of air pollution (the advantage it shares with all other nuclear plants). Figure 16-24 shows an artist's conception of a D-T fusion power plant and its major components.

A conceptual design for a fusion-power plant using a magnetic mirror reactor has been suggested by Frass [163].

16-13. TRITIUM-BREEDING AND NEUTRON MULTIPLICATION

One disadvantage of the D-T fuel cycle is the absence of large quantities of naturally occurring tritium, and the need, therefore, to breed tritium. Tritium is bred in the lithium blanket. All natural lithium is composed of 7.42 percent Li⁶ and 92.58 percent Li⁷. The breeding reactions are

$$\text{Li}^6 + n \longrightarrow \text{He}^4 + \text{T} + 4.8 \text{ Mev} \qquad (16\text{-}22)$$

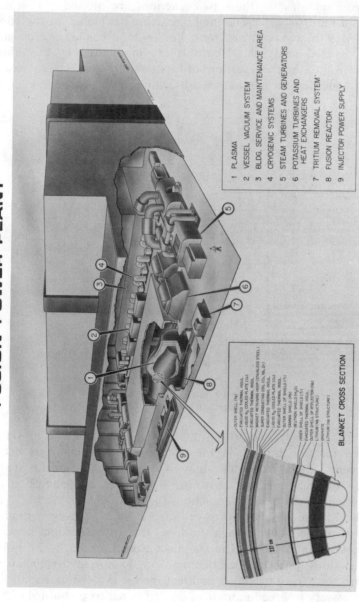

FUSION POWER PLANT

1 PLASMA
2 VESSEL VACUUM SYSTEM
3 BLDG. SERVICE AND MAINTENANCE AREA
4 CRYOGENIC SYSTEMS
5 STEAM TURBINES AND GENERATORS
6 POTASSIUM TURBINES AND HEAT EXCHANGERS
7 TRITIUM REMOVAL SYSTEM
8 FUSION REACTOR
9 INJECTOR POWER SUPPLY

BLANKET CROSS SECTION

FIG. 16-24. Artist's conception of a fusion power plant. (Courtesy Oak Ridge National Laboratory).

and

$$\text{Li}^7 + n + 2.5 \text{ Mev} \longrightarrow \text{He}^4 + \text{T} + n \qquad (16\text{-}23)$$

The cross sections for these two reactions are plotted against neutron energy in Fig. 16-25. While the Li^7 reaction has higher cross sections for the 14-Mev, newly born neutrons, Eq. 16-21, it has lower cross sections for moderated neutrons. Neutrons are moderated by the lithium to energies where the cross sections of Li^6 predominate. The Li^7 reaction is also an endothermic one that absorbs, rather than generates, energy.

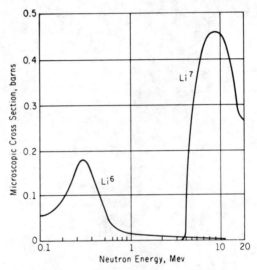

FIG. 16-25. Tritium-yielding reaction cross sections for Li^6 and Li^7 (Eqs. 16-22 and 16-23) as functions of neutron energy.

Since naturally occurring lithium is abundant in Li^7, consideration is given for enriching it in the isotope Li^6. Enrichments exceeding 20 percent are said to be desirable. The contributions of Li^7 are mainly in the downscattering of neutrons, but also it contributes to tritium production as well to $(n, 2n)$ reactions (below).

The lithium reactions, Eqs. 16-22 and 16-23, however, produce no neutrons of their own. Each produces one triton (tritium nucleus) for each neutron absorbed. The D-T fusion reaction, Eq. 16-21, unlike fission reactions, produces but one neutron. Some of these neutrons are lost through parasitic capture and leakage (as in a fission reactor). It is also necessary to breed more than one triton per triton burned because of the loss of tritium by radioactive (β) decay, and because of losses in plant holdup tanks, and in recovery and recycling operations. Tritium breeding ratios (tritons bred per triton burned) of more than 1.15 are said to be necessary [164]. This points to the necessity of finding a means of neutron multiplication through some $(n, 2n)$ reaction. Neutron multiplication can be accomplished, as we have become accustomed, in a fissionable process, or in nonfissionable processes.

In fissile neutron multiplication, the 14-Mev neutrons that are newly born in the D-T fusion reaction may be made to fission U^{238} which has a fission cross section of several barns at that energy. This would have the advantages of breeding some Pu^{239}, and the generation of fission heat which would keep the blanket hot. It thus results in a self-sustaining composite fuel cycle by the recycling of the plutonium and the production of tritium in the blanket. It has the disadvantage of turning the fusion reactor into a hybrid fusion-fission reactor which negates the main advantages of the fusion reactor in that it uses the abundant deuterium fuel only, and produces little radioactive waste. Fissile multiplication is also more costly than nonfissile multiplication.

Nonfissile neutron multiplication can be accomplished by bombarding several metals by the fast neutrons produced in the D-T fusion reaction. One promising metal is beryllium which occurs in nature as 100 percent Be^9. The reaction is

$$Be^9 + n \longrightarrow 2He^4 + 2n \tag{16-24}$$

The cross section for this reaction is about 0.6 barn for neutrons in the 5-15 Mev range, but drops to zero below 2 Mev. A beryllium layer may then be placed next to the vacuum chamber surrounding the plasma to take advantage of the energetic neutrons. Other metals with high $(n, 2n)$ cross sections are W, Pb, Ta, Mo, Fe, Cu, as well as Li^7 itself. The presence of the latter in the blanket may make it unnecessary to add a layer of beryllium or other material [165].

16-14. SOME TECHNOLOGICAL PROBLEMS

The technological problems to be solved before a commercial fusion-reactor power plant becomes feasible are, as should be apparent by now, enormous. Only a few of these problems are outlined below.

(a) *The interaction between plasma and vacuum chamber wall.* The wall is subjected to bombardment of ions, neutrons, electrons, and photons from the plasma. The effects of ions and neutrons are the most serious.

Ion bombardment causes *sputtering* of the wall. The sputtering ratio of deuterium ions on even niobium, the currently favored wall material, is about 0.004 atom per ion.* (Niobium is favored over molybdenum because of its superior welding characteristics and lower cost. Beryllium, while a good moderator and contributor to neutron multiplication, suffers from severe radiation damage and is also expensive.) Sputtering could result in an unacceptably high concentration of heavy niobium ions which would significantly increase bremsstrahlung radiation losses (See Eq. 16-5). Furthermore, the resulting niobium ions cause further sputtering of the niobium. This may lead to build up of niobium in the plasma as well as an erosion of the wall to the extent of weakening its mechanical strength after a few years operation. The cure here is for good confinement that would allow only a few of the charged particles to bombard the wall, and in the development of an efficient divertor that would remove heavy impurities as they are formed.

Neutron bombardment, estimated at 10^{23} neutron/cm^2 fluence in a 20-year lifetime, affects the mechanical integrity of the wall mainly by displacement of atoms from their lattice sites, and also by transmutations. While there now are inadequate data on the damage caused by the 14-Mev neutrons of the D-T reaction under fusion reactor conditions, preliminary theoretical studies seem to indicate no particularly disastrous effects.

(b) *Heat flux on the vacuum chamber wall.* The wall receives heat from the plasma and from the blanket. Of the three components of plasma energy, 14-Mev neutrons, charged particles and radiation, the neutrons pass through the wall without imparting energy to it and, ideally, the charged particles are confined by the magnetic field and removed in an orderly manner. Radiation is therefore the main plasma contribution to wall heating. Assuming the plasma to be continuously and successfully purged of impurities, only D-T bremsstrahlung radiation will affect the wall. The ratio of this bremsstrahlung to total plasma energy in a pure plasma is about 1 percent. In a parametric study of a fusion reactor [162], the total heat flux in a plasma is 1.2-1.7 kw(t)/cm^2 of wall surface area. The radiation from the plasma, therefore would be 12-17 w(t)/cm^2 which is very manageable.

The lithium blanket contributes gamma backscattering to the wall. In the above study, this was shown to add another 100-200 w(t)/cm^2 to the wall. The total sum, while large, is within the capability of both liquid-metal and fused-salt coolants.†

(c) *Fuel injection and heating.* There are two categories of fusion reactors. One is the *steady-state* reactor which requires a continuing supply of new fuel at a rate which depends upon the confinement time

---

* The ratio depends upon ion energy. It varies from about 0.006 at 12 kev to about 0.003 at 28 kev.
† 100 w(t) cm^2 corresponds to 317,000 Btu/hr ft². This figure and the wall material should probably be checked for burnout conditions.

or the fractional burnup of the plasma. The other is the *unsteady-state,* or *pulsed,* reactor, in which there may be no refueling during a power pulse and plasma existing only for the duration of the confinement time. The newly injected atoms stand the danger of being ionized and trapped in the outer few centimeters of the plasma, unless they are injected with uneconomically high velocities. A suggestion, now under consideration, was made by Spitzer for the injection of frozen D-T pellets of sufficient mass and velocity (about 2-cm diameter at about 10^4 m/sec for 10^{14} cm^{-3} density plasma). Such pellets would penetrate deep into the plasma before being completely ablated.

The energy required to heat the incoming plasma is equal to $3\,nkT$. The power output of the plasma is given by Eq. 16-8. For the D-T plasma, $n_1 n_2 = n^2/4$, and the ratio of the two (Joule/watt) is given by

$$\frac{\text{Plasma heating energy}}{\text{Fusion power}} = \frac{12kT}{n(\overline{\sigma v})\,\Delta E} \tag{16-25}$$

For the D-T reaction with kinetic temperature $kT = 15$ kev and $\Delta E = 17.6$ Mev, the above ratio would be about $3 \times 10^{13}/n$. For values of $n = 10^{14} - 10^{15}$ ion/cm^3, the energy required to heat the plasma to operating temperature would therefore be of the order of 0.1 Joule/watt. This energy need be supplied in a time less than the confinement time τ, and therefore in short-duration pulses. With confinement time of the order of 1 sec, the *peak* heating power would be of the same order as the fusion output power.

For steady-state closed (toroidal) systems, this poses no unsurmountable problems. Open (mirror) systems require temperatures much in excess of 15 kev with energy supplied from the outside. The ratio, therefore, is much higher.

16-15. A DIRECT-CONVERSION FUSION POWER PLANT

A direct-conversion fusion-reactor power plant using either the D-D, Eqs. 16-2, or D-He^3, Eq. 16-4, reactions and being developed at the Lawrence Radiation Laboratory at Livermore, California, is shown diagramatically in Fig. 16-26. The reactor in this system would be of the magnetic mirror type, Sec. 16-8. The D-He^3 reaction, Eq. 16-4, here repeated,

$$D^2 + He^3 \longrightarrow He^4 + H^1 + 18.3 \text{ Mev} \qquad [16\text{-}4]$$

has as products positively charged helium and hydrogen ions. The "exhaust" from the mirror machine would thus be composed of these products plus some of the original fuel ions and electrons. The total

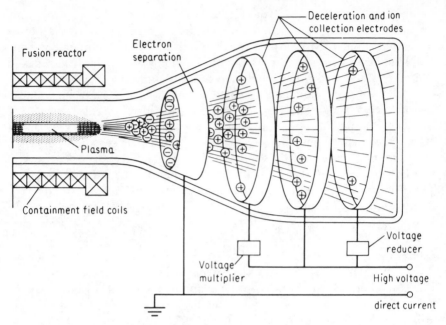

FIG. 16-26. Schematic representation of a direct-conversion fusion-power plant.

products will escape from the end of the magnetic mirror and allowed to expand through a diffuser-like channel to a larger chamber. This expansion causes further decrease in the already low density of the products; and like the expansion of hot gases in a rocket nozzle, the random motion of the charged particles is converted into an orderly unidirectional motion.

The direct conversion unit is in the exhaust chamber. It operates on the principle of direct collection of charged particles. The electrons are first separated electromagnetically from the positive ions and are allowed to flow to the ground of the electrical system. The stream of positive ions, carrying most of the energy of the fusion reaction, is composed of ions of different energies, and therefore electrical potential. These ions are caught by a series of electrostatic collectors, each one kept at slightly higher potential than the preceding one. A common average potential will then be obtained by the use of suitable dc voltage multipliers or reducers. It is estimated that only about 30 percent of the energy is different from average and need be subjected to such voltage transformation. The result is a high voltage dc supply.

Like direct collection energy converters (Chapter 15), a power plant of this type is not a heat engine, and therefore is not Carnot-efficiency limited. Consequently high efficiencies, possibly as high as 90 percent,

FIG. 16-27. Artist's conception of a direct conversion mirror machine fusion power plant. (Courtesy University of California Lawrence Radiation Laboratory and the U. S. Atomic Energy Commission.)

barring now unforeseen difficulties, may be possible. The concept is simple, resulting in expected low capital costs for large units.

Preliminary small-scale tests of the concept are expected in 1971 or 1972. The full-scale realization of the concept, if successful, is probably decades away, however. Figure 16-27 shows an artist's conception of a 1.000 Mw (e) direct-conversion fusion power plant.

16-16. THE FUSION TORCH

An interesting concept, though also not expected to be a reality for sometime, is one proposed by Eastlund and Gough of the US Atomic Energy Commission [166]. Called the *fusion torch,* the concept now involves two uses of the high-temperature plasma. One is to convert any material back to its basic elements, and the other to generate large quantities of ultraviolet radiation, Table 16-3.

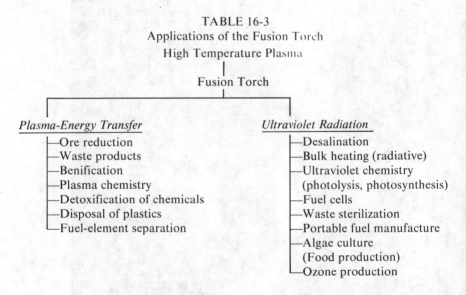

TABLE 16-3
Applications of the Fusion Torch

High Temperature Plasma

Fusion Torch

Plasma-Energy Transfer	*Ultraviolet Radiation*
—Ore reduction	—Desalination
—Waste products	—Bulk heating (radiative)
—Benification	—Ultraviolet chemistry
—Plasma chemistry	(photolysis, photosynthesis)
—Detoxification of chemicals	—Fuel cells
—Disposal of plastics	—Waste sterilization
—Fuel-element separation	—Portable fuel manufacture
	—Algae culture
	(Food production)
	—Ozone production

In the first use, also called *plasma-energy transfer*, the plasma energy is used to vaporize solid and other materials, including those with complex chemical structure, and convert them into ionized gases consisting only of basic elements. The gases may then be separated into these basic constituent elements. Some of the uses of plasma-energy transfer, Table 16-3, are in ore reduction, separation of alloys and handling of waste products. The latter use is of most importance in an age when many natural resources are being rapidly depleted and when industrial and domestic wastes are rapidly accumulating, with no place to throw them away. Plasma-energy transfer would make possible the reuse of the basic elements of these wastes and recycling them back into the economy. The second use of the fusion torch, that of *ultraviolet radiation,* also has vast potentialities. Among them are desalination of seawater on a large scale, the sterilization of food, food production by algae culture and by the synthesis of carbohydrates from CO_2 and H_2O, and the sterilization of sewage and other wastes. Daniels has suggested that ultraviolet radiation can be used to produce ozone on a large scale which could then be used to reduce industrial air pollution, sterilize domestic waters, revive dead lakes and rivers by reducing their excessive organic matter, and many other uses.

PROBLEMS

16-1. The *solar constant* is the intensity of solar energy falling on the earth, outside the atmosphere. It has a value of 2 langley/min (1 langley = 1 $g_m$cal/cm^2) or 442.4 Btu/hrft^2. Estimate the mass of hydrogen utilized in the sun in tons (metric) per hour. The earth rotates around the sun with a mean radius of 149.5 million km.

16-2. The total energy received from the sun at the surface of the earth is 1.2×10^{18} $g_m$cal/min. Should this energy be generated on earth by D-D and D-T reactions, find (a) the mass of deuterium needed, (b) the mass of natural water that must be processed to produce that deuterium, and (c) the masses of tritium and helium generated, in tons/day.

16-3. A hypothetical plasma composed of deuterium, tritium and helium-3 of equal densities is at a temperature of 930 million °K and a pressure of 1000 torr (mm Hg at 0°C). Calculate (a) the densities, and (b) the fusion cross sections of the three species.

16-4. A pure deuterium plasma at a temperature of 812 million °K generates 1 watt/cm^3. Calculate the plasma density, cm^{-3}, and pressure, torr.

16-5. A plasma at 696 million °K temperature and 10^4 torr pressure is composed of deuterium and tritium of equal ion densities. Calculate the rate of energy generated per unit volume (volumetric thermal source strength) due to fusion reactions in Mev/sec cm^3 and Btu/hr ft^3.

16-6. A deuterium-tritium plasma operates at a temperature of 232 million °K. Calculate the plasma density and pressure that would result in a net output of 10 watt/cm^3. Ignore all losses except bremsstrahlung.

16-7. By setting an energy balance for the D-T fusion reaction, show that the critical ignition temperature is independent of pressure or ion densities, and calculate that temperature. Take the D and T ion densities to be the same.

16-8. Consider a magnetic pinch confinement machine having a tube 18 cm in diameter. It is required to pinch a plasma of 10^{15} cm^{-3} density and 300 million °K temperature using an available magnetic field strength of 3.87×10^4 gauss.

Calculate the necessary current and the diameter, density and pressure of the pinched plasma.

16-9. It is required to pinch a D-T plasma so that it would generate 8 watt/ cm³ (before losses) in a magnetic pinch machine at a temperature of 232 million °K. Calculate the necessary magnetic field strength in gauss.

16-10. A circular torus of 20 m diameter and 30 cm cross-sectional radius contains a D-T plasma at a temperature of 162.5 million °K. It is required to generate 400 Mw (before losses) with a power density of 25 watt/cm³ of confined plasma. Calculate the necessary magnetic field strength and current.

16-11. Calculate the maximum possible power generation density, watt/cm³, obtainable from D-T and D-D fusion reactors and the plasma densities and temperatures at which these take place, if the maximum allowable pressure is 1000 atmosphere in either case.

16-12. D-T and D-D fusion reactors are operating at temperatures corresponding to maximum power densities and generating 100 and 10 watt/cm³ respectively. Estimate the minimum confinement times and the percent of the power generated that must be retained to heat the incoming fuel gas.

16-13. A D-T fusion reactor is made of a torus 40 m in overall diameter. The confined plasma has a radius of 1 m, operates at 162.5 million °K and has a density of 10^{15} cm⁻³. Calculate the necessary minimum confinement time, the total energy generated, the energy used up in heating the incoming fuel gas, and the energy lost as bremsstrahlung radiation, in Mw.

16-14. A D-T fusion reactor power plant has a vacuum chamber with a cross sectional diameter of 5 m. The plasma generates 8.8×10^{14} Mev/sec cm³. The container wall is made of a material that contains beryllium dispersed within it (for neutron multiplication). Assuming a minimum tritium multiplication of 1.15 to overcome tritium losses and decay, estimate the nuclear dispersion of beryllium in the wall material as percent of its normal nuclear density. Take density of beryllium as 1.8 g_m/cm³.

16-15. A fusion power plant similar to the one shown in Fig. 16-23 has a molten lithium blanket that is pumped to the potassium boiler at 2800 °R, leaving it at 2600 °R. Potassium vapor leaves the potassium boiler at 2600 °R and 164.5 psia. It enters the potassium condenser-steam boiler at 0.2364 psia. Steam is generated at 1200°F and 1000 psia. The steam condenser is at 1 psia. The steam turbogenerator generates 2000 Mw(e) gross. Both turbines have adiabatic efficiencies of 88 percent. Both generators have efficiencies of 95 percent. Using information from Chapter 2 and Appendix B, assuming no heat losses in the power plant and, for simplicity, no feed heating, find (a) the mass flow rates of steam, potassium and lithium, lb_m/hr, (b) the total station gross output, Mw(e), and (c) the station gross thermal efficiency.

chapter **17**

Nuclear Power Economics

17-1. INTRODUCTION

The nuclear power industry has shown a remarkable growth in the 1960's. This was attributed not only to such factors as proven technology, demonstrated safety, availability of nuclear fuels and freedom of air pollution, but also to the strong economic competitive position that the nuclear power industry has proven capable of.

An indication of nuclear power activity is the doubling of drilling for uranium during the year 1967 and the doubling again in 1968. The United States Atomic Energy Commission predicts that 150,000 Mw(e) of nuclear power will be on the line by 1980 in the United States and that the requirements for yellow cake (mostly U_3O_8) will grow from the 5,000 tons/year figure of 1968 to about 40,000 tons/year in 1980 for a total of 250,000 tons by that date. With an eight-year forward reserve as recommended by the AEC, the total requirements plus reserve will amount to 650,000 tons by 1980, for a value of about $ 10.4 billion (based on $ 8/lb_m). Ordinary-water reactors (PWR and BWR) will constitute the bulk of power in the United States by then. However, it is expected that extensive programs of plutonium recycling will begin in the late 1970's, that fast-breeder reactors and thorium molten-salt thermal breeders will make their appearance commercially in the 1980's, and that fusion reactors will be demonstrated some ten years after. The requirements of the rest of the world are expected to roughly equal those of the United States. Figure 17-1 shows the present and predicted share of nuclear power in the total power picture up to the year 2000 in the United States.

The engineer is often relied upon to make economic evaluations of his engineering projects. For this he should possess a knowledge of the various economic aspects of these projects. The problem is more challenging with newer projects. Nuclear power is no exception. This chapter will be devoted to a discussion of those factors that enter into shaping the nuclear-power economic picture. Power costs, composed of capital and operating costs, including the fuel cycle, will be discussed.

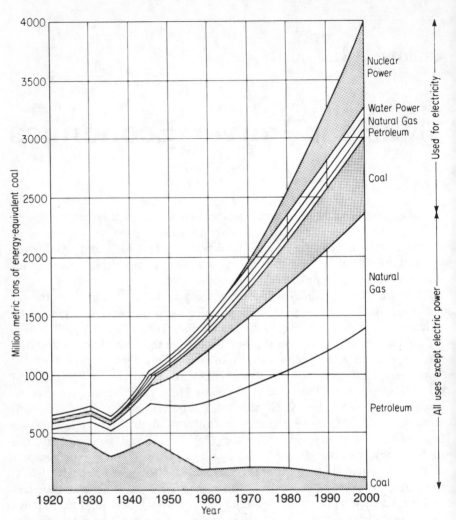

FIG. 17-1.   U.S. energy pattern.   Prepared by the ASEE.   From a pamphlet, "Nuclear Engineering in Your Future," 1968.

17-2.   NUCLEAR POWER COSTS

Nuclear power costs are evaluated on the basis of general ground rules that (a) vary from time to time, (b) vary from one country to another, and (c) even vary from one reactor type to another. The discussion that follows, therefore, must not be regarded as a precise evaluation of a particular type or situation. Rather, what the following discussion hopes to attain is to help the reader appreciate the different aspects that enter into making economic evaluations, and possibly make simple comparative studies between different reactor types.

What is required, of course, is a nuclear power plant that is cheap to build and produces electrical power at the lowest possible cost. There are, therefore, two figures of merit in discussing nuclear power costs: (a) the *capital costs* in \$/kw installed, and (b) the *power production costs* in mills/kwhr produced. A mill is 1/1000th of a U.S. dollar, and the kw is kw(e) with the electric (e) designation often dropped. Table 17-1 shows the major components of these two costs.

The capital costs denote the capital outlay necessary to build a nuclear power plant. The early experimental power plants of the 1950's cost \$600-700/kw to build. In the late 1960's, however, water-cooled reactor power plants (of the PWR and BWR types) cost in the neighborhood of \$120/kw depending upon size (the larger the cheaper), location, and other factors. A goal of \$100/kw has often been mentioned for both nuclear and fossil-fueled power plants, although costs below \$90/kw were realized in some gas-fired power plants near natural-gas fields in the southwest United States.

The capital costs are divided into *direct* and *indirect* capital costs. Direct capital costs are composed of land and land rights, special materials, and the physical plant, and are divided into depreciating and non-depreciating assets.

The *depreciating capital* is all capital costs, less land and D_2O inventory, while the *nondepreciating capital* includes the cost of land and land rights, the cost of the D_2O inventory (or certain other coolants and moderators), if any, and the working capital. The *working capital* is the money required for day-to-day operation and for the maintenance of stock. It is generally defined as the excess of current assets over current liabilities.

TABLE 17-1
Nuclear Power Costs

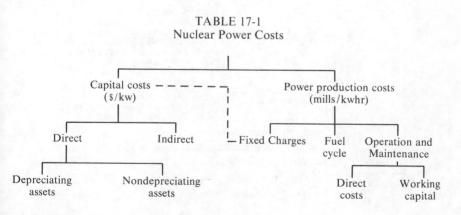

The indirect capital costs are expenses of a general nature which apply to the overall project of building a plant ready for operation, rather than to the actual material and labor of the installation of the complete facility.

Power production costs are the true measure of how economical a power plant really is. While the experimental nuclear power plants of

TABLE 17-2
Breakdown of Capital Costs

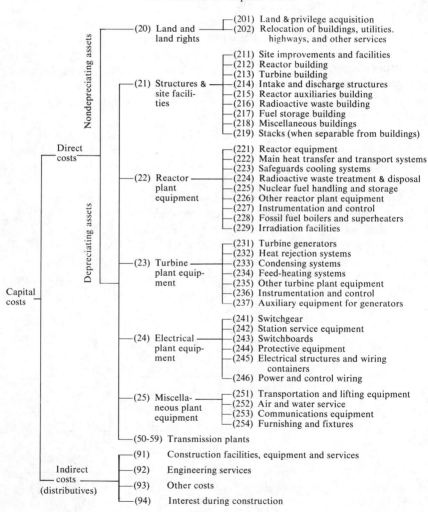

		(20) Land and land rights	(201) Land & privilege acquisition (202) Relocation of buildings, utilities. highways, and other services
		(21) Structures & site facilities	(211) Site improvements and facilities (212) Reactor building (213) Turbine building (214) Intake and discharge structures (215) Reactor auxiliaries building (216) Radioactive waste building (217) Fuel storage building (218) Miscellaneous buildings (219) Stacks (when separable from buildings)
		(22) Reactor plant equipment	(221) Reactor equipment (222) Main heat transfer and transport systems (223) Safeguards cooling systems (224) Radioactive waste treatment & disposal (225) Nuclear fuel handling and storage (226) Other reactor plant equipment (227) Instrumentation and control (228) Fossil fuel boilers and superheaters (229) Irradiation facilities
		(23) Turbine plant equipment	(231) Turbine generators (232) Heat rejection systems (233) Condensing systems (234) Feed-heating systems (235) Other turbine plant equipment (236) Instrumentation and control (237) Auxiliary equipment for generators
		(24) Electrical plant equipment	(241) Switchgear (242) Station service equipment (243) Switchboards (244) Protective equipment (245) Electrical structures and wiring containers (246) Power and control wiring
		(25) Miscellaneous plant equipment	(251) Transportation and lifting equipment (252) Air and water service (253) Communications equipment (254) Furnishing and fixtures
		(50-59) Transmission plants	
	Indirect costs (distributives)	(91)	Construction facilities, equipment and services
		(92)	Engineering services
		(93)	Other costs
		(94)	Interest during construction

Capital costs — Direct costs (Nondepreciating assets / Depreciating assets)

the early 1950's produced power at an understandably high cost (as high as 65 mills/kwhr), the figures have come down to around and below 4 mills/kwhr in the late 1960's. Present (1970) forecasts of technical developments, based on improved plant efficiencies, put the power production costs at 1.5 mills/kwhr or less around the year 2000.

The power production costs are divided into *fixed charges, fuel-cycle* costs and *operation and maintenance* (O&M) costs. Of these the fixed charges are largely dependent on the capital costs. They are calculated separately for depreciating and nondepreciating assets.

Tables 17-2 and 17-4 show breakdowns of capital and power production costs respectively. They are numbered according to an AEC accounting code [167] in a system of accounts designed to provide a standard distribution of costs and facilitate comparative economic analyses of various plants. The system parallels the Electric Plant Accounts, established by the Federal Power Commission, FPC. The items in Table 17-2 are, in turn, broken down into subcategories, subsubcategories, and so on. As an example:

(22) Reactor Plant Equipment
   (221) Reactor Equipment
      (221.1) Reactor vessel
         (221.11) Vessel supports and foundations
         (221.12) Vessel structure
         (221.13) Vessel internals, etc.
      (221.2) Reactor controls
         (221.21) Absorber/fuel positioning
         (221.22) Moderator characteristics, varying
         (221.23) Moderator/reflector, varying

The remainder of (221) is made up of (221.3), Moderator/reflector system, and (221.4), Shielding.

The following sections discuss some of the capital and power production cost items.

17-3. DIRECT CAPITAL COSTS

Land and land rights (20) are the costs of acquiring the power-plant site and include land acquisition and relocating roads and communication lines. Except for unusual sites, such as artificial islands or when attempts are made to build within heavily inhabited areas (possibly with double containment) this is usually a relatively small item, of the order of $1/kw. The *structures and site facilities* (21) item represents the costs of preparing the site, such as grading, lighting, etc., and constructing the reactor building, including containment and stacks, and other buildings and structures such as for the generating units, warehouses, etc. The costs here vary among other things with the particular reactor type and design, and may be in the neighborhood of $ 25/kw.

Reactor equipment (221) includes all equipment directly associated with the reactor, such as the reactor itself (less fuel), its vessel, foundation, controls, shielding, auxiliary heat transfer systems, floors, drainage, ventilation, cranes as well as the costs of the moderator and reflector (except if they also serve as coolant). Again the cost here depends largely on the particular reactor type and design and could vary between $10 and $50/kw.

The main *heat-transfer and transport systems* (222) include the reactor coolant and intermediate coolant systems, coolant charging, discharging, sampling and purification systems, the steam generators and superheaters, inert-gas systems, and the cost of the initial charge of the coolant, and fossil-fueled superheaters, if any. The cost here could vary, depending upon design (such as the number of primary coolant loops), between $15 and $30/kw.

Radioactive waste treatment and disposal equipment (224) includes filters and separators, storage tanks, ion exchangers, heat exchangers, concentrators, incinerators, and other equipment used in treating, storing and disposing of liquid, gaseous and solid radioactive wastes. The cost here is again minor, being less than $0.5/kw.

The nuclear *fuel-handling and storage systems* (225) includes cranes, fuel-assembly and disassembly equipment, spent-fuel decay-cooling facilities, storage and specialized machines and tools. The cost here is low, being in the neighborhood of $1/kw.

Instrumentation and control (227) includes devices for controlling the reactor, the heat-transfer system, the nuclear steam generators, the fuel-handling and storage systems, and the radioactive waste systems, as well as radiation monitoring devices. The cost here is around $3/kw.

Other reactor-plant equipment may include facilities for maintenance of radioactive equipment, traps, portable shielding, special tools, storage for contaminated equipment, etc., if not included in the above items. The total capital cost of reactor plant equipment (22), for PWR or BWR reactors, in the late 1960's has been of the order of $40-50/kw.

The *turbine-plant equipment* (23) includes the turbine generator itself, its foundations, valves, coolants, main and pilot exciters, gear, hydrogen cooling, instruments when furnished with the units, as well as condensers and their accessories, such as pumps, circulating-water lines, lubricating system, and such other equipment as glands, vacuum systems, panels, fire extinguishing, etc. This is a major item in the capital cost picture, and in the greater than 1000 Mw(e) plant capacity range, has been one that greatly influenced construction timetables. The cost, in the above large capacity range, is around $25-30/kw.

The *electrical plant equipment* (24) includes switchgear, switchboards, protective equipment, conduits, wiring, and service equipment (such as station transformers, emergency generators, etc.). The cost here is about $4/kw.

Shown with the direct capital costs, Table 17-2, are *transmission plant* costs (50-59), which include the main power transformers.

In general, all capital costs, except the turbine-generator costs, vary roughly as the plant capacity to the 0.6 power. The turbine-generator cost varies more or less in direct proportion to the plant capacity.

In making capital cost estimates, it is usually wise to allow for the *escalation* of costs during the period of procurement and construction. This escalation, in the form of a percentage of direct cost, is naturally uncertain, depends upon market and labor trends, delays, etc., and may be of the order of 10-20 percent of direct costs   A useful formula for calculating escalation during construction, E, is

$$E(\$10^6) = \frac{(1 - C)}{100} eTQ \qquad (17-1)$$

where e is the escalation rate in percent/yr, T the time of construction in years, Table 17-3, and Q is the direct capital investment in $ 10^6$. C is

TABLE 17-3
Typical Design and Construction Periods

Plant Rating Mw(e)	Design and Construction Period, months		
	Proven Design	Extrapolated Design	Novel Design
150	50	56	62
300	56	61	67
500	61	65	72
750	64	68	75
1,000	66	70	77
1,500	68	72	78

an averaging factor which places emphasis on the period of construction during which prices escalated. More money is usually spent in the latter half of the construction period and the value of C, therefore, is customarily taken as 0.45. The practice is by no means universal, however, and some utilities make no allowance for escalation.

17-4. INDIRECT CAPITAL COSTS

Indirect capital costs (91-94), Table 17-2, are usually figured on the basis of percentage of the total direct costs. Indirect construction costs, for example, are shown as a function of direct construction costs in Fig. 17-2. Indirect capital costs are broken down into four categories:

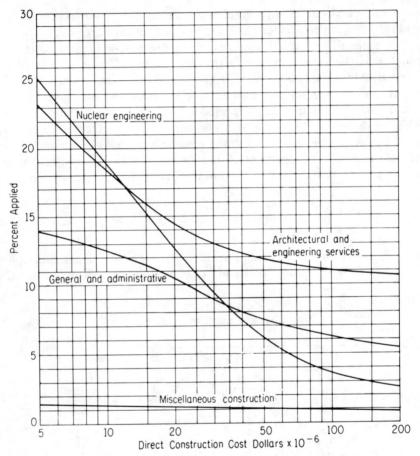

FIG. 17-2. Indirect construction-cost curves.

Construction facilities, equipment and services (91) include temporary facilities which are removed or dismantled after completion of construction. Facilities which become a permanent part of the plant are included under accounts (21) through (25). However, incremental costs associated with using them during the construction period must be included here. To be included under items (91) also, are construction equipment and construction services, provided the equipment is rented or bought for use during construction and subsequently removed from the site. Services such as insurance must exclude workmen's compensation insurance which is included in the labor charges associated with direct costs.

Engineering services (92) are composed of *reactor engineering* and *plant engineering.* Reactor-engineering services include core physics analysis, reactor-system design, procurement, inspection, expediting of materials and equipment in the reactor system, reactor-hazards calculations, design specifications, as well as plant startup, licensing activities and staff training. Plant engineering services comprise all other engineering services associated with the project, such as site selection, architect engineers and construction management services, and preparation of preliminary and final design documents.

Other costs (93) cover a variety of expense items towards the building of an operable nuclear power station, and which are not sensitive to reactor type. Costs under this account can be separated into three components: *Taxes and insurance* (931), *staff training and startup* (932), and *owner's general and accounting* (G&A) costs (933). Taxes and insurance (931) covers both state and local property taxes on the site and improvements during the construction period, sales taxes on purchased materials and equipment, and property and all-risk insurance with nuclear rider. There are no fixed rates of taxes and insurance of this type, but rather they vary widely even within a given state, and can sometimes be settled by negotiations. Taxes during construction, *TDC*, can be calculated from

$$TDC(\$) = \frac{tT}{100} (L + CQ_1) \qquad (17\text{-}2)$$

where t is the annual tax rate in percent, T the construction time in years (Table 17-3), L the land cost, and Q_1 the portion of the plant cost to which the property tax is applicable. C again is an averaging factor. Equation 17-2 is applicable on the assumption that the land is fully paid for at the beginning of construction and that construction expenditures follow estimations.

Interest during construction (94) is the sum of interest charges for each individual expenditure. It covers the net cost of funds utilized to finance the design and construction of the plant. Interest charges under this account are a function of the amount of expenditures, the time period for which funds are borrowed, and the interest rate, Figs. 17-3 and 17-4. The time reference is usually taken as that when the plant commences full-power commercial operation and the design and construction period begins with the start of the reactor design effort.

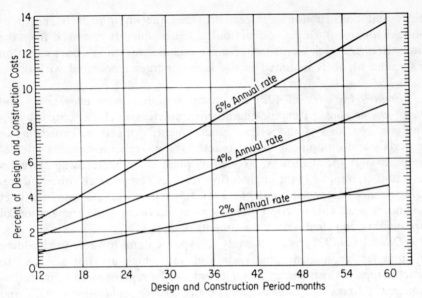

FIG. 17-3. Interest during construction (simple interest).

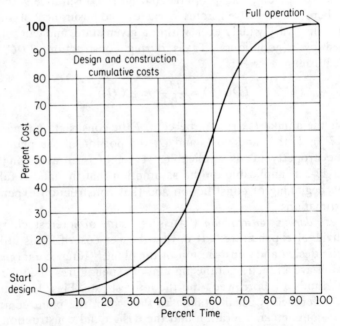

FIG. 17-4. Design and construction cumulative costs.

Interest during construction is included in the indirect capital cost even if the money used during construction is from the utility's own funds. It is in fact a return that could have been realized if that money were invested during construction. Consequently the interest during construction should yield a return equal to that of other capital items. Some utilities do not use this principle and allowance for this item is taken between that based on the utility rate and interest on short term borrowings. In general, the total interest during construction depends upon the construction time and on the way in which construction expenditures build up. A useful formula for estimating the total interest during construction, *IDC*, is

$$IDC(\$) = \frac{iT}{100}(W + L + CQ_2)\qquad(17\text{-}3)$$

where i is the annual interest rate in percent/year, W the average amount of working capital available during construction time T, and Q_2 the total construction cost of the plant including all overheads except interest during construction. Some utilities prefer to omit the working capital resulting in some reduction in *IDC*.

17-5. FIXED CHARGES

Fixed charges are the first contributors to power production costs, Table 17-4. They are those power costs (per annum) which are largely dependent on the plant capital investment and the working capital necessary for its operation. They do not depend on the extent to which the facilities and services provided by the investment are used. They are not dependent upon the actual electrical output of the plant, with the exception of taxes. They are therefore initially figured in \$/kw installed. Their contribution to the cost of power in mills/kwhr, therefore, depends upon the *plant operating factor*. The plant operating factor is defined as the ratio of the total kwhr (gross) actually generated in a given time, to the total kwhr of gross generating capacity during that period. It is often assumed as 80 percent for estimation purposes, Sec. 17-10.

The *interest charge* and the *minimum return required* items of the fixed charges represent the *cost of money* and are determined by the financial setup of the utility as well as the cost of money itself. The dollar cost of money decreases with time and hence the average cost of money over the plant lifetime. This average cost, expressed as a percentage of the initial plant capitalization, is less than the initial percentage used. The rate applied to the nondepreciable portion, on the other hand, remains the same. The average cost of money over the plant lifetime for the

TABLE 17-4
Breakdown of Power Production Costs

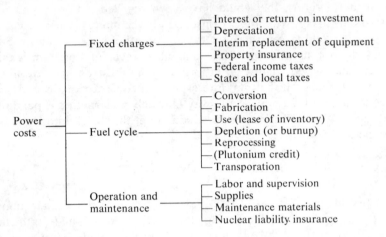

depreciable portion of the plant cost is given by the approximate formula

$$\langle r \rangle = r(0.547 + 0.000602\, rn) \qquad (17\text{-}4)$$

where $\langle r \rangle$ is the effective cost of money, r the actual or initial cost of money and n is the plant lifetime in years. Equation 17-4 assumes that the plant loses its value in equal increments over its lifetime.

Depreciation rates are based on an assumed average plant service life of 30 years, though they may vary between 25 and 40 years as is the experience with fossil-fueled plants. Depreciation is defined as a reasonable allowance for the exhaustion, wear and tear of the plant. In effect, it is a bookkeeping type of expense and is not a cash outflow during the operating period. Depreciation methods include the *sinking-fund* method (SF) which is often used because of its ease, the *straight-line* method, and others.

The depreciation does not provide for the replacement of equipment with service life shorter than the assumed average life (30 years). The *interim replacement* charge is a fixed charge which allows for such relatively unpredictable replacements. Regular and reasonably predictable replacement of equipment is usually considered in operation and maintenance (O&M) costs.

Insurance costs represent an item that has figured in much public discussion that, because of public historical fears of anything nuclear (the bomb was the first nuclear device to become public knowledge), has much emotion built into it. There are two types of insurance. The first is a conventional *all-risk property* insurance of the property (fire, storm, accidents to personnel, etc.), roughly estimated at 0.25-0.60

percent of the capital cost of the nuclear plant, including a nuclear rider for nuclear power plants. This type of insurance is figured under fixed charges. The second type is called *nuclear liability* insurance. It is independent of plant investment and is a relatively constant fixed cost. Consequently it is not included as part of the annual fixed charges, but is included under O&M costs, Sec. 17-9.

The annual fixed charges include all *taxes* as a percentage of the capital investment.

Table 17-5 shows typical nationwide fixed rates of the various items outlined above. The lower rates for the municipal and REA (Rural Electrification Administration)–financed plants are due to the lower interest rates and no return on investment.

TABLE 17-5
Nationwide Approximate Fixed Rates (Nonleveled)

Utility Type	Investor-owned Public Utility		Municipally Owned		REA*	
	Dep.	Nondep.	Dep.	Nondep.	Dep.	Nondep.
Interest charged	—	—	4.00	4.00	2.00	2.00
Minimum return required	7.20	7.20	—	—	—	—
Depreciation (30-yr SF†).	1.02	—	1.78	—	3.10	—
Interim replacement	0.35	—	0.35	—	0.35	—
Property insurance	0.25	—	0.25	—	1.00	—
Federal income taxes ...	2.04	4.80	—	—	—	—
State and local taxes	2.84	0.80	1.00	1.00	0.50	0.50
Total	13.70	12.80	7.38	5.00	6.95	2.50

* Public utilities financed by the Rural Electrification Administration.
† Sinking fund.

In calculating the fixed charges on depreciating capital cost certain percentages are applied annually to the book value of the investment. The book value represents the worth of the item as of a specific date and is obtained by subtracting all the depreciation charges made to date from the original cost of the item under consideration. Sometimes the book value is called the unamortized cost. Since the book value of a specific plant decreases year after year, we expect the fixed charges calculated on this basis to decrease correspondingly. Thus all costs have to be considered in fact year-by-year over the life of the plant. In order to obtain the total present worth of the plant prior to its construction, the present worth of the total costs for each future year must be summed. The present worth of a future amount is the present principal which must be deposited at a given interest rate to yield the desired amount at the future date.

The procedure of calculating fixed charges year-by-year using present-worth values has the advantage of providing a standard base for judging relative economics of proposed plants. Since this method gives the annual cash flow it can be used in conjunction with the fuel cycle and the operation and maintenance costs. Such detailed analysis requires digital computations if the whole plant is evaluated on the basis of a cash flow formulation. Computer codes are available and have been used, especially in connection with the fuel cycle economy.

A different and simpler approach is to use *levelized* annual fixed charges if there is no concern about the year-by-year costs as given by the above detailed approach. A constant fixed charge rate over the plant lifetime is applied to the initial investment and the rate is selected such that the present worth of the grand total is the same as that obtained by summing the variable present worth for each year. Levelized fixed charges are important in economical analysis, especially in fuel management and fuel cycle economics.

In the absence of proper information about fixed charge rates in a specific situation, fixed charges are figured by applying, separately, certain percentages to the depreciating and nondepreciating portions of capital costs and to the working capital. These percentages vary according to the type of plant ownership, Table 17-5.

17-6. THE FUEL CYCLE

Before discussing fuel cycle costs, it will be necessary to discuss the *fuel cycle* itself. The fuel cycle is shown schematically in Table 17-6. The various processes will now be described.

The ore produced in the mining process contains a mixture of uranium oxides called *black oxide* and usually formulated as $U_3 O_8$. The content of the raw material may vary from 2 to as much as 20 pounds of black oxide per ton of ore, with an average of about 5 pounds. Most of the deposits are mined from the underground, though some are shallow and mined by open-pit techniques.

In the *milling* operation, the uranium is extracted from the ore by a process called *leaching*. In this, the ore is first pulverized and brought into contact with a reagent that dissolves the uranium resulting in what is called *leach liquor*. The dissolved uranium oxides are then recovered by solvent extraction (chemical separation based on preferential solubility in one of two immiscible liquids), or by ion exchange (chemical separation based on preferential absorption of solute ions in insoluble resins). The

TABLE 17-6
The Fuel Cycle

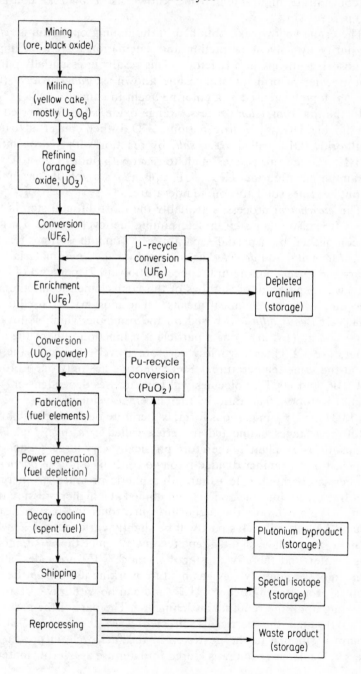

product is then calcined (roasted) to remove excess water. The result is a concentrate of uranium oxide known as *yellow cake* consisting of 70-90 percent U_3O_8.

The crude yellow cake obtained in the milling operation is *refined* to high purity by solvent extraction and calcination to remove impurities that absorb neutrons in a reactor. This results in essentially pure UO_3, a fine powder of brilliant orange hue known as *orange oxide*. (Orange oxide has long been used as a coloring agent in chinaware.)

In the first *conversion* process, orange oxide is first converted to uranium dioxide, UO_2, by hydrogenation. UO_2 is then converted to uranium tetrafluoride, UF_4, called *green salt*, by reaction with hydrogen fluoride gas, HF. Green salt is later made to react with fluorine gas to convert it to uranium hexafluoride, UF_6. UF_6 appears as a volatile gas in the process but becomes solid at room temperature.

The *enrichment* process is probably the most difficult step in the fuel cycle. It is also a key step in determining fuel-cycle costs. Enrichment is accomplished by a partial separation of uranium isotopes resulting in *enriched* uranium and *depleted* uranium (also called *tails* material), having higher- and lower-than-natural concentrations (0.71 percent) of U^{235} respectively. Separation of isotopes of the same element cannot, of course, be accomplished by chemical means. The most widely used technique is that of *gaseous diffusion,* based on the principle which states that the kinetic energy $(1/2\ mV^2)$ of a particle is a function only of the absolute temperature. Lighter particles therefore travel faster than the heavier ones at the same temperature. Since fluorine has only one isotope, F^{19}, the $U^{238}F_6$ and $U^{235}F_6$ molecules will have masses dependent only on the uranium isotope. The ratio of their average velocities is $\sqrt{352/349}$ or only 1.0043. UF_6, in gaseous form, is therefore passed through hundreds of diffusion stages connected in series, called a *cascade*, Fig. 17-5. A gaseous-diffusion plant is therefore physically very large. Each stage is composed of a chamber divided into two zones by a porous *barrier*. The chamber is operated at less than atmospheric pressure (to increase the mean free path), but one zone is maintained at a higher pressure than the other. UF_6 gas enters the higher-pressure zone and part of it diffuses through the barrier. This part will be slightly enriched in U^{235} while the remaining part will be slightly depleted in U^{235}, since the $U^{235}F_6$ molecules, being lighter and therefore faster than the $U^{238}F_6$ molecules, strike the barrier more frequently. A gas mixture working its way *up* the cascade becomes progressively richer in U^{235} and can be withdrawn at any stage, depending upon the required enrichment. The part of the gas that does not go through the barrier is directed *down* the cascade. Enriched uranium is shipped to the next step in the fuel cycle in pressurized cylinders. Depleted uranium is stored for later use as a fertile material and

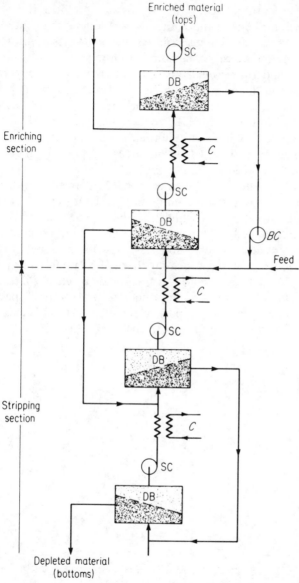

Enriched material
(tops)

Enriching
section

Feed

Stripping
section

Depleted material
(bottoms)

FIG. 17-5. Schematic of a gaseous diffusion cascade.
BC = booster compressor, C = cooler, DB = diffusion barrier, SC = stage compressor.

other uses. A gaseous-diffusion plant represents a tremendous capital
investment. If it were not for the fact that a good part of this investment
was borne by the military, the present reasonable enriching tolls could

not have been approached. In the Western World there are three gaseous-diffusion plants in the United States, one in the United Kingdom and one in France. Enriching facilities, using a series of ultracentrifuges for mass separation are receiving some attention.

In the second *conversion* step, UF_6 is chemically converted to the usable forms of uranium used in the fabrication process, such as metallic U, uranium oxide, UO_2, or uranium carbide, UC. UO_2, the most common form, is produced from UF_6 by leaching the latter first with water and then with a hydroxide salt resulting in a precipitate which is calcined to form orange oxide UO_3 (now enriched). UO_3 is then reduced with hydrogen to UO_2 powder.

In *fabrication*, the fuel elements to be used in the reactor core are manufactured. In the most common type, UO_2 powder is compacted into small cylindrical pellets which are then inspected for size with the rejects either scrapped or machined to size. The accepted pellets are then loaded into Zircaloy or stainless steel cladding tubes. Helium gas is then introduced into the tubes for thermal bonding and the tubes are end-capped. A fuel subassembly is then made by clustering a number of tubes into a bundle with top, intermediate and bottom spacers which allow the coolant to pass between the elements. Fuel element fabrication is the largest contributor to fuel cycle costs, Fig. 17-6.

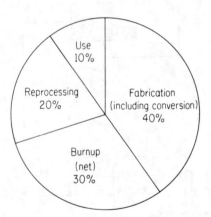

FIG. 17-6. Approximate breakdown
of fuel-cycle costs.

During *power generation,* the main effect on fuel cost is due to *burnup.* A high fuel burnup spreads the cost of the fuel cycle over a large number of kilowatthours power output. The fuel, removed for reprocessing, after it can no longer be economically used to sustain power generation because of the loss of reactivity due to fission product buildup and fissionable isotope depletion, or because of radiation damage, or both, is called the *spent,* or *irradiated*, fuel.

When removed from the reactor core, the spent fuel is intensely radioactive. Before shipping, therefore, it is stored for several months under water in order for some of the radioactivity to decay and the fuel can be safely shipped. This process is called *decay cooling.*

The spent fuel is then *shipped* for reprocessing in heavily shielded casks. The cost of shipping is more a function of the mass of the casks than the mass of the spent fuel itself. It is also a function of the distance between the reactor site and the reprocessing plant. Cask design is a function of burnup, size, and mass of fuel elements and the space and mass of the carrier. (Fuel is, of course, shipped from place to place at various times during the fuel cycle. The costs of such other shipping are usually included in the costs of the various fuel cycle processes). Shipping costs are composed of (a) *freight* costs (including returning empty containers), (b) *container* costs, covering annual fixed charges on container investment, or container rental, (c) *handling* costs, including loading, unloading, crane operations, decontamination and testing, and (d) *insurance* costs, against loss of containers and damage of contents. Liability insurance is included if shipment is outside the United States.

Shipment within the United States has to comply to Federal regulations (CFR, Title 10, Part 71) and to the Interstate Commerce Commission Order No. 70. For example, the dose rate 6 ft from the vehicle should not exceed 10 mr/hr, and one vehicle should not carry more than one cask. Other shipment regulations of interest are (a) spent-fuel subassemblies are shipped without disassembly or canning (except for fast reactor fuel), (c) shipment time is usually 16 days (20 days for fast reactors) per 1,000 miles distance between reactor site and reprocessing plant, (d) fresh fuel is shipped fully assembled, (e) maximum allowable cargo mass is 120 tons for railroad cars and 20 tons for trucks, and (f) spent fuel is usually shipped by rail while fresh fuel (except HTGR) is shipped by truck.

It is conceivable and expected that, when large nuclear power complexes are built, economically sized reprocessing plants may be built in conjunction with them, practically eliminating the cost of shipping.

Reprocessing of spent fuel is mainly done to reclaim unused uranium which is then recycled back as UF_6 to the first conversion step, and to recover plutonium. Plutonium accumulates due to conversion of some U^{238} in the fuel to fissionable Pu^{239} (Sec. 9-3). Some of the latter undergoes fission in place, contributing to reactor power, but the rest remains and is reclaimed in reprocessing. Reprocessing involves a series of operations, most of which are done by remote control in heavily shielded areas. The first operation is that of *decladding*. First, as much of the structural support as possible is cut away by mechanical means. Next, the cladding is removed. If it is not bonded to the fuel, as is usually the case in UO_2 pelletized fuel, it is removed mechanically and stored as solid active waste. If it is bonded to the fuel, separation is accomplished chemically. Fuel, and residual cladding, if any, are then dissolved in acid and the resulting solution made to undergo a series of chemical separations by solvent extraction. Most of the fission products are first removed,· followed by separation of uranium from plutonium. Subsequent steps remove residual fission products from the uranium and plutonium which leave as concentrated solutions ready for conversion to other forms.

Uranium, for example, may be converted to UF_6 and recycled to the enrichment step, Table 17-6, or converted to UO_2 which is then blended with UO_2 of higher enrichment. Pu may be converted to PuO_2 and blended with UO_2 to form PuO_2 -UO_2 fuel, or it may be stored as a byproduct for future use as reactor fuel or as an explosive. Special radioactive isotopes which have constructive uses such as fuel in isotopic power systems, or uses in medicine, industry and agriculture are also separated and stored for such uses.

The radioactive *waste products,* also in solution form, are boiled to reduce them in volume, and stored as a liquid concentrate in underground steel tanks, which are monitored against possible leakage. This method of storage is reliable but costly (2-3 percent of the cost of power) and, considering the fact that some isotopes are very long lived, it presents the prospect of continued monitoring. Several alternative methods of storing radioactive wastes that would minimize monitoring are being considered. One involves the conversion of the waste into a solid or glasslike material by calcination or incorporation in clays or ceramic mixtures. Such solids can then be stored in underground vaults without danger of leakage.

17-7. FUEL-CYCLE COSTS

Table 17-4 showed seven major contributors to fuel-cycle costs. The most important of these are shown in approximate relation to one another in Fig. 17-6. The various costs are usually based on a kg_m of *uranium charged into the reactor.* They will now be discussed.

Conversion of uranyl nitrate to UF_6 costs $\$5.60/kg_m$ U for U^{235} enrichments of 5 percent or less, and $\$32/kg_m$ U for enrichments greater than 5 percent. For the conversion of plutonium nitrate to Pu metal, the cost is $\$1.50/g_m$ Pu. Conversion costs are often included in fabrication costs, next.

Fabrication represents the major item in the fuel cycle costs, Fig. 17-6.

Fabrication costs are divided into direct and indirect costs. The *direct* costs often include conversion, above, and shipping, as well as actual fabrication. The indirect costs include materials, labor, and destructive and nondestructive testing. A representative direct fabrication cost in a water-cooled, zirconium-clad UO_2-fueled reactor with 15,000 Mwd/ton burnup is roughly $100/kg_m$ U, meaning several million dollars per core. Since nuclear fuel, unlike fossil fuel, represents a large working-capital investment during both fabrication and reactor operation, an annual charge against this working capital is a necessary indirect cost of the working cycle. This annual cost is about 6 percent of the direct fabrication costs.

Fabrication costs are composed of *capital* costs ($), *operating* costs ($/yr) of the fabrication plant, and the *hardware* material costs ($/yr). For a plant operating at full capacity (usually 500 to 10,000 kg_m/day), these fabrication costs are evaluated from

$$Y_{c,o,h} = AX^B + C \qquad (17\text{-}5)$$

where $Y_{c,o,h}$ = capital cost ($), operating cost ($/yr), or hardware cost ($/yr)

X = rated throughput of the fabrication facility, kg_m/day

A, B, C = constants given in Table 17-7 for two power-plant types

TABLE 17-7
Constants of Fabrication Cost, Eq. 17-5

Plant type	Fabrication Cost	A	B	C
PWR UO$_2$-fueled	Capital	0.1276	0.689	2.6768
	Operating	0.02258	0.7348	1.389
	Hardware	0.01181	0.9216	0.4702
HTGR (U, Th)O$_2$-fueled	Capital	0.0592	0.9096	8.421
	Operating	0.01119	0.9157	2.393
	Hardware	0.0304	0.7658	1.2161

For a plant operating at less than full capacity, the capital costs only are given by Eq. 17-5, the operating costs (based on 50 percent of full capacity being necessary to maintain plant in operation) are given by

$$Y_0 = \left(1 + \frac{Z}{X}\right)\left[0.5\,(AX^B + C)\right] \qquad (17\text{-}6)$$

and the hardware costs are given by

$$Y_h = AZ^B + C \qquad (17\text{-}7)$$

where Z is the actual throughput, kg_m/day. Equations 17-6 and 17-7 reduce to 17-5 when $Z = X$.

Use (or *lease*) charges, as the name suggests, are paid for the use or lease of the fuel. In the United States enriched fuel, up to about 1961, could only be leased from the AEC, and only natural and depleted fuels could be privately owned. Now, however, all fuel can be privately owned, and must be; beginning after 1971. When thus owned, the purchase price may be considered as capital. It is customary, however, to include such costs with the fuel cycle by using an annual rate, of 11 percent of the purchase price, as a use charge. Use charges are levied on fuel being fabricated, decay-cooled, reprocessed, shipped, and stored, as well as being used in the core. For fuel leased from the AEC, according to AEC price schedules, lease charges are figured on the basis of a percentage, currently 6.5 percent,* of the cost of the enriched uranium hexafluoride per annum. This latter is a function of enrichment and is given in Table 17-8 on the basis of $/kg$_m$ of contained uranium at various enrichments.

TABLE 17-8
Cost of Uranium Hexafluoride

Enrichment	$/kg$_m$U	Enrichment	$/kg$_m$U
0.71*	23.50	4.0	365.80
0.75	26.50		
0.80	30.50	5.0	479.40
0.90	38.90	7.0	710.50
1.00	47.70	10.0	1,002.00
1.50	95.30	30.0	3,456.00
2.00	146.50	70.0	8,329.00
2.50	200.00	90.0	10,808.00
3.00	254.30	93.0	11,188.00

* Natural uranium.

Fuel depletion costs are paid on the amount of fuel that has undergone burnup during reactor operation. This is composed of (a) fissionable fuel that gets consumed by fast and thermal fission and by radioactive capture, and (b) U^{238} that gets consumed by fast fission and by radioactive capture. The fuel depletion is therefore the difference between the cost of the fresh fuel loaded into the core and the cost of the spent fuel discharged from it, as determined by their respective masses and enrichments.

Reprocessing charges involve AEC plant rental at about $20,000/day. The number of days, D, are obtained from

$$D = \frac{M}{R} + T \qquad (17\text{-}8)$$

* Effective April 1, 1969, AEC release No. M-45, Feb. 24, 1969.

where M = mass of fuel in kg_m (annual throughput)

R = reprocessing rate, depending on *initial* enrichment r_1

= 1,000 kg_m/day if r_1 < 3 percent

$T = 2$ if $M/R < 2$, M/R if $2 < M/R < 8$, and 8 if $M/R > 8$

In addition to plant rental, reprocessing costs include UN to UF_6 conversion costs at $ 5.60/kg_m U if r < 5 percent, or $ 32.00/kg_m U if r > 5 percent, and PuN to Pu metal conversion costs at $ 1.50/g_m Pu. Losses during reprocessing must be accounted for. They may be estimated as 1 percent of U and Pu in converting to nitrates, 0.3 percent of U in converting UN to UF_6, and 1 percent of Pu in converting PuN to Pu metal.

The *plutonium credit* is an allowance (negative cost), fixed by Federal law, to be deducted from fuel cycle costs, which is made on reactor-grade plutonium in the spent fuel. The allowance is fixed by the U.S. Atomic Energy Commission at $ 9.28/g_m of fissile plutonium, until 1971. After that the value of Pu is to be decided by supply and demand in a competitive market. Plutonium credit may appreciably affect fuel cycle costs if the breeding ratio is substantially greater than unity. Also, change in the plutonium-credit allowance may have similar effects. The plutonium credit must be reduced by the charges of conversion of Pu nitrate to Pu metal ($ 1.50/g_m) and by the losses (1 percent) during conversion to the nitrate and the losses (1 percent) during conversion from the nitrate to the metal. The plutonium credit would therefore amount to $(9.28-1.50)0.98$ or $ 7.78/g_m.

Transportation costs have already been discussed in connection with the description of the fuel cycle. Estimates of these costs (1964), based on an average case, are as follows: AEC to fabricator, $ 1.50/kg_m U; fabricator to reactor, $ 1.50/kg_m U; reactor to reprocessing plant, $ 16.00/kg_m U; reprocessing plant to AEC, $ 1.00/kg_m U; total $ 20/kg_m U.

Before leaving fuel costs, it would be of interest to show the current (1969) average fossil fuel costs, Table 17-9.

TABLE 17-9
Average Fossil Fuel Costs (1978)

Fuel	Cost, Cents/10^6 Btu, as burned
Coal	126.0
Oil	270.0
Gas	170.0 (if available)
Nuclear	30.0 (old contracts)

17-8. COMPUTATIONAL TECHNIQUES OF THE FUEL CYCLE

The information presented in the previous section is adequate for general estimations of fuel-cycle costs. In this section, a more detailed breakdown of those costs will be presented.   A detailed analysis is, of course, necessary during the selection and design stages of different type power plants; but also for an operating plant in which such things as operational changes and changes in fuel-management schemes often cause changes in fuel-cycle costs.   To evaluate the various causes of cost change, fuel-cycle costs have been categorized as shown in Table 17-10 which makes a distinction between direct and indirect costs.

Table 17-10 essentially details the information on fuel-cycle costs in Table 17-4.   For example "Use" and "Plutonium credit" in Table 17-4 are detailed in items 1 through 4 in Table 17-10, and U^{235}, U^{233} and Th credits are added.   Conversion, fabrication, reprocessing, and transportation are included in items 5 through 7.   These are essentially inversely proportional to fuel burnup (if the same dollar expenditure is made

TABLE 17-10
Breakdown of Fuel-Cycle Costs

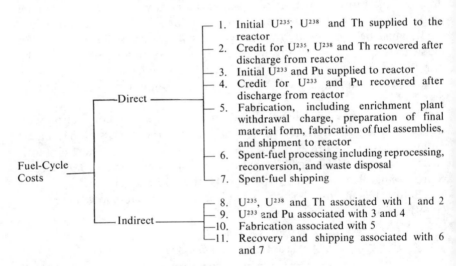

1. Initial U^{235}, U^{238} and Th supplied to the reactor
2. Credit for U^{235}, U^{238} and Th recovered after discharge from reactor
3. Initial U^{233} and Pu supplied to reactor
4. Credit for U^{233} and Pu recovered after discharge from reactor
5. Fabrication, including enrichment plant withdrawal charge, preparation of final material form, fabrication of fuel assemblies, and shipment to reactor
6. Spent-fuel processing including reprocessing, reconversion, and waste disposal
7. Spent-fuel shipping

8. U^{235}, U^{238} and Th associated with 1 and 2
9. U^{233} and Pu associated with 3 and 4
10. Fabrication associated with 5
11. Recovery and shipping associated with 6 and 7

irregardless of the energy obtained).   Items 5 and 7 represent the unit energy costs of process and service.

The indirect costs are those associated with the time value of money. They result from the time lag between fuel expenditures and revenues. The time lag forces payment of direct fuel expenses with borrowed money.   The interest on that borrowed money, in addition to taxes and profits related to fuel expenses make up the indirect costs.   Indirect costs can be further broken to out-of-core and in-core costs.   The latter are proportional to the plant-capacity factor.

A general approach to fuel-cost analysis is to obtain the timing and amount of direct expenditures and credits for each item of the fuel cost over a certain period of time, and the timing and amount of energy produced during the same period. A balance between the consumption and ownership costs of the fuel and the revenue from the energy produced yields the unit fuel cost. In the computations care must be taken to include all expenses of all steps in the fuel cycle, including shipping between various processes.

The cost of each item is usually calculated in units of dollars per unit mass of Pu, Th and/or U inserted in the reactor, but the final results are expressed in terms of the thermal or electrical energy of the reactor. To make this conversion the following factors are used

$$p_f(\text{¢/MBtu}) = \frac{p_f(\$/\text{kg}_{mi}, \text{ or } \$/\text{MT}_i) \times (100\text{¢}/\$)}{(24 \text{ hr/d}) \times (3413 \times 10^{-6} \text{MBtu/kwhr}) \times \overline{BU}(\text{kwd/kg}_i \text{ or Mwd/MT}_i)}$$

(17-9)

and

$$p_f(\text{mills/Kwhr}) = \frac{p_f(\$/\text{kg}_i \text{ or } \$/\text{MT}_i) \times (1{,}000 \text{ mills}/\$-)}{(24 \text{ hr/d}) \times \overline{BU}(\text{kwd/kg}_i \text{ or Mwd/MT}_i) \times \eta}$$

(17-10)

where $\overline{BU}$ is the average burnup and η is the net plant efficiency. All mass units above refer to Pu, Th, or U inserted initially in the reactor.

The total fuel cost P_f is given by

$$P_f = \sum_{j=1}^{7} p_f(j) \text{ direct} + \sum_{j=8}^{11} p_f(j) \text{ indirect}$$

(17-11)

where the summation is over the components of Table 17-10.

17-9. OPERATION AND MAINTENANCE COSTS

Operation and maintenance costs are the third main item in power production costs, Table 17-4, and are believed to account for something less than 10 percent of the total power costs. They include direct and indirect costs of labor and supervisory personnel, continuously used supplies such as office supplies, personnel needs, etc., necessary for the operation of the power plant, and maintenance materials such as repair parts and inventory, and nuclear liability insurance.

Manpower requirements for nuclear power plants are less than for fossil-fueled power plants. They, however, require more highly trained personnel, such as reactor operators and health physicists. It is therefore, estimated that manpower requirements for nuclear plants are probably only a little bit higher than those in fossil-fuel plants of the same output.   While not much experience has yet been accumulated on commercially operable, nonexperimental, power plants, it is estimated that manpower requirements are fairly large for small-sized plants but do not increase rapidly with plant size.   For PWR and BWR plants, they may be of the order of 73 for a 300 Mw(e) plant, increasing to only 78 and 81 for 1,000 and 1,500 Mw(e) plants respectively.   For advanced converter and breeder plants these figures should be increased by about 10.

TABLE 17-11
Computation of Nuclear Accident Insurance

Insurance Required, $	Premium, $/$ 10⁶ yr	Premium, $/yr
First 10^6	40,000	40,000
Next 4×10^6	20,000	80,000
Next 5×10^6	8,000	40,000
Next 10×10^6	4,000	40,000
Next 20×10^6	2,000	40,000
Next 20×10^6	1,000	20,000
Total 60×10^6		260,000

The *nuclear liability insurance* is computed at $ 150 per kw(t) of maximum power times a *location* factor which accounts for the proximity of the reactor plant site to population and the total population of the area surrounding the plant, with a minimum of $ 3.5 × 10⁶ and a maximum of $ 60 × 10⁶ [corresponding to plant size of 23.3 Mw(t) and 400 Mw(t) respectively].   Mw(t), rather than Mw(e), represents the "nuclear" capacity of the plant.   The cost of insurance is given in Table 17-11.   It can be easily computed that the minimum premium is $ 90,000/year, and the maximum, as shown, is $ 260,000/year.   In addition to the above

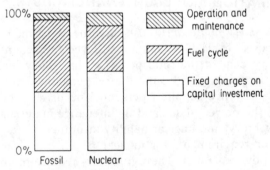

FIG. 17-7. Approximate breakdown of produc
tion costs in fossil and nuclear power plants.

there is a *Government indemnification* which pays any claims beyond the above, up to $\$500 \times 10^6$, at a cost per annum of $\$30/\text{Mw(t)}$.

An approximate breakdown of production costs in nuclear and fossil-fueled power plants, both normalized to 100 percent, is shown in Fig. 17-7 for comparison purposes.

17-10. SOME BASIC DEFINITIONS

Before presenting a computation of nuclear power costs, some useful basic definitions will be presented.

A power plant *plant operating factor* (also called the *plant capacity factor*) is the ratio of the total kwhr(net) generated during a period of time (usually a year) to the total kwhr(net) *rated* generated capacity of the plant during the same period, Fig. 17-8. For estimation purposes, a

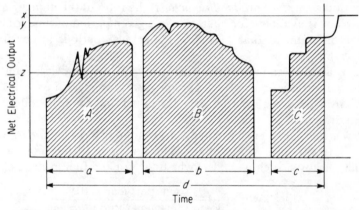

FIG. 17-8. Hypothetical diagram of operation of a power plant: x = rated net capacity of plant, y = peak load during period d, z = average load during period d. *Plant operating factor* = $(A + B + C)/xd$. *Plant load factor* = z/y. *Plant availability factor* = $(a + b + c)/d$. d is usually taken as 1 year.

plant operating factor of 0.8, or 80 percent is often used and the AEC has adopted this value for evaluation of nuclear-plant concepts. This factor is usually attained, and often exceeded, by new power plants. It decreasses, however, with plant life as newer and more economical plants are added to the electrical network, Table 17-12. On a lifetime basis, fossil-fueled plants have shown an average plant operating factor of 50 percent. This is expected to be exceeded by nuclear power plants. It may be added here, that many conservatively designed nuclear power plants, after a few months operation, exceeded (with AEC permission) their initial rated capacity, and a new rated capacity was established.

By definition, a power plant with a plant operating factor of 0.8 pro-

TABLE 17-12
Typical Plant Operating Factors

Years of Commercial Operation	Plant Operating Factor
1-5	0.8
6-15	0.9
16-20	0.8
21-25	0.7
26-30	0.6

duces rated capacity during 0.8 year/year, or $0.8(365 \times 24) = 7{,}008$ hours/year. This last figure is usually rounded to 7,000 hours/year, a convenient number to remember.

By contrast a *plant load factor* is the ratio of the average load carried by an electric power plant during a given period to its peak load during that period. A *plant availability factor* is the ratio of the number of hours that a plant is available for operation in a given period of time to the total number of hours in that period. Again these periods are customarily taken as one year.

As is already known, power costs are of two categories; capital costs and power production costs. *Capital costs* are given in total $. *Unit capital costs* are given in terms of $/kw(e) net and the (e) net designation is often dropped.

Production costs are obtained on an annual basis in *mills/kwhr*. A mill is 1/1000th of a U.S. dollar. The unit is used for total or partial production costs, such as mills/kwhr total, for the fuel cycle, etc. Thus

$$\text{Mills/kwhr} = \frac{\$ \text{ spent per year} \times 10^3}{\text{kwhr (net) generated per year}} \qquad (17\text{-}12)$$

where kwhr(net) generated per year = plant net rating kw(e) × 8766 hr/year × plant operating factor.

For a plant operating factor of 80 percent, Eq. 17-12 becomes

$$\text{Mills/kwhr} = \frac{\$ \text{ spent per year} \times 10^3}{\text{plant net rating kw(e)} \times 7000} \qquad (17\text{-}13)$$

Example 17-1.* A 1,000-Mw(e) nuclear power plant cost $130 × 10^6 to build and $12.6 × 10^6 for the fuel cycle in its third year of operation. During that year, the plant operating factor was 80 percent. What are the known power costs?

Solution.

$$\text{Unit capital cost} = \frac{130 \times 10^6}{1{,}000{,}000} = \$130/\text{kw}$$

$$\text{Fuel-cycle cost} = \frac{12.6 \times 10^6 \times 10^3}{1{,}000{,}000 \times 7000} = 1.8 \text{ mills/kwhr}$$

*These costs are too low by 1978 standards.

Fuel *burnup* refers to the quantity of material consumed by fission or neutron capture per unit quantity of fuel. Units may be atom percent burnup, or fissions per cc. Since the energy released per fission is approximately constant [2], burnup is also expressed in terms of the *total* energy released per unit mass of fuel. The common unit is Mwd/ton. The megawatts are thermal and the ton is metric (1,000 kg_m), based on U metal basis. Burnup affects the average residence time of the fuel in the core in the following manner.

$$\text{Residence time} = \frac{\text{burnup, Mwd/ton}}{[\text{Mw(t)/ton U}] \times \text{plant operating factor}}$$

(17-14)

It should be recalled that *enrichment* is a *mass* percentage of all fissile isotopes in the fuel [2], that *fuel* is defined as all U, Pu, and Th isotopes in a fuel element, not including chemical or alloying materials (such as oxygen in UO_2), but that the term *fuel material* includes the latter.

While the *fuel cycle* is defined as the series of steps which include obtaining raw material, fabricating fuel elements and subassemblies, storage at the reactor, removing, storing, shipping and processing the spent fuel, the term *fuel management* applies to the processes of refueling and the fuel movement program utilized in operating the nuclear power plant. An *out-in* fuel-management program is one in which new fuel is loaded into the outer region of the core, then subsequently shifted radially inward, and finally discharged from the center region. This is usually associated with a *zoned enrichment* core, a fuel-management program in which new fuel is loaded in lots of graded enrichment and fuel subassemblies are positioned in the core according to enrichment.

Working capital is that required (in addition to the capital investment in the plant) to provide funds for acquisition of fuel and other operating supplies and to meet payroll and other operating expenditures.

17-11. COMPUTATION OF POWER COSTS

A computation of nuclear power costs (capital and production) is best shown by example. The reader is here cautioned, however, that the computations presented here are only a condensed version of what is done by the utilities in actual cases. For a more complete method, the reader should refer to [167 and 168]. The reader is also cautioned that many of the figures used below are, of necessity, only representative and may become dated.

Example 17-2.*. Estimate the expected power costs of a proposed 1,300 Mw(e) investor-owned public utility nuclear power plant (of the PWR or BWR types). Power transmission equipment, lines and losses are not to be included. The plant has:

Direct capital costs (including labor and allowances for cost escalation):

(20)   Land and land rights = $0.3 × 10⁶

(21)   Structures and improvement = $10 × 10⁶

(22)   Reactor plant equipment:

    (22.1)   Reactor equipment, $15 × 10⁶

    (22.2)   Major heat transfer and transport system, $18 × 10⁶

    (22.3)   Safeguards cooling systems, $1.3 × 10⁶

    (22.4)   Radioactive waste treatment and disposal, $0.2 × 10⁶

    (22.5)   Nuclear fuel handling and storage systems, $1.1 × 10⁶

    (22.6)   Other reactor equipment, $5 × 10⁶

    (22.7)   Instrumentation and control, $2.5 × 10⁶

    (22.8)   Fossil-fueled boilers and superheaters, $2 × 10⁶

    (22.9)   Irradiation facilities, none

        Total (22) = $45.3 × 10⁶

(23)   Turboplant equipment = $28 × 10⁶

(24)   Electrical plant equipment = $4.0 × 10⁶

(25)   Miscellaneous power plant equipment = 1 × 10⁶

        Total items 20 through 25 = $88.6 × 10⁶

Working capital = $700,000 for plant operation and maintenance, $7.5 × 10⁶ for fuel-cycle **operation and materials and supplies in inventory**.

Fuel loading: Zr-clad UO_2, containing 114,286 kg_m U initial, 97,000 kg_m U at discharge.

Fuel ownership: Private

Fuel enrichment: 2.50 percent U^{235}, initial, 0.71 percent at discharge. Discharged fuel also contains 4.5g_m fissile Pu/kg_m U discharged.

Fuel burnup: 24,500 Mw-day/ton

Plant thermal efficiency = 32.5 percent

Solution.

I. Capital costs

Direct construction costs

    (20)   Land and land rights . . . . $0.300 × 10⁶

    (21-25)   Other   . . . . $ 88.300 × 10⁶

Indirect construction costs

    (91)   General & Administrative, 9.8 percent   8.653 × 10⁶

            Subtotal . . . .   96.953 × 10⁶

    (93)   Miscellaneous, 1.2 percent . . . . . .   1.163 × 10⁶

            Subtotal . . . .   98.116 × 10⁶

    (92)   Engineering services

        (922)   Plant engineering,

            15.4 percent   . . . .   15.110 × 10⁶

            Subtotal   . . . . 113.226 × 10⁶

*These costs are too low by 1978 standards.

(921) Reactor engineering,
12.4 percent . . . . 14.040×10^6

Subtotal . . . . $127.266 \times 10^6 127.266 \times 10^6$

(94) Interest during construction, 6.7 percent . . . . . 8.527×10^6

Total . . . . 135.793×10^6

Contingencies, 10 percent . . . . 13.579×10^6

Total depreciating capital cost . . . . $\$149.372 \times 10^6$

Total capital cost $= (149.372 + 0.3 + 0.7 + 7.5)10^6 = \$157.872 \times 10^6$

Thus, unit capital cost $= \dfrac{157.872 \times 10^6}{1300 \times 10^3} = \$121.440/kw$

II. Fuel Costs

Fuel costs are computed separately for each item: (a) processing, (b) shipping, (c) use charges, (d) uranium losses or consumption, and (e) plutonium losses, consumption or production. For each of these items, the costs are grouped under three headings: Charges incurred during fabrication, at the reactor, and in chemical processing. This is shown in Table 17-13. All charges are computed on the basis of $\$/kg_m$ U *charged* into the reactor.

TABLE 17-13
Summary of Fuel Costs, $\$/kg_m$ U

Process	(a)	(b)	(c)	(d)	(e)	Totals
Fabrication						
Transit to conversion		1.68	0.82			2.50
Conversion & fabrication	100.00		5.12	5.62		110.74
Transit to reactor		1.65	0.80			2.45
Reactor						
Preirradiation inventory . . .			2.20			2.20
Irradiation			21.38	124.40	(82.32)*	63.46
Decay			2.45			2.45
Chemical Processing						
Transit to processor		15.68	0.41			16.09
Separation	24.82		0.94	1.57	0.82	28.15
Uranium conversion	5.43		0.42	0.46		6.31
Plutonium conversion	10.18				0.81	10.99
Transit to receiving		0.97				0.97
Totals	140.43	19.98	34.54	132.05	(80.69)*	246.31

* Values shown in parentheses are credits for net Pu production.

Mw(t) of plant $=$ Mw(e)/efficiency $= 1300/0.325 = 4,000$ Mw(t)

Mw(t)/ton fuel $= 4000/114.286 = 35$ Mw(t)/ton U

Average fuel residence time in core (Using Eq. 17-14)

$$= \frac{24,500}{35 \times 0.8} = 875 \text{ days} = 2.4 \text{ years}$$

$$\text{Annual fuel throughput} = \frac{114{,}286}{2.4} = 47{,}619 \text{ kg}_m \text{ U/year}$$

Annual fuel cost = total fuel cost (from Table 17-13) × annual throughput
$$= \$246.31 \times 47{,}619 = \$11.729 \times 10^6$$
Fuel cycle costs, using Eq. 17-13:

$$= \frac{11.729 \times 10^6 \times 10^3}{1{,}300{,}000 \times 7{,}000} = 1.289 \text{ mills/kwhr}$$

III. Operating and Maintenance Costs

These are estimated on the basis of 78 personnel (5 in management, 12 technical staff, 43 operating staff, and 18 in maintenance) at an annual average wage of $10,000. Thus

Total labor expenses   = $10,000 × 78   = $0.780 × 10⁶
Fringe benefits, at 20% = 0.2 × 0.78 × 10⁶   = $0.156 × 10⁶
Maintenance material and operating supplies   = $1.200 × 10⁶

Nuclear liability insurance, based on maximum coverage up to $500 × 10⁶ (Sec. 17-9).

$$= \$0.260 \times 10^6 + \$30/\text{Mw(t)}, \text{ Government indemnity}$$
$$= 0.260 \times 10^6 + \times 30 \times \frac{1300}{0.2857} = \$0.397 \times 10^6/\text{year}$$

Total annual operating and maintenance costs = $2.136 × 10⁶
$$\text{unit} = \frac{2.136 \times 10^6 \times 10^3}{1{,}300{,}000 \times 7000} = \$0.235 \text{ mills/kwhr}$$

IV. Fixed Charges

Depreciating capital   = $149.372 × 10⁶
Nondepreciating capital = Land and land rights + working
                          capital
                        = 0.300 × 10⁶ + 8.200 × 10⁶
                        = $8.500 × 10⁶

Annual fixed charges:

For depreciating capital at 13.7 percent (Table 17-5) = 0.137 × 149.372 × 10⁶ = $20.464 × 10⁶/yr.

For nondepreciating capital at 12.8 percent = 0.128 × 8.5 × 10⁶ = $1.088 × 10⁶/yr.

Unit costs for fixed charges:

$$\text{For depreciating capital} = \frac{20.464 \times 10^6 \times 10^3}{1{,}300{,}000 \times 7000} = 2.248$$

$$\text{For nondepreciating capital} = \frac{1.088 \times 10^6 \times 10^3}{1{,}300{,}000 \times 7000} = 0.120$$

V. Production Costs

The total production costs are now obtained from the fixed charges, and from the fuel cycle and operating and maintenance costs as shown in Table 17-14.

TABLE 17-14
Power Production Costs of Example 17-2

Cost item	Capital Cost $\times 10^6$	Rate, percent	Annual Cost, $\$10^6$	Unit Cost mills/kwhr
Fixed charges				
Depreciating capital........	149.372	13.7	20.464	2.248
Nondepreciating capital	8.500	12.8	1.088	0.120
Subtotal......................	157.872		21.552	2.368
Operating costs				
Fuel cycle..................			11.729	1.289
Operating and maintenance.			2.136	0.235
Nuclear liability insurance ..			0.397	0.044
Subtotal......................			14.262	1.568
TOTAL......................			35.814	3.936

PROBLEMS

17-1. A 1,000-Mw(e) nuclear power plant of novel design has been estimated to cost a total of $122 million including $86 million in direct capital costs. The land costs are $0.333 million. Estimate (a) the allowance for escalation, if escalation is at the rate of 2 percent per year, (b) the taxes during construction if the local tax rate is 1.3 percent per annum, and (c) the interest during construction if the working capital available to the utility is $5 million and the annual interest rate is 6.5 percent.

17-2. If the initial cost of money for a nuclear power plant is 7.20 percent, estimate the effective cost of money for a nuclear power plant if the life of the power plant is (a) 25 years, (b) 35 years.

17-3. A fuel-fabrication plant has a rated throughput of 1000 kg_m/day. Calculate the unit fuel fabrication cost, $\$/kg_m$ for the fuel used in a UO_2-fueled reactor when the actual throughput is 800 kg_m/day and the fixed charge rate is 25 percent. Assume 260 working days per year.

17-4. Shipment of spent fuel between reactor site and reprocessing plant, a distance of 1,000 miles, takes 16 days round trip by rail at a rate of 2.04¢/lb_m for loaded casks and 1.87¢/lb_m for empties. Casks cost $1.25/$lb_m$ with an annual fixed charge of 15 percent (including taxes, return and recovery of investment and maintenance). Handling plus insurance (against damage and loss) costs are $1350/trip. The 10-fuel-element cask weighs 110 tons. Calculate the total shipment costs in $ and $\$/kg_m$ U when 80 fuel elements weighing 0.8 tons each must be shipped every 40 days from a reactor complex, and the cost of a railroad car (carrying one cask) is $25,000/yr plus 5 percent of car and fuel cost per year for maintenance and insurance.

17-5. Calculate the Pu credit in Mills/kwhr for a low-enriched uranium fuel if 5.5 g_m Pu are produced per kg_m initial U. No Pu or U^{233} is supplied in the initial loading. Pu suffers a 1 percent loss in reprocessing. The thermal efficiency is 30 percent and the average fuel burnup is 21,000 Mwd/ton.

17-6. For the same O&M costs, find if plant A or B is more economical to operate if both plants have $\eta = 0.32$ and the fixed charge rate is (a) 10 percent, (b) 15 percent.

Plant	Capital cost, $/kw(e)	Fuel cycle cost, ¢/10⁶ Btu
A	115	30
B	105	35

17-7. Compare the costs of nuclear insurance in mills/kwhr for (a) a 768-Mw(e) nuclear power plant that has a 32 percent thermal efficiency, and (b) a 50-Mw(e) plant that has a thermal efficiency of 25 percent.

17-8. A 1,000-Mw(e) nuclear power plant operates with a thermal efficiency of 30 percent. A major piece of equipment costing a total of $10⁶ (capital, installation, etc.) would, if installed, improve the efficiency to 30.5 percent. Should that equipment be added if the fuel cycle costs before installation were 2.2 mills/kwhr and the fixed charge rate remains at 15 percent? Assume other power production costs are unaffected.

17-9. A, 1,000-Mw(e) plant is to be installed in a particular locality because of increased load demand. (a) Which of the following two plants would be chosen on economic grounds if the plant operating factor is 80 percent? (b) What plant operating factor would cause a reversal of that decision?

	Fossil-Fueled Plant	Nuclear Plant
Unit capital cost, $/kw(e)	120	145
O&M, mills/kwhr	0.28	0.33
Fuel cycle costs, ¢/10⁶ Btu	34	18
Thermal efficiency, percent	40	30
Fixed charge rate, percent	15	15

17-10. A 1,300-Mw(e) nuclear power plant has a thermal efficiency of 32 percent. The core initial loading contained 70,000 kg_m U of 3 percent enrichment. In a batch-core fuel management scheme (entire core inserted and removed-no zoning), the discharged core contained 68,000 kg_m U of 1 percent enrichment. The average core burnup was 20,000 Mwd/ton U (initial). Estimate (a) the fuel-residence time in the core in years, (b) the fuel-depletion cost, and (c), the fuel use charge for the in-core period in mills/kwhr.

17-11. 20,000 kg_m U are inserted in a 1,000-Mw(e) reactor core with an enrichment of 3.00 percent, and removed (batch management) with an enrichment of 1.50 percent. Two percent of all U will be consumed in burnup. Fuel takes 1 year in fabrication, 1 year in the core and 6 months in reprocessing. What are the depletion and use costs in mills/kwhr. Use rate = 6.5 percent.

17-12. A 1,000-Mw(e) reactor plant has a thermal efficiency of 32 percent. 40,500 kg_m U of 2.5 percent enriched fuel are inserted in the core. The entire core was removed (batch core management) one year later at which time it contained 39,375 kg_m U of 1 percent enrichment plus 80 kg_m Pu. The fuel was obtained 6 months prior to insertion and was returned 6 months after removal from the core. The fuel fabrication and conversion costs were $100/kg$_m$ U, and the shipping costs were $15/kg$_m$ U. Compute the fuel cycle cost in mills/kwhr.

17-13. Fuel is inserted in a reactor core with an enrichment of 3.00 percent and removed (batch management) with an enrichment of 1.50 percent. When acquired, it remains 1 year in fabrication, 1 year in use in the core and 6 months in reprocessing. Two percent of all uranium will be consumed in burnup. What would the open market cost of a kg$_m$ natural U have to be to compete with government-supplied fuel? Government use rate = 6.5 percent, open-market interest rate = 8 percent. Assume reprocessing costs are the same and that only naturally enriched fuel can be obtained on the open market. Enriching costs are as set by the government.

17-14. It is desired to evaluate the effect of fuel burnup, Mwd/ton, on fuel-cycle costs, mills/kwhr, for a 1,000-Mw(e) nuclear power plant that has a thermal efficiency of 32 percent. The following data apply: Fuel loading, 50,000 kg$_m$ U; fuel fabrication time, 6 months; irradiated fuel storage, 4 months; reprocessing time, 2 months; shipping costs, $20/kg$_m$ U; fabrication and conversion costs, $100/kg$_m$ U; reprocessing costs, $60/kg$_m$ U; annual use charge, 6.5 percent, plutonium credit, none; enrichments given in table below. Plot the fuel cycle costs vs. burnup.

Burnup, Mwd/ton	10,000	20,000	30,000	40,000
Initial enrichment, percent	2.8	3.3	3.8	4.3
Final enrichment, percent	2.2	2.3	2.4	2.5

appendix **A**

Alphabetical
List of the Elements

ALPHABETICAL LIST OF THE ELEMENTS

Element	Symbol	Atomic number, Z	Element	Symbol	Atomic number, Z
Actinium	Ac	89	Mendelevium	Md	101
Aluminum	Al	13	Mercury	Hg	80
Americium	Am	95	Molybdenum	Mo	42
Antimony	Sb	51	Neodymium	Nd	60
Argon	A	18	Neon	Ne	10
Arsenic	As	33	Neptunium	Np	93
Astatine	At	85	Nickel	Ni	28
Barium	Ba	56	Niobium (Columbium)	Nb	41
Berkelium	Bk	97	Nitrogen	N	7
Beryllium	Be	4	Nobelium	No	102
Bismuth	Bi	83	Osmium	Os	76
Boron	B	5	Oxygen	O	8
Bromine	Br	35	Palladium	Pd	46
Cadmium	Cd	48	Phosphorus	P	15
Calcium	Ca	20	Platinum	Pt	78
Californium	Cf	98	Plutonium	Pu	94
Carbon	C	6	Polonium	Po	84
Cerium	Ce	58	Potasium	K	19
Cesium	Cs	55	Praseodymium	Pr	59
Chlorine	Cl	17	Promethium	Pm	61
Chromium	Cr	24	Protactinium	Pa	91
Cobalt	Co	27	Radium	Ra	88
Copper	Cu	29	Radon	Rn	86
Curium	Cm	96	Rhenium	Re	75
Dysprosium	Dy	66	Rhodium	Rh	45
Einsteinium	Es	99	Rubidium	Rb	37
Erbium	Er	68	Ruthenium	Ru	44
Europium	Eu	63	Samarium	Sm	62
Fermium	Fm	100	Scandium	Sc	21
Fluorine	F	9	Selenium	Se	34
Francium	Fr	87	Silicon	Si	14
Gadolinium	Gd	64	Silver	Ag	47
Gallium	Ga	31	Sodium	Na	11
Germanium	Ge	32	Strontium	Sr	38
Gold	Au	79	Sulfur	S	16
Hafnium	Hf	72	Tantalum	Ta	73
Helium	He	2	Technetium	Tc	43
Holmium	Ho	67	Tellurium	Te	52
Hydrogen	H	1	Terbium	Tb	65
Indium	In	49	Thallium	Tl	81
Iodine	I	53	Thorium	Th	90
Iridium	Ir	77	Thulium	Tm	69
Iron	Fe	26	Tin	Sn	50
Krypton	Kr	36	Titanium	Ti	22
Khurchatorium	Ku	104	Tungsten (Wolfram)	W	74
Lanthanum	La	57	Uranium	U	92
Lawrencium	Lw	103	Vanadium	V	23
Lead	Pb	82	Xenon	Xe	54
Lithium	Li	3	Ytterbium	Yb	70
Lutecium	Lu	71	Yttrium	Y	39
Magnesium	Mg	12	Zinc	Zn	30
Manganese	Mn	25	Zirconium	Zr	40

appendix **B**

Some
Thermodynamic Properties

TABLE B-1*
Properties of Dry Saturated Steam†
Pressure

Abs. press., psia	Temp., °F	Specific volume		Enthalpy			Entropy		
		Sat. liquid	Sat. vapor	Sat. liquid	Evap.	Sat. vapor	Sat. liquid	Evap.	Sat. vapor
p	t	v_f	v_g	h_f	h_{fg}	h_g	s_f	s_{fg}	s_g
1.0	101.74	0.01614	333.6	69.70	1036.3	1106.0	0.1326	1.8456	1.9782
2.0	126.08	0.01623	173.73	93.99	1022.2	1116.2	0.1749	1.7451	1.9200
3.0	141.48	0.01630	118.71	109.37	1013.2	1122.6	0.2008	1.6855	1.8863
4.0	152.97	0.01636	90.63	120.86	1006.4	1127.3	0.2198	1.6427	1.8625
5.0	162.24	0.01640	73.52	130.13	1001.0	1131.1	0.2347	2.6094	1.8441
6.0	170.06	0.01645	61.98	137.96	996.2	1134.2	0.2472	1.5820	1.8292
7.0	176.85	0.01649	53.64	144.76	992.1	1136.9	0.2581	1.5586	1.8167
8.0	182.86	0.01653	47.34	150.79	988.5	1139.3	0.2674	1.5383	1.8057
9.0	188.28	0.01656	42.40	156.22	985.2	1141.4	0.2759	1.5203	1.7962
10	193.21	0.01659	38.42	161.17	982.1	1143.3	0.2835	1.5041	1.7876
14.696	212.00	0.01672	26.80	180.07	970.3	1150.4	0.3120	1.4446	1.7566
15	213.03	0.01672	26.29	181.11	969.7	1150.8	0.3135	1.4415	1.7549
20	227.96	0.01683	20.089	196.16	960.1	1156.3	0.3356	1.3962	1.7319
25	240.07	0.01692	16.303	208.42	952.1	1160.6	0.3533	1.3606	1.7139
30	250.33	0.01701	13.746	218.82	945.3	1164.1	0.3680	1.3313	1.6993
35	259.28	0.01708	11.898	227.91	939.2	1167.1	0.3807	1.3063	1.6870
40	267.25	0.01715	10.498	236.03	933.7	1169.7	0.3919	1.2844	1.6763
45	274.44	0.01721	9.401	243.36	928.6	1172.0	0.4019	1.2650	1.6669
50	281.01	0.01727	8.515	250.09	924.0	1174.1	0.4110	1.2474	1.6585
55	287.07	0.01732	7.787	256.30	919.6	1175.9	0.4193	1.2316	1.6509
60	292.71	0.01738	7.175	262.09	915.5	1177.6	0.4270	1.2168	1.6438
65	297.97	0.01743	6.655	267.50	911.6	1179.1	0.4342	1.2032	1.6374
70	302.92	0.01748	6.206	272.61	907.9	1180.6	0.4409	1.1906	1.6315
75	307.60	0.01753	5.816	277.43	904.5	1181.9	0.4472	1.1787	1.6259
80	312.03	0.01757	5.472	282.02	901.1	1183.1	0.4531	1.1676	1.6207
85	316.25	0.01761	5.168	286.39	897.8	1184.2	0.4587	1.1571	1.6158
90	320.27	0.01766	4.896	290.56	894.7	1185.3	0.4641	1.1471	1.6112
95	324.12	0.01770	4.652	294.56	891.7	1186.2	0.4692	1.1376	1.6068
100	327.81	0.01774	4.432	298.40	888.8	1187.2	0.4740	1.1286	1.6026
110	334.77	0.01782	4.049	305.66	883.2	1188.9	0.4832	1.1117	1.5948

* For list of symbols used in Appendix B, see Table 2-2, p. 24.
† Abridged from Joseph H. Keenan and Frederick G. Keyes, *Thermodynamic Properties of Steam*, John Wiley & Sons., Inc., New York. Copyright, 1937, by Joseph H. Keenan and Frederick G. Keyes.

TABLE B-1
Properties of Dry Saturated Steam *(continued)*

Abs. press., psia	Temp., °F	Specific volume		Enthalpy			Entropy		
		Sat. liquid	Sat. vapor	Sat. liquid	Evap.	Sat. vapor	Sat. liquid	Evap.	Sat. vapor
p	t	v_f	v_g	h_f	h_{fg}	h_g	s_f	s_{fg}	s_g
120	341.25	0.01789	3.728	312.44	877.9	1190.4	0.4916	1.0962	1.5878
130	347.32	0.01796	3.455	318.81	872.9	1191.7	0.4995	1.0817	1.5812
140	353.02	0.01802	3.220	324.82	868.2	1193.0	0.5069	1.0682	1.5751
150	358.42	0.01809	3.015	330.51	863.6	1194.1	0.5138	1.0556	1.5694
160	363.53	0.01815	2.834	335.93	859.2	1195.1	0.5204	1.0436	1.5640
170	368.41	0.01822	2.675	341.09	854.9	1196.0	0.5266	1.0324	1.5590
180	373.06	0.01827	2'532	346.03	850.8	1196.9	0.5325	1.0217	1.5542
190	377.51	0.01833	2.404	350.79	846.8	1197.6	0.5381	1.0116	1.5497
200	381.79	0.01839	2.288	355.36	843.0	1198.4	0.5435	1.0018	1.5453
250	400.95	0.01865	1,8438	376.00	825.1	1201.1	0.5675	0.9588	1.5263
300	417.33	0.01890	1.5433	393.84	809.0	1202.8	0.5879	0.9225	1.5104
350	431.72	0.01913	1.3260	409.69	794.2	1203.9	0.6056	0.8910	1.4966
400	444.59	0.0193	1.1613	424.0	780.5	1204.5	0.6214	0.8630	1.4844
450	456.28	0.0195	1.0320	437.2	767.4	1204.6	0.6356	0.8378	1.4734
500	467.01	0.0197	0.9278	449.4	755.0	1204.4	0.6487	0.8147	1.4634
550	476.94	0.0199	0.8424	460.8	743.1	1203.9	0.6608	0.7934	1.4542
600	486.21	0.0201	0.7698	471.6	731.6	1203.2	0.6720	0.7734	1.4454
650	494.90	0.0203	0.7083	481.8	720.5	1202.3	0.6826	0.7548	1.4374
700	503.10	0.0205	0.6554	491.5	709.7	1201.2	0.6925	0.7371	1.4296
750	510.86	0.0207	0.6092	500.8	699.2	1200.0	0.7019	0.7204	1.4223
800	518.23	0.0209	0.5687	509.7	688.9	1198.6	0.7108	0.7045	1.4153
850	525.26	0.0210	0.5327	518.3	678.8	1197.1	0.7194	0.6891	1.4085
900	531.98	0.0212	0.5006	526.6	668.8	1195.4	0.7275	0.6744	1.4020
950	538.43	0.0214	0.4717	534.6	659.1	1193.7	0.7355	0.6602	1.3957
1000	544.61	0.0216	0.4456	542.4	649.4	1191.8	0.7430	0.6467	1.3897
1100	556.31	0.0220	0.4001	557.4	630.4	1187.7	0.7575	0.6205	1.3780
1200	567.22	0.0223	0.3619	571.7	611.7	1183.4	0.7711	0.5956	1.3667
1300	577.46	0.0227	0.3293	585.4	593.2	1178.6	0.7840	0.5719	1.3559
1400	587.10	0.0231	0.3012	598.7	574.7	1173.4	0.7963	0.5491	1.3454
1500	596.23	0.0235	0.2765	611.6	556.3	1167.9	0.8082	0.5269	1.3351
2000	635.82	0.0257	0.1878	671.7	463.4	1135.1	0.8619	0.4230	1.2849
2500	668.13	0.0287	0.1307	730.6	360.5	1091.1	0.9126	0.3197	1.2322
3000	695.36	0.0346	0.0858	802.5	217.8	1020.3	0.9731	0.1885	1.1615
3206.2	705.40	0.0503	0.0503	902.7	0	902.7	1.0580	0	1.0580

TABLE B-2
Properties of Dry Saturated Steam
Temperature

Temp., °F	Abs. press., psia	Specific volume Sat. liquid	Specific volume Sat. vapor	Enthalpy Sat. liquid	Enthalpy Evap.	Enthalpy Sat. vapor	Entropy Sat. liquid	Entropy Evap.	Entropy Sat. vapor
t	p	v_f	v_g	h_f	h_{fg}	h_g	s_f	s_{fg}	s_g
32	0.08854	0.01602	3306	0.00	1075.8	1075.8	0.0000	2.1877	2.1877
35	0.09995	0.01602	2947	3.02	1074.1	1077.1	0.0061	2.1709	2.1770
40	0.12170	0.01602	2444	8.05	1071.3	1079.3	0.0162	2.1435	2.1597
45	0.14752	0.01602	2036.4	13.06	1068.4	1081.5	0.0262	2.1167	2.1429
50	0.17811	0.01603	1703.2	18.07	1065.6	1083.7	0.0361	2.0903	2.1264
60	0.2563	0.01604	1206.7	28.06	1059.9	1088.0	0.0555	2.0393	2.0948
70	0.3631	0.01606	867.9	38.04	1054.3	1092.3	0.0745	1.9902	2.0647
80	0.5069	0.01608	633.1	48.02	1048.6	1096.6	0.0932	1.9428	2.0360
90	0.6982	0.01610	468.0	57.99	1042.9	1100.9	0.1115	1.8972	2.0087
100	0.9492	0.01613	350.4	67.97	1037.2	1105.2	0.1295	1.8531	1.9826
110	1.2748	0.01617	265.4	77.94	1031.6	1109.5	0.1471	1.8106	1.9577
120	1.6924	0.01620	203.27	87.92	1025.8	1113.7	0.1645	1.7694	1.9339
130	2.2225	0.01625	157.34	97.90	1020.0	1117.9	0.1816	1.7296	1.9112
140	2.8886	0.01629	123.01	107.89	1014.1	1122.0	0.1984	1.6910	1.8894
150	3.718	0.01634	97.07	117.89	1008.2	1126.1	0.2149	1.6537	1.8685
160	4.741	0.01639	77.29	127.89	1002.3	1130.2	0.2311	1.6174	1.8485
170	5.992	0.01645	62.06	137.90	996.3	1134.2	0.2472	1.5822	1.8293
180	7.510	0.01651	50.23	147.92	990.2	1138.1	0.2630	1.5480	1.8109
190	9.339	0.01657	40.96	157.95	984.1	1142.0	0.2785	1.5147	1.7932
200	11.526	0.01663	33.64	167.99	977.9	1145.9	0.2938	1.4824	1.7762
210	14.123	0.01670	27.82	178.05	971.6	1149.7	0.3090	1.4508	1.7598
212	14.696	0.01672	26.80	180.07	970.3	1150.4	0.3120	1.4446	1.7566
220	17.186	0.01677	23.15	188.13	965.2	1153.4	0.3239	1.4201	1.7440
230	20.780	0.01684	19.382	198.23	958.8	1157.0	0.3387	1.3901	1.7288
240	24.969	0.01692	16.323	208.34	952.2	1160.5	0.3531	1.3609	1.7140
250	29.825	0.01700	13.821	218.48	945.5	1164.0	0.3675	1.3323	1.6998
260	35.429	0.01709	11.763	228.64	938.7	1167.3	0.3817	1.3043	1.6860
270	41.858	0.01717	10.061	238.84	931.8	1170.6	0.3958	1.2769	1.6727
280	49.203	0.01726	8.645	249.06	924.7	1173.8	0.4096	1.2501	1.6597
290	57.556	0.01735	7.461	259.31	917.5	1176.8	0.4234	1.2238	1.6472
300	67.013	0.01745	6.466	269.59	910.1	1179.7	0.4369	1.1980	1.6350
310	77.68	0.01755	5.626	279.92	902.6	1182.5	0.4504	1.1727	1.6231
320	89.66	0.01765	4.914	290.28	894.9	1185.2	0.4637	1.1478	1.6115
330	103.06	0.01776	4.307	300.68	887.0	1187.7	0.4769	1.1233	1.6002
340	118.01	0.01787	3.788	311.13	879.0	1190.1	0.4900	1.0992	1.5891

TABLE B-2
Properties of Dry Saturated Steam *(continued)*
Temperature

Temp., °F	Abs. press., psia	Specific volume		Enthalpy			Entropy		
		Sat. liquid	Sat. vapor	Sat. liquid	Evap.	Sat. vapor	Sat. liquid	Evap.	Sat. vapor
t	p	v_f	v_g	h_f	h_{fg}	h_g	s_f	s_{fg}	s_g
350	134.63	0.01799	3.342	321.63	870.7	1192.3	0.5029	1.0754	1.5783
360	153.04	0.01811	2.957	332.18	852.2	1194.4	0.5158	1.0519	1.5677
370	173.37	0.01823	2.625	342.79	853.5	1196.3	0.5286	1.0287	1.5573
380	195.77	0.01836	2.335	353.45	844.6	1198.1	0.5413	1.0059	1.5471
390	220.37	0.01850	2.0836	364.17	835.4	1199.6	0.5539	0.9832	1.5371
400	247.31	0.01864	1.8633	374.97	826.0	1201.0	0.5664	0.9608	1.5272
410	276.75	0.01878	1.6700	385.83	816.3	1202.1	0.5788	0.9386	1.5174
420	308.83	0.01894	1.5000	396.77	806.3	1203.1	0.5912	0.9166	1.5078
430	343.72	0.01910	1.3499	407.79	796.0	1203.8	0.6035	0.8947	1.4982
440	381.59	0.01926	1.2171	418.90	785.4	1204.3	0.6158	0.8730	1.4887
450	422.6	0.0194	1.0993	430.1	774.5	1204.6	0.6280	0.8513	1.4793
460	466.9	0.0196	0.9944	441.4	763.2	1204.6	0.6402	0.8298	1.4700
470	514.7	0.0198	0.9009	452.8	751.5	1204.3	0.6523	0.8083	1.4606
480	566.1	0.0200	0.8172	464.4	739.4	1203.7	0.6645	0.7868	1.4513
490	621.4	0.0202	0.7423	476.0	726.8	1202.8	0.6766	0.7653	1.4419
500	680.8	0.0204	0.6749	487.8	713.9	1201.7	0.6887	0.7438	1.4325
520	812.4	0.0209	0.5594	511.9	686.4	1198.2	0.7130	0.7006	1.4136
540	962.5	0.0215	0.4649	536.6	656.6	1193.2	0.7374	0.6568	1.3942
560	1133.1	0.0221	0.3868	562.2	624.2	1186.4	0.7621	0.6121	1.3742
580	1325.8	0.0228	0.3217	588.9	588.4	1177.3	0.7872	0.5659	1.3532
600	1542.9	0.0236	0.2668	610.0	548.5	1165.5	0.8131	0.5176	1.3307
620	1786.6	0.0247	0.2201	646.7	503.6	1150.3	0.8398	0.4664	1.3062
640	2059.7	0.0260	0.1798	678.6	452.0	1130.5	0.8679	0.4110	1.2789
660	2365.4	0.0278	0.1442	714.2	390.2	1104.4	0.8987	0.3485	1.2472
680	2708.1	0.0305	0.1115	757.3	309.9	1067.2	0.9351	0.2719	1.2071
700	3093.7	0.0369	0.0761	823.3	172.1	995.4	0.9905	0.1484	1.1389
705.4	3206.2	0.0503	0.0503	902.7	0	902.7	1.0580	0	1.0580

TABLE B-3
Properties of Superheated Steam*

Abs. press., psia (Sat. temp., °F.)		Temperature, °F											
		200	300	400	500	600	700	800	900	1000	1100	1200	1400
1 (101.74)	v	392.6	452.3	512.0	571.6	631.2	690.8	750.4	809.9	869.5	929.1	988.7	1107.8
	h	1150.4	1195.8	1241.7	1288.3	1335.7	1383.8	1432.8	1482.7	1533.5	1585.2	1637.7	1745.7
	s	2.0512	2.1153	2.1720	2.2233	2.2702	2.3137	2.3542	2.3923	2.4283	2.4625	2.4952	2.5566
5 (162.24)	v	78.16	90.25	102.26	114.22	126.16	138.10	150.03	161.95	173.87	185.79	197.71	221.6
	h	1148.8	1195.0	1241.2	1288.0	1335.4	1383.6	1432.7	1482.6	1533.4	1585.1	1637.7	1745.7
	s	1.8718	1.9370	1.9942	2.0456	2.0927	2.1361	2.1767	2.2148	2.2509	2.2851	2.3178	2.3792
10 (193.21)	v	38.85	45.00	51.04	57.05	63.03	69.01	74.98	80.95	86.92	92.88	98.84	110.77
	h	1146.6	1193.9	1240.6	1287.5	1335.1	1383.4	1432.5	1482.4	1533.2	1585.0	1637.6	1745.6
	s	1.7927	1.8595	1.9172	1.9689	2.0160	2.0596	2.1002	2.1383	2.1744	2.2068	2.2413	2.3028
14.696 (212.00)	v		30.53	34.68	38.78	42.86	46.94	51.00	55.07	59.13	63.19	67.25	75.37
	h		1192.8	1239.9	1287.1	1334.8	1383.2	1432.3	1482.3	1533.1	1584.8	1637.5	1745.5
	s		1.8160	1.8743	1.9261	1.9734	2.0170	2.0576	2.0958	2.1319	2.1662	2.1989	2.2603
20 (227.96)	v		22.36	25.43	28.46	31.47	34.47	37.46	40.45	43.44	46.42	49.41	55.37
	h		1191.6	1239.2	1286.6	1334.4	1382.9	1432.1	1482.1	1533.0	1584.7	1637.4	1745.4
	s		1.7808	1.8396	1.8918	1.9392	1.9829	2.0235	2.0618	2.0978	2.1321	2.1648	2.2263
40 (267.25)	v		11.040	12.628	14.168	15.688	17.198	18.702	20.20	21.70	23.20	24.69	27.68
	h		1186.8	1236.5	1284.8	1333.1	1381.9	1431.3	1481.4	1532.4	1584.3	1637.0	1745.1
	s		1.6994	1.7608	1.8140	1.8619	1.9058	1.9467	1.9850	2.0212	2.0555	2.0883	2.1498
60 (292.71)	v		7.259	8.357	9.403	10.427	11.441	12.449	13.452	14.454	15.453	16.451	18.446
	h		1181.6	1233.6	1283.0	1331.8	1380.9	1430.5	1480.8	1531.9	1583.8	1636.6	1744.8
	s		1.6492	1.7135	1.7678	1.8162	1.8605	1.9015	1.9400	1.9762	2.0106	2.0434	2.1049
80 (312.03)	v			6.220	7.020	7.797	8.562	9.322	10,077	10.830	11.582	12.332	13.830
	h			1230.7	1281.1	1330.5	1379.9	1429.7	1480.1	1531.3	1583.4	1636.2	1744.5
	s			1.6791	1.7346	1.7836	1.8281	1.8694	1.9079	1.9442	1.9787	2.0115	2.0731
100 (327.81)	v			4.937	5.589	6.218	6.835	7.446	8.052	8.656	9.259	9.860	11.060
	h			1227.6	1279.1	1329.1	1378.9	1428.9	1479.5	1530.8	1582.9	1635.7	1744.2
	s			1.6518	1.7085	1.7581	1.8029	1.8443	1.8829	1.9193	1.9538	1.9867	2.0484
120	v			4.081	4.636	5.165	5.683	6.195	6.702	7.207	7.710	8.212	9.214
	h			1224.4	1277.2	1327.7	1377.8	1428.1	1478.1	1530.2	1582.4	1635.3	1743.9

Abs. Press. Lb./Sq. In. (Sat. Temp.)		Sat.									
140 (353.02)	v	....	....	....	....	....	....	....	....	....	....
	h	1221.1	1275.2	1326.4	1376.8	1427.3	1478.2	1529.7	1581.9	1634.9	1743.5
	s	1.6087	1.6683	1.7190	1.7645	1.8063	1.8451	1.8817	1.9163	1.9493	2.0110
160 (363.53)	v	3.008	3.443	3.849	4.244	4.631	5.015	5.396	5.775	6.152	6.906
	h	1217.6	1273.1	1325.0	1375.7	1426.4	1477.5	1529.1	1581.4	1634.5	1743.2
	s	1.5908	1.6519	1.7033	1.7491	1.7911	1.8301	1.8667	1.9014	1.9344	1.9962
180 (373.06)	v	2.649	3.044	3.411	3.764	4.110	4.452	4.792	5.129	5.466	6.136
	h	1214.0	1271.0	1323.5	1374.7	1425.6	1476.8	1528.6	1581.0	1634.1	1742.9
	s	1.5745	1.6373	1.6894	1.7355	1.7776	1.8167	1.8534	1.8882	1.9212	1.9831
200 (381.79)	v	2.361	2.726	3.060	3.380	3.693	4.002	4.309	4.613	4.917	5.521
	h	1210.3	1268.9	1322.1	1373.6	1424.8	1476.2	1528.0	1580.5	1633.7	1742.6
	s	1.5594	1.6240	1.6767	1.7232	1.7655	1.8048	1.8415	1.8763	1.9094	1.9713
220 (389.86)	v	2.125	2.465	2.772	3.066	3.352	3.634	3.913	4.191	4.467	5.017
	h	1206.5	1266.7	1320.7	1372.6	1424.0	1475.5	1527.5	1580.0	1633.3	1742.3
	s	1.5453	1.6117	1.6652	1.7120	1.7545	1.7939	1.8308	1.8656	1.8987	1.9607
240 (397.37)	v	1.9276	2.247	2.533	2.804	3.068	3.327	3.584	3.839	4.093	4.597
	h	1202.5	1264.5	1319.2	1371.5	1423.2	1474.8	1526.9	1579.6	1632.9	1742.0
	s	1.5319	1.6003	1.6546	1.7017	1.7444	1.7839	1.8209	1.8558	1.8889	1.9510
260 (404.42)	v	....	2.063	2.330	2.582	2.827	3.067	3.305	3.541	3.776	4.242
	h	....	1262.3	1317.7	1370.4	1422.3	1474.2	1526.3	1579.1	1632.5	1741.7
	s	....	1.5897	1.6447	1.6922	1.7352	1.7748	1.8118	1.8467	1.8799	1.9420
280 (411.05)	v	....	1.9047	2.156	2.392	2.621	2.845	3.066	3.286	3.504	3.938
	h	....	1260.0	1316.2	1369.4	1421.5	1473.5	1525.8	1578.6	1632.1	1741.4
	s	....	1.5796	1.6354	1.6834	1.7265	1.7662	1.8033	1.8383	1.8716	1.9337
300 (417.33)	v	....	1.7675	2.005	2.227	2.442	2.652	2.859	3.065	3.269	3.674
	h	....	1260.0	1316.2	1368.3	1420.6	1472.8	1525.2	1578.1	1631.7	1741.0
	s	....	1.5701	1.6268	1.6751	1.7184	1.7582	1.7954	1.8305	1.8638	1.9260
350 (431.72)	v	....	1.4923	1.7036	1.8980	2.084	2.266	2.445	2.622	2.798	3.147
	h	....	1251.5	1310.9	1365.5	1418.5	1471.1	1523.8	1577.0	1630.7	1740.3
	s	....	1.5481	1.6070	1.6563	1.7002	1.7403	1.7777	1.8130	1.8463	1.9086
400 (444.59)	v	....	1.2851	1.4770	1.6508	1.8161	1.9767	2.134	2.290	2.445	2.751
	h	....	1245.1	1306.9	1362.7	1416.4	1469.4	1522.4	1575.8	1629.6	1739.5
	s	....	1.5281	1.5894	1.6398	1.6842	1.7247	1.7623	1.7977	1.8311	1.8936

* Abridged from Joseph H. Keenan and Frederick G. Keyes, *Thermodynamic Properties of Steam*, John Wiley and Sons, Inc., New York. Copyright, 1937, by Joseph H. Keenan and Frederick G. Keyes.

TABLE B-3
Properties of Superheated Steam (continued)

Abs. pres., psia (Sat. temp., °F,)		500	550	600	620	640	660	680	700	800	900	1000	1200	1400
450 (456.28)	v	1.1231	1.2155	1.3005	1.3332	1.3652	1.3967	1.4278	1.4584	1.6074	1.7516	1.8928	2.170	2.443
	h	1238.4	1272.0	1302.8	1314.6	1326.2	1337.5	1348.8	1359.9	1414.3	1467.7	1521.0	1628.6	1738.7
	s	1.5095	1.5437	1.5735	1.5845	1.5951	1.6054	1.6153	1.6250	1.6699	1.7108	1.7486	1.8177	1.8803
500 (467.01)	v	0.9927	1.0800	1.1591	1.1893	1.2188	1.2478	1.2763	1.3044	1.4405	1.5715	1.6996	1.9504	2.197
	h	1231.3	1266.8	1298.6	1310.7	1322.6	1334.2	1345.7	1357.0	1412.1	1466.0	1519.6	1627.6	1737.9
	s	1.4919	1.5280	1.5588	1.5701	1.5810	1.5915	1.6016	1.6115	1.6571	1.6982	1.7363	1.8056	1.8683
550 (476.94)	v	0.8852	0.9686	1.0431	1.0714	1.0989	1.1259	1.1523	1.1783	1.3038	1.4241	1.5414	1.7706	1.9957
	h	1223.7	1261.2	1294.3	1306.8	1318.9	1330.8	1342.5	1354.0	1409.9	1464.3	1518.2	1626.6	1737.1
	s	1.4751	1.5131	1.5451	1.5568	1.5680	1.5787	1.5890	1.5991	1.6452	1.6868	1.7250	1.7946	1.8575
600 (486.21)	v	0.7947	0.8753	0.9463	0.9729	0.9988	1.0241	1.0489	1.0732	1.1899	1.3013	1.4096	1.6208	1.8279
	h	1215.7	1255.5	1289.9	1302.7	1315.2	1327.4	1339.3	1351.1	1407.7	1462.5	1516.7	1625.5	1736.3
	s	1.4586	1.4990	1.5323	1.5443	1.5558	1.5667	1.5773	1.5875	1.6343	1.6762	1.7147	1.7846	1.8476
700 (503.10)	v	......	0.7277	0.7934	0.8177	0.8411	0.8639	0.8860	0.9077	1.0108	1.1082	1.2024	1.3853	1.5641
	h	......	1243.2	1280.6	1294.3	1307.7	1320.3	1332.8	1345.0	1403.2	1459.0	1513.9	1623.5	1734.8
	s	......	1.4722	1.5084	1.5212	1.5333	1.5449	1.5559	1.5665	1.6147	1.6573	1.6963	1.7666	1.8299
800 (518.23)	v	......	0.6154	0.6779	0.7006	0.7223	0.7433	0.7635	0.7833	0.8763	0.9633	1.0470	1.2088	1.3662
	h	......	1229.8	1270.7	1285.4	1299.4	1312.9	1325.9	1338.6	1398.6	1455.3	1511.0	1621.4	1733.2
	s	......	1.4467	1.4863	1.5000	1.5129	1.5250	1.5366	1.5476	1.5972	1.6407	1.6801	1.7510	1.8146
900 (531.98)	v	......	0.5264	0.5873	0.6089	0.6294	0.6491	0.6680	0.6863	0.7716	0.8506	0.9262	1.0714	1.2124
	h	......	1215.0	1260.1	1275.9	1290.9	1305.1	1318.8	1332.1	1393.9	1451.8	1508.1	1619.3	1731.6
	s	......	1.4216	1.4653	1.4800	1.4938	1.5066	1.5187	1.5303	1.5814	1.6257	1.6656	1.7371	1.8009
1000 (544.61)	v	......	0.4533	0.5140	0.5350	0.5546	0.5733	0.5912	0.6084	0.6878	0.7604	0.8294	0.9615	1.0893
	h	......	1198.3	1248.8	1265.9	1281.9	1297.0	1311.4	1325.3	1389.2	1448.2	1505.1	1617.3	1730.0
	s	......	1.3961	1.4450	1.4610	1.4757	1.4893	1.5021	1.5141	1.5670	1.6121	1.6525	1.7245	1.7886
1100 (556.31)	v	......	......	0.4532	0.4738	0.4929	0.5110	0.5281	0.5445	0.6191	0.6866	0.7503	0.8716	0.9885
	h	......	......	1236.7	1255.3	1272.4	1288.5	1303.7	1318.3	1384.3	1444.5	1502.2	1615.2	1728.4
	s	......	......	1.4251	1.4425	1.4583	1.4728	1.4862	1.4989	1.5535	1.5995	1.6405	1.7130	1.7775

Temperature, °F

Pressure (Sat. temp)		(1)	(2)	(3)	(4)	(5)	(6)	(7)	(8)	(9)	(10)	(11)
1200 (567.22)	v	0.4016	0.4222	0.4410	0.4586	0.4752	0.4909	0.5617	0.6250	0.6843	0.7967	0.9046
	h	1223.5	1243.9	1262.4	1279.6	1295.7	1311.0	1379.3	1440.7	1499.2	1613.1	1726.9
	s	1.4052	1.4243	1.4413	1.4568	1.4710	1.4843	1.5409	1.5879	1.6293	1.7025	1.7672
1400 (587.10)	v	0.3174	0.3390	0.3580	0.3753	0.3912	0.4062	0.4714	0.5281	0.5805	0.6789	0.7727
	h	1193.0	1218.4	1240.4	1260.3	1278.5	1295.5	1369.1	1433.1	1493.2	1608.9	1723.7
	s	1.3639	1.3877	1.4079	1.4258	1.4419	1.4567	1.5177	1.5666	1.6093	1.6386	1.7489
1600 (604.90)	v	.....	0.2733	0.2936	0.3112	0.3271	0.3417	0.4034	0.4553	0.5027	0.5906	0.6738
	h	.....	1187.8	1215.2	1238.7	1259.6	1278.7	1358.4	1425.3	1487.0	1604.6	1720.5
	s	.....	1.3489	1.3741	1.3952	1.4137	1.4304	1.4964	1.5476	1.5914	1.6669	1.7328
1800 (621.03)	v	.....	.....	0.2407	0.2597	0.2760	0.2907	0.3502	0.3986	0.4421	0.5218	0.5968
	h	.....	.....	1185.1	1214.0	1238.5	1260.3	1347.2	1417.4	1480.8	1600.4	1717.3
	s	.....	.....	1.3377	1.3638	1.3855	1.4044	1.4765	1.5301	1.5752	1.6520	1.7185
2000 (635.82)	v	.....	.....	0.1936	0.2161	0.2337	0.2489	0.3074	0.3532	0.3935	0.4668	0.5352
	h	.....	.....	1145.6	1184.9	1214.8	1240.0	1335.5	1409.2	1474.5	1596.1	1714.1
	s	.....	.....	1.2945	1.3300	1.3564	1.3783	1.4576	1.5139	1.5603	1.6384	1.7055
2500 (668.13)	v	.....	.....	.....	.....	0.1484	0.1686	0.2294	0.2710	0.3061	0.3678	0.4244
	h	.....	.....	.....	.....	1132.3	1176.8	1303.6	1387.8	1458.4	1585.3	1706.1
	s	.....	.....	.....	.....	1.2687	1.3073	1.4127	1.4772	1.5273	1.6088	1.6775
3000 (695.36)	v	.....	.....	.....	.....	.....	0.0984	0.1760	0.2159	0.2476	0.3018	0.3505
	h	.....	.....	.....	.....	.....	1060.7	1267.2	1365.0	1441.8	1574.3	1698.0
	s	.....	.....	.....	.....	.....	1.1966	1.3690	1.4439	1.4984	1.5837	1.6540
3206.2 (705.40)	v	.....	.....	.....	.....	.....	.....	0.1583	0.1981	0.2288	0.2806	0.3267
	h	.....	.....	.....	.....	.....	.....	1250.5	1355.2	1434.7	1569.8	1694.6
	s	.....	.....	.....	.....	.....	.....	1.3508	1.4309	1.4874	1.5742	1.6452
3500	v	.....	.....	.....	.....	.....	0.0306	0.1364	0.1762	0.2058	0.2546	0.2977
	h	.....	.....	.....	.....	.....	780.5	1224.9	1340.7	1424.5	1563.3	1689.8
	s	.....	.....	.....	.....	.....	0.9515	1.3241	1.4127	1.4723	1.5615	1.6336
4000	v	.....	.....	.....	.....	.....	0.0287	0.1052	0.1462	0.1743	0.2192	0.2581
	h	.....	.....	.....	.....	.....	763.8	1174.8	1314.4	1406.8	1552.1	1681.7
	s	.....	.....	.....	.....	.....	0.9347	1.2757	1.3827	1.4482	1.5417	1.6154

TABLE B-4

Temperature, °R (Sat. press., psia)		Sat. liquid	Sat. vapor	800	900	1000	1100	1200
700 (8.7472 × 10⁻⁹)s	v	1.7232 × 10⁻²	>10¹⁰	.....	....	....	....	....
	h	219.7	2180.5	2203.1	2224.7	2246.4	2268.0	2289.5
	s	0.6854	3.4866	3.5169	3.5424	3.5652	3.5857	3.6043
800 (5.0100 × 10⁻⁷)s	v	1.7548 × 10⁻²	7.4375 × 10⁸	......	8.3835 × 10⁸	9.3168 × 10⁸	1.0249 × 10⁹	1.1180 × 10⁹
	h	252.3	2200.1	......	2224.4	2246.3	2267.9	2289.5
	s	0.7290	3.1637	......	3.1925	3.2155	3.2360	3.2546
900 (1.1480 × 10⁻⁵)s	v	1.7864 × 10⁻²	3.6411 × 10⁷	......	....	4.0267 × 10⁷	4.4718 × 10⁷	4.8789 × 10⁷
	h	284.3	2217.6	......	....	2245.2	2267.7	2289.4
	s	0.7667	2.9148	......	....	2.9440	2.9653	2.9841
1000 (1.3909 × 10⁻⁴)s	v	1.8180 × 10⁻²	3.323 × 10⁶	......	....	....	3.6834 × 10⁶	4.0254 × 5
	h	325.9	2232.7	......	....	....	2264.8	2288.6
	s	0.7999	2.7168	......	....	....	2.7474	2.7680
1100 (1.0616 × 10⁻³)s	v	1.8496 × 10⁻²	4.7592 × 10⁵	......	....	....	....	5.2512 × 10⁵
	h	347.0	2245.1	......	....	....	....	2282.7
	s	0.8296	2.5551	......	....	....	....	2.5878
1200 (5.7398 × 10⁻³)s	v	1.8812 × 10⁻²	9.5235 × 10⁴	......	....	....	....	....
	h	377.7	2254.9					
	s	0.8563	2.4207					
1300 (2.3916 × 10⁻²)s	v	1.9128 × 10⁻²	2.4520 × 10⁴	......				
	h	408.2	2262.8					
	s	0.8807	2.3073					
1400 (8.1347 × 10⁻²)s	v	1.9444 × 10⁻²	7.6798 × 10³	......				
	h	438.4	2269.3					
	s	0.9031	2.2109					
1500 (2.3351 × 10⁻¹)s	v	1.9760 × 10⁻²	2.8334 × 10³	......				
	h	468.5	2274.9					
	s	0.9239	2.1282					
1600 (5.8425 × 10⁻¹)s	v	2.0076 × 10⁻²	1.1935 × 10³	......				
	h	498.5	2280.0					
	s	0.9433	2.0567					
1700 (1.3170)	v	2.0392 × 10⁻²	5.5585 × 10²	....				
	h	528.5	2285.3					
	s	0.9615	1.9948					
1800 (2.7164)	v	2.0708 × 10⁻²	2.8200 × 10²	....				
	h	558.6	2291.1					
	s	0.9786	2.9411					
1900 (5.1529)	v	2.1024 × 10⁻²	1.5512 × 10²	....				
	h	588.8	2297.2					
	s	0.9949	1.8941					
2000 (9.1533)	v	2.1340 × 10⁻²	90.914	....				
	h	619.1	2304.1					
	s	1.0105	1.8530					
2100 (15.392)	v	2.1656 × 10⁻²	56.185	....				
	h	649.7	2312.1					
	s	1.0255	1.8171					
2200 (24.692)	v	2.1972 × 10⁻²	36.338	....				
	h	680.7	2321.0					
	s	1.0399	1.7855					
2300 (38.013)	v	2.2288 × 10⁻²	24.454	....				
	h	712.0	2330.7					
	s	1.0538	1.7576					
2400 (56.212)	v	2.2604 × 10⁻²	17.109	....				
	h	743.8	2341.2					
	s	1.0673	1.7329					
2500 (80.236)	v	2.2920 × 10⁻²	12.388	....				
	h	776.2	2352.6					
	s	1.0805	1.7111					
2600 (1.1116 × 10²)s	v	2.3236 × 10⁻²	9.2328	....				
	h	809.1	2365.1					
	s	1.0934	1.6919					
2700 (1.052 × 10²)s	v	2.3552 × 10⁻²	7.0380					
	h	842.7	2378.8					
	s	1.1061	1.6751					

* From Ref. 6.

Thermodynamic Properties of Sodium *

Superheated Vapor, °R

1400	1600	1800	2000	2200	2400	2600	2700
....	....	....	....	....	....	....	....
2332.7	2375.9	2419.1	2462.3	2505.4	2548.6	2591.8	2613.4
3.6381	3.6665	3.6924	3.7148	3.7354	3.7545	3.7713	3.7796
1.3044×10^9	1.4907×10^9	1.6771×10^9	1.8634×10^9	2.0498×10^9	2.2361×10^9	2.4224×10^9	2.5156×10^9
2332.7	2375.9	2419.1	2462.3	2505.4	2548.6	2591.8	2613.4
3.2884	3.3169	3.3428	3.3652	3.3858	3.4048	3.5217	3.4299
5.6924×10^7	6.5056×10^7	7.3188×10^7	8.1320×10^7	8.9452×10^7	9.7584×10^7	1.0572×10^8	1.0978×10^8
2332.7	2375.9	2419.1	2462.3	2505.4	2548.6	2591.8	2613.4
3.0179	3.0464	3.0723	3.0947	3.1153	3.1343	3.1511	3.1594
4.6978×10^6	5.3693×10^6	6.0406×10^6	6.7118×10^6	7.383×10^6	8.0541×10^6	8.7253×10^6	9.0609×10^6
2332.6	2375.9	2419.1	2462.3	2505.4	2548.6	2591.8	2613.4
2.8024	2.8309	2.8568	2.8792	2.8998	2.9188	2.9357	2.9439
6.1515×10^5	7.0339×10^5	7.9139×10^5	8.7935×10^5	9.6729×10^5	1.0552×10^6	1.1432×10^6	1.1871×10^6
2331.7	2375.7	2419.0	2462.3	2505.4	2548.6	2591.8	2613.4
2.6263	2.6552	2.6812	2.7036	2.7243	2.7433	2.7601	2.7684
1.1345×10^5	1.3001×10^5	1.4635×10^5	1.6263×10^5	1.7891×10^5	1.9517×10^5	2.1144×10^5	2.1957×10^5
2327.6	2374.7	2418.7	2462.2	2505.4	2548.6	2591.8	2613.4
2.4778	2.5089	2.5353	2.5578	2.5785	2.5975	2.6143	2.6226
2.6936×10^4	3.1124×10^4	3.5095×10^4	3.9019×10^4	4.2931×10^4	4.6838×10^4	5.0743×10^4	5.2695×10^4
2312.2	2371.1	2417.6	2461.7	2505.1	2548.5	2591.8	2613.4
2.3445	2.3836	2.4115	2.4343	2.4551	2.4742	2.4911	2.4993
....	9.0793×10^3	1.0292×10^4	1.1460×10^4	1.2616×10^4	1.3767×10^4	1.4916×10^4	1.5491×10^4
....	2359.9	2414.0	2460.3	2504.5	2548.1	2591.6	2613.2
....	2.2715	2.3040	2.3280	2.3491	2.3683	2.3853	2.3935
....	3.1025×10^3	3.5625×10^3	3.9820×10^3	4.3896×10^3	4.7929×10^3	5.1944×10^3	5.3948×10^2
....	2332.6	2404.7	2456.5	2502.7	2547.2	2591.0	2612.8
....	2.1651	2.2083	2.2352	2.2573	2.2769	2.2940	2.3023
....	....	1.4040×10^3	1.5823×10^3	1.7496×10^3	1.9128×10^3	2.0743×10^3	2.1548×10^3
....	....	2384.7	2448.0	2498.7	2545.1	2589.7	2611.8
....	....	2.1192	2.1523	2.1765	2.1969	2.2144	2.2228
....	....	6.0659×10^2	6.9378×10^2	7.7180×10^2	8.4601×10^2	9.1858×10^2	9.5458×10^2
....	....	2347.7	2431.3	2490.5	2540.7	2587.1	2609.8
....	....	2.0309	2.0747	2.1031	2.1252	2.1433	2.1519
....	....	....	3.2952×10^2	3.7033×10^2	4.0785×10^2	4.4385×10^2	4.6158×10^2
....	....	....	2402.3	2475.6	2532.5	2582.2	2605.9
....	....	....	1.9996	2.0347	2.0597	2.0792	2.0882
....	....	....	1.6838×10^2	1.9197×10^2	2.1298×10^2	2.3265×10^2	2.4224×10^2
....	....	....	2359.5	2451.8	2518.8	2573.9	2599.3
....	....	....	1.9259	1.9701	1.9996	2.0212	2.0209
....	....	....	....	1.0543×10^2	1.1816×10^2	1.2980×10^2	1.3539×10^2
....	....	....	....	2417.4	2498.0	2560.9	2588.9
....	....	....	....	1.9072	1.9426	1.9673	1.9780
....	....	....	....	60.665	68.825	76.167	79.656
....	....	....	....	2372.9	2469.0	2451.9	2573.5
....	....	....	....	1.8455	1.8876	1.9164	1.9284
....	....	....	....	....	41.754	46.622	48.920
....	....	....	....	....	2431.8	2516.2	2552.3
....	....	....	....	....	1.8340	1.8674	1.8811
....	....	....	....	....	26.244	29.585	31.163
....	....	....	....	....	2388.2	2484.0	2525.0
....	....	....	....	....	1.7820	1.8201	1.8356
....	....	....	....	....	....	19.460	20.580
....	....	....	....	....	....	2446.8	2492.6
....	....	....	....	....	....	1.7748	1.7922
....	....	....	....	....	....	13.219	14.032
....	....	....	....	....	....	2406.5	2456.5
....	....	....	....	....	....	1.7321	1.7501
....	....	....	....	....	....	....	9.8326
....	....	....	....	....	....	....	2418.1
....	....	....	....	....	....	....	1.7120

TABLE B-5

Temperature, °R (Sat. press., psia)		Sat. liquid	Sat. vapor	1900	2000	2100
1800 (1.4713 × 10⁻²)	v	3.4650×10^{-2}	1.8423×10^5	1.9703×10^5	2.0870×10^5	2.1983×10^5
	h	1717.9	10966.1	11127.8	11242.8	11336.2
	s	2.3840	7.5219	7.6088	7.6672	7.7127
1900 (3.8545 × 10⁻²)	v	3.4972×10^{-2}	7.3717×10^4	....	7.8783×10^4	8.3378×10^4
	h	1817.0	10990.5	....	11165.6	11291.7
	s	2.4388	7.2669	....	7.3563	7.4178
2000 (9.2135 × 10⁻²)	v	3.5294×10^{-2}	3.2213×10^4	....	....	3.4410×10^4
	h	1916.1	11009.2	....	....	11197.5
	s	2.4892	7.0357	....	....	7.1276
2100 (2.0300 × 10⁻¹)	v	3.5617×10^{-2}	1.5226×10^1	....	....	....
	h	2015.2	11024.8			
	s	2.5369	6.8272			
2200 (4.1594 × 10⁻¹)	v	3.5939×10^{-2}	7.7194×10^3	....		
	h	2114.2	11039.6			
	s	2.5828	6.6398			
2300 (7.9884 × 10⁻¹)	v	2.6261×10^{-2}	4.1671×10^3	....		•
	h	2213.0	11055.4			
	s	2.6274	6.4719			
2400 (1.4530)	v	3.6583×10^{-2}	2.3691×10^3	....		
	h	2311.8	11066.8			
	s	2.6694	6.3173			
2500 (2.5189)	v	3.6905×10^{-2}	1.4098×10^3	....		
	h	2410.4	11074.4			
	s	2.7089	6.1745			
2600 (4.1705)	v	3.7228×10^{-2}	8.7756×10^2	....		
	h	2508.9	11086.6			
	s	2.7480	6.0471			
2700 (6.6395)	v	3.7550×10^{-2}	5.6763×10^2	....		
	h	2607.2	11103.1			
	s	2.7861	5.9327			
2800 (10.282)	v	3.7872×10^{-2}	3.7684×10^2	....		
	h	2705.6	11118.8			
	s	2.8211	5.8258			
2900 (15.438)	v	3.8194×10^{-2}	2.5779×10^2	....		
	h	2803.5	11136.5			
	s	2.8550	5.7284			
3000 (22.498)	v	3.8517×10^{-2}	1.8153×10^{-2}	....		
	h	2900.9	11156.8			
	s	2.8884	5.6403			
3100 (31.989)	v	3.8839×10^{-2}	1.3090×10^2	....		
	h	2998.2	11178.9			
	s	2.9207	5.5596			
3200 (44.498)	v	3.9161×10^{-2}	96.404	....		
	h	3095.8	11202.7			
	s	2.9518	5.4852			
3300 (60.843)	v	3.9483×10^{-2}	72.187	....		
	h	3193.2	11229.1			
	s	2.9812	5.4163			
3400 (81.759)	v	3.9805×10^{-2}	54.970	....		
	h	3290.5	11257.9			
	s	3.0092	5.3526			
3500 (1.07715 × 10²)	v	4.0128×10^{-2}	42.669	....		
	h	3387.6	11288.6			
	s	3.0367	5.2942			
3600 (1.3845 × 10⁷)	v	4.0450×10^{-2}	33.925	....		
	h	2484.6	11320.4			
	s	3.0651	5.2417			

* From Ref. 6

Thermodynamic Properties of Lithium*

Temperature of Superheated Vapor, °R							
2200	2400	2600	2800	3000	3200	3400	3600
2.3068×10^5	2.5202×10^5	2.7315×10^5	2.9423×10^5	3.1528×10^5	3.3631×10^5	3.5734×10^5	3.7836×10^5
11419.4	11572.6	11719.2	11863.9	12007.7	12151.2	12294.5	12438.3
7.7518	7.8192	7.8768	7.9303	7.9804	8.0273	8.0716	8.1118
8.7721×10^4	9.6053×10^4	1.0419×10^5	1.1227×10^5	1.2032×10^5	1.2836×10^5	1.3639×10^5	1.4442×10^5
11392.8	11562.1	11714.4	11861.4	12006.4	12150.4	12294.0	12438.0
7.4652	7.5397	7.5995	7.6540	7.7044	7.7515	7.7959	7.8361
3.6397×10^4	4.0051×10^4	4.3522×10^4	4.6832×10^4	5.0314×10^4	5.3685×10^4	5.7050×10^4	6.0411×10^4
11335.3	11538.7	11703.5	11855.9	12003.3	12148.6	12292.9	12437.2
7.1921	7.2815	7.3464	7.4028	7.4541	7.5016	7.5463	7.5866
1.6257×10^4	1.8056×10^4	1.9691×10^4	2.1266×10^4	2.2816×10^4	2.4353×10^4	2.5884×10^4	2.7412×10^4
11224.9	11491.7	11681.2	11844.5	11996.9	12144.8	12290.5	12435.6
6.9207	7.0379	7.1127	7.1732	7.2262	7.2745	7.3196	7.3601
....	8.7048×10^3	9.5534×10^3	1.0347×10^4	1.1116×10^4	1.1873×10^4	1.2625×10^4	1.3373×10^4
....	11406.6	11639.6	11822.9	11984.9	12137.5	12285.9	12432.5
....	6.8008	6.8931	6.9611	7.0174	7.0673	7.1132	7.1541
....	4.4415×10^3	4.9236×10^3	5.3588×10^3	5.7705×10^3	6.1709×10^3	6.5659×10^3	6.9576×10^3
....	11268.3	11568.3	11785.0	11963.4	12124.6	12277.8	12427.0
....	6.5626	6.6820	6.7624	6.8244	6.8770	6.9244	6.9661
....	....	2.6631×10^3	2.9202×10^3	3.1565×10^3	3.3823×10^3	3.6028×10^3	3.8202×10^3
....	....	11456.2	11723.1	11927.7	12102.9	12264.0	12417.7
....	....	6.4727	6.5717	6.6428	6.6999	6.7497	6.7927
....	....	1.4999×10^3	1.6617×10^3	1.8063×10^3	1.9415×10^3	2.0718×10^3	2.1990×10^3
....	....	11295.1	11628.9	11871.8	12068.3	12241.8	12402.7
....	....	6.2608	6.3848	6.4691	6.5331	6.5866	6.6317
....	....	....	9.8441×10^2	1.0782×10^3	1.1641×10^3	1.2454×10^3	1.3239×10^3
....	....	....	11497.3	11790.2	12016.6	12208.3	12379.8
....	....	....	6.1995	6.3011	6.3749	6.4339	6.4820
....	....	....	6.0268×10^2	6.6631×10^2	7.2353×10^2	7.7692×10^2	8.2767×10^2
....	....	....	11326.8	11678.5	11943.5	12159.9	12346.4
....	....	....	6.0140	6.1359	6.2221	6.2886	6.3410
....	....	....	....	4.2099×10^2	4.6047×10^2	4.9683×10^2	5.3085×10^2
....	....	....	....	11532.3	11843.7	12092.1	12298.9
....	....	....	....	5.9691	6.0703	6.1464	6.2047
....	....	....	....	2.7294×10^2	3.0098×10^2	3.2664×10^2	3.5034×10^2
....	....	....	....	11355.8	11716.6	12002.7	12234.7
....	....	....	....	5.8031	5.9203	6.0078	6.0733
....	....	....	....	....	2.0183×10^2	2.2048×10^2	2.3755×10^2
....	....	....	....	....	11564.4	11891.0	12152.5
....	....	....	....	....	5.7726	5.8724	5.9464
....	....	....	....	....	1.3819×10^2	1.5199×10^2	1.6460×10^2
....	....	....	....	....	11391.0	11757.4	12050.8
....	....	....	....	....	5.6272	5.7390	5.8221
....	....	....	....	....	....	1.0674×10^2	1.1622×10^2
....	....	....	....	....	....	11604.6	11930.2
....	....	....	....	....	....	5.6077	5.7000
....	....	....	....	....	....	76.021	83.221
....	....	....	....	....	....	11435.9	11791.4
....	....	....	....	....	....	5.4783	5.5792
....	....	....	....	....	....	....	60.476
....	....	....	....	....	....	....	11638.5
....	....	....	....	....	....	....	5.4607
....	....	....	....	....	....	....	44.746
....	....	....	....	....	....	....	11478.4
....	....	....	....	....	....	....	5.3470

TABLE B-6

Temperature, °R (Sat. press., psia)		Sat. liquid	Sat. vapor	Temperature of				
				800	900	1000	1100	1200
700 (1.2525×10⁻⁶)	v	1.9490×10^{-2}	1.5338×10^8	1.7531×10^8	1.9772×10^8	2.1914×10^8	2.4105×10^8	2.6297×10^8
	h	134.7	1084.1	1096.8	1109.6	1122.3	1135.0	1147.6
	s	0.4831	1.8394	1.8565	1.8715	1.8847	1.8967	1.9078
800 (3.4995×10⁻⁵)	v	1.9890×10^{-2}	6.2721×10^6	....	7.0584×10^6	7.8430×10^6	8.6275×10^6	9.4119×10^6
	h	153.9	1096.6		1109.5	1122.3	1135.0	1147.6
	s	0.5087	1.6872		1.7023	1.7155	1.7275	1.7387
900 (4.5743×10⁻⁴)	v	2.0290×10^{-2}	5.3950×10^5			5.9983×10^5	6.5994×10^5	7.2000×10^5
	h	172.8	1109.0			1122.1	1134.9	1147.6
	s	0.5310	1.5712			1.5848	1.5969	1.6081
1000 (3.5203×10⁻³)	v	2.0690×10^{-2}	7.7766×10^4				8.5672×10^4	9.3519×10^4
	h	191.4	1120.8				1134.3	1147.4
	s	0.5507	1.4800				1.4928	1.5043
1100 (1.8552×10⁻²)	v	2.1090×10^{-2}	1.6182×10^4					1.7711×10^4
	h	209.9	1131.7					1146.3
	s	0.5683	1.4062					1.4190
1200 (7.3860×10⁻²)	v	2.1490×10^{-2}	4.4179×10^3					
	h	228.3	1142.3					
	s	0.5842	1.3459					
1300 (0.2364)	v	2.1890×10^{-2}	1488.7					
	h	246.5	1152.3					
	s	0.5988	1.2956					
1400 (0.6374)	v	2.2290×10^{-2}	591.02					
	h	264.7	1161.6					
	s	0.6123	1.2530					
1500 (1.5006)	v	2.2690×10^{-2}	267.15					
	h	283.0	1170.4					
	s	0.6249	1.2166					
1600 (3.1657)	v	2.3090×10^{-2}	134.03					
	h	301.2	1178.7					
	s	0.6367	1.1851					
1700 (6.1050)	v	2.3490×10^{-2}	73.252					
	h	319.6	1186.8					
	s	0.6478	1.1580					
1800 (10.918)	v	2.3890×10^{-2}	43.011					
	h	338.0	1194.8					
	s	0.6584	1.1344					
1900 (18.281)	v	2.4290×10^{-2}	26.880					
	h	356.7	1202.7					
	s	0.6685	1.1137					
2000 (29.024)	v	2.4690×10^{-2}	17.667					
	h	375.6	1210.5					
	s	0.6782	1.0956					
2100 (44.067)	v	2.5090×10^{-2}	12.113					
	h	394.8	1218.4					
	s	0.6875	1.0797					
2200 (64.389)	v	2.5490×10^{-2}	8.6138					
	h	414.3	1226.5					
	s ...	0.6966	1.0658					
2300 (90.965)	v	2.5890×10^{-2}	6.3248					
	h	434.2	1234.9					
	s	0.7054	1.0536					
2400 (124.29)	v	2.6290×10^{-2}	4.7934					
	h	454.4	1243.4					
	s	0.7141	1.0428					
2500 (164.49)	v	2.6690×10^{-2}	3.7349					
	h	475.2	1252.0					
	s	0.7225	1.0333					
2600 (213.60)	v	2.7090×10^{-2}	2.9786					
	h	496.4	1260.9					
	s	0.7308	1.0249					
2700 (271.59)	v	2.7490×10^{-2}	2.4180					
	h	518.2	1270.4					
	s	0.7391	1.0177					

* From Ref. 6

Thermodynamic Properties of Potassium*

Superheated Vapor, °R

1400	1600	1800	2000	2200	2400	2600	2700
3.0679×10^8 1173.0 1.9275	3.5062×10^8 1198.4 1.9444	3.9445×10^8 1223.8 1.9594	4.3828×10^8 1249.2 1.9726	4.8211×10^8 1274.6 1.9847	5.2593×10^8 1300.0 1.9959	5.6976×10^8 1325.4 2.0058	5.9168×10^8 1338.1 2.0107
1.0981×10^7 1173.0 1.7583	1.2549×10^7 1198.4 1.7753	1.4118×10^7 1223.8 1.7903	1.5686×10^7 1249.2 1.8034	1.7255×10^7 1274.6 1.8156	1.8824×10^7 1300.0 1.8268	2.0392×10^7 1325.4 1.8367	2.1177×10^7 1338.1 1.8415
8.4404×10^5 1173.0 1.6278	9.6005×10^5 1198.4 1.6447	1.0801×10^6 1223.8 1.6597	1.2001×10^6 1249.2 1.6729	1.3201×10^6 1274.6 1.6850	1.4401×10^6 1300.0 1.6962	1.5601×10^6 1325.4 1.7061	1.6201×10^6 1338.1 1.7110
1.0914×10^5 1173.0 1.5241	1.2474×10^5 1198.4 1.5411	1.4034×10^5 1223.8 1.5561	1.5593×10^5 1249.2 1.5693	1.7153×10^5 1274.6 1.5814	1.8712×10^5 1300.0 1.5926	2.0272×10^5 1325.4 1.6025	2.1051×10^5 1338.1 1.6073
2.0700×10^4 1172.7 1.4395	2.3666×10^4 1198.3 1.4566	2.6628×10^4 1223.8 1.4717	2.9588×10^4 1249.2 1.4848	3.2548×10^4 1274.6 1.4970	3.5507×10^4 1300.0 1.5081	3.8466×10^4 1325.4 1.5181	3.9946×10^4 1338.1 1.5229
5.1901×10^3 1171.6 1.3687	5.9406×10^3 1197.9 1.3862	6.6864×10^3 1223.6 1.4014	7.4307×10^3 1249.1 1.4146	8.1745×10^3 1274.5 1.4268	8.9180×10^3 1300.0 1.4380	9.6614×10^3 1325.4 1.4479	1.0033×10^4 1338.1 1.4527
1613.4 1168.6 1.3077	1852.8 1196.8 1.3266	2087.6 1223.1 1.3421	2320.9 1248.8 1.3554	2553.7 1274.4 1.3676	2786.2 1299.9 1.3789	3018.6 1325.3 1.3888	3134.8 1338.0 1.3936
....	683.89 1194.1 1.2747	772.51 1221.9 1.2911	859.69 1248.2 1.3048	946.35 1274.0 1.3171	1032.8 1299.6 1.3284	1119.1 1325.1 1.3383	1162.2 1337.9 1.3432
....	287.70 1188.6 1.2283	326.67 1219.3 1.2464	364.29 1246.8 1.2607	401.40 1273.1 1.2733	438.28 1299.1 1.2847	475.03 1324.8 1.2947	493.39 1337.6 1.2996
....	....	153.57 1214.5 1.2063	171.90 1244.1 1.2217	189.76 1271.5 1.2348	207.39 1298.0 1.2465	224.90 1324.0 1.2566	233.63 1336.9 1.2615
....	....	78.535 1206.5 1.1692	88.449 1239.6 1.1865	97.938 1268.8 1.2004	107.22 1296.2 1.2125	116.37 1322.8 1.2229	120.93 1335.9 1.2279
....	....	....	48.863 1232.7 1.1541	54.355 1264.4 1.1693	59.657 1293.4 1.1820	64.847 1320.7 1.1927	67.423 1334.2 1.1979
....	....	....	28.687 1223.0 1.1240	32.111 1258.2 1.1409	35.371 1289.1 1.1545	38.529 1317.7 1.1657	40.091 1331.6 1.1710
....	....	....	....	19.928 1249.7 1.1144	22.055 1283.3 1.1291	24.092 1313.4 1.1409	25.095 1327.9 1.1465
....	....	....	....	12.878 1239.0 1.0894	14.333 1275.6 1.1054	15.714 1307.8 1.1181	16.390 1323.0 1.1239
....	....	....	....	....	9.6477 1266.2 1.0832	10.621 1300.6 1.0967	11.097 1316.8 1.1029
....	....	....	....	....	6.6955 1255.3 1.0623	7.4050 1292.1 1.0767	7.7514 1309.3 1.0833
....	....	....	....	....	....	7.4050 1292.1 1.0767	7.7514 1309.3 1.0833
....	....	....	....	....	....	3.9373 1271.9 1.0410	4.1379 1291.1 1.0482
....	....	....	....	....	....	....	3.1363 1281.0 1.0325

TABLE B-7

Temperature, °R (Sat. press., psia)		Sat. liquid	Sat. vapor	Temperature of				
				800	900	1000	1100	1200
700 (1.2061×10⁻²)	v	0.0012	3.107×10³	3.551×10³	3.995×10³	4.439×10³	4.883×10³	5.327×10³
	h	29.26	156.94	159.21	161.49	163.77	166.06	168.36
	s	0.09998	0.2824	0.2854	0.2882	0.2906	0.2928	0.2948
800 (1.2541×10⁻¹)	v	0.00122	3.4150×10²	....	38.42	42.69	46.96	51.23
	h	32.51	159.75	....	162.03	164.31	166.60	168.89
	s	0.1043	0.2634	....	0.2661	0.2685	0.2707	0.2727
900 (0.76637)	v	0.00123	62.870			69.86	76.85	83.83
	h	35.75	162.54			164.82	167.11	169.41
	s	0.1081	0.2490			0.2514	0.2536	0.2557
1000 (3.2319)	v	0.00124	16.570				18.22	19.88
	h	38.98	165.32				167.61	169.91
	s	0.1115	0.2379				0.2041	0.2421
1100 (10.415)	v	0.00125	5.654					6.168
	h	42.21	168.10					170.40
	s	0.1146	0.2291					0.2311
1200 (27.450)	v	0.00127	2.340					
	h	45.45	170.88					
	s	0.1174	0.2220					
1300 (62.007)	v	0.00128	1.122					
	h	48.70	173.69					
	s	0.1200	0.2162					
1400 (1.2413×10²)	v	0.00129	0.6038					
	h	51.97	176.51					
	s	0.1225	0.2114					
1500 (2.2566×10²)	v	0.00131	0.3559					
	h	55.27	179.36					
	s	0.1247	0.2075					
1600 (3.7943×10²)	v	0.00132	0.2258					
	h	58.61	182.25					
	s	0.1269	0.2042					
1700 (5.9840×10²)	v	0.00133	0.1521					
	h	62.00	185.19					
	s	0.1289	0.2014					
1800 (8.9481×10²)	v	0.00135	0.1077					
	h	65.44	188.18					
	s	0.1309	0.1991					
1900 (1.2795×10³)	v	0.00136	0.0795					
	h	68.94	191.23					
	s	0.1328	0.1972					
2000 (1.7617×10³)	v	0.00137	0.06078					
	h	72.51	194.35					
	s	0.1346	0.1956					
2100 (2.3483×10³)	v	0.00139	0.04788					
	h	76.16	197.55					
	s	0.1364	0.1942					
2200 (3.0443×10³)	v	0.00140	0.03869					
	h	79.89	200.83					
	s	0.1382	0.1931					
2300 (3.8523×10³)	v	0.00141	0.03196					
	h	83.71	204.21					
	s	0.1399	0.1922					
2400 (4.7732×10³)	v	0.00143	0.02692					
	h	87.64	207.68					
	s	0.1415	0.1915					
2500 (5.8059×10³)	v	0.00144	0.02305					
	h	91.67	211.27					
	s	0.1432	0.1910					
2600 (6.9479×10³)	v	0.00145	0.02003					
	h	95.82	214.97					
	s	0.1448	0.1906					
2700 (8.1953×10³)	v	0.00147	0.01764					
	h	100.10	218.79					
	s	0.1464	0.1904					

* From Ref. 6

Thermodynamic Properties of Mercury*

Superheated Vapor, °R

1400	1600	1800	2000	2200	2400	2600	2700
6.214×10^3 172.97 0.2984	7.102×10^3 177.61 0.3015	7.990×10^3 182.27 0.3042	8.878×10^3 186.97 0.3067	9.765×10^3 191.69 0.3090	1.0650×10^4 196.44 0.3111	1.1540×10^4 201.21 0.3130	1.1980×10^4 203.61 0.3139
5.977×10^2 173.51 0.2763	6.831×10^2 178.15 0.2794	7.684×10^2 182.81 0.2822	8.538×10^2 187.51 0.2847	9.392×10^2 192.23 0.2869	1.025×10^3 196.97 0.2890	1.110×10^3 201.75 0.2909	1.153×10^3 204.14 0.2918
97.80 174.02 0.2592	1.118×10^2 178.66 0.2633	1.257×10^2 183.33 0.2651	1.397×10^2 188.02 0.2676	1.537×10^2 192.74 0.2698	1.677×10^2 197.49 0.2719	1.816×10^2 202.26 0.2738	1.886×10^2 204.66 0.2747
23.19 174.52 0.2457	26.51 179.16 0.2488	29.82 183.82 0.2515	33.13 188.52 0.2540	36.44 193.24 0.2563	39.76 197.98 0.2584	43.07 202.76 0.2603	44.73 205.14 0.2612
7.197 175.01 0.2347	8.225 179.65 0.2378	9.253 184.31 0.2405	10.28 189.01 0.2430	11.31 193.73 0.2453	12.34 198.47 0.2473	13.37 203.25 0.2493	13.88 205.64 0.2502
2.731 175.50 0.2255	3.121 180.14 0.2287	3.511 184.80 0.2314	3.901 189.50 0.2339	4.291 194.22 0.2362	4.681 198.96 0.2382	5.071 203.74 0.2402	5.266 206.13 0.2411
1.209 175.99 0.2179	1.381 180.63 0.2210	1.554 185.30 0.2238	1.727 189.99 0.2263	1.900 194.71 0.2285	2.072 199.46 0.2306	2.245 204.23 0.2325	2.331 206.63 0.2334
....	0.6901 181.15 0.2145	0.7764 185.81 0.2173	0.8626 190.51 0.2198	0.9489 195.23 0.2220	1.035 199.97 0.2241	1.121 204.75 0.2260	1.165 207.14 0.2269
....	0.3796 181.68 0.2090	0.4271 186.35 0.2117	0.4745 191.04 0.2142	0.5220 195.76 0.2165	0.5694 200.51 0.2185	0.6169 205.28 0.2205	0.6406 207.68 0.2214
		0.2540 186.92 0.2069	0.2822 191.61 0.2094	0.3104 196.33 0.2117	0.3386 201.08 0.2137	0.3669 205.85 0.2157	0.3810 208.25 0.2166
		0.1610 187.52 0.2028	0.1789 192.22 0.2052	0.1968 196.94 0.2075	0.2147 201.68 0.2096	0.2326 206.46 0.2115	0.2416 208.86 0.2124
			0.1197 192.87 0.2016	0.1316 197.59 0.2038	0.1436 202.34 0.2059	0.1556 207.11 0.2078	0.1615 209.51 0.2087
			0.08368 193.58 0.1984	0.09205 198.30 0.2006	0.1004 203.05 0.2027	0.1088 207.82 0.2046	0.1130 210.22 0.2055
				0.06686 199.07 0.1978	0.07294 203.82 0.1999	0.07901 208.59 0.2018	0.08205 210.99 0.2027
				0.05016 199.91 0.1953	0.05472 204.66 0.1974	0.05928 209.43 0.1993	0.06156 211.83 0.2002
					0.04221 205.58 0.1952	0.04572 210.35 0.1971	0.04748 212.75 0.1980
					0.03335 206.58 0.1933	0.03613 211.36 0.1952	0.03752 213.75 0.1961
						0.02804 210.07 0.1925	0.03028 214.85 0.1944
						0.02398 213.66 0.1919	0.02490 216.06 0.1929
							0.02081 217.37 0.1915

TABLE B-8

Temperature, °R (Sat. press., psia)		Sat. liquid	Sat. vapor	Temperature of				
				800	900	1000	1100	1200
700 (2.1536×10⁻⁵)	v	8.7648×10^{-3}	2.6223×10^{6}	2.9990×10^{6}	3.3742×10^{6}	3.7492×10^{6}	4.1242×10^{6}	4.4991×10^{6}
	h	41.5	283.8	287.8	291.4	295.2	298.9	302.6
	s	0.1796	0.5258	0.5310	0.5354	0.5394	0.5429	0.5461
800 (3.8740×10⁻⁴)	v	8.8412×10^{-3}	1.6634×10^{5}	....	1.8747×10^{5}	2.0839×10^{5}	2.2925×10^{5}	2.5010×10^{5}
	h	47.2	287.3	....	291.3	295.1	298.9	302.6
	s	0.1873	0.4874	....	0.4922	0.4962	0.4997	0.5029
900 (3.6022×10⁻³)	v	8.9176×10^{-3})	2.0065×10^{4}	....	....	2.2377×10^{4}	2.4641×10^{4}	2.6890×10^{4}
	h	52.9	290.6	....	....	294.9	298.8	302.6
	s	0.1940	0.4581	....	....	0.4626	0.4663	0.4696
1000 (2.1173×10⁻²)	v	8.9940×10^{-3}	3.7755×10^{3}	....	....	....	4.1781×10^{3}	4.5678×10^{3}
	h	58.6	293.7	....	....	....	298.3	302.3
	s	0.2000	0.4351	....	....	....	0.4395	0.4430
1100 (8.9240×10⁻²)	v	9.0704×10^{-3}	9.7916×10^{2}	....	....	....	....	1.0774×10^{3}
	h	64.3	296.5	....	....	....	....	301.5
	s	0.2055	0.4165	....	....	....	....	0.4208
1200 (2.9353×10⁻¹)	v	9.1468×10^{-3}	3.2217×10^{2}	....	....	....	....	....
	h	70.1	299.0	....	....	....	....	....
	s	0.2105	0.4013	....	....	....	....	....
1300 (8.0137×10⁻¹)	v	9.2232×10^{-3}	1.2669×10^{2}					
	h	75.8	301.5					
	s	0.2150	0.3886					
1400 (1.8925)	v	9.2996×10^{-3}	5.7220×10					
	h	81.5	303.8					
	s	0.2193	0.3781					
1500 (3.9520)	v	9.3760×10^{-3}	2.9060×10					
	h	87.2	306.1					
	s	0.2232	0.3691					
1600 (7.4718)	v	9.4524×10^{-3}	1.6222×10					
	h	92.9	308.3					
	s	0.2269	0.3615					
1700 (13.1253)	v	9.5288×10^{-3}	9.7128					
	h	98.7	310.6					
	s	0.2304	0.3550					
1800 (21.6481)	v	9.6052×10^{-3}	6.1762					
	h	104.4	313.0					
	s	0.2337	0.3495					
1900 (33.5849)	v	9.6816×10^{-3}	4.1618					
	h	110.1	315.3					
	s	0.2367	0.3448					
2000 (40.7219)	v	9.7580×10^{-3}	2.9322					
	h	115.8	317.8					
	s	0.2397	0.3407					
2100 (70.8207)	v	9.8344×10^{-3}	2.1437					
	h	121.5	320.4					
	s	0.2425	0.3371					
2200 (95.5743)	v	9.9108×10^{-3}	1.6179					
	h	127.3	323.1					
	s	0.2451	0.3341					
2300 (1.3059×10²)	v	9.9872×10^{-3}	1.2554					
	h	133.0	325.9					
	s	0.2477	0.3315					
2400 (1.6985×10²)	v	1.0064×10^{-2}	1.0005					
	h	138.7	328.7					
	s	0.2501	0.3293					
2500 (2.1545×10²)	v	1.0140×10^{-2}	8.1633×10^{-1}					
	h	144.4	331.6					
	s	0.2524	0.3273					
2600 (2.6768×10²)	v	1.0216×10^{-2}	6.7936×10^{-1}					
	h	150.1	334.5					
	s	0.2547	0.3256					
2700 (3.2739×10²)	v	1.0293×10^{-2}	5.7438×10^{-1}					
	h	155.9	337.7					
	s	0.2568	0.3242					

* From Ref. 6.

Thermodynamic Properties of Cesium*

Superheated Vapor, °R

1400	1600	1800	2000	2200	2400	2600	2700
5.2490×10⁶ 310.1 0.5520	5.9988×10⁶ 317.6 0.5569	6.7487×10⁶ 325.0 0.5614	7.4985×10⁶ 332.5 0.5652	8.2484×10⁶ 340.0 0.5688	8.9982×10⁶ 347.4 0.5721	9.7481×10⁶ 354.9 0.5750	1.0123×10⁷ 358.7 0.5764
-2.9180×10⁵ 310.1 0.5088	3.3348×10⁵ 317.6 0.5137	3.7517×10⁵ 325.0 0.5182	4.1685×10⁵ 332.5 0.5221	4.5854×10⁵ 340.0 0.5256	5.0022×10⁵ 347.4 0.5289	5.4191×10⁵ 354.9 0.5318	5.6275×10⁵ 358.7 0.5333
3.1379×10⁴ 310.1 0.4755	3.5864×10⁴ 317.6 0.4804	4.0348×10⁴ 325.0 0.4849	4.4831×10⁴ 332.5 0.4887	4.9315×10⁴ 340.0 0.4923	5.3798×10⁴ 347.4 0.4956	5.8281×10⁴ 354.9 0.4985	6.0523×10⁴ 358.7 0.4999
5.3361×10³ 310.0 0.4490	6.1004×10³ 317.5 0.4539	6.8637×10³ 325.0 0.4584	7.6267×10³ 332.5 0.4623	8.3895×10³ 340.0 0.4659	9.1523×10³ 347.4 0.4691	9.9151×10³ 354.9 0.4721	1.0296×10⁴ 358.7 0.4735
1.2639×10³ 309.8 0.4273	1.4464×10³ 317.4 0.4324	1.6280×10³ 325.0 0.4369	1.8092×10³ 332.5 0.4408	1.9903×10³ 339.9 0.4444	2.1713×10³ 347.4 0.4476	2.3524×10³ 354.9 0.4506	2.4429×10³ 358.6 0.4520
3.8231×10² 309.0 0.4091	4.3884×10² 317.1 0.4144	4.9445×10² 324.8 0.4190	5.4972×10² 332.4 0.4230	6.0488×10² 339.9 0.4265	6.5998×10² 347.4 0.4298	7.1504×10² 354.9 0.4328	7.4257×10² 358.6 0.4342
1.3836×10² 307.2 0.3930	1.5994×10² 316.4 0.3990	1.8067×10² 324.5 0.4039	2.0108×10² 332.2 0.4079	2.2137×10² 339.8 0.4115	2.4161×10² 347.3 0.4148	2.6180×10² 354.8 0.4177	2.7190×10² 358.6 0.4192
....	6.7030×10 314.8 0.3854	7.6115×10 323.7 0.3907	8.4899×10 331.7 0.3948	9.3569×10 339.5 0.3986	1.0218×10² 347.1 0.4019	1.1076×10² 354.7 0.4049	1.1505×10² 358.5 0.4063
....	3.1523×10 312.2 0.3730	3.6113×10 322.3 0.3790	4.0438×10 330.9 0.3835	4.4657×10 339.0 0.3874	4.8822×10 346.8 0.3908	5.2954×10 354.4 0.3938	5.5016×10 358.2 0.3952
....	....	1.8819×10 320.1 0.3686	2.1199×10 329.6 0.3735	2.3487×10 338.1 0.3776	2.5725×10 346.2 0.3811	2.7931×10 354.0 0.3841	2.9030×10 357.9 0.3856
....	....	1.0482×10 317.0 0.3587	1.1907×10 327.6 0.3643	1.3254×10 336.8 0.3687	1.4557×10 345.3 0.3724	1.5831×10 353.4 0.3755	1.6464×10 357.3 0.3771
....	....	....	7.0856 324.9 0.3558	7.9367 335.0 0.3606	8.7497 344.0 0.3645	9.5368 352.4 0.3678	9.9270 356.5 0.3694
....	....	....	4.4618 321.6 0.3479	5.0337 332.6 0.3532	5.5752 342.3 0.3574	6.0942 351.1 0.3609	6.3509 355.4 0.3625
....	....	....	....	3.3331 329.8 0.3464	3.7113 340.2 0.3509	4.0709 349.5 0.3546	4.2483 353.9 0.3562
....	....	....	....	2.2867 326.6 0.3400	2.5605 337.7 0.3449	2.8194 347.5 0.3487	2.9471 352.2 0.3505
....	....	....	....	....	1.8216 334.9 0.3393	2.0139 345.2 0.3433	2.1089 350.1 0.3452
....	....	....	....	....	1.3315 331.8 0.3341	1.4780 342.6 0.3383	1.5505 347.8 0.3403
....	....	....	....	....	....	1.1147 340.0 0.3337	1.1715 345.3 0.3357
....	....	....	....	....	....	8.6137×10⁻¹ 337.2 0.3295	9.0671×10⁻¹ 342.8 0.3316
....	....	....	....	....	....	....	7.1615×10⁻¹ 340.3 0.3278

TABLE B-9
Thermodynamic Properties of Helium *

Pressure, psia		Temperature, °F					
		100	200	300	400	500	600
14.696	v	102.23	120.487	138.743	157.00	175.258	193.515
	ρ	0.0097820	0.0082997	0.0072076	0.006394	0.0057059	0.0051676
	h	707.73	827.56	952.38	1077.20	1202.02	1326.83
	s	6.8421	7.0472	7.2233	7.3776	7.5149	7.6386
50	v	30.085	35.451	40.817	46.183	51.549	56.915
	ρ	0.033239	0.028208	0.024500	0.021653	0.019399	0.017570
	h	703.08	827.90	952.72	1077.54	1202.36	1327.28
	s	6.2342	6.4393	6.6153	6.7697	6.9070	7.0307
150	v	10.063	11.8522	13.6407	15.4293	17.2183	19008
	ρ	0.099372	0.084372	0.073310	0.064812	0.058078	0.052610
	h	704.08	828.91	953.73	1078.55	1203.37	1328.19
	s	5.6886	5.8937	6.0698	6.2241	6.3614	6.4852
400	v	3.8062	4.4775	5.1487	5.8197	6.4905	7.1616
	ρ	0.26273	0.22334	0.194225	0.171831	0.154072	0.139633
	h	706.58	831.42	956.24	1081.06	1205.88	1330.70
	s	5.2013	5.4065	5.5827	5.7371	5.8744	5.9981
600	v	2.5546	3.0023	3.44995	3.8973	4.3449	4.7923
	ρ	0.39146	0.33308	0.28986	0.25658	0.23016	0.20867
	h	708.49	833.33	958.15	1082.97	1207.79	1332.61
	s	4.9998	5.2050	5.3813	5.5357	5.6730	5.7968
900	v	1.7200	2.0187	2.3173	2.6157	2.91399	3.2124
	ρ	0.58139	0.49537	0.43154	0.38230	0.34317	0.33129
	h	710.29	835.38	960.40	1085.42	1210.42	1335.36
	s	4.7981	5.0035	5.1797	5.3342	5.4715	5.5953
1,500	v	1.05192	1.2314	1.4108	1.58994	1.7690	1.9483
	ρ	0.95064	0.81207	0.70880	0.62897	0.56528	0.51328
	h	715.54	840.77	965.88	1090.92	1215.93	1340.97
	s	4.5437	4.7475	4.9257	5.0801	5.2176	5.3414
2,500	v	0.65044	0.75847	0.86635	0.97410	1.08176	1.18947
	ρ	1.53741	1.31845	1.15427	1.02659	0.92442	0.84071
	h	724.37	849.73	974.95	1100.10	1225.22	1350.29
	s	4.2887	4.4928	4.6712	4.8258	4.9634	5.0873
4,000	v	0.42377	0.49161	0.55932	0.62694	0.69444	0.76191
	ρ	2.3598	2.0341	1.78789	1.59503	1.44000	1.31248
	h	736.48	862.24	987.70	1113.12	1238.46	1363.73
	s	4.0531	4.2576	4.4363	4.5912	4.7287	4.8530

* From Ref. 103.

TABLE B-10
Thermodynamic Properties of Carbon Dioxide*

Pressure, psia		Temperature, °F												
		50	100	150	200	300	400	600	800	1000	1200	1400	1600	1800
10.0	v	12.38	13.61	14.84	16.06	18.51	20.96	25.85	30.73	35.61	40.49	45.36	50.24	55.11
	h	307.3	317.7	328.2	339.0	361.3	384.6	434.4	487.1	542.4	599.6	658.6	718.8	780.0
	s	1.4277	1.4467	1.4645	1.4813	1.5126	1.5412	1.5930	1.6384	1.6790	1.7158	1.7494	1.7799	1.8084
20.0	v	6.119	6.778	7.407	8.016	9.247	10.47	12.92	15.36	17.80	20.24	22.68	25.11	27.55
	h	306.8	317.3	327.9	338.8	361.1	384.5	434.3	487.1	542.4	599.6	658.6	718.8	780.0
	s	1.3964	1.4154	1.4332	1.4500	1.4813	1.5099	1.5617	1.6071	1.6477	1.6845	1.7181	1.7486	1.7771
40.0	v	3.053	3.363	3.688	3.993	4.615	5.230	6.458	7.688	8.901	10.12	11.37	12.56	13.78
	h	305.9	316.5	327.4	338.4	360.9	384.3	434.2	487.0	542.4	599.6	658.6	718.8	780.0
	s	1.3642	1.3834	1.4014	1.4184	1.4499	1.4787	1.5305	1.5759	1.6165	1.6533	1.6869	1.7174	1.7459
80.0	v	1.498	1.657	1.828	1.982	2.298	2.608	3.226	3.839	4.448	5.060	5.670	6.281	6.887
	h	304.1	315.1	326.4	337.7	360.2	383.9	434.0	486.9	542.3	599.5	658.6	718.8	780.0
	s	1.3284	1.3490	1.3679	1.3855	1.4177	1.4468	1.4991	1.5446	1.5852	1.6220	1.6556	1.6861	1.7146
120	v	0.9799	1.088	1.208	1.311	1.525	1.734	2.148	1.559	2.966	3.373	3.781	4.188	4.592
	h	302.2	313.6	325.4	337.0	359.7	383.5	433'8	486.8	542.3	599.5	658.6	718.8	780.0
	s	1.3086	1.3297	1.3488	1.3666	1.3993	1.4285	1.4808	1.5263	1.5669	1.6037	1.6373	1.6678	1.6963
160.	v	0.7207	0.8033	0.8986	0.9760	1.139	1.297	1.610	1.918	2.224	2.530	2.836	3.141	3.445
	h	300.4	312.1	324.4	336.3	359.1	383.1	433.6	486.6	542.2	599.5	658.6	718.8	780.0
	s	1.2928	1.3154	1.3350	1.3529	1.3857	1.4151	1.4675	1.5133	1.5539	1.5907	1.6243	1.6548	1.6833
200	v	0.5652	0.6376	0.7125	0.7748	0.9075	1.035	1.287	1.534	1.779	2.024	2.269	2.513	2.757
	h	298.6	310.6	323.4	335.6	358.5	382.7	433.4	486.5	542.2	599.5	658.5	718.8	780.0
	s	1.2805	1.3038	1.3239	1.3421	1.3753	1.4049	1.4574	1.5033	1.5439	1.5807	1.6143	1.6448	1.6733
240	v	0.4614	0.5237	0.5886	0.6407	0.7532	0.8604	1.071	1.273	1.482	1.687	1.891	2.095	2.297
	h	296.7	309.1	322.4	334.9	358.0	382.3	433.1	486.4	542.1	599.5	658.5	718.8	780.0
	s	1.2694	1.2940	1.3145	1.3330	1.3671	1.3963	1.4490	1.4948	1.5356	1.5724	1.6060	1.6365	1.6650
300	v	0.3563	0.4100	0.4636	0.5065	0.5985	0.6868	0.8556	1.021	1.186	1.349	1.513	1.676	1.838
	h	294.0	306.9	320.9	333.9	357.1	381.6	432.8	486.2	542.0	599.4	658.5	718.7	780.0
	s	1.2562	1.2813	1.3029	1.3219	1.3560	1.3862	1.4389	1.4848	1.5256	1.5624	1.5960	1.6265	1.6550
360	v	0.2858	0.3341	0.3780	0.4171	0.4958	0.5693	0.7212	0.8502	0.9874	1.125	1.261	1.397	1.533
	h	291.2	304.6	319.4	332.8	356.3	381.0	432.5	486.0	541.9	599.4	658.5	718.7	779.9
	s	1.2436	1.2699	1.2925	1.3124	1.3475	1.3779	1.4307	1.4766	1.5174	1.5542	1.5878	1.6183	1.6468
440	v	0.2216	0.2652	0.3040	0.3358	0.4022	0.4633	0.5817	0.6950	0.8079	0.9201	1.032	1.142	1.255
	h	287.6	301.6	317.4	331.4	355.1	380.2	432.1	485.8	541.7	599.3	658.4	718.6	779.9
	s	1.2282	1.2559	1.2797	1.3006	1.3370	1.3681	1.4215	1.4675	1.5083	1.5451	1.5787	1.6092	1.6377
520	v	0.1772	0.2174	0.2513	0.2795	0.3374	0.3901	0.4912	0.5881	0.6832	0.7785	0.8733	0.9672	1.062
	h	283.9	298.7	315.4	330.0	354.0	379.4	431.7	485.5	541.5	599.2	658.3	718.6	779.9
	s	1.2148	1.2438	1.2687	1.2905	1.3281	1.3599	1.4138	1.4599	1.5007	1.5375	1.5711	1.6010	1.6301
600	v	0.1452	0.1823	0.2123	0.2383	0.2898	0.3363	0.4250	0.5093	0.5921	0.6747	0.7571	0.8385	0.9202
	h	280.3	295.7	313.4	328.6	352.8	378.6	431.1	485.3	541.4	599.0	658.2	718.6	779.8
	s	1.2020	1.2323	1.2583	1.2809	1.3198	1.3525	1.4071	1.4534	1.4942	1.5310	1.5646	1.5951	1.6236
800	v		0.1196	0.1483	0.1712	0.2126	0.2489	0.3173	0.3812	0.4436	0.5060	0.5680	0.6292	0.6906
	h		288.2	308.4	325.1	350.0	376.5	430.1	484.7	541.0	598.8	658.0	718.4	779.7
	s		1.2111	1.2391	1.2631	1.3041	1.3380	1.3935	1.4404	1.4812	1.5180	1.5516	1.5821	1.6106
1000	v		0.1101	0.1310	0.1663	0.1966	0.2526	0.3048	0.3547	0.4049	0.4545	0.5037	0.5526	
	h		303.4	321.6	347.1	374.5	429.1	484.0	540.6	598.5	657.8	718.3	779.6	
	s		1.2218	1.2472	1.2903	1.3258	1.3828	1.4302	1.4712	1.5080	1.5416	1.5721	1.6006	

* From Ref. 103

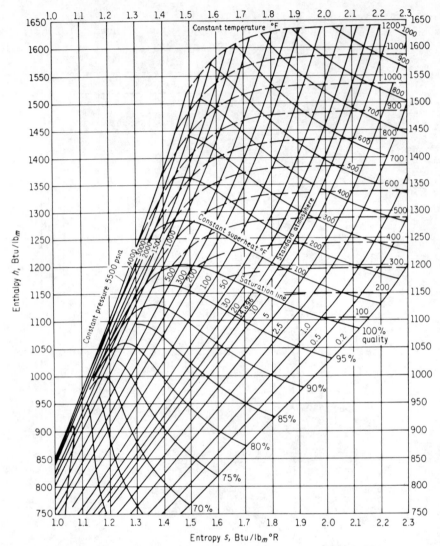

FIG. B-1. Mollier diagram for steam. (Based on data in Ref. 170.)

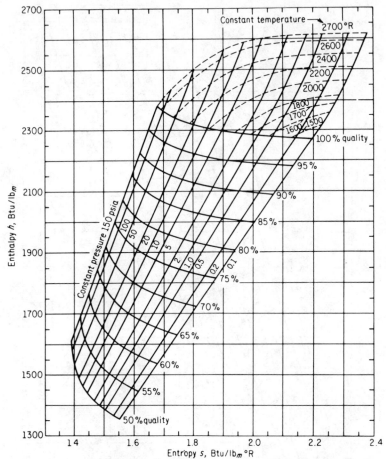

FIG. B-2. Mollier diagram for sodium. (Based on charts, courtesy Flight Propulsion Laboratory, General Electric Company.)

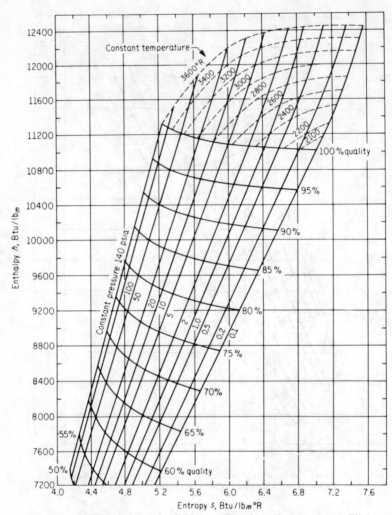

FIG. B-3. Mollier diagram for lithium. (Based on charts, courtesy Flight
Propulsion Laboratory, General Electric Company.)

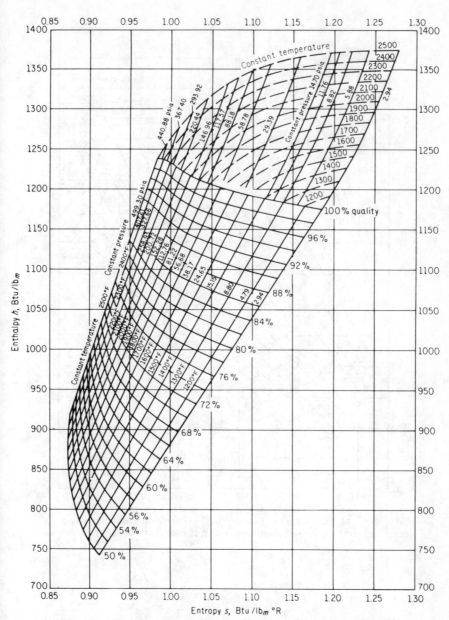

FIG. B-4. Mollier diagram for potassium. (Based on chart, courtesy U.S. Naval Research Laboratory, Ref. 171.)

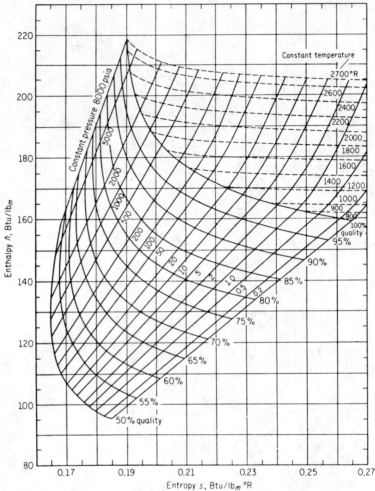

FIG. B-5. Mollier diagram for mercury. (Based on charts, courtesy
Flight Propulsion Laboratory, General Electric Company.)

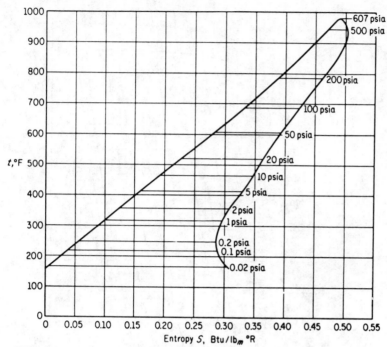

FIG. B-6. Temperature-entropy diagram for diphenyl. (Based on data from, Ref. 117.)

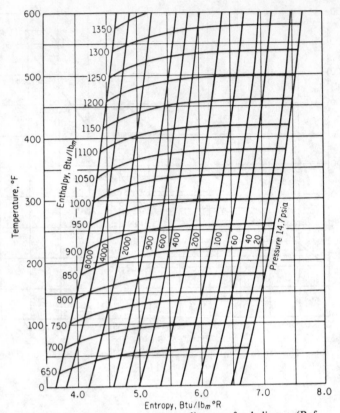

FIG. B-7. Temperature-entropy diagram for helium. (Ref. 103.)

appendix **C**

Some Physical Properties

TABLE C-1
Physical Properties of Ordinary Liquid Water*

Temp., °F	Specific heat c_p, Btu/lb$_m$ °R			Thermal conductivity k, Btu/hr ft °F			Viscosity μ, lb$_m$/hr ft			Density ρ, lb$_m$/ft³		
	Sat. liquid	1,000 psia	2,000 psia	Sat. liquid	1,000 psia	2,000 psia	Sat. liquid	1,000 psia	2,000 psia	Sat. liquid	1,000 psia	2,000 psia
32	1.0083	1.0032	1.0004	0.3185	0.3198	0.3211	4.340	4.309	4.279	62.422	62.637	62.846
40	1.0048	1.0014	0.9986	0.3245	0.3260	0.3275	3.742	3.721	3.699	62.422	62.657	62.854
60	0.9990	0.9968	0.9939	0.3397	0.3414	0.3433	2.731	2.722	2.714	62.344	62.539	62.755
80	0.9975	0.9943	0.9912	0.3532	0.3537	0.3570	2.084	2.084	2.083	62.189	62.383	62.586
100	0.9976	0.9932	0.9897	0.3641	0.3659	0.3680	1.650	1.654	1.658	61.996	62.185	62.371
120	0.9977	0.9934	0.9895	0.3733	0.3751	0.3771	1.353	1.360	1.366	61.728	61.920	62.104
140	0.9988	0.9940	0.9897	0.3810	0.3828	0.3847	1.137	1.145	1.154	61.387	61.576	61.767
160	1.0004	0.9959	0.9913	0.3861	0.3880	0.3902	0.970	0.979	0.988	61.013	61.200	61.395
180	1.0022	0.9980	0.9931	0.3905	0.3924	0.3945	0.839	0.849	0.858	60.569	60.753	60.953
200	1.0047	1.0008	0.9958	0.3935	0.3957	0.3980	0.738	0.748	0.757	60.132	60.314	60.511
210	1.0064	1.0024	0.9974	0.3944	0.3972	0.3998	0.687	0.697	0.706	59.809	60.006	60.205
220	1.0079	1.0039	0.9988	0.3950	0.3977	0.4003	0.660	0.670	0.680	59.630	59.830	60.031
240	1.0119	1.0075	1.0023	0.3961	0.3988	0.4016	0.595	0.604	0.614	59.102	59.305	59.506
260	1.0165	1.0117	1.0061	0.3964	0.3992	0.4021	0.542	0.551	0.560	58.514	58.727	58.938
280	1.0222	1.0163	1.0102	0.3959	0.3987	0.4018	0.494	0.502	0.511	57.937	58.156	58.377
300	1.0289	1.0232	0.0166	0.3952	0.3981	0.4013	0.452	0.460	0.468	57.307	57.537	57.767
320	1.0354	1.0307	1.0235	0.3944	0.3969	0.3998	0.420	0.426	0.433	56.657	56.883	57.136
340	1.0455	1.0999	1.0322	0.3921	0.3947	0.3977	0.391	0.396	0.404	55.960	56.211	56.465
360	1.0564	1.0496	1.0411	0.3891	0.3919	0.3951	0.366	0.372	0.378	55.218	55.463	55.710
380	1.0669	1.0611	1.0510	0.3857	0.3885	0.3919	0.346	0.351	0.356	54.466	54.720	55.012
400	1.0794	1.074	1.062	0.3809	0.3840	0.3880	0.327	0.330	0.335	53.648	53.903	54.218
420	1.0941	1.087	1.075	0.3753	0.3787	0.3833	0.310	0.312	0.317	52.798	53.042	53.396
440	1.1114	1.105	1.091	0.3693	0.3728	0.3776	0.294	0.296	0.301	51.921	52.154	52.546
460	1.1319	1.124	1.109	0.3640	0.3664	0.3713	0.280	0.282	0.286	51.020	51.230	51.661
480	1.1345	1.149	1.131	0.3575	0.3595	0.3642	0.267	0.270	0.273	50.000	50.191	50.659
500	1.1861	1.176	1.154	0.3494	0.3510	0.3562	0.256	0.257	0.260	49.020	49.097	49.618
520	1.23	1.21	1.188	0.3397	0.3410	0.3475	0.246	0.246	0.249	47.847	......	48.527
540	1.28	......	1.225	0.3298	......	0.3371	0.235	.....	0.239	46.512	......	47.181
560	1.34	......	1.278	0.3189	......	0.3256	0.225	.....	0.231	45.249	......	45.905
580	1.41	......	1.341	0.3064	......	0.3118	0.217	.....	0.222	43.860	......	44.492
600	1.51	......	1.448	0.2919	......	0.2962	0.210	.....	0.212	42.373	......	42.913
620	1.65	......	1.62	0.2753	......	0.2778	0.200	.....	0.202	40.486	......	40.950
640	1.88	......	......	0.2565	......	......	0.190	.....	.....	38.462		
660	2.34	......	......	0.2335	......	......	0.177	.....	.....	35.971		
680	3.5	......	......	0.2056	......	......	0.161	.....	.....	32.787		
690	5.5	......	......	0.1854	......	......	0.48	.....	.....	30.488		

* c_p, μ, and k data from Ref. 103. ρ data computed from Keenan and Keyes [4].

TABLE C-2
Physical Properties of Helium*
(at 10 atm pressure)

T, °F	Density ρ, lb_m/ft^3	Viscosity μ, $lb_m/ft\ hr$	Specific heat [†] c_p, $Btu/lb_m\ °F$	Thermal conductivity [‡] k, $Btu/ft\ hr\ °F$	Prandtl no. Pr, $c_p\mu/k$
32	0.1117	0.0457	1.248	0.083	0.687
100	0.0974	0.0495	1.248	0.090	0.687
200	0.0827	0.0555	1.248	0.100	0.687
300	0.0718	0.0605	1.248	0.110	0.686
400	0.0635	0.0653	1.248	0.119	0.684
500	0.0569	0.0700	1.248	0.128	0.682
600	0.0516	0.0743	1.248	0.136	0.679
700	0.0475	0.0780	1.248	0.145	0.675
800	0.0430	0.0821	1.248	0.153	0.671
900	0.0399	0.0859	1.248	0.160	0.667
1000	0.0373	0.0889	1.248	0.167	0.662
1100	0.0351	0.0918	1.248	0.175	0.656

* From Ref. 169.
[†] Extrapolated.
[‡] Atmospheric pressure.

TABLE C-3
Physical Properties of Carbon Dioxide*

T, °F	Density [†] ρ, lb_m/ft^3	Viscosity [‡] μ, $lb_m/ft\ hr$	Specific heat [‡] c_p, $Btu/lb_m\ °F$	Thermal conductivity [‡] k, $Btu/ft\ hr\ °F$	Prandtl no. [‡] Pr, $c_p\mu/k$
32	1.3190	0.03318	0.2187	0.008415	0.782
100	1.1277	0.03739	0.2202	0.009962	0.768
200	0.9373	0.04332	0.2262	0.01261	0.749
300	0.8051	0.04892	0.2342	0.01533	0.729
400	0.7071	0.05419	0.2423	0.01818	0.710
500	0.6310	0.05920	0.2503	0.02117	0.691
600	0.5701	0.06397	0.2476	0.02425	0.672
700	0.5202	0.06851	0.2643		
800	0.4784	0.07288	0.2704		
900	0.4428	0.07709	0.2760		
1000	0.4125	0.0811	0.2812		
1100	0.3860	0.08511	0.2858		
1200	0.3626	0.8891	0.2901		

* From Ref. 169
[†] At 10 atm pressure.
[‡] At atmospheric pressure.

appendix **D**

Moody Friction Factor Chart

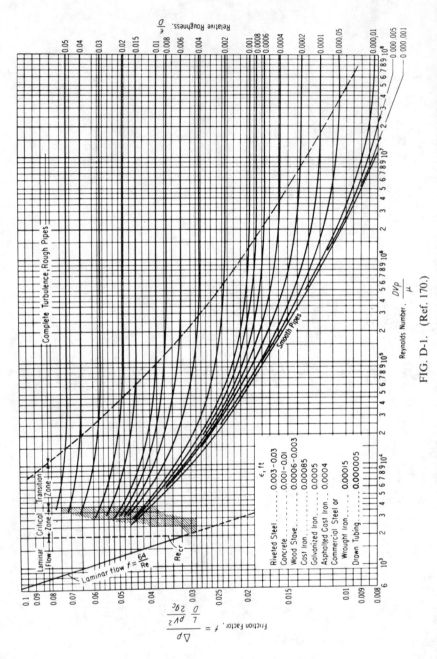

FIG. D-1. (Ref. 170.)

appendix **E**

Some Useful Constants

Avogadro's number, Av . . . . . 0.602252 × 10^{24} molecules/g_m mole
                                                2.731769 × 10^{26} molecules/lb_m mole
Barn . . . . 10^{-24} cm², 1,0765 × 10^{-27} ft²
Boltzmann's constant, $k = \bar{R}/Av$. 1.38054 × 10^{-16} erg/°K,
                                                8,61747 × 10^{-5} ev/°K
Curie . . . . 3.70 × 10^{10} dis/sec
Electron charge . . . . 4.80298 × 10^{-10} esu, 1.60210 × 10^{-20} emu
g_c conversion factor . . . . 1.0 g_m cm²/erg sec², 32.2 lb. ft/lb_fsec² 4.17 ×
                                                10^8 lb_m ft/lb_f hr², 0.9648 × 10^{18} amu cm²/
                                                Mev sec².
Gravitational acceleration . . . . 32.1739 ft/sec², 980.665 cm/sec²
    (standard)
Joule's equivalent . . . . 778.16 ft-lb_f/Btu
Mass-energy conversion . . . . . 1 amu = 931.478 Mev = 1.41492 × 10^{-13} Btu
                                                    = 4.1471 × 10^{-17} kwhr
                                                lb_m = 5.60984 × 10^{26} Mev
                                                    = 2.49760 × 10^7 kwhr = 1.04067 Mwd
                                                1 lb_m = 2.54458 × 10^{32} Mev
                                                    = 3.86524 × 10^{16} Btu
Molecular volume . . . . 2,2413.6 cm³/g_m mole, 359.0371 ft³/lb_m mole,
                                                at 1 atm and 0°C
Neutron energy . . . . 0.0252977 ev at 2,200 m/sec
                                                $\frac{1}{40}$ ev at 2,187.017 m/sec
Planck's constant, h . . . . 6.6256 × 10^{-27} erg sec
                                                4.13576 × 10^{-15} ev sec
Rest masses:
    Electron, m_e . . . . 5.48597 × 10^{-4} amu, 9.10909 × 10^{-28} g_m
                                                2.00819 × 10^{-30} lb_m
    Neutron, m_n . . . . 1.0086654 amu, 1.6748228 × 10^{-24} g_m
                                                3.692314 × 10^{-27} lb_m
    Proton, m_p . . . . 1.0072766 amu, 1.672499 × 10^{-24} g_m
                                                3.687192 × 10^{-27} lb_m
Universal gas constant, $\bar{R}$ . . . . . 1545.08 ft-lb_f/lb_m mole °R
                                                1.98545 cal/g_m mole °K
                                                1.98545 Btu/lb_m mole °R
                                                8.31434 × 10^7 erg/g_m mole °K
Velocity of light, c . . . . 2.997925 × 10^{10} cm/sec
                                                9.83619 × 10^8 ft/sec

appendix **F**

Some Conversion Factors

TABLE F-1
Length

Centimeters, cm	Meters, m	Inches in.	Feet, ft	Miles	Microns, μ	Angstroms, Å	Light-year
1	0.01	0.3937	0.03281	6.214×10^{-6}	10^4	10^8	1.0564×10^{-18}
100	1	39.37'	3.281	6.214×10^{-4}	10^6	10^{10}	1.0564×10^{-16}
2.540	0.0254	1	0.08333	1.578×10^{-5}	2.54×10^4	2.54×10^8	2.683×10^{-18}
30.48	0.3048	12	1	1.894×10^{-4}	0.3048×10^6	0.3048×10^{10}	3.2199×10^{-17}
1.6003×10^5	1609.3	6.336×10^4	5280	1	1.6093×10^9	1.6093×10^{13}	1.7001×10^{-13}
10^{-4}	10^{-6}	3.937×10^{-5}	3.281×10^{-6}	6.2139×10^{-10}	1	10^4	1.0564×10^{-22}
10^{-8}	10^{-10}	3.937×10^{-9}	3.281×10^{-10}	6.2139×10^{-14}	10^{-4}	1	1.0564×10^{-26}
9.4663×10^{17}	9.4663×10^{15}	3.727×10^{17}	3.1068×10^{16}	5.8822×10^{12}	9.4663×10^{21}	9.4663×10^{25}	1

TABLE F-2
Area

cm²	m²	in²	ft²	mile²	acre	barn	millibarn
1	10^{-4}	0.155	1.0764×10^{-3}	3.861×10^{-11}	2.4711×10^{-8}	10^{24}	10^{27}
10^4	1	1550	10.764	3.861×10^{-7}	2.4711×10^{-4}	10^{28}	10^{31}
6.4516	6.4516×10^{-4}	1	6.944×10^{-3}	2.491×10^{-10}	1.5944×10^{-7}	64517×10^{24}	6.4517×10^{27}
929	0.0929	144	1	3.587×10^{-8}	2.2957×10^{-5}	9.29×10^{26}	9.2903×10^{29}
2.59×10^{10}	2.59×10^6	4.0144×10^{11}	2.7878×10^7	1	640	2.59×10^{34}	2.59×10^{37}
4.0469×10^7	4.0469×10^3	6.2726×10^6	4.356×10^4	1.5625×10^{-3}	1	4.0469×10^{31}	4.0469×10^{34}
10^{-24}	10^{-28}	1.55×10^{-25}	1.0764×10^{-27}	3.861×10^{-35}	2.4711×10^{-32}	1	10^3
10^{-27}	10^{-31}	1.55×10^{-28}	1.0764×10^{-30}	3.861×10^{-38}	2.4711×10^{-35}	10^{-3}	1

TABLE F-3
Volume

cm³	liters	m³	in³	ft³	cubic yards	U.S. (liq.) gallons	Imperial gallons	acre-feet
1	10^{-3}	10^{-6}	0.06102	3.532×10^{-5}	1.308×10^{-6}	2.642×10^{-4}	2.20×10^{-4}	8.105×10^{-10}
10^3	1	10^{-3}	61.02	0.03532	1.308×10^{-3}	0.2642	0.220	8.105×10^{-7}
10^6	10^3	1	6.102×10^4	35.31	1.308	264.2	220.0	8.105×10^{-4}
16.39	0.01639	1.639×10^{-5}	1	5.787×10^{-4}	2.143×10^{-5}	4.329×10^{-3}	3.605×10^{-3}	1.328×10^{-8}
2.832×10^4	28.32	0.02832	1728	1	0.03704	7.481	6.229	2.295×10^{-5}
7.646×10^5	764.6	0.7646	4.666×10^4	27.0	1	202.0	168.2	6.196×10^{-4}
3785	3.785	3.785×10^{-3}	231.0	0.1337	4.951×10^{-3}	1	0.8327	3.068×10^{-6}
4546	4.546	4.546×10^{-3}	277.4	0.1605	5.946×10^{-3}	1.201	1	3.684×10^{-6}
1.234×10^9	1.234×10^6	1.234×10^3	7.529×10^7	4.357×10^4	1.614×10^3	3.259×10^5	2.714×10^5	1

TABLE F-4
Mass

grams, g_m	kilograms, kg_m	pounds, lb_m	tons (short)	tons (long)	tons (metric)	atomic mass units, amu
1	0.001	2.2046×10^{-3}	11.102×10^{-6}	9.842×10^{-7}	10^{-6}	0.60225×10^{24}
1,000	1	2.2046	0.001102	9.842×10^{-4}	10^{-3}	6.0225×10^{26}
453.6	0.4536	1	5.0×10^{-4}	4.464×10^{-4}	4.536×10^{-4}	2.7318×10^{26}
9.072×10^5	907.2	2,000	1	0.8929	0.9072	5.4636×10^{29}
1.016×10^6	1,016	2,240	1.12	1	1.016	6.1192×10^{29}
10^6	1,000	2,204.7	1.1023	0.9843	1	6.0225×10^{29}
1.6604×10^{-24}	1.6604×10^{-27}	3.6606×10^{-27}	1.8303×10^{-30}	1.6342×10^{-30}	1.6604×10^{-30}	1

TABLE F-5
Density

g_m/cm^3	kg_m/m^3	lb_m/in^3	lb_m/ft^3	$lb_m/U.S.$ gal.	$lb_m/Imp.$ gal.
1	10^3	0.03613	62.43	8.345	10.02
10^{-3}	1	3.613×10^{-5}	0.06243	8.345×10^{-3}	0.01002
27.68	2.768×10^4	1	1728	231	277.4
0.01602	16.02	5.787×10^{-4}	1	0.1337	0.1605
0.1198	119.8	4.329×10^{-3}	7.481	1	1.201
0.09978	99.78	3.605×10^{-3}	6.229	0.8327	1

TABLE F-6
Time

Microseconds, μsec	Seconds, sec	Minutes, min	Hours, hr	Days	Years, yr
1	10^{-6}	1.667×10^{-8}	2.778×10^{-10}	1.157×10^{-11}	3.169×10^{-14}
10^6	1	1.667×10^{-2}	2.778×10^{-4}	1.157×10^{-5}	3.169×10^{-8}
6×10^7	60	1	1.667×10^{-2}	6.944×10^{-4}	1.901×10^{-6}
3.6×10^9	3,600	60	1	0.04167	1.141×10^{-24}
8.64×10^{10}	8.64×10^4	1,440	24	1	2.737×10^{-3}
3.1557×10^{13}	3.1557×10^7	5.259×10^5	8,766	365.24	1

TABLE F-7
Flow

cm^3/sec	ft^3/min	U.S. gal/min	Imperial gal/min
1	0.002119	0.01585	0.01320
472.0	1	7.481	6.229
63.09	0.1337	1	0.8327
75.77	0.1605	1.201	1

TABLE F-8
Force

Dyne $(g_m cm/sec^2)$	Newton $(kg_m m/sec^2)$	Kilogram-force or kilopond (kg_f)	Pound-force (lb_f)	Poundal $(lb_m ft/sec^2)$
1	10^{-5}	1.01972×10^{-6}	2.248089×10^{-6}	7.2330137×10^{-5}
10^5	0.1019716	0.2248089	7.2330137
9.80665×10^5	9.80665	1	2.2046226	70.931647
4.44822×10^5	4.44822	0.453592	1	32.174054
1.38255×10^5	0.138255	0.014098	0.031081	1

TABLE F-9
Pressure

kg_f/cm^2	$lb_f/in.^2$	lb_f/ft^2	cm Hg (0°C)	in. Hg (32°F)	in. H$_2$O (60°F)	atm
1	14.22	2,048	73.56	28.96	394.1	0.9678
0.07031	1	144	5.171	2.036	27.71	0.06805
4.882×10^{-4}	0.006944	1	0.03591	0.01414	0.1924	4.725×10^{-4}
0.01360	0.1934	27.85	1	0.3937	5.358	0.01316
0.03453	0.4912	70.73	2.540	1	13.61	0.03342
0.002538	0.03609	5.197	0.1866	0.07348	1	0.002456
1.033	14.70	2,116	76.0	29.92	407.2	1

TABLE F-10
Energy

Ergs	Joules	kwhr	g_m-cal	ft-lb$_f$	hp-hr	Btu	ev	Mev
1	10^{-7}	2.778×10^{-14}	2.388×10^{-8}	7.376×10^{-8}	3.725×10^{-14}	9.478×10^{-11}	6.2421×10^{11}	6.242×10^{5}
10^{7}	1	2.778×10^{-7}	0.2388	0.7376	3.725×10^{-7}	9.478×10^{-4}	6.2421×10^{18}	6.242×10^{12}
3.6×10^{13}	3.6×10^{6}	1	8.598×10^{5}	2.655×10^{6}	1.341	3412	2.25×10^{25}	2.25×10^{19}
4.187×10^{7}	4.187	1.163×10^{-6}	1	3.088	1.56×10^{-6}	3.968×10^{-3}	2.616×10^{19}	2.613×10^{13}
1.356×10^{7}	1.356	3.766×10^{-7}	0.3238	1	5.051×10^{-7}	1.285×10^{-3}	8.462×10^{18}	8.462×10^{12}
2.685×10^{13}	2.685×10^{6}	0.7457	6.412×10^{5}	1.98×10^{6}	1	2545	1.677×10^{25}	1.677×10^{19}
1.055×10^{10}	1055	2.931×10^{-4}	252	778.2	3.93×10^{-4}	1	6.584×10^{21}	6.584×10^{15}
1.6021×10^{-12}	1.6021×10^{-19}	4.44×10^{-26}	3.826×10^{-20}	1.178×10^{-19}	5.95×10^{-26}	1.519×10^{-22}	1	10^{-6}
1.6021×10^{-6}	1.6021×10^{-13}	4.44×10^{-20}	3.826×10^{-14}	1.178×10^{-13}	5.95×10^{-20}	1.519×10^{-16}	10^{6}	1

TABLE F-11
Power

Ergs/sec	Joule/sec watt	kw	Btu/hr	hp	ev/sec
1	10^{-7}	10^{-10}	3.412×10^{-7}	1.341×10^{-10}	6.2421×10^{11}
10^7	1	10^{-3}	3.412	0.001341	6.2421×10^{18}
10^{10}	10^3	1	3412	1.341	6.2421×10^{21}
2.931×10^6	0.2931	2.931×10^{-4}	1	3.93×10^{-4}	1.8294×10^{18}
7.457×10^9	745.7	0.7457	2545	1	4.6548×10^{21}
1.6021×10^{-12}	1.6021×10^{-19}	1.6021×10^{-22}	5.4664×10^{-19}	2.1483×10^{-22}	1

TABLE F-12
Specific Power

w/g_m, kw/kg_m	$cal/sec\ g_m$	$Btu/hr\ lb_m$	$ft\ lb_f/hr\ lb_m$	Hp/lb_m
1	0,2390	1.5488×10^1	1.2044×10^6	6.0827×10^{-2}
4.184	1	6.48×10^3	5.0392×10^6	2.545
6.4568×10^{-4}	1.5432×10^{-4}	1	7.7765×10^2	3.9275×10^{-4}
8.303×10^{-7}	1.9845×10^{-7}	1.2859×10^{-3}	1	5.0505×10^{-7}
1.64399	0.39292	2.5461×10^2	1.98×10^6	1

TABLE F-13
Power Density, Volumetric Thermal Source Strength

$watt/cm^3$	$cal/sec\ cm^3$	$Btu/hr\ in.^3$	$Btu/hr\ ft^3$	$Mev/sec\ cm^3$
1	0.2388	55.91	9.662×10^4	6.2420×10^{12}
4.187	1	234.1	4.045×10^5	2.613×10^{13}
0.01788	4.272×10^{-3}	1	1728	1.1164×10^{11}
1.035×10^{-5}	2.472×10^{-6}	5.787×10^{-4}	1	6.4610×10^7
1.602×10^{-13}	3.826×10^{-14}	8.9568×10^{-12}	1.5477×10^{-8}	1

TABLE F-14
Heat Flux

watt/cm²	cal/sec cm²	Btu/hr ft²	Mev/sec cm²
1	0.2388	3170.2	6.2420×10^{12}
4.187	1	1.3272×10^4	2.6134×10^{13}
3.155×10^{-4}	7.535×10^{-5}	1	1.9691×10^9
1.602×10^{-13}	3.826×10^{-14}	5.0785×10^{-10}	1

TABLE F-15
Thermal Conductivity

watt/cm °C	cal/sec cm °C	Btu/hr ft °F	Btu in./hr ft² °F	Mev/sec cm °C
1	0.2388	57.78	693.3	6.2420×10^{12}
4.187	1	241.9	2903	2.6134×10^{13}
0.01731	4.134×10^{-3}	1	12	1.0805×10^{11}
1.441×10^{-3}	3.445×10^{-4}	0.08333	1	9.004×10^9
1.602×10^{-13}	3.8264×10^{-14}	9.2551×10^{-12}	1.111×10^{-10}	1

TABLE F-16
Viscosity

Centipoise	Poise	kg_m/sec m	lb_m/sec ft	lb_m/hr ft	lb_fsec/ft²
1	0.01	0.001	6.720×10^{-4}	2.419	2.089×10^{-5}
100	1	0.1	0.06720	241.9	2.089×10^{-3}
1,000	10	1	0.6720	2,419	0.02089
1,488	14.88	1.488	1	3,600	0.03108
0.4134	4.134×10^{-3}	4.134×10^{-4}	2.778×10^{-4}	1	8.634×10^{-6}
4.788×10^4	478.8	47.88	32.17	1.158×10^5	1

TABLE F-17
Electric Charge

Electron Charge	Coulomb (amp sec)	Amp hr	Faraday	Erg/volt
1	1.6021×10^{-19}	4.45028×10^{-23}	1.66043×10^{-24}	1.6021×10^{-12}
6.24180×10^{18}	1	2.77778×10^{-4}	1.03641×10^{-5}	10^7
2.24705×10^{22}	3600	1	3.73107×10^{-2}	3.6×10^{10}
6.02253×10^{23}	9.6487×10^4	26.80194	1	9.64870×10^{11}
6.24180×10^{11}	10^{-7}	2.77778×10^{-11}	1.03641×10^{-12}	1

References

1. El-Wakil, M. M. *Nuclear Power Engineering*, McGraw-Hill, 1962.
2. El-Wakil, M. M. *Nuclear Heat Transport*, International Textbook, Scranton, Pa., 1971.
3. Obert, E. F. *Concepts of Thermodynamics*, McGraw-Hill, New York, 1960.
4. Keenan, J. H., and F. G. Keyes. *Thermodynamic Properties of Steam– Including Data for the Liquid and Solid Phases,* 32nd printing. Wiley, New York, 1959.
5. *Organic Coolant Databook*, compiled by M. McEwen, Monsanto Chemical Co. Tech. Publ. AT-1, July 1958.
6. Meisl, C. J., and A. Shapiro. *Thermodynamic Properties of Alkali Metal Vapors and Mercury*–2nd Revision, Gen. Elec. Flight Propulsion Lab. Rept. R60FPD358-A, November, 1960.
7. Carnot, N. L. S. "Reflexions sur la puissance mortice du feu et sur les machines propres a developper cette puissance," Paris, 1824; English translation by R. H. Thurston. American Society of Mechanical Engineers, New York, 1943.
8. Novikov, I. I. "The Efficiency of Atomic Power Stations (a Review)," *J. Nuclear Energy*, Vol. 7, (1958), pp. 125-128.
9. Trub, I. A. "The Temperature of Regenerative Preheating of Water in an Atomic Power Station With a Water-Cooled Reactor," *J. Nuclear Energy,* Vol. 9 (1959), pp. 185-189.
10. English, R. E., H. O. Slone, D. T. Bernatowicz, E. H. Davison, and S. Lieblein. "A 2000-kilowatt Nuclear Turboelectric Power Supply for Manned Space Vehicles," Natl. Aeronaut. and Space Admin., Memo 2-20-59E, March 1959.
11. Orr, J. "Mercury Vapor Unit Operates Successfully at South Meadow," *Power,* Vol. 72, No. 1 (July 1, 1930), pp. 4-9.
12. "Mercury Vapor at Kearney," *Power,* Vol. 79 (July 1935), pp. 348-350.
13. Dwight, H. B. *Tables of Integrals and Other Mathematical Data*, Macmillan, New York, 1947.
14. Weil, John W. "Void-Induced Power Distortion in Boiling Water Reactors," *Nucleonics* (June 1958), pp. 90-94.
15. Tennessee Valley Authority. "Browns Ferry Nuclear Power Station." *Design and Analysis Report*, 1968.
16. "EBWR, The Experimental Boiling Water Reactor," *U.S. Atomic Energy Comm. Nuclear Technol. Ser. Rept.* ANL-5607, prepared by Argonne National Laboratory, May, 1957.
17. Foster, L. E. "Technical Design of the Dresden Nuclear Power Station," paper presented at Semiannual Meeting, American Society of Mechanical Engineers, June 1958.
18. Pathfinder Atomic Power Plant. Final Feasibility Report. "Boiling Water Reactor with Internal Superheater," *Allis-Chalmers Manufacturing Company Rept.* ACNP-5917, U.S. Atomic Energy Commission Contract AT (11-1)-589, August 1959.
19. Dollezhal, N. A., A. K. Krasin, et al. "Uranium Graphite Reactor with Superheated High Pressure Steam." *Proc. Second United Nations Intern.*

Conf. on Peaceful Uses of Atomic Energy, Geneva, September, 1958, Paper P/2139, pp. 398-414.

20. "A Power Reactor Moderator Level Controlled." American Standard Atomic Energy Div. Bull. 107, March 1958.

21. Elston, C. W. "First Large Steam Turbine for Operation with a Boiling Water Reactor," paper presented at the Twentieth Annual Meeting of the American Power Conference, Chicago, Ill., March, 1958; General Electric Co. reprint, GER-1505.

22. Cataldi, H. A., C. F. Cheng, and V. S. Musick. "Investigation of Erosion and Corrosion of Turbine Materials in Wet Oxygenated Steam." *Trans. ASME,* Vol. 80, No. 7 (October 1958), pp. 1465-1478.

23. Tipton, C. R. Jr. (ed.). *Reactor Handbook,* Vol. 1, *Materials,* 2d ed., Prepared under contract with U.S. Atomic Energy Commission. Interscience Publishers, New York, 1960.

24. Byerley, W. M. "Nuclear Plant Steam Generators," *Power Engineering,* Vol. 72, No. 10 (October 1968), pp. 44-47.

25. Glasser, T. H. "Basic Equations for Predicting Performance of a Nuclear Power Plant Pressurizer," presented at the Second Nuclear Engineering and Science Conference, Philadelphia, Pa., *ASME Paper* 57-NESC-95 (March 1957).

26. Gallagher, J. M., Jr., et al. "Startup Experiment Program for the Yankee Reactor," *YAEC-184,* June 1961.

27. Thie, J. A. "Fluid Poison Control of Boiling Water Reactors," *Nucleonics,* Vol. 16, No. 5 (1958), p. 82.

28. U. S. Atomic Energy Commission. Wisconsin-Michigan Power Company, Point Beach Nuclear Plant, Preliminary Facility and Safety Analysis Report, 1968.

29. Cohen, P., and H. W. Graves, Jr. "Chemical Shim Control for Power Reactors, *Nucleonics,*" Vol. 22, No. 5 (May 1964), p. 75.

30. Lacey, P. G. "Fine Structure Power Peaking in a Critical Experiment Mockup of a Chemical Shim Core," USAEC Report WCAP-3723, March 1963.

31. Aisu, H., R. F. Barry, and P. G. Lacey. "Load Variation Restrictions in a Chemically Poisoned Large PWR Core," USAEC Report WCAP-3724, March 1963.

32. Pawliv, J. "Canadian HWR Fuel," *Nuclear Engineering,* Vol. 11, No. 121 (June 1966), pp. 464-466.

33. Ross-Ross, P. A., and K. L. Smith. "Pressure-Tube Development for Canada's Power Reactors", ASME Paper No. 64-WA/NE-5 (1964).

34. Nilson, R. "Heavy-Water Steam in Direct Cycle-Marviken," *Nuclear Engineering,* Vol. 11, No. 121 (June 1966), pp. 456.

35. Edlund, M. C., and G. T. Rhode. "Spectral Shift Control," *Nucleonics,* Vol. 16, No. 5 (1958), p. 80.

36. Blokhintsev, D. I., N. A. Dollezhal, and A. K. Krasin. "Atomic Reactor Electric Station of the U.S.S.R.," *Atomnaya Energia,* Vol. 1 (1956).

37. "The Significance of Shippingport," *Nucleonics,* Vol. 16, No. 4 (April 1958), pp. 53-72.

38. Gumprich, W. C. The Consolidated Edison Reactor, paper presented at Annual Meeting of Atomic Industrial Forum, Inc., on Commercial and International Developments in Atomic Energy, Washington, D.C., September, 1955 (Babcock & Wilcox Company Bull. AR-19).

39. U. S. Atomic Energy Commission. *Final Facility Description and Safety Analysis Report*, Wisconsin Electric Power Company and Wisconsin-Michigan Power Company Point Beach Nuclear Plant, Units No. 1 and 2, 1969.

40. Morison, W. G. "Pickering Generating Station-Design," The Eleventh AECL Symposium on Atomic Power, Ontario, Canada, October 1966.

41. van Heerden, C., A. P. P. Nobel, and D. W. van Krevelen. "Studies on Fluidization. I-The Critical Mass Velocity," *Chem. Eng. Sci.,* Vol. 1, (1952), pp. 37-49.

42. Mickley, H. S., and C. A. Trilling. "Heat Transfer Characteristics of Fluidized Beds," *Ind. Eng. Chem.*, Vol. 41 (1949), p. 1135.

43. Toomey, R. D., and H. F. Johnstone. "Gaseous Fluidization of Solid Particles," *Chem. Eng. Progr.*, Vol. 48, No. 5 (1952), pp. 220-226.

44. Weintrab, M., and M. Leva. "A Review of the Literature on Fluidization," *Ind. Eng. Chem.,* Vol. 45, No. 1 (January 1953), pp. 76-78.

45. Leva, M., M. Weintraub, M. Grummer, M. Pollchik, and H. H. Storch. "Fluid Flow Through Packed and Fluidized Systems," U. S. Bur. Mines, Bull. 504, 1951.

46. Lewis, W. K., E. R. Gilliland, and W. C. Bauer. "Characteristics of Fluidized Particles," *Ind. Eng. Chem.,* Vol. 41 (1949), p. 1104.

47. Scheve, M. R. "Liquid Fluidized Bed Reactor Program," *Proc. American Power Conf.,* Illinois Institute of Technology, Vol. 22 (March 1960), pp. 138-146.

48. Dahlberg, R. C., E. G. Beasley, T. K. DeBoer, T. C. Evans, D. F. Molino, W. S. Rothwell, and W. R. Sivka. "Gas-cooled, Natural Uranium, D$_2$O-moderated Power Reactor; Reactor Design and Feasibility Problem," U. S. Atomic Energy Comm. Rept. CF-S6-8-207 (Del.), August 1956.

49. Alikhanov, A. I., V. V. Vladimirsky, P. A. Petrov, and P. I. Khristenko. "A Heavy-water Power Reactor with Gas-cooling," *J. Nuclear Energy* Vol. 3, Pt. 2 (August 1956), pp. 77-82.

50. Sutton, O. G. "The Theoretical Distribution of Airborne Pollution from Factory Chimneys," *Quart. J. Roy. Meteorol. Soc.*, Vol. 73 (1947), pp. 426-436.

51. Kreith, F. *Principles of Heat Transfer*, 2d. ed., International Textbook, Scranton, Pa., 1965.

52. Nichols, R. W. "Uranium and Its Alloys," *Nuclear Engineering*, Vol. 2, No. 18 (September 1957), pp. 355-365.

53. Bell, I. P. "Thorium, Its Properties and Characteristics," *Nuclear Engineering*, Vol. 2, No. 19 (October 1957), pp. 418-422.

54. Technical Aspects of the Gas-cooled, Graphite-moderated Reactor, Report summarizing a series of six U.S. Atomic Energy Commission Reports, Hearing before Subcommittee on Legislation of the Joint Committee on

Atomic Energy, Congress of the United States. Government Printing Office, Washington, D.C., April 1958.

55. Choudhury, W. U., and M. M. El-Wakil. "On the Use of Porous Fuel Elements in Nuclear Reactors," *ASME Paper* No. 68-WA/NE-7 (1968).

56. Harteck, P., and S. Dondes. "Glass Fibers–A New Form for Reactor Fuels," *Nucleonics*, Vol. 15, No. 8 (August 1957), p. 94.

57. Dahlberg, R. C., and T. C. Evans. "Spiral Fuel Elements for Gas-cooled Reactors," *Nucleonics*, Vol. 16, No. 4 (April 1958), p. 106.

58. Wootton, W. R., A. J. Taylor, and N. G. Worley. "Steam Cycles for Gas-cooled Reactors," *Proc. Second United Nations Intern. Conf. on Peaceful Uses of Atomic Energy*, Geneva, Paper P/273 UK, Vol. 7 (September 1958), pp. 827-834.

59. "Calder Hall, A Special Report," *Nucleonics*, Vol. 14, No. 12 (December 1956).

60. Arms, H. S., C. Bottrell, and P. H. Wolff. "The Hinkley Point Power Station," *Proc. Second United Nations Intern. Conf. on Peaceful Uses of Atomic Energy*, Geneva, Paper P/75 UK, Vol. 8 (September 1958), pp. 434-449.

61. Ritz, H. L. "The Polyzonal Spiral Fuel Element," *Proc. Second United Nations Intern. Conf. on Peaceful Uses of Atomic Energy*, Geneva, Paper P/48 UK, Vol. 7 (September 1958), pp. 725-737.

62. "Hinkley Point B, A Special Survey," *Nuclear Engineering*, Vol. 13, No. 147 (August 1968), pp. 652-668.

63. Jaye, S., and W. V. Goeddel. "High-temperature Gas-cooled Reactor Fuels and Fuel Cycles–Their Progress and Promise," General Atomic Report No. GA-748, December 1966.

64. "HTGR-The High Temperature, Gas-cooled Power Reactor System," General Atomic Report, May 1962.

65. Habush, A. L., and R. F. Walker. "The Fort Saint Vrain HTGR," General Atomic Report No. GA-7817, April 1967.

66. Schulten, R. "The Development of High-temperature Reactors," *Proc. Symposium on Gas-cooled Reactors*, sponsored jointly by Franklin Institute and American Nuclear Society, Delaware Valley Section (February 1960), pp. 109-126.

67. Robinson, S. T. "The Pebble Bed Reactor," *Proc. Symposium on Gas-cooled Reactors*, sponsored jointly by Franklin Institute and American Nuclear Society, Delaware Valley Section (February 1960), pp. 87-108.

68. Carman, P. C. "Fluid Flow Through Granular Beds," *Trans. Inst. Chem. Engrs. (London)*, Vol. 15 (1937), pp. 150-166.

69. "The Pebble Bed Reactor Program," Progress Report, *U. S. Atomic Energy Comm. Rept.*, NYO 2373, June 1, 1958–May 31, 1959.

70. Lancashire, R. B., E. A. Lezberg, and J. F. Morris. "Heat-transfer Coefficients for a Fullscale Pebble Bed Heater," *Natl. Air and Space Adm. Tech. Note* N-83423X, May 1960.

71. "The Pebble Bed Reactor," *Nuclear Engineering*, Vol. 10, No. 112 (September 1965), pp. 333-334.

72. Sanderson & Porter, Inc. "Design and Feasibility Study of a Pebble Bed

Reactor Steam Power Plant," *U. S. Atomic Energy Comm. Rept.* NYO-8753, Vol. 1 (May 1, 1958).

73. Newby, G. A. "The Army Gas-cooled Reactor Systems Program," Proc. Symposium on Gas-cooled Reactors, sponsored jointly by Franklin Institute and American Nuclear Society, Delaware Valley Section (February 1960), pp. 43-51.

74. Daniels, Farrington. "Small Gas-cycle Reactor Offers Economic Power," *Nucleonics* (March 1956), pp. 34-41.

75. Bernsen, S. A., et al. "The Marine Gas-cooled Reactor Program," information meeting on gas-cooled power reactors, October, 1958; unclassified U. S. Atomic Energy Comm. Rept. TID-7564 (December 1958), p. 243.

76. Conklin, D. L., et al. "Economics of Nuclear and Conventional Merchant Ships, unclassified," U. S. Atomic Energy Comm. Rept. TID-7563 (June 1958), p. 33.

77. Roy, G. M., and M. F. Valerino: "Design Concept for a Gas-cooled, Pressure Tube D_2O-Moderated Reactor," information meeting on gas-cooled power reactors, October, 1958; U. S. Atomic Energy Comm. Rept. TID-7564 (December 1958), p. 216.

78. Perry, A. M. "Advanced Design Studies," Information meeting on gas-cooled reactors, October, 1958; U. S. Atomic Energy Comm. Rept. TID-7564 (December 1958), p. 334.

79. Rengel, J. C. "Only High-Gain Breeder Reactors Can Stabilize Uranium Fuel Requirements," *Westinghouse Engineer,* Vol. 28, No. 1 (1968), p. 3.

80. Katcoff, S. "Fission-Product Yields from U, Th, and Pu," *Nucleonics,* Vol. 16, No. 4 (April 1958), p. 78.

81. Glasstone, S., and A. Sesonske. *Nuclear Reactor Engineering,* D. Van Nostrand Company, Inc., Princeton N.J., 1963.

82. Meghreblian, R. V., and D. K. Holmes. *Reactor Analysis*, McGraw-Hill, New York, 1960.

83. Okrent, D. "Neutron Physics Consideration in Large Fast Reactors," *Power Reactor Technology,* Vol. 7, No. 2 (Spring 1964).

84. Nicholson, R. B. "The Doppler Effect in Fast Neutron Reactors," U. S. Atomic Energy Comm. Report APDA-139, 1960.

85. U.S. AEC, *Reactor Physics Constants*, ANL-5800, 2nd ed., 1962.

86. Koch, L. S., and H. C. Paxton. *American Review of Nuclear Science*, Vol. 9 (1959), p. 437.

87. R. N. Lyon (ed.). *Liquid Metals Handbook,* sponsored by Committee on the Basic Properties of Liquid Metals, Office of Naval Research, Department of the Navy, in collaboration with U. S. Atomic Energy Commission and Bureau of Ships, Department of the Navy, June, 1952. See also *Liquid Metals Handbook, Na-Nak Supplement,* U.S. Atomic Energy Commission, July 1955.

88. Civilian Power Reactor Program Part III, Status Report on Sodium Graphite Reactors as of 1959, U.S. Atomic Energy Comm. Rept. TID-8518(6), 1960.

89. David, M., and A. Draycott. "Compatibility of Reactor Materials in

Flowing Sodium," Proc. Second United Nations Intern. Conf. on Peaceful Uses of Atomic Energy, Geneva, Vol. 7 (1958), pp. 94-110.

90. Fortescue, P., J. Broido, G. Schultz, and R. Shanstrom. "The Gas-Cooled Fast-Reactor Experiment," paper presented at the Fast-Reactor Topical Meeting, San Francisco, California, April 1967.

91. Huebotter, P. R. "Material Problems and Mechanical Design," Paper presented at the Eleventh Annual ANL-AUA Nuclear Engineering Education Conference, Argonne National Laboratory, February 1970.

92. Civilian Power Reactor Program, Part III, TID-8515 (1), Status Report on Fast Reactors as of 1959, U.S. Atomic Energy Commission, 1960.

93. "Dounreay," *Nuclear Eng.*, Vol. 2, No. 15, (June 1957), pp. 229-245.

94. "Enrico Fermi Atomic Power Plant," Atomic Power Development Associates, Inc., Rept. APDA-124, January 1959.

95. Billuris, G., K. Hikido, E. E. Olich, and A. B. Reynolds. "SEFOR Plant Design," paper presented at the Fast Reactor Topical Meeting, San Francisco, California, April 1967.

96. Design Studies of a 1000-Mw(e) Fast Reactor, *Power Reactor Technology*, Vol. 8, No. 2 (Spring 1965), pp. 147-155.

97. Fortescue, P. G. Melese-d'Hospital, and W. I. Thompson. Status of the Technology of Gas Cooling for Fast Reactors, GA-7888, March 1967.

98. Fortescue, P., and G. M. Schultz. A 1000 Mw(e) Gas-cooled Fast Breeder Reactor, GA-7823, paper presented at the 29th Annual Meeting of the American Power Conference, Chicago, III., April 1967.

99. Leitz, F. J., P. M. Murphy, P. R. Pluta, and H. J. Rubinstein. Status of Steam Cooled Fast Reactor Technology, paper presented at the Fast Reactor Topical Meeting, San Francisco, California, April 1967.

100. Sutherland, W. A., and C. W. Miller. "Heat Transfer to Superheated Steam-II Improved Performance with Turbulence Promoters," GEAP 4749, November 1964.

101. Jaffey, A. H. "Long Term Variation in Composition and Neutron Yield in Pile Plutonium," *Nuclear Sci. and Eng.*, Vol. 1 (1956), pp. 204-215.

102. Civilian Power Reactor Program, Part III, Status Report on Aqueous Homogeneous Reactors as of 1959, U.S. Atomic Energy Comm. Rept. TID-8518 (3), 1960.

103. Hoegerton, J. F., and R. C. Grass (eds.). "Reactor Handbook: Vol. 3, Engineering," selected reference material, U.S. Atomic Energy Commission, August 1955.

104. Hays, E. E., and P. Gordon. "The Solubility of Uranium and Thorium in Liquid Metals and Alloys," U.S. Atomic Energy Comm. Rept. TID-65, July 1948.

105. Bryner, J. S. "The Thorium-Bismuth-Lead System," U.S. Atomic Energy Comm. Rept. TID-7502, Pt. I, 1958.

106. Teitel, R. J. "The Uranium-Bismuth System," *J. Metals*, Vol. 9 (1957), pp. 131-140.

107. Bettis, E. S., et al. "The Aircraft Reactor Experiment–Design and Construction," *Nuclear Sci. and Eng.*, Vol. 2 (1957), p. 804.

108. Deissler, R. G., and J. S. Boegli. "An Investigation of Effective Thermal

Conductivities of Powders in Various Gases," *Trans. ASME*, Vol. 80, No. 7 (October 1958), pp. 1417-1425.

109. Weinberg, A. M. "Some Aspects of Fluid Fuel Reactor Development," *Nuclear Sci. and Eng.*, Vol. 8 (1960), pp. 346-360.

110. Liquid Metal Fuel Reactor Experiment, The Economic Comparison of a Large Scale LMFR Breeder and Burner, Babcock & Wilcox Company, Atomic Energy Div. Paper BAW-1012, May 1957.

111. Miles, F. T., et al. "Liquid Metal Fuel Reactor with Recycled Plutonium," *Proc. Second United Nations Intern. Conf. on Peaceful Uses of Atomic Energy*, Geneva, Vol. 9, Pt. 2 (1958), pp. 180-187.

112. Rosenthal, M. W., R. B. Briggs, and P. R. Kasten. "Molten Salt Reactor Program," Semiannual Progress Report, Oak Ridge National Laboratory, ORNL-4254, August 1968.

113. Proceedings of Sodium Reactor Experiment–Organic Moderated Reactor Experiment (SRE-OMRE) Forum, Los Angeles, Calif., November, 1956; U.S. Atomic Energy Comm. Rept. TID-7525 (NAA-SR-1804), January 1957.

114. Trilling, C. A. "OMRE Operating Experience," *Nucleonics*, Vol. 17, No. 11, p. 113, November, 1959; see also C. A. Trilling, "The OMRE–A Test of the Organic Moderator-Coolant Concept," *Proc. Second United Nations Intern, Conf. on Peaceful Uses of Atomic Energy*, Paper P/421, Vol. 9, Pt. 2 (1958), pp. 468-489.

115. Weisner, E. F. "Engineering Design of Piqua OMR," *Nuclear Eng.*, Vol. 5, No. 45 (February 1960), pp. 68-71.

116. Compilation of Organic Moderator and Coolant Technology, U.S. Atomic Energy Comm. Rept. TID-7007, Pts. 1 and 2, January 1957.

117. Organic Coolant Databook, compiled by M. McEwen, Monsanto Chemical Company, Organic Chemicals Div. Tech. Publ. AT-1, July 1958.

118. U.S. Atomic Energy Commission: AEC Unclassified Programs January-December, 1967, Annual Tech. Rept. NAA-SR-2400, Pt. I, March 1958.

119. Gilroy, H. M. "Analysis of OMRE Waste Gases," U.S. Atomic Energy Comm. Rept. NAA-TDR-4130, June 1959.

120. Keen, R. T. "Composition of Reactor Coolant," Annual Technical Progress Report on Fiscal Year 1959, U.S. Atomic Energy Comm. Rept. NAA-SR-3850, June 1959.

121. Bley, W. N. "An In-Pile Loop Study of the Performance of Polyphenyls Reactor Coolants," U.S. Atomic Energy Comm. Rept. NAA-SR-2470, September 1958.

122. Silberberg, M., and D. A. Huber. "Forced-convection Heat-transfer Characteristics of Polyphenyl Reactor Coolants," U.S. Atomic Energy Comm. Rept. NAA-SR-2796, 1959.

123. Perlow, M. A. "Heat Transfer and Fouling," Annual Technical Progress Report for Fiscal Year 1959, U.S. Atomic Energy Comm. Rept. NAA-SR-3850, June 1959.

124. Siegel, S., and R. F. Wilson. "OMCR Power Plants, Their Status and Promise," *Nucleonics*, Vol. 17, No. 11 (November 1959), p. 118.

125. Jordan, D. P., and G. Leppert. "Nucleate Boiling Characteristics of

Organic Reactor Coolants," *Nuclear Sci. and Eng.*, Vol. 4, No. 6 (June 1959), pp. 349-359.

126. Forster, K. E., and R. Greif. "Heat Transfer to a Boiling Liquid–Mechanism and Correlations," *Trans. ASME*, Vol. 81, Ser. C, No. 1, (February 1959), pp. 43-53.

127. Core, T. C., and K. Sato. "Determination of Burnout Limits of Polyphenyl Coolants," U.S. Atomic Energy Comm. Rept. IDO-28007, February 1958.

128. Balent, R., G. H. Bosworth, and J. Plawchan. Design Study of a 300-Mw Organic-cooled Reactor, Paper 59-A-178, presented at the Annual Meeting of the American Society of Mechanical Engineers, Atlantic City, N. J., November 1959.

129. McNelly, M. H. "A Heavy-water Moderated Power Reactor Employing an Organic Coolant," *Proc. Second United Nations Intern. Conf. on Peaceful Uses of Atomic Energy*, Geneva, Vol. 9 (September 1958), pp. 78-87.

130. Halg, W., and T. Schaub. "Diphenyl Cooled, Heavy-water Moderated, Natural Uranium Reactor Prototype," *Proc. Second United Nations Intern. Conf. on Peaceful Uses of Atomic Energy*, Geneva, Vol. 9 (September 1958), pp. 88-98.

131. Bridoux, C., H. Foulquier, J. Kaufman, and P. Thome. "Sintered Aluminum Powder as a Canning Material," *Nuclear Eng.*, Vol. 6, No. 60 (May 1961), pp. 189-192.

132. Schlichter, W. *Dissertation,* Cottingen, 1915.

133. Kittel, C. *Introduction to Solid State Physics*, John Wiley and Sons, Inc., New York, 1956.

134. Langmuir, I. Effect of Space Change and Initial Velocities on the Potential Distribution and Thermionic Current Between Parallel Plane Electrodes, *Phy. Rev.*, Vol. 21, p. 426, 1923.

135. Walsh, E. M. *Energy Conversion*, The Ronald Press Company, New York, 1967.

136. Breitwieser, R. "Direct Energy Conversion Programs at NASA," *Conference on Special Topics on Nuclear Education and Research*, Oak Ridge Associated Universities, Inc., held at Gatlinburg, Tennessee, August 1967.

137. Perry, L. W., and W. G. Homeyer. "Synthesis of Thermionic Power Conversion to Nuclear Reactors for Space Power Applications," *Thermionic Electrical Power Generation Symposium*, ENEA-IEE, September 1965.

138. Rouklove, P. "Thermionic Converters and Generators for Space Application, Part I: Thermionic Converter Performance and Material Problems," *Thermionic Electrical Power Generation Symposium*, ENEA-IEE, September 1965.

139. Dunn, P. D., and J. Adam. "Fission-Heated Thermionic Diode," Paper No. P/132, *Proceedings of the Third International Conference on the Peaceful Uses of Atomic Energy*, Geneva, 31 August-9 September, 1964, Vol. 15, pp. 114-122, United Nations, 1965.

140. Knapp, D. E., et al. "Nuclear Electric Power Sources for Biomedical Applications," Paper presented at the *Fourth Annual Intersociety Energy Conversion Engineering Conference*, Washington, D. C., September 1969.

141. Sutton, G. W. *Direct Energy Conversion*, McGraw-Hill Book Co., New York, 1966.

142. Ure, R. W. Jr., and R. R. Heiks. "Theoretical Calculations of Device Performance," Chapter 15 in Heikes and Ure, *Thermoelectricity: Science and Engineering*, Interscience Publishers, New York, 1961.

143. Sherman, B. R., R. Heikes, and R. Ure, Jr. *Journal of Appl. Phys.*, Vol. 31, 1960.

144. Green, W. B. *Thermoelectric Handbook*, Westinghouse Electric Corp., Youngswood, Pa., 1962.

145. Wilson, R. F., and H. M. Dieckamp. "What Happened to SNAP 10A," *Astronautics and Aeronautics*, pp. 60-65, Oct., 1965.

146. Ohmart, P. E. "Method and Apparatus for Converting Ionic Energy into Electrical Energy," U.S. Patent No. 3, 152, 254, Oct. 6, 1964.

147. Murphy, E. L. "Direct Electricity Generation from Ionized Vapors and Flames," *Tech. Report ONR* 1-2-65, Office of Naval Research, American Embassy, London, Feb. 1965.

148. Steinberg, M. "Chemonuclear and Radiation Chemical Process Research and Development," *Isotopes and Radiation Technology*, Vol. 4, No. 2, p. 142, 1966.

149. Braun, J. "Linear Constant-Mach-Number MHD Generator with Nuclear Ionization, A Parameter Study," *J. Nucl. Energy*, Pt. C., Vol. 7, p. 525, 1965.

150. Olsen, L. C., S. E. Seeman, and B. I. Griffin. "Betavoltaic Nuclear Electric Power Sources," Paper presented at the *Winter Meeting of the American Nuclear Society*, San Francisco, California, Nov. 30-Dec. 4, 1969.

151. Samonov, G. "Direct Conversion of Fission to Electric Energy in Low Temperature Reactors," Report No. RM-1870, Rand Research Corp., Jan. 1957.

152. Heindl, C. J., W. F. Krieve, and V. Meghreblian. "The Fission Electric Cell Reactor Concept," *Nucleonics,* Vol. 21, No. 4 (April 1963), p. 80.

153. Anno, J. N., and. S. L. Fawcett. "The Triode Concept of Direct Conversion," *Battelle Technical Review*, pp. 3-9, 1962.

154. Plummer, A. M., and. J. N. Anno. "Battelle Studies on the Triode Concept of Direct Energy Conversion," *Proc. AEC-ASEE Summer Institute on Direc Energy Conversion,* University of Illinois, Urbana, 1963.

155. Krieve, W. F. "JPL Fission-Electric Cell Experiment," Report JPL Tech. No. 32-981, *Jet Propulsion Laboratory,* Now. 1966.

156. Rose, D. J., and M. Clark, Jr. *Plasmas and Controlled Fusion,* M.I.T. Press Wiley, New York, 1961.

157. Bishop, A. S. *Project Sherwood, the U.S. Program in Controlled Fusion,* Addison-Wesley, Reading, Mass., 1958.

158. Post, R. F. "Controlled Fusion Research: An Application of the Physics of High Temperature Plasmas," *Revs. Modern Phys.*, Vol. 28, No. 3 (July 1956), pp. 338-362; see also R. F. Post, "Fusion Power," *Sci. American*, Vol. 197, No. 6 (December 1957), pp. 73-88.

159. Mills, R. G. "Some Engineering Problems of Thermonuclear Reactors," *Nuclear Fusion*, Vol. 7 (1967), pp. 223-236.

160. Artsimovich, L. A., et al. "Experimènts in Tokamak Devices," *Proc. I.A.E.A. Conf. on Plasma Physics and Controlled Nuclear Fusion Research,* Novosibirsk, 1968.
161. Artsimovich, L. A., et al. "Ion Heating in Tokamak-3," *Zh.E.T.F. Letters,* No. 10 (August 1969), pp. 130-133.
162. Carruthers, R. "Engineering Parameters of a Fusion Reactor," Nuclear Fusion Reactor Conference, British Nuclear Energy Society, September 17-19, 1969.
163. Frass, A. P. "Conceptual Design of a Fusion Plant to Meet the Total Energy Requirements of an Urban Complex," paper presented at the Nuclear Fusion Reactor Conference, British Nuclear Energy Society, September 17-19, 1969.
164. Rose, D. J. "On the Feasibility of Power by Nuclear Fusion," ORNL-TM-2204, May 1968.
165. Steiner, D. "Neutronic Calculations and Cost Estimates for Fusion Reactor Blanket Assemblies," USAEC Report ORNL-TM-2360, November, 1968.
166. Eastlund, B. J., and W. C. Gough. "The Fusion Torch, Closing the Cycle from Use to Reuse," USAEC Report WASH-1132, May 1969.
167. Kaiser Engineers: "Guide to Nuclear Power Cost Evaluation," AEC Report TID-7025; Vol. 1, Reference Data and Standards; Vol. 2, Land, Improvement, Buildings and Structure; Vol. 3, Equipment Costs; Vol. 4, Fuel Cycle Costs; Vol. 5, Production Costs, March 15, 1962.
168. NUS Corporation: "Guide for Economic Evaluations of Nuclear Reactor Plant Designs," Prepared for the U.S. Atomic Energy Commission, November 1967.
169. Tables of Thermal Properties of Gases, Natl. Bur. Standards (U.S.), Circ. 564, November 1955.
170. Moody, L. F. "Friction Factors for Pipe Flows," *Trans. ASME,* Vol. 66 (1944), pp. 671-694.
171. Ewing, C. T., J. P. Stone, J. R., J. R. Spann, and R. R. Miller. "High Temperature Properties of Potassium," NRL Report 6233, U.S. Naval Research Laboratory, September, 1965.

Index